Lecture Notes in Mathematics 1794

Springer
Berlin
Heidelberg
New York
Barcelona
Hong Kong
London
Milan
Paris
Tokyo

N. Pytheas Fogg

Substitutions in Dynamics, Arithmetics and Combinatorics

Editors: V. Berthé
S. Ferenczi
C. Mauduit
A. Siegel

Author

N. Pytheas Fogg
Marseille, France

Editors

Valérie Berthé
Sébastien Ferenczi
Christian Mauduit

Univ. de la Méditerranée
IML, Case 907
163 av. de Luminy
13288 Marseille Cedex 09, France

e-mail:
berthe@iml.univ-mrs.fr
ferenczi@iml.univ-mrs.fr
mauduit@iml.univ-mrs.fr

Anne Siegel

IRISA
Campus de Beaulieu
35042 Rennes Cedex
France
e-mail: Anne.Siegel@irisa.fr

Authors' name: N. stands for "no nomen". Pytheas was the Greek navigator and scientist (4th century B.C.) who became one of the historical heroes of the city of Marseille (Massalia). Phileas Fogg was the excentric and phlegmatic hero of Jules Verne who travelled around the world by "jumbing mathematically from train to boat, never forgetting his sense of humour."

Cataloging-in-Publication Data applied for.

Die Deutsche Bibliothek - CIP-Einheitsaufnahme

Pytheas Fogg, N.:
Substitutions in dynamics, arithmetics and combinatorics / N. Pytheas Fogg.
Ed.: V. Berthé .... - Berlin ; Heidelberg ; New York ; Barcelona ; Hong Kong
; London ; Milan ; Paris ; Tokyo : Springer, 2002
(Lecture notes in mathematics ; 1794)
ISBN 3-540-44141-7

Mathematics Subject Classification (2000): 11B85, 1A55, 11A63, 11J70, 11KXX, 11R06, 28A80, 28DXX, 37AXX, 37BXX, 40A15, 68Q45, 68R15

ISSN 0075-8434
ISBN 3-540-44141-7 Springer-Verlag Berlin Heidelberg New York

Springer-Verlag Berlin Heidelberg New York a member of BertelsmannSpringer
Science + Business Media GmbH

http://www.springer.de

Printed in Germany

Typesetting: Camera-ready TEX output by the editors

SPIN: 10890994 41/3142/DU - 543210 - Printed on acid-free paper

Preface

There are two basic ways of constructing dynamical systems. One approach is to take an already existing system from the vast reserve arising in biology, physics, geometry, or probability; such systems are typically rather complex and equipped with rigid structures (a "natural" system is generally a "smooth" system). Alternatively one can build a system by hand using some basic tools like strings of letters on a finite alphabet; the latter system will have a simple structure. Nevertheless, these "simple" systems prove to be very useful in many mathematical fields (number theory, harmonic analysis, combinatorics, ergodic theory, and so on), as well as in theoretical computer science and physics. This category also includes various "classical" systems.

To make more precise this intuitive concept of "simple" systems, we can use the combinatorial notion of complexity of a sequence of letters with values in a finite alphabet, which counts the number of factors of given length of this sequence, a factor being any string of consecutive letters appearing in the sequence. This gives an indication of the degree of randomness of the sequence: a periodic sequence has a bounded complexity, while the g-adic expansion of a normal number in base g has an exponential complexity. There are many examples of sequences having a reasonably low complexity function, the most famous being automatic sequences and Sturmian sequences, and the "reasonably simple" dynamical systems we like to consider are those which are canonically associated with this kind of sequences.

Among them, substitutive sequences play an important role. Substitutions are very simple combinatorial objects (roughly speaking, these are rules to replace a letter by a word) which produce sequences by iteration. Let us note that substitutions will be considered here as particular cases of free group morphisms, the main simplification being that we have no problem of cancellations. Substitutive dynamical systems have a rich structure as shown by the natural interactions with combinatorics on words, ergodic theory, linear algebra, spectral theory, geometry of tilings, theoretical computer science, Diophantine approximation, transcendence, graph theory, and so on.

Notice that the notion of substitution we consider here differs from that used for self-similar tilings; in this framework, substitutions produce matching rules acting on a finite set of prototiles and determining the ways in which the tiles are allowed to fit together locally; the best known example is the

Penrose tiling. Here, we consider substitutions acting on strings which are more elementary in nature.

There exist several books in the literature on related subjects. For instance [247, 265] consider symbolic dynamics, though mainly in the positive-entropy case, while [28] deals with automatic sequences, and [271, 272] with combinatorics on words; all these works are generalist books covering rather wide areas and decribing extensively the appropriate techniques. We decided to focus on a well-delimited subject (namely, substitutive dynamical systems) but try to give as many different viewpoints as possible, with emphasis on interactions between different areas; for example, Sturmian sequences are presented in details in both [272] and our book, the former giving a complete review on their combinatorial properties and the latter a description of their links with dynamical systems. Also, our book deals with zero-entropy symbolic dynamics, a subject where few such tools are available: the most famous one at this time is [340], and our book may be seen first as its updating, and then as a sequel together with an opening on wider perspectives.

The idea for this book stemmed from the collaboration between various groups of mathematicians mainly from France, Japan and China. It is largely based on courses performed by the authors in several universities and given during various Summer Schools in the past five years.

We chose to use the pseudonym N. Pytheas Fogg for two reasons. First, our work is the work of a group which is clearly more than the sum of its individual members, and a collective identity is a good way to stress this point. Second, this group is still active, and willing to produce more mathematical publications; though we do not claim to be a new Bourbaki, we do hope that there will be at least a second book, papers in mathematical journals, and a seminar bearing the name of Pytheas Fogg, so the present volume is just a beginning, together with a motivation, for more to come.

The overall structure of this book reflects our purpose which is twofold. We first want to provide an introduction to the theory of substitutive dynamical systems by focusing on several topics including various aspects of mathematics (as for instance geometry, combinatorics, ergodic theory and spectral analysis, number theory, numeration systems, fractals and tilings) but also computer science and theoretical physics. Secondly, we want to give a state of the art on this field, spotlighting representative aspects of the theory. More precisely, we focus on the following themes:

An introduction to elementary properties of **combinatorics on words** is given in the first three chapters. In particular Chaps. 2 and 3 provide an introduction to automatic sequences, which are produced by very natural algorithms coming from theoretical computer science. Chapters 6 and 9 give an analysis of the combinatorial properties of Sturmian words: these are the sequences (or infinite words) which have the smallest complexity function among non-ultimately periodic sequences.

Numeration systems appear in a natural way in the study of low complexity sequences. Beatty sequences and in particular the Fibonacci numeration system are introduced in Chap. 2 and dealt with in detail in Chap. 4. Chapter 6 introduces Ostrowski's numeration system, while the numeration systems defined in Chap. 8 are linked to the geometric properties of Pisot substitutive sequences.

An interesting field of application of automatic sequences deals with the **transcendence** of formal power series. Indeed, the study of automata provides a very fruitful combinatorial transcendence criterion for formal power series with coefficients in a finite field, which can be considered as a natural translation into algebraic terms of the properties of automatic sequences. This criterion is known as Christol, Kamae, Mendès France, and Rauzy's theorem. Chapter 3 presents this criterion and surveys the most recent transcendence results obtained via finite automata theory. Note that a real number whose g-adic expansion is an automatic sequence is conjectured to be either transcendental or rational, and has been proved to be transcendental when the sequence is Sturmian. Hence this presentation emphasizes the following philosophy: algebraicity strongly depends on the (generalized) "base" in which one works. Some connected results of transcendence are studied in Chap. 4; Chapter 8 reviews some Diophantine approximation properties issued from the study of substitutive systems.

Tools from **ergodic theory and spectral analysis** are introduced in Chap. 5, via a detailed study of systems associated with sequences having a low complexity function. An elementary introduction to correlation properties, and some examples of computations of correlation measures, are given in Chap. 2. Chapter 7 surveys the latest results in the spectral study of substitutive dynamical systems. Chapter 11 gives a special account of the ergodic properties of the Perron–Frobenius transfer operator.

The question of the **geometric representation** of substitutive sequences and more generally of low complexity sequences has given birth to a great amount of work. We give an account of the development and current state of this problem in Chap. 7. The study of the Sturmian case is particularly instructive (Chap. 6); it provides a well-known and fundamental interaction between ergodic theory, number theory, and symbolic dynamics, which comes from the study of irrational rotations on the one-dimensional torus $\mathbb{T}$. With an irrational real number α we associate a geometric dynamical system, the rotation $R : x \mapsto x + \alpha \bmod 1$, an arithmetic algorithm, the usual continued fraction approximation, and a set of Sturmian sequences which are codings of trajectories under R by a canonical partition; the continued fraction algorithm arises naturally as the link between the dynamical system and the symbolic sequences, and the study of the arithmetic and symbolic objects is very useful for the study of the dynamical system. Chapters 2 and 4 allude to Beatty sequences, whereas Chap. 9 studies in detail the connections between Sturmian and invertible substitutions over a two-letter alphabet.

Chapter 8 gives a detailed introduction to the tools and techniques used in this framework, in order to generalize this interaction to further systems and sequences.

The notion of **self-similarity** is illustrated by symbolic objects, namely substitutions, and by geometric objects, the fractal sets. Chapters 7 and 8 give examples of interactions between these two notions, as the most natural geometric representations of a substitutive sequence are sets with fractal boundaries. See also Chap. 11 for the study of fractal sets associated in a natural way with piecewise linear transformations of the unit interval.

We illustrate the connections between **physics** and low complexity sequences through the study of trace maps in Chaps. 8 and 9. Indeed free monoids, groups and their morphisms occur in a natural way in physics: finite automata and substitutive sequences are very useful to model and describe certain situations in solid state physics. In particular, one important question in quasicrystal theory is to compute the traces of matrices defined inductively according to a substitutive process. Trace maps are effective algorithms for constructing the recursion relations that the traces satisfy.

We have tried to allow the reader to read the different chapters as independently as possible, and to make each chapter essentially self-contained. Furthermore, the reader is not assumed to have a detailed knowledge in each of the fields covered by this book; we rather try to provide the necessary information, allowing it to be be used by graduate students. We describe hereafter the necessary preliminary knowledge, and which chapters are required for a better understanding of the following chapters.

Chapter 1 is needed as a prerequisite for all the other chapters. We recommend the lecture of Chaps. 2 and 3 for getting used to combinatorial manipulations on substitutions. Chapter 5 (as well as the spectral part of this book covered in some sections of Chap. 1, and in Chaps. 7 and 11) requires some basic knowledge on measure-theory and functional analysis, and is supposed to be self-contained as far as ergodic theory is concerned. This chapter will be needed for Chap. 7. We also recommend to read first Chaps. 6 and 7 before Chap. 8. Chapter 6 is essentially self-contained and is a good introduction for Chap. 9. Chapter 3 can be understood with no special algebraic knowledge except some familiarity with the notion of finite fields.

The book divides naturally into three parts, and is organized as follows.

The **introductory chapter** unifies the notation and contains the necessary background for the following chapters. Indeed we introduce in this chapter the basic introductory material that we will use throughout the book: words, languages, complexity function, substitutions, automatic sequences, substitutive dynamical systems, introduction to discrete dynamical systems and its spectral theory, group rotations, and so on.

In the **first part**, we focus on the aspects of substitutions which do not require any background on measure theory: these include combinatorial esults, but also deep problems of number theory.

- The aim of **Chap. 2** is to focus on the notions of substitutions and automatic sequences by showing some typical examples of arithmetic situations in which they occur in a quite natural way. For example, the Morse sequence was first introduce to answer a question in combinatorial number theory and rediscovered by many people in various other circumstances (including geometry, group theory, logic). The Rudin-Shapiro sequence, first introduced to answer a question in harmonic analysis asked by Salem, is nowadays a basic construction in number theory and ergodic theory. The Fibonacci sequence, introduced as a natural example of a generator of a symbolic dynamical system, is deeply connected with the continued fraction algorithm and gives rise to many applications in theoretical computer science (for example to obtain good algorithms for the drawing of a straight line on a computer screen). We will consider the statistical properties of these sequences through the study of their correlation measure. The tools developed here are as simple as possible providing an elementary introduction to these classical examples. The following chapters will study them with a heavier theoretic background.
- The aim of **Chap. 3** is to investigate the connections between automatic sequences and transcendence in fields of positive characteristic, based on the following criterion due to Christol, Kamae, Mendès France, and Rauzy: a formal power series is algebraic if and only if the sequence of its coefficients is automatic, that is it is the image by a letter-to-letter projection of a fixed point of a substitution of constant length. We also allude to the differences concerning transcendence between the real and the positive characteristic case. We then introduce some functions defined by Carlitz (exponential, logarithm, zeta) which are analogous to the corresponding real functions, and review the results of transcendence involving automata for these functions. We end this chapter by reviewing some techniques for disproving the automaticity of a sequence.
- **Chapter 4** is devoted to various partitions of the set of positive integers: Beatty sequences and connections with Sturmian sequences, partitions generated by substitutions, and similis partitions illustrated by linguistic properties of the Hungarian and Japanese languages. Special attention is devoted to non-periodic words which are shown to be fixed points of some combinatorial processes: above all the notion of log-fixed points is introduced. This study is illustrated by the Kolakoski word, which is shown to be not only a log-fixed point but also the unique fixed point of several maps: a map based on the Minkowski question-mark map, or maps defined by using the continued fraction expansion, the base-3 expansion, and so on. We then generalize these situations and present some open problems.

With any substitution, we can associate in a very natural way a dynamical system. In the **second part** we study these so-called substitutive dynamical systems, from the viewpoints of symbolic dynamics, ergodic theory and geometry.

- In **Chap. 5**, we introduce the fundamental notions of ergodic theory, also through the study of a few examples of symbolic dynamical systems, both in the topological and measure-theoretic framework. These include substitutions, which are maps on a finite alphabet, and which define naturally a class of infinite sequences, and the shift on this set. We study four examples of substitutions (Morse, Rudin-Shapiro, Fibonacci, already introduced in Chap. 2, and Chacon) and use them, together with symbolic notions (language, frequencies, complexity) to define basic ergodic notions (minimality, ergodicity), and begin the study of elementary spectral properties, geometric representation of symbolic systems, and the vast problem of joinings between systems.
- The main idea of **Chap. 6** is to show how it is possible to recover all the classical properties of rotations and continued fractions in a purely combinatorial way; we only use the combinatorial definition of Sturmian sequences, and obtain the existence of a combinatorial continued fraction acting on the set of Sturmian sequences. Some proofs, and also some geometric interpretations, become simpler and more natural in this setting; it also suggests non-trivial ways to generalize the usual continued fraction.
- **Chapter 7** presents an overview of the general spectral theory of substitutive dynamical systems. After recalling the tools and concepts required (including subshifts of finite type and adic systems, the notions of recognizability, and Markov partitions), a complete description of the related literature is given, including very recent work and some important conjectures in this subject.
- The problem of the geometric representation of substitutive dynamical systems is studied in detail in **Chap. 8**. Some basic tools and notions are introduced: stepped surface, generation by generalized substitutions, renormalization, study of the fractal boundary, and so on. Special attention is devoted to some important applications, as the quasi-periodicity of the tiling related to the stepped surface, the existence of Markov partitions of group automorphisms on $\mathbb{T}^3$ or Diophantine approximation properties in connection with the modified Jacobi-Perron algorithm.

In the **third part**, we extend the notion of substitution in two directions, the automorphisms of the free group and the piecewise linear maps. We also state a few of the many open problems related to substitutions.

- The purpose of **Chap. 9** is to study the properties of factors of sequences generated by invertible substitutions over a two-letter alphabet; this study is based on the following important result: invertible substitutions over a two-letter alphabet (that is, substitutions which are automorphisms of free groups) are shown to be exactly the Sturmian substitutions. We first discuss the local isomorphism between two invertible substitutions, reducing the problem to the study of some special classes of invertible substitutions. We then study some elementary properties of factors (including palindrome properties, powers of factors), which generalize some classical results for the Fibonacci sequence. By introducing the notion of singular factors, we establish a decomposition of the fixed points of invertible substitutions and we discuss the factor properties associated with this decomposition.

- **Chapter 10** investigates more deeply trace maps. Trace maps are dynamical systems attached to endomorphisms of free groups, which occur in a wide range of physical situations. By exploiting polynomial identities in rings of matrices, recursion formulas are produced between the traces of matrices defined by an induction using substitutions. We then study from an algebraic point of view endomorphisms of free monoids and free groups. Such endomorphisms are shown to induce a map of the ring of Fricke characters into itself. Particular emphasis is given to the group structure of trace maps and the Fricke-Vogt invariant.
- **Chapter 11** deals with Cantor sets generated by expanding piecewise linear maps. The main tool is the α-Fredholm matrix. This is the extension of the Fredholm matrix which is introduced to study the spectral properties of the Perron–Frobenius operator associated with one-dimensional maps. Using this α-Fredholm matrix, the Hausdorff dimension of the Cantor sets we consider is studied, as well as the ergodic properties of the dynamical system on it.
- This book ends with a survey in **Chap. 12** of some open problems in the subject. The first is the S-adic conjecture, about the equivalence for a minimal sequence of having sub-linear complexity and being generated by a finite number of substitutions. Then we look at possible generalizations of the interaction between rotations and Sturmian sequences through the usual continued fraction algorithm; these involve Arnoux-Rauzy sequences, interval exchanges and codings of the $\mathbb{Z}^2$-action of two rotations on the one-dimensional torus. This approach induces various open questions about geometric representations of substitutions, arithmetics in $SL(3,\mathbb{Z})$ and $SL(3,\mathbb{N})$, and about the definition in the two-dimensional case of some fundamental combinatorial objects as the complexity function or the notion of substitution. In **Appendix A**, J. Rivat states that infinitely many prime matrices exist in $SL(3,\mathbb{Z})$, contrary to what happens in the two-dimensional case.

Acknowledgements. We would like to express our gratitude to Jean-Paul Allouche, Guy Barat, Julien Cassaigne, Madelena Cropley-Gonzalez, Michel Dekking, Bernard Host, Michel Koskas, Éric Lozingot, Sebastian Mayer, Guy Mélançon, Michel Mendès France, François Parreau, Natalie Priebe, Alain Rémondière, Dierk Schleicher and Luca Q. Zamboni for their careful reading, and for their numerous suggestions, including pointing out mathematical and stylistic errors.

The editors

V. Berthé S. Ferenczi C. Mauduit A. Siegel

List of contributors

P. Arnoux (Chapters 6, 12)
Institut de Math. de Luminy
Case 907, 163 avenue de Luminy
F-13288 Marseille cedex 09
FRANCE
arnoux@iml.univ-mrs.fr

V. Berthé (Chapters 1, 3, 12)
Institut de Math. de Luminy
Case 907, 163 avenue de Luminy
F-13288 Marseille cedex 09
FRANCE
berthe@iml.univ-mrs.fr

S. Ferenczi (Chapter 5)
Institut de Math. de Luminy
Case 907, 163 avenue de Luminy
F-13288 Marseille cedex 09
FRANCE
ferenczi@iml.univ-mrs.fr

S. Ito (Chapter 8)
Department of Mathematics
Tsuda College
2-1-1 Tsuda Machi
Kodaira-shi, Tokyo 187
JAPON
ito@tsuda.ac.jp

C. Mauduit (Chapter 2)
Institut de Math. de Luminy
Case 907, 163 avenue de Luminy
F-13288 Marseille cedex 09
FRANCE
mauduit@iml.univ-mrs.fr

M. Mori (Chapter 11)
Department of Mathematics
College of Humanities and Sciences
Nihon University
3-25-40 Sakura Jyousui
Setagayaku Tokyo
JAPAN 156-8550
mori@math.chs.nihon-u.ac.jp,
mori@tsuda.ac.jp

J. Peyrière (Chapter 10)
Université Paris-Sud
Mathématiques, bât. 425
91405 Orsay Cedex
FRANCE
Jacques.PEYRIERE@math.u-psud.fr

A. Siegel (Chapters 1,7)
IRISA
Campus de Beaulieu
F-35042 Rennes Cedex
FRANCE
Anne.Siegel@irisa.fr

J.- I. Tamura (Chapter 4)
3-3-7-307 Azamino
Aoba-ku Yokohama
225-0011 JAPAN
jtamura@tsuda.ac.jp

Z.- Y. Wen (Chapter 9)
Tsinghua University
Department of Mathematics
Beijing 100084
P. R. CHINA
wenzy@tsinghua.edu.cn

Contents

Part III. Extensions to free groups and interval transformations

1. Basic notions on substitutions

The aim of this chapter is to introduce some concepts and to fix the notation we will use throughout this book. We will first introduce some terminology in combinatorics on words. These notions have their counterpart in terms of symbolic dynamics. We shall illustrate these definitions through the example of a particular sequence, the Morse sequence, generated by an algorithmic process we shall study in details in this book, namely a substitution. After recalling some basic notions on substitutions, we shall focus on the concept of automatic sequences. We then introduce the first notions of ergodic theory and focus on the spectral description of discrete dynamical systems.

Let us start by giving a short list of basic references concerning the concepts developed here. Let us first mention the indispensable [340] on substitution dynamical systems. For detailed introductions to the symbolic dynamics, see [59, 81, 247, 265]. More generally, there are a number of excellent books on dynamical systems and ergodic theory [62, 80, 122, 145, 185, 194, 234, 241, 308, 309, 324, 326, 335, 376, 445], and topological dynamics [50, 134, 160, 161]. For references on word combinatorics, automata and formal languages, see [28, 159, 212, 271, 272, 312, 325, 368] and the references therein. See also [55, 162] for an approach to fractals, and [113, 127, 381, 383] for connections with number theory and Diophantine approximation.

We shall use the following usual notation: $\mathbb{R}$ denotes the set of *real numbers*; $\mathbb{Q}$ denotes the set of *rational numbers*; $\mathbb{Z}$ denotes the set of *integer numbers*; $\mathbb{C}$ denotes the set of *complex numbers*; $\mathbb{N}$ denotes the set of *nonnegative integers* ($0 \in \mathbb{N}$); $\mathbb{N}^+$ denotes the set of positive integers ($0 \notin \mathbb{N}^+$). By x positive we mean $x > 0$, and by x nonnegative, we mean $x \geq 0$. The *one-dimensional torus* $\mathbb{R}/\mathbb{Z}$ is denoted by $\mathbb{T}$, the *d-dimensional torus* by $\mathbb{T}^d$, and the unit circle, that is, the set of complex numbers of modulus one, is denoted by $\mathbb{U}$. The cardinality of a set S shall be denoted by Card S.

Within each section, results (theorems, propositions, lemmas, formulas, exercises, and so on) are labelled by the section and then the order of occurrence.

[1] This chapter has been written by V. Berthé and A. Siegel

1.1 Word combinatorics

Let us first introduce some terminology in word combinatorics.

1.1.1 Sequences, words, languages

Letters. Let $\mathcal{A}$ be a finite set that shall be called *alphabet.* Throughout this book the letters (also called *symbols*) of the alphabet $\mathcal{A}$ shall either be denoted as digits ($\mathcal{A} = \{0, 1, \ldots, d-1\}$) or as letters ($\mathcal{A} = \{a_1, \ldots, a_d\}$).

Monoid of finite words. A *word* or *block* is a finite string of elements in $\mathcal{A}$. The set of all finite words over $\mathcal{A}$ is denoted by $\mathcal{A}^\star$. We denote by ε the empty word.

The *concatenation* of two words $V = v_1...v_r$ and $W = w_1...w_s$ is the word $VW = v_1...v_r w_1...w_s$. This operation is associative and has a unit element, the empty word ε.

The set $\mathcal{A}^\star$ is thus endowed with the structure of a monoid, and is called the *free monoid* generated by $\mathcal{A}$. The set $\mathcal{A}^\star - \{\varepsilon\}$ is endowed with the structure of a *free semi-group.*

If $W = w_1...w_s$, s is called the *length* of W and denoted by $|W|$. We denote by $|W|_a$ the number of occurrences of a letter $a \in \mathcal{A}$ appearing in a word $W \in \mathcal{A}^\star$.

Sequences. A (one-sided) *sequence* of elements of $\mathcal{A}$, also called *(right) infinite word* on $\mathcal{A}$, shall be here an element $u = (u_n)_{n\in\mathbb{N}}$ in $\mathcal{A}^{\mathbb{N}}$.

A *two-sided sequence* in $\mathcal{A}$, also called *biinfinite word,* shall be here an element $u = (u_n)_{n\in\mathbb{Z}}$ in $\mathcal{A}^{\mathbb{Z}}$. It is also denoted by $u = \ldots u_{-2}u_{-1}.u_0u_1\ldots$.

Language. A word $v_1...v_r$ is said to *occur* at *position* (or *index,* or *rank*) m in a sequence $u = (u_n)$ or in a finite word $u_1...u_s$, if there exists a rank m such that $u_m = v_1, \ldots, u_{m+r-1} = v_r$.

We say also that the word $v_1...v_r$ is a *factor* of the sequence u.

The *language* (respectively the language of length n) of the sequence u, denoted by $\mathcal{L}(u)$ (respectively $\mathcal{L}_n(u)$), is the set of all words (respectively of length n) in $\mathcal{A}^\star$ which occur in u.

Specific properties of sequences. A sequence u is said to be *recurrent* if every factor occurs infinitely often.

A sequence u is *periodic* (respectively *ultimately periodic*) if there exists a positive integer T such that

$$\forall n,\ u_n = u_{n+T} \quad \text{(respectively } \exists n_0 \in \mathbb{N},\ \forall |n| \geq n_0,\ u_n = u_{n+T}).$$

We will use in both cases the terminology *shift-periodic.*

1.1.2 Complexity function

There is a classical measure of disorder for sequences taking their values in a finite alphabet, the so-called complexity function (see [16, 169] for many results linked to the notion of complexity).

Complexity. Let u be a sequence; we call *complexity function* of u, and denote by $p_u(n)$, the function which with each positive integer n associates Card $\mathcal{L}_n(u)$, that is, the number of different words of length n occurring in u.

In other words, the complexity function counts the number of distinct factors that occur through a sliding window. The complexity function is obviously non-decreasing. For any positive integer n, one has furthermore $1 \leq p_u(n) \leq d^n$, where d denotes the cardinality of the alphabet $\mathcal{A}$. Note that the complexity function can be considered as a good measure of the disorder of a sequence as it is smallest for periodic sequences. Namely, a basic result of [124] is the following.

Proposition 1.1.1. *If u is a periodic or ultimately periodic sequence, $p_u(n)$ is a bounded function. If there exists an integer n such that $p_u(n) \leq n$, u is an ultimately periodic sequence.*

Proof. The first part is trivial. In the other direction, we have $p_u(1) \geq 2$ otherwise u is constant, so $p_u(n) \leq n$ implies that $p_u(k+1) = p_u(k)$ for some k. For each word W of length k occurring in u, there exists at least one word of the form Wa occurring in u, for some letter $a \in \mathcal{A}$. As $p_u(k+1) = p_u(k)$, there can be only one such word. Hence, if $u_i \ldots u_{i+k-1} = u_j \ldots u_{j+k-1}$, then $u_{i+k} = u_{j+k}$. As the set $\mathcal{L}_k(u)$ is finite, there exist $j > i$ such that $u_i \ldots u_{i+k-1} = u_j \ldots u_{j+k-1}$, and hence $u_{i+p} = u_{j+p}$ for every $p \geq 0$, one period being $j - i$. ■

Special word. Let W be a factor of the sequence $u \in \mathcal{A}^{\mathbb{N}}$. A *right extension* (respectively *left extension*) of the factor W is a word Wx (respectively xW), where $x \in \mathcal{A}$, such that Wx (respectively xW) is also a factor of the sequence u.

A factor is said to be a *left special factor* (respectively *right special factor*) if it has more than one left (respectively right) extension.

Let W^+ (respectively W^-) denote the number of right (respectively left) extensions of W. We have

$$p_u(n+1) - p_u(n) = \sum_{W \in \mathcal{L}_n(u)} (W^+ - 1) = \sum_{W \in \mathcal{L}_n(u)} (W^- - 1).$$

This equality is a very useful tool for the computation of the complexity: see for instance Exercise 5.4.11 in Chap. 5, and also Chap. 6.

Remark. One can also introduce the notion of bispecial factors: these are the factors which are simultaneously right and left special factors (for a more detailed exposition, see [107]).

Topological entropy. There is a natural notion of entropy associated with the complexity function. The *topological entropy* of the sequence u is defined as the exponential growth rate of the complexity of u as the length increases:

$$H_{top}(u) = \lim_{n \to +\infty} \frac{\log_d(p_u(n))}{n},$$

where d denotes the cardinality of the alphabet $\mathcal{A}$. The existence of the limit follows from the subadditivity of the function $n \mapsto \log_d(p_u(n))$:

$$\forall m, n, \ \log_d(p_u(n+m)) \leq \log_d(p_u(m)) + \log_d(p_u(n)).$$

This notion of topological entropy comes from topological dynamics. One can associate in a natural way (see Sec. 1.1.3 below) a topological dynamical system with a sequence. The topological entropy of a sequence is nothing else than the topological entropy of its associated dynamical system. For more details on the entropy of dynamical systems, see for instance [393, 445].

1.1.3 Symbolic dynamical systems

Let us introduce some basic notions in symbolic dynamics. For expository books on the subject, see [59, 81, 247, 265, 340]. A more detailed exposition of this subject shall be made in Chap. 5.

The set $\mathcal{A}^{\mathbb{N}}$ shall be equipped with the product topology of the discrete topology on each copy of $\mathcal{A}$. Thus, this set is a compact space. The topology defined on $\mathcal{A}^{\mathbb{N}}$ is the topology defined by the following distance:

$$\text{for } u \neq v \in \mathcal{A}^{\mathbb{N}}, \ d(u, v) = 2^{-\min\{n \in \mathbb{N};\ u_n \neq v_n\}},$$

Thus, two sequences are close to each other if their first terms coincide. Note that the space $\mathcal{A}^{\mathbb{N}}$ is complete as a metric compact space. Furthermore, it is a *Cantor set*, that is, a totally disconnected compact set without isolated points.

Let S denote the following map defined on $\mathcal{A}^{\mathbb{N}}$, called the *one-sided shift*:

$$S((u_n)_{n \in \mathbb{N}}) = (u_{n+1})_{n \in \mathbb{N}}.$$

The map S is uniformly continuous, onto but not one-to-one on $\mathcal{A}^{\mathbb{N}}$.

All these notions extend in a natural way to $\mathcal{A}^{\mathbb{Z}}$, the distance on $\mathcal{A}^{\mathbb{Z}}$ being defined as:

$$\text{for } u \neq v \in \mathcal{A}^{\mathbb{Z}}, \ d(u, v) = 2^{-\min\{n \in \mathbb{N};\ u_{|n|} \neq v_{|n|}\}}.$$

Here the shift S is one-to-one.

System associated with a sequence. The *symbolic dynamical system* associated with a one-sided (respectively two-sided) sequence u with values in $\mathcal{A}$ is the system $(\overline{\mathcal{O}(u)}, S)$, where $\overline{\mathcal{O}(u)} \subset \mathcal{A}^{\mathbb{N}}$ (respectively $\mathcal{A}^{\mathbb{Z}}$) is the closure of the *orbit* of the sequence u under the action of the shift S; the orbit is the set $\{S^n u, n \in \mathbb{N}\}$ (respectively $\{S^n u, n \in \mathbb{Z}\}$). Note that $\overline{\mathcal{O}(u)}$ is finite if and only if u is shift-periodic.

The set $X_u := \overline{\mathcal{O}(u)}$ is a compact space and S is a continuous map from X_u to X_u. Indeed, the set X_u is a closed subset of the compact set $\mathcal{A}^{\mathbb{N}}$, and hence compact; $d(Sx, Sy) \leq 2d(x, y)$, therefore the continuity. This implies that if $x = \lim_{n \to +\infty} S^{k_n} u$, then $Sx = \lim_{n \to +\infty} S^{k_n + 1} u$ and $Sx \in X_u$.

Lemma 1.1.2. *For every sequence $w \in \mathcal{A}^{\mathbb{N}}$, the following statements are equivalent:*

1. *the sequence $w \in X_u$;*
2. *there exists a sequence $(k_n)_{n \in \mathbb{N}}$ such that $w_0 ... w_n = u_{k_n} ... u_{k_n + n}$ for every $n \geq 0$;*
3. *$\mathcal{L}_n(w) \subset \mathcal{L}_n(u)$ for all n.*

Proof. Statement 1. is equivalent to $d(w, S^{k_n} u) < 2^{-n}$ for some sequence k_n, and that is exactly 2.; 2. implies 3. as any word occurring in w must occur in some $w_0 ... w_n$, and 3. implies 2. because $w_0 ... w_n$ is in $\mathcal{L}_{n+1}(u)$. ∎

Exercise 1.1.3. Prove that a sequence u is recurrent if and only if there exists a strictly increasing sequence $(n_k)_{k \in \mathbb{N}}$ such that $u = \lim_{k \to +\infty} S^{n_k} u$ (that is, u is a cluster point of X_u). Deduce that the sequence u is recurrent if and only if S is onto on $\overline{\mathcal{O}(u)}$ (see also Lemma 5.1.11).

Cylinders. For a word $W = w_0 ... w_r$, the *cylinder set* $[W]$ is the set $\{v \in X_u; v_0 = w_0, ..., v_r = w_r\}$.

The cylinder sets are *clopen* (open and closed) sets and form a basis of open sets for the topology of X_u. Indeed, if the cylinder $[W]$ is nonempty and v is a point in it, $[W]$ is identified with both the open ball $\{v'; d(v, v') < 2^{-n}\}$ and the closed ball $\{v'; d(v, v') \leq 2^{-n-1}\}$.

Exercise 1.1.4. Prove that a clopen set is a finite union of cylinders.

Remark. Note that the topology extends in a natural way to $\mathcal{A}^{\mathbb{N}} \cup \mathcal{A}^{\star}$. Indeed, let $\mathcal{B}$ be a new alphabet obtained by adding a further letter to the alphabet $\mathcal{A}$; words in $\mathcal{A}^{\star}$ can be considered as sequences in $\mathcal{B}^{\mathbb{N}}$, by extending them by the new letter in $\mathcal{B}$. The set $\mathcal{A}^{\mathbb{N}} \cup \mathcal{A}^{\star}$ is thus metric and compact, as a closed subset of $\mathcal{B}^{\mathbb{N}}$. This will be needed in particular in Section 1.2.1.

The cylinders are defined over $\mathcal{A}^{\mathbb{Z}}$ as the following clopen sets, where $W_1, W_2 \in \mathcal{A}^{\star}$:

$$[W_1.W_2] = \{(v_i)_i \in X_u | v_{-|W_1|} \ldots v_{-1}.v_0 \ldots v_{|W_2|-1} = W_1 W_2\}.$$

Depending on the chapter, we will work either with $\mathcal{A}^{\mathbb{N}}$ or with $\mathcal{A}^{\mathbb{Z}}$.

Consider now the notion of *minimal* sequence (also called *uniform recurrence*):

Minimality. A sequence $u = (u_n)$ is minimal (or uniformly recurrent) if every word occurring in u occurs in an infinite number of positions with bounded gaps, that is, if for every factor W, there exists s such that for every n, W is a factor of $u_n \dots u_{n+s-1}$.

Exercise 1.1.5. Build an example of a sequence which is recurrent but not minimal.

Let us note that the term "minimality" comes from symbolic dynamics: a sequence u is minimal if and only if the dynamical system $(\overline{\mathcal{O}(u)}, S)$ is minimal. For more details, see Sec. 1.4.1, Sec. 5.1.4, and for instance, [340].

1.1.4 Sturmian sequences

Let us illustrate the preceding section by evoking an important family of sequences, the so-called Sturmian sequences.

Sturmian sequences. A *Sturmian sequence* is defined as a (one-sided) sequence u the complexity function p_u of which satisfies:

$$\forall n \in \mathbb{N}, \ p_u(n) = n + 1.$$

In particular, a Sturmian sequence is defined over a two-letter alphabet. Chaps. 6 and 9 shall study in detail the properties of Sturmian sequences. The more classical example of a Sturmian sequence is the Fibonacci sequence, defined in Sec. 1.2.1 below; a proof of the fact that the Fibonacci sequence is Sturmian is given in Exercise 5.4.11.

Exercise 1.1.6. For any $s \geq 1$, construct an example of a sequence u with complexity $p_u(n) = s + n$. *(Hint: start from a Sturmian sequence and add new letters.)*

Remark. One can define Sturmian sequences over a larger size alphabet as recurrent sequences of complexity $n + s$: see [108, 147, 202] and Chap. 4.

Arnoux-Rauzy sequences. One can also consider a generalization of Sturmian sequences over a three-letter alphabet, namely the *Arnoux-Rauzy sequences* introduced in [49]. These are recurrent sequences defined over a three-letter alphabet with complexity $2n+1$ with the following extra combinatorial property: for every n, there is exactly one right special factor and one left special factor of length n, and these special factors can be extended in three different ways. This condition is called the * condition in the seminal paper [49], and more generally in the literature on the subject. A further generalization of this notion is given in [149].

Exercise 1.1.7. Prove that the Arnoux-Rauzy sequences are uniformly recurrent. *(Hint: prove that the right special factors appear with bounded gaps.)*

One shall consider these sequences in Chap. 12.

1.2 Substitutions

Substitutions are maps defined over the set of words $\mathcal{A}^\star$, which generate in a natural way infinite sequences with low complexity function.

1.2.1 Definition

Substitution. A *substitution* σ is an application from an alphabet $\mathcal{A}$ into the set $\mathcal{A}^\star - \{\varepsilon\}$ of nonempty finite words on $\mathcal{A}$; it extends to a morphism of $\mathcal{A}^\star$ by concatenation, that is, $\sigma(WW') = \sigma(W)\sigma(W')$ and $\sigma(\varepsilon) = \varepsilon$. It also extends in a natural way to a map defined over $\mathcal{A}^{\mathbb{N}}$ or $\mathcal{A}^{\mathbb{Z}}$.

It is called *of constant length* k if $\sigma(a)$ is of length k for any $a \in A$.

Fixed point. A *fixed point* of the substitution σ is an infinite sequence u with $\sigma(u) = u$.

A *periodic point* of σ is an infinite sequence u with $\sigma^k(u) = u$ for some $k > 0$.

An n*-word* for σ is any one of the words $\sigma^n(a)$ for $a \in \mathcal{A}$.

Remark. Substitutions are very efficient tools for producing sequences. Let σ be a substitution over the alphabet $\mathcal{A}$, and a be a letter such that $\sigma(a)$ begins with a and $|\sigma(a)| \geq 2$. Then there exists a unique fixed point u of σ beginning with a. This sequence is obtained as the limit in $\mathcal{A}^\star \cup \mathcal{A}^{\mathbb{N}}$ (when n tends toward infinity) of the sequence of n-words $\sigma^n(a)$, which is easily seen to converge.

Exercise 1.2.1. Prove that every substitution σ such that $|\sigma^n(a)|$ tends to infinity with n, for every $a \in \mathcal{A}$, has at least one periodic point.

Let us consider a classical example of a substitution of constant length, namely the Morse substitution. For a nice survey of the many properties of this substitution, see [27]; see also [284]. This sequence will be studied in detail in Chaps. 2 and 5.

The Morse sequence. The *Morse sequence* u is the fixed point beginning with a of the *Morse substitution* σ_1 defined over the alphabet $\{a, b\}$ by $\sigma_1(a) = ab$ and $\sigma_1(b) = ba$.

$$u = abbabaabbaababbabaababbaabbabaabbaababbaabbabaababbabaabbaabab...$$

Another classical example, namely the Fibonacci substitution, provides the more natural example of Sturmian sequence (see Chaps. 2, 6, 8 and 9).

The Fibonacci sequence. The *Fibonacci sequence* is the fixed point v beginning with a of the the *Fibonacci substitution* σ_2 defined over the two-letter alphabet $\{a, b\}$ by $\sigma_2(a) = ab$ and $\sigma_2(b) = a$.

$$v = abaababaabaababaababaabaababaabaababaababaabaababaababaaba...$$

Invertible substitution. Let Γ_d denote the free group over d-letters. A substitution over a d-letter alphabet can be naturally extended to Γ_d by defining $\sigma(s^{-1}) = (\sigma(s))^{-1}$. It is said to be *invertible* if there exists a map $\eta : \Gamma_d \to \Gamma_d$ such that $\sigma\eta(a) = \eta\sigma(a) = a$ for every $a \in \Gamma_d$.

The Fibonacci substitution is invertible: its inverse is $a \mapsto b$, $b \mapsto b^{-1}a$. For a nice introduction to the free groups and invertible substitutions, see [69]. See also [275, 278, 416].

1.2.2 Abelianization

There is a convenient way to associate with a substitution a matrix, namely the incidence matrix of the substitution σ.

Incidence matrix. Let σ be a substitution defined over the alphabet $\mathcal{A} = \{a_1, \ldots, a_d\}$ of cardinality d. The *incidence matrix* of the substitution σ is, by definition, the $d \times d$ matrix $\mathbf{M}_\sigma$ the entry of index (i, j) of which is $|\sigma(a_j)|_{a_i}$, that is, the number of occurrences of a_i in $\sigma(a_j)$.

Let us note that for every $(i, j) \in \{1, 2, \ldots, d\}^2$ and for every $n \in \mathbb{N}$, $|\sigma^n(a_j)|_{a_i}$ is equal to the coefficient of index (i, j) of the matrix $\mathbf{M}_\sigma^n$.

Example 1.2.2. The incidence matrices of the Morse substitution σ_1 and the Fibonacci substitution σ_2 are respectively the matrices:

$$\mathbf{M}_{\sigma_1} = \begin{pmatrix} 1 & 1 \\ 1 & 1 \end{pmatrix}, \quad \mathbf{M}_{\sigma_2} = \begin{pmatrix} 1 & 1 \\ 1 & 0 \end{pmatrix}.$$

Unimodular substitution. A substitution is said to be *unimodular* if the determinant of its incidence matrix is ± 1. The Fibonacci substitution is unimodular. The Morse substitution is not.

Canonical homomorphism. Let σ be a substitution defined over the alphabet $\mathcal{A} = \{a_1, \ldots, a_d\}$ of cardinality d. Let $\mathbf{l} : \mathcal{A}^\star \to \mathbb{Z}^d$ denote the *canonical homomorphism*, also called *homomorphism of abelianization*, defined as follows:

$$\forall W \in \mathcal{A}^\star, \quad \mathbf{l}(W) = (|w|_{a_i})_{1 \le i \le d} \in \mathbb{N}^d.$$

As a consequence, the incidence matrix $\mathbf{M}_\sigma$ satisfies:

$$\mathbf{M}_\sigma = (\mathbf{l}(\sigma(a_1)), \ldots, \mathbf{l}(\sigma(a_d)))\,.$$

Furthermore, we have the following commutative relation:

$$\forall W \in \mathcal{A}^\star, \ \mathbf{l}(\sigma(W)) = \mathbf{M}_\sigma \mathbf{l}(W).$$

This notion will be needed in Chaps. 7, 8 and 10.

1.2.3 Primitivity

Let us introduce the notion of primitivity.

Definition. A substitution σ over the alphabet $\mathcal{A}$ is *primitive* if there exists a positive integer k such that, for every a and b in $\mathcal{A}$, the letter a occurs in $\sigma^k(b)$.

Proposition 1.2.3. *If σ is primitive, any of its periodic points is a minimal sequence.*

Proof. Let $u = \sigma^p(u)$ be a periodic point of σ. We have $u = (\sigma^p)^k(u) = (\sigma^p)^k(u_0)(\sigma^p)^k(u_1)...$; for any $b \in \mathcal{A}$, a occurs in $(\sigma^p)^k(b)$, hence a occurs in u infinitely often with bounded gaps; but then so does every $(\sigma^p)^n(a)$ in $u = (\sigma^p)^n(u)$, hence so does any word occurring in u. ■

Remark. The following substitution is not primitive but its fixed point beginning by 0 is a minimal infinite sequence:

$$0 \mapsto 0010, \; 1 \mapsto 1.$$

This sequence is called the *Chacon sequence.* Indeed, it is not difficult to check that Chacon's sequence begins with the following sequence of words (b_n):

$$b_0 = 0, \text{ and } \forall n \in \mathbb{N}, \; b_{n+1} = b_n b_n 1 b_n.$$

One thus can deduce from this that Chacon's sequence is minimal; for more details, see Chap. 5. Notice that a rule of the form $b_{n+1} = b_n b_n 1 b_n$ is called a *catenative rule* (see [365] for introducing the notion of locally catenative rules and [390] for a generalization of this notion).

A characterization of primitivity. Let us add now the following extra assumptions on substitutions:

1. there exists a letter $a \in \mathcal{A}$ such that $\sigma(a)$ begins with a (this condition guarantees the existence of a fixed point);
2. $\lim_{n \to \infty} |\sigma^n(b)| = \infty$, for every letter $b \in \mathcal{A}$.

Let us note that these two conditions imply the existence of an (infinite) fixed point beginning with a. We will furthermore suppose:

3. all the letters in $\mathcal{A}$ actually occur in this fixed point.

Let us note that, given a primitive substitution, then there exists a power of this substitution for which these three conditions hold.

Under these assumptions, we have the following equivalence:

Proposition 1.2.4. *Let σ be a substitution satisfying the above conditions. The substitution σ is primitive if and only if the fixed point of σ beginning by a is minimal.*

The proof is left as an exercise (see also [340]).

1.2.4 The Perron–Frobenius theorem

The primitivity of a substitution can easily be expressed in terms of its incidence matrix. Indeed the substitution σ is primitive if and only if there exists an integer k such that the k-th power of its incidence matrix $\mathbf{M}_\sigma$ has positive entries. We thus say that a matrix is *primitive* if its entries are nonnegative and if there exists an integer k such that the k-th power of the matrix has positive entries. A matrix $\mathbf{M}_\sigma$ is said *irreducible* if it has no nontrivial invariant space of coordinates.

Exercise 1.2.5 ([189]).

Let $\mathbf{M}$ be an irreducible matrix (of size n).

1. Prove that $(\mathbf{I}+\mathbf{M})^{n-1} > 0$. *(Hint: prove that for any nonnegative vector* $\mathbf{y}$*, the vector* $(\mathbf{I}+\mathbf{M})\mathbf{y}$ *has a number of zero coordinates which is strictly smaller than that of* $\mathbf{y}$*.)*
2. Deduce that $\mathbf{M}$ satisfies the following: for any (i,j), there exists a positive integer k such that the entry of index (i,j) of $\mathbf{M}^k$ is positive. *(Hint: consider the binomial expansion of* $(\mathbf{I}+\mathbf{M})^{n-1}$*.)*

Primitive matrices are also called *irreducible and aperiodic matrices.* See [63, 189, 387] for more details on matrices with nonnegative coefficients.

Irreducible matrices (and hence primitive matrices) satisfy the Perron–Frobenius theorem:

Theorem 1.2.6 (Perron–Frobenius' theorem). *Let* $\mathbf{M}$ *be a nonnegative* irreducible *matrix. Then* $\mathbf{M}$ *admits a strictly positive eigenvalue* α *which dominates in modulus the other eigenvalues* λ: $\alpha \geq |\lambda|$. *The eigenvalue* α *is a simple eigenvalue and there exists an eigenvector with positive entries associated with* α.

Furthermore, if $\mathbf{M}$ *is* primitive, *then the eigenvalue* α *dominates strictly in modulus the other eigenvalues* λ: $\alpha > |\lambda|$.

This important property of primitive matrices implies the existence of frequencies for every factor of a fixed point of a primitive substitution.

Frequencies. Let u be a sequence. The *frequency* f_W of a factor W of u is defined as the limit (when n tends towards infinity), if it exists, of the number of occurrences of the factor W in $u_0u_1\ldots u_{n-1}$ divided by n.

Theorem 1.2.7. *Let* σ *be a primitive substitution satisfying the conditions mentioned in Sec. 1.2.3. Let* u *be a fixed point of* σ. *Then every factor of* u *has a frequency. Furthermore, all the frequencies are positive. The frequencies of the letters are given by the coordinates of the positive eigenvector associated with the dominating eigenvalue, renormalized in such a way that the sum of its coordinates equals* 1.

For a proof of this result, see Chap. 5 or [340].

1.2.5 Substitutions of Pisot type

The case where the dominant eigenvalue is a Pisot number is of particular interest.

Pisot-Vijayaraghavan number. Let us recall that an algebraic integer $\alpha > 1$ is a *Pisot-Vijayaraghavan number* or a *Pisot number* if all its algebraic conjugates λ other than α itself satisfy $|\lambda| < 1$. This class of numbers has been intensively studied and has some special Diophantine properties (see for instance [113]).

Pisot type substitution. A substitution is *of Pisot type* if the eigenvalues of the incidence matrix satisfy the following: there exists a dominant eigenvalue α such that for every other eigenvalue λ, one gets

$$\alpha > 1 > |\lambda| > 0.$$

Proposition 1.2.8 ([103]). *Let σ be a Pisot substitution of Pisot type. Then the characteristic polynomial χ_σ of the incidence matrix $\mathbf{M}_\sigma$ is irreducible over $\mathbb{Q}$. Hence, the dominant eigenvalue α is a Pisot number and the matrix $\mathbf{M}_\sigma$ is diagonalizable (over $\mathbb{C}$), the eigenvalues being simple. Furthermore, σ cannot be of constant length.*

Proof. Suppose there are two polynomials Q and R with integer coefficients such that $QR = \chi_\sigma$. As 0 is not a root, each of the polynomials Q and R has a root in $\mathbb{C}$ with modulus larger than or equal to 1, which implies that at least two eigenvalues have a modulus larger than or equal to 1, hence the irreducibility over $\mathbb{Q}$. We deduce from this that the roots in $\mathbb{C}$ are simple. If σ is of constant length l, then l is an eigenvalue for the eigenvector $(1, 1, \cdots, 1)$, which is a contradiction for the irreducibility of the characteristic polynomial. ∎

Remark. Following the previous definition, 0 cannot be an eigenvalue of the incidence matrix of a Pisot substitution. Hence, the Morse substitution is not Pisot.

It is worth noticing the following result which is a direct consequence of the theory of irreducible matrices:

Theorem 1.2.9 ([103]). *Any Pisot type substitution is primitive.*

Proof. We deduce from the irreducibility of χ_σ, that the matrix $\mathbf{M}_\sigma$ is irreducible. The proof of the theorem is thus a direct consequence of the following result (see for instance [63]): let $\mathbf{M}$ be a matrix with nonnegative entries; $\mathbf{M}$ is primitive if and only if $\mathbf{M}$ is irreducible and the spectral radius of $\mathbf{M}$ is greater in magnitude than any other eigenvalue. ∎

We shall focus on substitutions of Pisot type in Chaps. 7, 8 and 12.

1.2.6 Substitutive dynamical systems

We deduce from Proposition 1.2.3, that if σ is primitive, then all its periodic points have the same language. In other words, all the symbolic dynamical systems associated with all its periodic points do coincide.

Substitutive system. Hence, we can associate in a natural way with a primitive substitution a symbolic dynamical system following Sec. 1.1.3. Indeed, if σ is primitive, the *symbolic dynamical system associated with* σ is that, denoted by (X_σ, S), generated by any of its periodic points. It is finite if and only if there is a periodic point for σ which is also shift-periodic. In this case, the substitution is also called *shift-periodic.* We are mainly interested in primitive and not shift-periodic substitutions.

Note that C. Holton and L. Q. Zamboni prove that a substitution of Pisot type cannot be shift-periodic:

Proposition 1.2.10 ([208]). *If σ is a primitive substitution the matrix of which has a nonzero eigenvalue of modulus less than 1, then no fixed point u of σ is shift-periodic.*

In particular, all the usual hypotheses (primitivity, non-periodicity) are satisfied by Pisot type substitutions. In Chaps. 7 and 8, we will see that Pisot type substitutive systems have very special spectral and geometrical properties.

Topological properties. The dynamical system associated with a primitive substitution can be endowed with a Borel probability measure μ. Furthermore this measure is *invariant* under the action of the shift S, that is, $\mu(S^{-1}B) = \mu(B)$, for every Borel set B. Indeed, this measure is uniquely defined by its values on the cylinders. The measure of the cylinder $[W]$ is defined as the frequency of the finite word W in any element of X_σ, which does exist following Theorem. 1.2.7.

Let us note that the system (X_σ, S) is *uniquely ergodic*: there exists a unique shift-invariant measure. For a proof, see [340], and see also Chap. 5. We will come back to this notion later in this chapter.

1.3 Automata

Among substitutions, substitutions of constant length play an important role. The aim of this section is to prove that they are connected in a fundamental way with automatic sequences.

1.3.1 Definition

Let k be an integer greater than or equal to 2. Let us introduce the notion of k-automaton: such an automaton is a finite complete deterministic automaton (also called 2-*tape automaton or transducer*). For more details, the reader is referred to Chaps. 2 and 5; see also [121, 159, 212] and the surveys [11, 22].

Definition of a k-automaton. A k-automaton is represented by a directed graph defined by:

- a finite set $\mathcal{S}$ of vertices called *states*. One of these states is called the *initial state* and is denoted by i;
- k oriented edges called *transition maps* from the set of states $\mathcal{S}$ into itself, labelled by integers between 0 and $k-1$;
- a set Y and a map φ from $\mathcal{S}$ into Y, called *output function* or *exit map.*

Automaticity. A sequence $(u(n))_{n\in\mathbb{N}}$ with values in Y is called k-*automatic* if it is generated by a k-automaton as follows: let $\sum_{i=0}^{j} n_i k^i$ $(n_j \neq 0)$ be the base k expansion of the integer n; starting from the initial state one feeds the automaton with the sequence $n_0, n_1 \dots n_j$ (the digits being read in increasing order of powers); after doing this the automaton is in the state $a(n)$; then put $u(n) := \varphi(a(n))$. The automaton is then said to generate the k-automatic sequence in *reverse reading.*

One can similarly give another definition of k-automaticity by reading the digits in the reverse order, i.e., by starting with the most significant digit (that is, starting with n_j), but these two notions are easily seen to be equivalent (see Proposition 1.3.4 below). The automaton is then said to generate the k-automatic sequence in *direct reading.* We will use the terminology k-*automatic in direct reading* or *in reverse reading,* until Proposition 1.3.4 will be proved.

Remark. An automaton is considered here as a machine producing a sequence, and not as usual as a machine recognizing a language.

1.3.2 Cobham's theorem

There is an important connection between automatic sequences and fixed points of substitutions of constant length. The following characterization of automatic sequences is due to Cobham (see [121]).

Letter-to-letter projection. Consider a map from a finite alphabet $\mathcal{A}$ to a finite alphabet, say $\mathcal{B}$. This map extends in a natural way by concatenation to a map from $\mathcal{A}^\star \cup \mathcal{A}^\mathbb{N}$ to $\mathcal{B}^\star \cup \mathcal{B}^\mathbb{N}$. Such a map is called a *letter-to-letter projection.*

Proposition 1.3.1. *A sequence u is k-automatic in direct reading if and only if u is the image by a letter-to-letter projection of a fixed point of a substitution of constant length k.*

Proof. Let us give a constructive proof of this equivalence by showing that the image by a letter-to-letter projection of the fixed point of a substitution of constant length is generated by an automaton in direct reading with as set of states the alphabet of the substitution, with transition maps given by the substitution, and with as output function, the projection p.

- Let us first start with the fixed point $u = (u(n))_{n\in\mathbb{N}}$ of the substitution σ of constant length k defined over the alphabet $\mathcal{A}$. Define $\sigma_i : \mathcal{A} \to \mathcal{A}$, which sends the letter a to the $(i+1)$-th letter of $\sigma(a)$. We then have $\sigma(a) = \sigma_0(a) \cdots \sigma_{k-1}(a)$. Furthermore, for any integer n,
$$\sigma(u_n) = u_{kn}u_{kn+1} \ldots u_{kn+k-1} = \sigma_0(u_n) \cdots \sigma_{k-1}(u_n),$$
that is, $\sigma_i(u(n)) = u(kn+i)$, for any integer n and for any $i \in [0, k-1]$.
Let us construct a k-automaton A in direct reading which recognizes the sequence u: consider as set of states the alphabet $\mathcal{A}$, and as set of edges the maps σ_i. There is an edge from a to b if b occurs in $\sigma(a)$, labelled by i if b is the $i+1$-th letter of $\sigma(a)$. Put as initial state u_0 the first letter of u. Define as output function the identity.
It is easily seen that the k-automaton A generates in direct reading the sequence u. Indeed, write $n = \sum_{i=0}^{j} n_i k^i$: we start from $u(0)$, then go to the $n_t + 1$-th letter of $\sigma(u(0))$, denoted by a_1, then go to the $n_{t-1} + 1$-th letter of $\sigma(a_1)$, which is also the $kn_t + n_{t-1} + 1$-th letter of $\sigma^2(u(0))$, and, after n steps, we arrive to the $n+1$-th letter of $\sigma^{t+1}(u(0))$, which is $u(n)$.
If now, v is the image by a letter-to-letter projection $p : \mathcal{A} \to \mathcal{B}$ of the fixed point $u = (u(n))_{n\in\mathbb{N}} \in \mathcal{A}^{\mathbb{N}}$ of a substitution σ of constant length k defined on the alphabet $\mathcal{A}$, then v is generated by the automaton previously defined for u, but with output function the projection p.
- Conversely, let u be a sequence generated by a k-automaton A in direct reading. Let $\mathcal{S}$ be the set of states of the automaton A and let $f_0, \cdots, f_{k-1}$ be the transition maps. Define the substitution of constant length $\sigma = f_0 \cdots f_{k-1}$ over $\mathcal{S}$. Let v be the fixed point of σ beginning with the initial state i (note that the edge f_0 maps i onto itself by definition, hence the existence of a fixed point). It is easily checked that the sequence u is the image by the output function φ of the fixed point v. ■

Remark. Following Prop. 1.3.1, k-automatic sequences are also called *k-uniform tag sequences* in [121].

Example 1.3.2. One deduces from the proposition above that the 2-automaton (Fig. 1.1) with initial state a and exit map $Id_{\{a,b\}}$ generates the Morse sequence in direct reading.

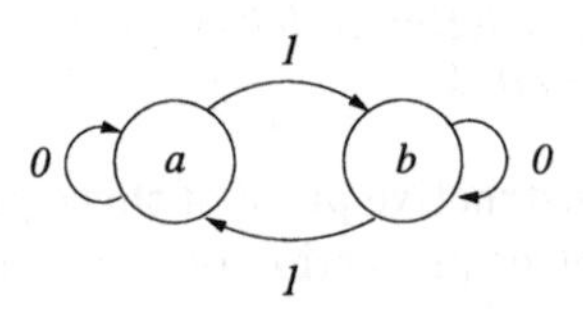

Fig. 1.1. Automaton associated with the Morse sequence.

1.3.3 The kernel of a sequence

The aim of this section is to give a further characterization of automaticity in terms of a family of subsequences.

Proposition 1.3.3. *A sequence $u \in \mathcal{A}^{\mathbb{N}}$ is k-automatic in reverse reading if and only if the* k-kernel $N_k(u)$ *of the sequence u is finite, where $N_k(u)$ is the set of subsequences of the sequence $(u(n))_{n\in\mathbb{N}}$ defined by*

$$N_k(u) = \{(u(k^l n + r))_{n\in\mathbb{N}};\ l \geq 0;\ 0 \leq r \leq k^l - 1\}.$$

Proof.
Let us first suppose that the k-kernel $N_k(u)$ of the sequence u is finite. We denote by $u = u_1, u_2, \cdots, u_d$ the sequences of $N_k(u)$. Consider a finite set of d states $\mathcal{S} = \{a_1, \cdots, a_d\}$ in bijection with $N_k(u)$ (a_j denotes the state in bijection with the sequence u_j). Define, for any r in $[0, k-1]$, the map $r : \mathcal{S} \to \mathcal{S}$ which associates with any state a_i the state $r.a_i$ in bijection with the sequence of the k-kernel $(u_i(kn+r))_{n\in\mathbb{N}}$. Let $n = \sum_{i=0}^{j} n_i k^i$ be the base k expansion of the integer n, with $n_i \in [0, k-1]$ and $n_j \neq 0$. We define the map n from $\mathcal{S}$ into itself by $n(a_i) = n_j(n_{j-1}(\cdots(n_0(a_i))))$, if $n \neq 0$, otherwise the map 0 equals the identity. It is easily seen that $n.a_1$ is the state in bijection with $(u(k^{j+1}l + n))_{l\in\mathbb{N}}$. Hence if $n.a_1 = m.a_1$, then $u(n) = u(m)$, by considering the first term of the two corresponding subsequences. We thus can define as output function the map φ (which is well defined) which associates with a state a_i the value $u(n)$ for any integer such that $n.a_1 = a_i$ (such an integer always exists by construction). The sequence u is hence generated by the automaton in reverse reading with states $\mathcal{S}$, initial state a_1, transition maps $0, 1, \cdots, k-1$ and output function φ.

Let us show that a k-automatic sequence u in reverse reading has a finite k-kernel. Let A be a finite automaton which generates the sequence u of initial state i. The subsequence $(u(k^l n + r))_{n\geq 0}$, where $l \geq 0$ and $0 \leq r \leq k^l - 1$, is generated by A with the initial state $\overline{r}.i$, where $\overline{r}$ is the word of k letters obtained by concatenating in front of the base q expansion of r as many zeros as necessary. As the automaton A has a finite number of states, we hence obtain a finite number of subsequences. ∎

Proposition 1.3.4. *A sequence is k-automatic in direct reading if and only if it is k-automatic in reverse reading.*

Proof. Following Proposition 1.3.1 and Proposition 1.3.3, it is sufficient to prove the equivalence between being the image by a letter-to-letter projection of the fixed point of a substitution of constant length k and having a finite k-kernel.

Let us prove that the image by a letter-to-letter projection of the fixed point of a substitution of constant length k has a finite k-kernel. It suffices to prove this result for the fixed point $(u(n))_{n\in\mathbb{N}}$ of a substitution σ of constant length k defined over the alphabet $\mathcal{A}$. For $0 \leq i \leq k-1$, define as in the preceding proof $\sigma_i : \mathcal{A} \to \mathcal{A}$ which associates with a letter x the $(i+1)$-th letter in its image under σ. We have $\sigma_i(u(n)) = u(kn+i)$, for any integer n and for any $i \in [0, k-1]$. Let $l \geq 0$, $0 \leq r \leq k^l - 1$. Write $r = \sum_{i=0}^{l-1} r_i k^i$, where $0 \leq r_i \leq k-1$. We thus have $u(k^l n + r) = \sigma_{r_0} \circ \sigma_{r_1} \circ \cdots \circ \sigma_{r_{l-1}}(u(n))$. As there are at most $(\mathrm{Card}(\mathcal{A}))^{\mathrm{Card}(\mathcal{A})}$ such maps, this concludes the proof.

Conversely, suppose that the k-kernel $N_k(u)$ of the sequence u is finite. We denote by $u = u_1, u_2, \cdots, u_d$ the sequences of $N_k(u)$. Let $U = (U(n))_{n \in \mathbb{N}}$ be the sequence with values in $\mathcal{A}^d$ defined by $U(n) = (u_1(n), \cdots, u_d(n))$. Let us construct a substitution σ of constant length k defined on $\mathcal{A}^d$ with U as a fixed point. As $N_k(u)$ is stable by the maps A_r, where $A_r(v(n))_{n\in\mathbb{N}} = v(kn+r))_{n\in\mathbb{N}}$, then, for any $0 \leq r \leq k-1$, $U(kn+r) = U(km+r)$, if $U(n) = U(m)$. Let us define, for $0 \leq r \leq k-1$, σ_r: $\mathcal{A}^d \to \mathcal{A}^d$, $U(n) \mapsto U(kn+r)$ if there exists n such that $(a_1, \cdots, a_r) = U(n)$, and otherwise $(a_1, \cdots, a_r) \mapsto (0, \cdots, 0)$. The sequence U is thus the fixed point of the substitution of constant length $\sigma : x \mapsto \sigma_0(x) \cdots \sigma_{k-1}(x)$ and the sequence u is the image of U by the projection on the first coordinate from $\mathcal{A}^d$ into $\mathcal{A}$.■

Remark. These proofs are constructive, i.e., given the k-kernel of a sequence, then one can compute a k-automaton in reverse reading, generating the sequence u, the states of which are in bijection with the subsequences of the k-kernel. Furthermore a k-automaton in direct reading can be derived from a substitution of constant length generating u from the proof of Proposition 1.3.1. We give other characterizations of automaticity in Chap. 3, known as the Christol, Kamae, Mendès France and Rauzy theorem (see [118] and also [119]).

Exercise 1.3.5. Give a description of the 2-kernel of the Thue-Morse sequence. Prove that the automaton in Fig. 1.1 also generates the Thue-Morse sequence in reverse reading.

1.4 An introduction to topological dynamics and measure-theoretic dynamical systems

We have introduced the notion of a symbolic dynamical system associated with an infinite sequence. Such a discrete system belongs to the larger class of topological dynamical systems, which have been intensively studied (for instance, see [122]).

The aim of this section is to provide a brief introduction to basic notions in topological and measure-theoretic dynamics, that may be useful for a better understanding of Part 2. Note that most of the notions introduced here will be studied in more details through the study of representative examples in Chap. 5.

1.4.1 Topological dynamical systems: basic examples

Definition. A *topological dynamical system* is defined as a compact metric space X together with a continuous map T defined onto the set X.

Remark. Note that topological systems may also be defined without assuming that the map T is onto. In all the examples we consider here, the maps are onto.

Minimality. A topological dynamical system (X, T) is *minimal* if for all x in X, the orbit of x (that is, the set $\{T^n x;\ n \in \mathbb{N}\}$) is dense in X.

Exercise 1.4.1. 1. Prove that minimality is equivalent to each of the following assertions:
a) X does not contain a nontrivial closed subset E such that $T(E) \subset E$;
b) for every non-empty subset U of X, then $\bigcup_{n \geq 0} T^{-n}(U) = X$.
2. A Borel function f on X is said T*-invariant* if $\forall x \in X,\ f(Tx) = f(x)$. Suppose that (X, T) is minimal. Prove that if $f : X \to X$ is continuous, then f is T-invariant implies that f is constant.

Existence of an invariant measure. Let X be a compact metrizable space. Recall that the set $\mathcal{M}_X$ of Borel probability measures on X is identified with a convex subset of the unit ball of the dual space of the set $\mathcal{C}(X)$ of the continuous complex-valued functions on X for the weak-star topology. The set $\mathcal{M}_X$ is metrizable, and compact. In the weak-star topology, one gets

$$\mu_n \to \mu \iff \forall f \in \mathcal{C}(X),\ \int f d\mu_n \to \int f d\mu.$$

Recall that a Borel measure μ defined over X is said T-invariant if $\mu(T^{-1}(B)) = \mu(B)$, for every Borel set B. This is equivalent to the fact that for any continuous function $f \in \mathcal{C}(X)$, then $\int f(Tx) d\mu(x) = \int f(x) d\mu(x)$.

Lemma 1.4.2. *A topological system (X, T) always has an invariant probability measure.*

Proof. Indeed, given a point $x \in X$, any cluster point for the weak-star topology of the sequence of probability measures:

$$\frac{1}{N} \sum_{n<N} \delta_{T^n x}$$

is a T-invariant probability measure, where δ_y denotes the Dirac measure at point y. ∎

Since the space $\mathcal{C}(X)$ is metrizable, it is equivalent to say that there exists an increasing sequence of integers $(N_k)_{k \in \mathbb{N}}$ such that

$$\forall f \in \mathcal{C}(X),\ \frac{1}{N_k} \sum_{n<N_k} f(T^n x) \to \int f d\mu.$$

Unique ergodicity. The case where there exists only one T-invariant measure is of particular interest. A topological dynamical system (X, T) is *uniquely ergodic* if there exists one and only one T-invariant Borel probability measure over X.

Conjugacy. A natural question is to try to "compare" two dynamical systems. Two dynamical systems (X, S) and (Y, T) are said to be *topologically conjugate* (or *topologically isomorphic*) if there exists an homeomorphism f from X onto Y which conjugates S and T, that is:

$$f \circ S = T \circ f.$$

Semi-conjugacy. Two dynamical systems (X, S) and (Y, T) are said to be *semi-topologically conjugate* if there exist two sets X_1 and Y_1 which are at most countable such that $B_1 = X \setminus X_1$, $B_2 = Y \setminus Y_1$, and f is a bicontinuous bijection from B_1 onto B_2 which conjugates S and T. The map f is said to be a *semi-topological conjugacy.* For an example of semi-topological conjugacy, see Sec. 5.2.3.

Topological factor. A topological dynamical system (Y, T) is a *topological factor* of (X, S) if there exists a continuous map from X onto Y which conjugates the maps S and T. The system (X, T) is then called an *extension* of (Y, S).

Exercise 1.4.3. Let u, v be two sequences with values in finite alphabets, and let X_u, X_v denote respectively the associated symbolic dynamical systems.

1. Suppose that (X_v, S) is a topological factor of (X_u, S), where S denotes the shift. Let ϕ denote the conjugation map from X_u onto X_v. Prove that the map ϕ satisfies the following: there exists a positive integer q such that for every i, the coordinate of index i of $\phi(x)$ depends only on $(x_i, \ldots x_{i+q})$. *(Hint: let $\phi_0 : X_u \to \mathcal{B}$, $x = (x_n) \mapsto |\phi(x)|_0$, where $\mathcal{B}$ denotes the alphabet of v; prove that $\phi_0^{-1}([a])$ is a clopen set, for every letter a; see also Lemma 5.1.14.)*
2. Deduce that if (X_u, S) and (X_v, S) are topologically conjugate, then they have the same topological entropy (see Sec. 1.1.2), and if (X_v, S) is a topological factor of (X_u, S), then $H_{top}(v) \leq H_{top}(u)$.

Rotations. The simplest example of a topological dynamical system consists of *toral translations.* Toral translations belong to a larger class of dynamical systems, namely, the *translations over a compact group* G, the invariant probability measure being the Haar measure. These are usually called *rotations* of the group G. This class contains in particular all the toral rotations, the additions over a finite group, translations over p-adic integer groups or translations over p-adic solenoids (see Sec. 1.6.2).

Exercise 1.4.4. Let G be a compact metric group. Let $T : G \to G$, $x \mapsto ax$ be a rotation of G. Prove that (G, T) is minimal if and only if $\{a^n,\ n \in \mathbb{N}\}$ is dense in X.

Following Kronecker's theorem, the minimality of toral rotations can be expressed as follows (for a proof, see for instance [234]):

Proposition 1.4.5. *The rotation by the vector* $\alpha = (\alpha_1, \ldots, \alpha_d) \in \mathbb{R}^d$ *on the d-dimensional torus* $\mathbb{T}^d = \mathbb{R}^d/\mathbb{Z}^d$ *is minimal if and only if* $\alpha_1, \ldots, \alpha_d$ *and 1 are rationally independent.*

Remark. Note that the minimality of a rotation (G, T) (G being a compact metric group) is equivalent to its unique ergodicity. For a proof, see for instance [445]. In particular, a rotation with irrational angle on the one-dimensional torus $\mathbb{T} = \mathbb{R}/\mathbb{Z}$ is minimal and uniquely ergodic, the invariant measure being the Haar measure.

1.4.2 Measure-theoretic dynamical systems

We have considered here the notion of dynamical system, that is, a map acting on a given set, in a topological context. This notion can be extended to measurable spaces: we thus get measure-theoretic dynamical systems. For more details about all of the notions defined in this section, one can refer to [122, 194, 445].

Definition. A *measure-theoretic dynamical system* is defined as a system $(X, T, \mu, \mathcal{B})$, where μ is a probability measure defined on the σ-algebra $\mathcal{B}$ of subsets of X, and $T : X \to X$ is a measurable map which preserves the measure μ.

Remark. Despite unique ergodicity is a concept from topological dynamics, one should note that any uniquely ergodic topological dynamical system (X, T) is a measure-theoretic dynamical system with respect to the unique probability measure which is invariant under T on the σ-algebra of Borel sets of X.

Example 1.4.6. A uniquely ergodic rotation on a compact metric abelian group or a primitive substitutive dynamical system are measure-theoretic dynamical systems.

Ergodicity. A measure-theoretic dynamical system (X, T, μ) is *ergodic* if every Borel subset B of X such that $T^{-1}(B) = B$ has zero measure or full measure.

Exercise 1.4.7. Prove that ergodicity is equivalent to each of the following assertions:

- every Borel set B such that $\mu(T^{-1}B \Delta B) = 0$ has zero or full measure;
- every Borel set B with $\mu(B) > 0$ satisfies $\mu(\cup_{n=0}^{\infty} T^{-n}B) = 1$.

Let us recall that a Borel function f on X is said T-invariant if $f \circ T = f$. If the system (X, T, μ) is ergodic, then every Borel function which is T-invariant is almost everywhere constant. Otherwise, if f is not constant almost everywhere, then one can cut its image into two disjoint sets, whose inverse images have a nontrivial measure and are invariant sets.

Remark. The T-invariant measure of a uniquely ergodic system (X, T) is ergodic. Otherwise, if E satisfies $T^{-1}(E) = E$ and $\mu(E) > 0$, then the measure defined over the Borel sets by: $\mu(B) = \frac{\mu(B \cap E)}{\mu(E)}$ is another T-invariant probability measure.

Ergodic transformations have the following interesting property, which is a measure-theoretical counter-part of the property stated in Exercise 1.4.1 (for a proof, see for instance [445], Sec. 1.5):

Theorem 1.4.8. *Let $(X, T, \mu, \mathcal{B})$ be a measure-theoretic dynamical system. The map T is ergodic if and only if every T-invariant measurable function $f : X \to X$ is constant almost everywhere.*

Exercise 1.4.9. Prove that if $f : X \to \mathbb{R}$ is measurable and $f(Tx) \geq f(x)$ for almost every $x \in X$, then f is almost everywhere constant. *(Hint: introduce sets of the form $\{x \in X,\ f(x) \geq c\}$, with $c \in \mathbb{R}$.)*

Group rotations. In the case of rotations on compact metric groups, one has furthermore the following equivalence (for a proof, see for instance [445]):

Theorem 1.4.10. *Let G be a compact metric group, and let $T : G \to G$, $x \mapsto ax$ be a rotation of G. The following properties are equivalent:*

1. *T is minimal;*
2. *T is ergodic;*
3. *T is uniquely ergodic;*
4. *$\{a^n,\ n \in \mathbb{N}\}$ is dense in G.*

Remark. We thus deduce by using the density of $\{a^n,\ n \in \mathbb{N}\}$ that the ergodicity of T implies that G is abelian.

Isomorphism. Let us introduce an equivalent of the notion of topological conjugacy for measure-theoretic dynamical systems. The idea here is to remove sets of measure zero in order to conjugate the spaces via an invertible measurable transformation. For a nice exposition of connected notions of isomorphism, see [445].

Two measure-theoretic dynamical systems $(X_1, T_1, \mu_1, \mathcal{B}_1)$ and $(X_2, T_2, \mu_2, \mathcal{B}_2)$ are said to be *measure-theoretically isomorphic* if there exist two sets of full measure $B_1 \in \mathcal{B}_1$, $B_2 \in \mathcal{B}_2$, a measurable map $f : B_1 \to B_2$ called *conjugacy map* such that

- the map f is one-to-one,
- the reciprocal map of f is measurable,
- f conjugates T_1 and T_2 over B_1 and B_2,
- μ_2 is the image $f_* \mu_1$ of the measure μ_1 with respect to f, that is,

$$\forall B \in \mathcal{B}_2,\ \mu_1(f^{-1}(B)) = \mu_2(B).$$

If the map is f is only onto, then $(X_2, T_2, \mu_2, \mathcal{B}_2)$ is said to be a *measure-theoretic factor* of $(X_1, T_1, \mu_1, \mathcal{B}_1)$.

A nice way to obtain explicit factors is to study the spectral properties of the system. The aim of the next section is to introduce this spectral aspect.

1.5 Spectral theory

With any measure-theoretic dynamical system (X, T, μ), one can associate a Hilbert space: the space $\mathcal{L}^2(X, \mu)$. The aim of this section is to introduce a unitary operator acting on this Hilbert space and show how to use *spectral theory* to give us insight into the dynamics of (X, T, μ). Good references on the subject are [122, 194, 309, 340, 445]. See also for more details, Sec. 5.2.2 in Chap. 5.

1.5.1 First properties

Unitary operator. Let (X, T, μ) be a measure-theoretic dynamical system, where T is *invertible* (that is, T^{-1} is also measurable and measure-preserving). One can associate with it in a natural way an *operator* U acting on the Hilbert space $\mathcal{L}^2(X, \mu)$ defined as the following map:

$$\begin{aligned} U : \mathcal{L}^2(X, \mu) &\to \mathcal{L}^2(X, \mu) \\ f &\mapsto f \circ T. \end{aligned} \tag{1.1}$$

Since T preserves the measure, the operator U is easily seen to be a *unitary operator*. Note that the surjectivity of the operator U comes from the invertibility of the map T.

The *eigenvalues* of (X, T, μ) are defined as being those of the map U. By abuse of language, we will call *spectrum* the set of eigenvalues of the operator U. It is a subgroup of the unit circle. The *eigenfunctions* of (X, T, μ) are defined to be the eigenvectors of U. Let us note that the map U always has 1 as an eigenvalue and any non-zero constant function is a corresponding eigenfunction.

The spectrum is said to be *irrational* (respectively *rational*) if it is included in $\exp(2i\pi\mathbb{R} \setminus \mathbb{Q})$ (respectively in $\exp(2i\pi\mathbb{Q})$).

Eigenvalues and ergodicity. One can deduce ergodic information from the spectral study of the operator U.

In particular, T is ergodic if and only if the eigenvalue one is a simple eigenvalue, that is, if all eigenfunctions associated with 1 are constant almost everywhere. Indeed, this follows from Theorem 1.4.8 and from the following remark: if a Borel set E of nontrivial measure satisfies $T^{-1}E = E$, then 1_E is a nonconstant eigenfunction associated with one.

Furthermore, if T is ergodic, every eigenfunction is simple and every eigenfunction is of constant modulus. Indeed, if f is an eigenfunction for the eigenvalue β, $|f|$ is an eigenfunction for the eigenvalue $|\beta| = 1$ and hence is a constant. If f_1 and f_2 are eigenfunctions for β, $|f_2|$ is a nonzero constant, and f_1/f_2 is an eigenfunction for 1 and hence a constant (see also Chap. 5, Sec. 5.2.2).

Structure of the spectrum. The spectrum is said to be *discrete* if $\mathcal{L}^2(X,\mu)$ admits an Hilbert basis of eigenfunctions, that is, if the eigenfunctions span $\mathcal{L}^2(X,\mu)$. Hence, if $\mathcal{L}^2(X,\mu)$ is separable (this is the case for instance if X is a compact metric set), then there are at most a countable number of eigenvalues.

If the spectrum contains only the eigenvalue 1, with multiplicity 1, the system is said to be *weakly mixing* or to have a *continuous spectrum.* This implies in particular that T is ergodic. We will see in Sec. 1.5.2 where the terminology "continuous spectrum" comes from.

We say that the system has a *partially continuous spectrum* if the system neither has discrete spectrum, nor is weakly mixing.

Spectrum of group rotations. Consider a rotation $T: x \mapsto ax$ on a compact abelian group G equipped with the Haar measure μ. Let $\hat{G}$ denote the group of *characters* of the group G, that is, the continuous group morphisms $\gamma: G \to \mathbb{C}^*$. Note that a character γ takes its values in a compact set of the unit circle $\mathbb{U} := \{z \in \mathbb{C},\ |z| = 1\}$ of the complex plane.

Example 1.5.1. We have $\hat{\mathbb{U}} = \{z \mapsto z^n,\ n \in \mathbb{Z}\}$ and $\hat{\mathbb{U}}$ is isomorphic to $\mathbb{Z}$. The group $\hat{\mathbb{T}}^d = \{(\alpha_1, \ldots, \alpha_d) \mapsto \sum_{j=1}^d k_j\alpha_j,\ (k_1, \ldots, k_d) \in \mathbb{Z}^d\}$ and $\hat{\mathbb{T}}^d$ is isomorphic to $\mathbb{Z}^d$. For a proof, see [445].

Note that if G is a compact metric space (this is the situation we will be in throughout this book), then $\mathcal{L}^2(G,\mu)$ is separable, and $\hat{G}$ is at most countable. One thus can prove that the set of characters form an Hilbert basis of $\mathcal{L}^2(G,\mu)$. We deduce that λ is an eigenvalue for U if and only if there exists $\gamma \in \hat{G}$ such that $\lambda = \gamma(a)$ (recall that a denotes the translation vector). Hence, any group rotation (G,T,μ) has discrete spectrum.

Exercise 1.5.2. Show that (see also Chap. 5 Prop. 5.2.18):

1. the spectrum of the rotation R_α on the one-dimensional torus $\mathbb{T}$ with irrational angle α is $\exp(2i\pi\mathbb{Z}\alpha) = \{e^{2i\pi\, k\alpha};\ k \in \mathbb{Z}\}$;
2. the spectrum of the addition of 1 on the group $\mathbb{Z}/r\mathbb{Z}$ ($r \geq 1$), is $\exp(2i\pi\frac{\mathbb{Z}}{r}) = \{e^{2i\pi\, k/r};\ k \in \mathbb{Z}\}$;
3. the spectrum of the rotation of angle $(\alpha_1, \ldots \alpha_d)$ on the d-dimensional torus $\mathbb{T}^d$ is $\exp(2i\pi\sum_j \mathbb{Z}\alpha_j) = \{e^{2i\pi\, \sum_j k_j\alpha_j};\ k_j \in \mathbb{Z}\}$.

An invariant for measure-theoretical isomorphism. One of the main questions in ergodic theory is to decide when two dynamical systems are isomorphic. A first answer to this problem can be provided by looking at some natural conjugacy invariants. A property of a topological (respectively measure-theoretic) dynamical system is a topological (respectively measure-theoretic) *invariant* if given (X_1,T_1) which has this property and (X_2,T_2) which is topologically conjugate (respectively measure-theoretically isomorphic) to (X_1,T_1), then (X_2,T_2) has this property. An invariant is said *complete* if (X_1,T_1) and (X_2,T_2) are conjugate if and only if they have this same invariant.

Exercise 1.5.3. Which of the following properties is a topological invariant or a measure-theoretic invariant: unique ergodicity, minimality, weak-mixing, ergodicity?

It is usually difficult to find complete invariants. Nevertheless, Von Neumann proved that, restricting to invertible ergodic system with discrete spectrum, the spectrum is a complete measure-theoretic invariant (see a proof in [194]).

Theorem 1.5.4 (Von Neumann). *Two invertible and ergodic transformations with identical discrete spectrum are measure-theoretically isomorphic.*

Any invertible and ergodic system with discrete spectrum is measure theoretically isomorphic to a rotation on a compact abelian group, equiped with the Haar measure.

The proof of the second assertion of the theorem is based on the following idea: consider the group Λ of eigenvalues of U endowed with the discrete topology; the group of the rotation will be the character group of Λ, which is compact and abelian.

Topological discrete spectrum. This theorem has its counterpart in topological terms. Let (X, T) be a topological dynamical system, where T is an homeomorphism. A nonzero complex-valued continuous in $\mathcal{C}(X)$ is an *eigenfunction* for T if there exists $\lambda \in \mathbb{C}$ such that $\forall x \in X,\ f(Tx) = \lambda f(x)$. The set of the eigenvalues corresponding to those eigenfunctions is called the *topological spectrum* of the operator U. If two systems are topologically conjugate they have the same group of eigenvalues. The operator U is said to have *topological discrete spectrum* if the eigenfunctions span $\mathcal{C}(X)$.

Example 1.5.5. An ergodic rotation on a compact metric abelian group has topological discrete spectrum.

The theorem of Von Neumann becomes (see for instance [445]):

Theorem 1.5.6. *Two minimal topological dynamical systems* (X_1, T_1) *and* (X_2, T_2) *with discrete spectrum, where* T_1 *and* T_2 *are homeomorphisms, are topologically conjugate if and only if they have the same eigenvalues.*

Any invertible and minimal topological dynamical system with topological discrete spectrum is topologically conjugate to a minimal rotation on a compact abelian metric group.

1.5.2 Spectral type

For systems which do not have a discrete spectrum, more precise invariants solve the problem of isomorphism: the most important of them are the spectral type introduced below and the spectral multiplicity. These notions are illustrated in Sec. 5.2.2.

Fourier coefficients. Let μ be a probability measure defined over $\mathbb{T}$. The *Fourier coefficients* $(\hat{\mu}(n))_{n\in\mathbb{Z}} \in \mathbb{C}^{\mathbb{Z}}$ of μ are defined by:

$$\forall n \in \mathbb{Z},\ \hat{\mu}(n) = \int_{\mathbb{T}} exp(2i\pi nt)d\mu(t).$$

Positive definite sequence. A sequence $(a_n)_{n\in\mathbb{Z}} \in \mathbb{C}^{\mathbb{Z}}$ is *positive definite* if for any complex sequence $(z_i)_{i\geq 1}$, we have:

$$\forall n \geq 1,\ \sum_{1\leq i,j\leq n} z_i\overline{z_j}a_{i-j} \geq 0.$$

In particular, the sequence $(\hat{\mu}(n))_{n\in\mathbb{Z}}$ is positive definite. Conversely, one can associate with any positive definite sequence a positive Borel measure on the torus (see for instance [235]).

Theorem 1.5.7 (Bochner-Herglotz theorem). *Any positive definite sequence $(a_n)_{n\in\mathbb{Z}} \in \mathbb{C}^{\mathbb{Z}}$ is the Fourier transform of a finite positive Borel measure on $\mathbb{T}$.*

Consider now a measure-theoretical dynamical system (X, T, μ) and let U be the unitary operator associated with it.

Exercise 1.5.8. Let $f \in \mathcal{L}^2(X, \mu)$. Prove that the sequence $(U^n f, f)_{n\in\mathbb{Z}}$ is positive definite (where $(U^n f, f) = \int_X U^n f.\overline{f}d\mu$).

Spectral type. Let $f \in \mathcal{L}^2(X, \mu)$. The *spectral type* ϱ_f of f is the finite positive Borel measure ϱ_f on the torus $\mathbb{T}$ defined by

$$\forall n \in \mathbb{Z},\ \hat{\varrho_f}(n) = (U^n f, f) = \int_X f \circ T^n.\overline{f}d\mu.$$

Its total mass is $\hat{\varrho_f}(0) = ||f||_2$. For examples of methods of computation of spectral types, see Chap. 5.

Exercise 1.5.9. Let f be an eigenfunction with norm 1 associated with the eigenvalue $e^{2i\pi\lambda}$. Prove that $\varrho_f = \delta_\lambda$.

Remark. Let us recall that the spectrum is said continuous if 1 is the only eigenvalue of T and the only eigenfunctions are the (almost everywhere) constants. In this case if $f \in \mathcal{L}^2(X, \mu)$ is orthogonal to the constant function 1, then ϱ_f is continuous.

Maximal spectral type. Let μ, ν be two Borel measures. Recall that μ is said *absolutely continuous* with respect to ν ($\mu << \nu$) if $\nu(E) = 0$ implies $\mu(E) = 0$, for any Borel set E. Two measures μ and ν are said to be *equivalent* if $\mu << \nu$ and $\nu << \mu$.

Theorem 1.5.10. *Let (X, T, μ) be a measure-theoretic dynamical system. There exists a probability measure ϱ defined over $\mathbb{T}$ characterized by the following property:*

$$\forall f, \ \varrho_f << \varrho.$$

Let ν be a positive measure on $\mathbb{T}$, if $\nu << \varrho$, then there exists f such that $\nu = \varrho_f$.

By definition, this measure ϱ is unique up to equivalence. The measure ϱ is called the maximal spectral type *of U.*

Remark. We thus have that $e^{2i\pi\lambda}$ is an eigenvalue of U if and only if $\varrho\{\lambda\} \neq 0$. For more details, see [194, 340, 445]. Note that $\varrho\{1\} \neq 0$.

The above theorem comes from a more general theorem stating that the Hilbert space $\mathcal{L}^2(X, \mu)$ can be uniquely decomposed as an orthogonal sum of cyclic spaces H_i, where the spaces H_i are stable under the action of U. A *cyclic space* is the closure of a set $\{U^n(f), n \in \mathbb{Z}\}$, for an element f in $\mathcal{L}^2(X, \mu)$. We say that U has a spectrum of *multiplicity* at most k if $\mathcal{L}^2(X, \mu)$ is the direct sum of k cyclic spaces; if $k = 1$, we say that the spectrum is *simple*. For more details, see [340] or Sec. 5.2.2. In particular, Proposition 5.2.21 deals with the spectrum of the Morse system which is proved to be nondiscrete and simple. An example of system having a spectrum with multiplicity at most four, and functions having a spectral type equivalent to the Lebesgue measure is given in Sec. 5.3.2.

1.5.3 Correlation measures of a sequence

There is an interesting way to associate with an infinite sequence u with values in a finite alphabet, a family of finite positive measures on $\mathbb{T}$. We will see that such measures can be considered as spectral types of the associated dynamical system (X_u, S).

Correlation measure. Let u be a sequence with values in the finite alphabet $\mathcal{A} \subset \mathbb{C}$. Let us note that the following definitions and results also hold for bounded complex sequences.

A *correlation sequence* of u is any cluster point of the sequence $(\gamma_N)_{N \geq 1}$ with values in $\mathbb{C}^{\mathbb{N}}$, where γ_N is defined as follows for a fixed N:

$$\gamma_N : \mathbb{N} \to \mathbb{C}, \ k \mapsto \frac{1}{N} \sum_{n<N} u_{n+k} \overline{u_n}.$$

By compacity, if γ is a correlation sequence of u, there exists a sequence $(N_j)_{j \geq 1}$ such that

$$\forall k \in \mathbb{N}, \ \gamma_k = \lim_{j \to +\infty} \frac{1}{N_j} \sum_{n<N_j} u_{n+k} \overline{u_n}.$$

Let us extend the definition of γ to the negative numbers by introducing $\tilde{\gamma}$ defined over $\mathbb{Z}$ as follows: $\forall k \in \mathbb{N}$, $\tilde{\gamma}_{-k} = \overline{\gamma_k}$. The sequence γ is positive definite.

A *correlation measure* of u associated with the correlation sequence γ is the finite positive measure defined on $\mathbb{T}$, the Fourier transform of which is given by the positive definite sequence γ.

Let us note that a correlation measure μ is not necessarily a probability measure. Its total mass is equal to $\lim_{j\to+\infty} \frac{1}{N_j} \sum_{n<N_j} |u_n|^2$.

Remark. If $\forall k \in \mathbb{N}$, $\lim_{N\to+\infty} \frac{1}{N} \sum_{n<N} u_n \overline{u_{n+k}}$ exists, then the sequence u admits a unique correlation measure.

Exercise 1.5.11. Consider the dynamical system (X_u, S) associated with the sequence u. Let μ be an S-invariant measure which is a cluster point for the weak-star topology of the sequence of probability $(\frac{1}{N} \sum_{n<N} \delta_{S^n u})_{N\geq 1}$. Consider, for a fixed $m \in \mathbb{N}$, the map $\pi_m : X_u \to \mathbb{C}$, $x \mapsto x_m$. Prove that the spectral type of π_m is a correlation measure, and that it does not depend on m.

For examples of correlation measures, see Chap. 2, where the correlation measure for the Morse (Sec. 2.1.3) and the Rudin-Shapiro (Sec. 2.2.2) sequence are explicitly computed. See also [79, 340, 348] and the nice surveys [22, 342].

1.6 Factors of substitutive dynamical systems

In this section, we illustrate the important connection between the eigenvalues of a measure theoretical dynamical system and its translation factors. We will then apply these results to substitutive dynamical systems, by alluding to the fact that the spectrum of a dynamical system appears to be an efficient way to determine the largest rotation component. These results will be needed in particular in Chap. 7.

1.6.1 Translation factors on a torus or on a finite group

First, let us present the relationship between the eigenvalues of a measure-theoretical dynamical system and its translation factors on a torus or a finite group. We will next study in the following section examples of factor rotations on inverse limit groups, such as the group of p-adic numbers or the 2-adic solenoid.

Lemma 1.6.1. *The spectrum of a measure-theoretic dynamical system contains the spectrum of any of its measure-theoretic factors.*

Proof. Let (X_1, T_1, μ_1) be a factor of (X, T, μ). Let f be the conjugacy map. Let g_1 be an eigenfunction of (X_1, T_1, μ_1) for the eigenvalue λ. We have $g_1 \circ T_1 = \lambda\, g_1$. Let $g = g_1 \circ f$. Then $g \circ T = g_1 \circ f \circ T = g_1 \circ T_1 \circ f = \lambda\, g_1 \circ f = \lambda\, g$. Thus, g is an eigenfunction of (X, T, μ) for λ. ∎

In the other direction, the following lemmas state that the knowledge on the existence of some eigenvalues of a dynamical system allows the determination of some rotation factors:

Lemma 1.6.2. *A rotation R_α of irrational angle α on the one-dimensional torus $\mathbb{T}$, that we denote $(\mathbb{T}, R_\alpha)$, is a measure-theoretic factor of an ergodic dynamical system (X, T, μ) if and only if $e^{2i\pi\,\alpha}$ is an eigenvalue of (X, T, μ).*

Proof. The necessary condition is a consequence of Lemma 1.6.1.

Let g be an eigenfunction of (X, T, μ) for the eigenvalue $e^{2i\pi\,\alpha}$. We will prove that the rotation $T_\alpha : \mathbb{U} \to \mathbb{U}$, $x \mapsto e^{2i\pi\,\alpha} x$ is a measure-theoretic factor of (X, T, μ). This is equivalent (since $(\mathbb{U}, T_\alpha)$ and $(\mathbb{T}, R_\alpha)$ are conjugate) to the fact that $(\mathbb{T}, R_\alpha)$ is a factor of (X, T, μ).

By ergodicity, g is of constant modulus, which can be chosen equal to 1. We thus have $g : X \to \mathbb{U}$, with $g \circ T = e^{2i\pi\,\alpha} g = T_\alpha \circ g$. It remains to prove that g is onto.

We have $\mu(g^{-1}\mathbb{U}) = \mu(X) = 1 \neq 0$, and the measure $\mu_* g$ on $\mathbb{U}$ is nonzero and invariant under T_α. By unique ergodicity of T_α, this measure is nothing else than the Haar measure. Since $g(X)$ is invariant under T_α and of nonzero measure, we get $g(X) = \mathbb{U}$, by ergodicity of T_α, and $(\mathbb{U}, T_\alpha)$ is a measure-theoretic factor of (X, T, μ). ∎

Exercise 1.6.3. Prove that a minimal rotation R_α on the torus $\mathbb{T}^d$, with $\alpha = (\alpha_1, \ldots, \alpha_d)$, is a measure-theoretic factor of an ergodic dynamical system (X, T, μ) if and only if for every $1 \leq i \leq d$, $e^{2i\pi\alpha_i}$ is an eigenvalue of (X, T, μ).

Similar results are obtained for the addition of 1 on the finite group $\mathbb{Z}/r\mathbb{Z}$, where r is a positive integer, this map being uniquely ergodic.

Lemma 1.6.4. *The addition of 1 on the finite group $\mathbb{Z}/r\mathbb{Z}$, that we denote $(\mathbb{Z}/r\mathbb{Z}, 1)$, is a measure-theoretic factor of the ergodic dynamical system (X, T, μ) if and only if $e^{2i\pi/r}$ is an eigenvalue of (X, T, μ).*

Proof. The necessary condition comes from Lemma 1.6.1.

Suppose that g is an eigenfunction of (X, T, μ) for $\exp(2i\pi/r)$, normalized so that it is of modulus 1. For all $0 \leq \varepsilon \leq 1/p^n$, the set $X_\varepsilon = g^{-1}(\cup_{k\in\mathbb{Z}} \exp\{2i\pi(k/r + [0, \varepsilon[)\})$ is measurable and invariant through the action of T. Consequently, the measure of this set is either 0 or 1. Moreover, the sequence $(X_\varepsilon)_{0\leq\varepsilon\leq r}$ is an increasing sequence of sets, X_0 has measure 0, and $X_{1/r}$ has measure 1. Thus, there exists a lowerbound for the set of the reals ε such that X_ε has measure 1. This real is denoted by ε_0.

Then $\cup_{\varepsilon<\varepsilon_0} X_\varepsilon = g^{-1}(\cup_{k\in\mathbb{Z}} \exp\{2i\pi(k/r+[0,\varepsilon_0[)\})$ has measure 0, whereas $\cup_{\varepsilon>\varepsilon_0} X_\varepsilon = g^{-1}(\cup_{k\in\mathbb{Z}} \exp\{2i\pi(k/r+[0,\varepsilon_0])\})$ has measure 1. Let $z_0 = e^{2i\pi\varepsilon_0}$. Then $g^{-1}\{z_0 e^{2i\pi(k/r)}, k\in\mathbb{Z}\}$ has measure 1.

This implies that there exists a unique function $f : X \to \mathbb{Z}/r\mathbb{Z}$ such that $g = z_0 \exp(2i\pi/r\, f)$ in $\mathcal{L}^2(X,\mu)$. This map satisfies $f\circ T = 1+f$. By unique ergodicity of $\mathbb{Z}/p^n\mathbb{Z}$, we prove as in the preceding lemma that f preserves the measure and is onto. ■

Remark. Let (X,T,μ) be an ergodic dynamical system, and let β be an eigenvalue of the operator U. According to the fact that the argument of β is rational or not, we deduce that the system (X,T,μ) admits as a factor a rotation on a finite group or on the torus $\mathbb{T}$. Let us consider in the next section, the case of a p-adic or of a solenoidal factor.

1.6.2 Solenoidal and p-adic translation factors

The aim of this section is to introduce two classes of translations over a compact abelian group which totally differ from that of the translations studied in the previous section, that is, translations over p-adic groups and p-adic solenoids. Such objects appear in a natural way in the study of the spectral properties of substitutive dynamical systems: for instance, for a precise study of the Morse system, see Chap. 5; we also refer to p-adic rotations and geometric representation of substitutive dynamical systems in Chap. 7. Nevertheless, the reader may consider the following as a digression more than a fundamental notion on substitutions.

p-adic translations. Let us first consider additions on p-adic groups. We will suppose that the reader is already familiar with the notion of p-adic numbers. One can refer to [34, 83, 192, 360] for more details on this notion. See also Chaps. 3 and 5.

Let us recall that the *ring of p-adic integers* $\mathbb{Z}_p$ can be realized as the *inverse limit* of the finite groups $\mathbb{Z}/p^n\mathbb{Z}$. More precisely, for $i\geq 1$, let p^i_{i+1} denote the canonical projection from $\mathbb{Z}/p^{i+1}\mathbb{Z}$ onto $\mathbb{Z}/p^i\mathbb{Z}$. Then we have

$$\mathbb{Z}_p = \varprojlim \mathbb{Z}/p^n\mathbb{Z} = \{(a_i)_{i\geq 1};\ a_i\in\mathbb{Z}/p^i\mathbb{Z},\ \forall i\geq 1,\ p^i_{i+1}(a_{i+1}) = a_i\}.$$

We get an expansion of the elements of $\mathbb{Z}_p$ with respect to the reference sequence $(p^i)_{i\in\mathbb{Z}}$, which tends to zero in $\mathbb{Q}_p$. This expansion, similar to the decimal expansion in $\mathbb{R}$, is called *Hensel expansion*:

$$\forall z\in\mathbb{Z}_p,\ z\neq 0,\ \exists\,(b_i)_{i\geq 0}\in\{0,\ldots,p-1\},\ z=\sum_{i\geq 1} b_i p^i.$$

The *field of p-adic numbers* $\mathbb{Q}_p$, that is, the fraction field $\mathbb{Q}_p$ of $\mathbb{Z}_p$ is the completion of $\mathbb{Q}$ for the *p-adic absolute value* defined on $\mathbb{Q}$ as $|z|_p = p^{-v_p(z)}$, where $v_p(z)$ is the largest power of p which appears in the decomposition of p in prime factors. This absolute value is *non-Archimedean*, that is, for any two elements x, y with $|x|\neq|y|$, then $|x+y| = \max(|x|,|y|)$.

Symbolic representation as an odometer. From a symbolic point of view, the Hensel expansion means that the addition of 1 on $\mathbb{Z}_p$ is topologically conjugate to the following symbolic system called *p-odometer* (see Chap. 5 and [193] for a definition of a generalized odometer associated with a numeration system).

Definition 1.6.5. *The p-odometer is the topological dynamical system* $(\{0,\ldots,p-1\}^{\mathbb{N}}, \chi)$ *where* χ *is defined by:*

$$\begin{aligned} \chi : \{0,\ldots,p-1\}^{\mathbb{N}} &\to \{0,\ldots,p-1\}^{\mathbb{N}} \\ w = w_0 w_1 \ldots \quad &\mapsto v = v_0 v_1 \ldots \end{aligned}$$

with

$$\text{if } \forall i \geq 0,\ w_i = p-1, \quad \text{then } \forall i \geq 0, v_i = 0;$$

$$\text{if } \begin{cases} \forall i < i_0,\ w_i = p-1 \\ w_{i_0} \neq p-1, \end{cases} \quad \text{then } \begin{cases} \forall i < i_0,\ v_i = 0 \\ v_{i_0} = w_{i_0} + 1, \\ \forall i > i_0,\ v_i = w_i. \end{cases}$$

By Theorem 1.4.10, we obtain that the addition of 1 on the group $\mathbb{Z}_p$, that we denote $(\mathbb{Z}_p, 1)$ is a minimal and uniquely ergodic topological dynamical system.

A basis for the topology of $\mathbb{Q}_p$ consists in the clopen balls $z + p^k\mathbb{Z}_p$, where $k \in \mathbb{N}$ and $z \in \mathbb{Q}_p$ ($\mathbb{Z}_p$ is the closed unit ball of $\mathbb{Q}_p$). In particular, one can note that the Hensel map sends the clopen sets $z + p^k\mathbb{Z}_p$ in $\mathbb{Z}_p$ onto the cylinders in $\{0,\ldots,p-1\}^{\mathbb{N}}$. On this space, the Bernoulli measure is invariant under the action of the odometer. A consequence is that the Haar measure μ_p on $\mathbb{Z}_p$ is explicit; it is defined on the clopen sets by:

$$\forall z \in \mathbb{Z}_p, \quad \forall k \geq 0, \qquad \mu_p(z + k^n\,\mathbb{Z}_p) = 1/p^k.$$

Spectral theory. The determination of the spectrum of the addition of 1 on $\mathbb{Z}_p$ can be realized as a generalization of that of the eigenvalues of the addition of 1 on a finite group. We left it as an exercise:

Exercise 1.6.6. The spectrum of the addition of 1 on the p-adic group $\mathbb{Z}_p$ is the group generated by

$$\bigcup_{n\geq 1} \exp(2i\pi \frac{\mathbb{Z}}{p^n}) = \{e^{2i\pi\, k/p^n};\ n \geq 1,\ k \in \mathbb{Z}\}.$$

Conversely, the following lemma states that having the elements $\{e^{2i\pi\, k/p^n};\ n \geq 1,\ k \in \mathbb{Z}\}$ as eigenvalues is equivalent with having $\mathbb{Z}_p$ as a factor. See also the proof of Proposition 5.2.18.

Lemma 1.6.7. *The addition of 1 on the p-adic group* $\mathbb{Z}_p$ *is a measure-theoretic factor of the ergodic dynamical system* (X, T, μ) *if and only if, for every* $n \geq 1$, $e^{2i\pi/p^n}$ *is an eigenvalue of* (X, T, μ).

Proof. The condition is necessary from Lemma 1.6.1.

From Lemma 1.6.4, if $e^{2i\pi/p^n}$ is an eigenvalue of (X, T, μ), there exists an eigenfunction $g_n : X \to \mathbb{C}$ (of constant modulus 1) for the eigenvalue $e^{2i\pi/p^n}$, and a constant $z_n \in \mathbb{C}$ such that $g_n = z_n \exp(2i\pi/p^n\, f_n)$, with $f_n : X \to \mathbb{Z}/p^n\mathbb{Z}$ and $f_n \circ T = 1 + f_n$.

Since g_{n+1} is an eigenfunction for $e^{2i\pi/p^{n+1}}$, the map $(g_{n+1})^p$ is also an eigenfunction for $e^{2i\pi/p^n}$. By ergodicity of (X, T, μ) , the eigenvalues are simple, and $(g_{n+1})^p$ and g_n are proportional. So that we get $z_n \exp(2i\pi/p^n\, f_n) = {z_{n+1}}^p \exp(2i\pi/p^n\, f_{n+1})$. In particular, $f_n - f_{n+1}$ is constant modulo p^n.

It will be recalled that p^i_{i+1} is the canonical projection of $\mathbb{Z}/p^{i+1}\mathbb{Z}$ onto $\mathbb{Z}/p^i\mathbb{Z}$. Modifying the map f_{n+1} by subtracting the constant (we thus get $f_n = f_{n+1}$ modulo p^n), we obtain a sequence of maps $f_n : X \to \mathbb{Z}/p^n\mathbb{Z}$ such that $f_n \circ T = 1 + f_n$ and $p^n_{n+1} \circ f_{n+1} = f_n$, for every positive integer n.

Let $f : X \to \prod_{n\geq 1} \mathbb{Z}/p^n\mathbb{Z}$ be the map defined by $f(x) = (f_n(x))_{n\geq 1}$. According to what precedes, we have for all $i \geq 1$, $p^i_{i+1}(f_{i+1}(x)) = f_i(x)$. Thus, $f(X)$ is included in the inverse limit of $\mathbb{Z}/p^n\mathbb{Z}$, that is, $\mathbb{Z}_p$. Moreover, we have $f \circ T = 1 + f$. By unique ergodicity of the addition of 1 on $\mathbb{Z}_p$, we conclude just as in Lemma 1.6.2 that $\mathbb{Z}_p$ is a measure-theoretic factor of (X, T, μ). ■

The 2-adic solenoid. Solenoids usually appear in dynamical systems as examples of attractors (see for instance the Smale attractor in [234], or [145]). See also [247] for a purely symbolic approach in connection with the notion of dimension group. We will study some properties of the 2-adic solenoid solenoid through an exercise. Everything extends in a natural way to the p-adic solenoid.

The 2-*adic solenoid* $\mathcal{S}_2$ can be realized as the following inverse limit space:

$$\mathcal{S}_2 = \{\mathbf{x} = (x_k)_{k\geq 1} \mid \forall k \geq 1,\ \ x_k \in \mathbb{T},\ \ x_k = 2x_{k+1} \bmod 1\}\,.$$

Exercise 1.6.8. Let us first prove some basic properties of the 2-adic solenoid:

1. Prove that $\mathcal{S}_2$ is nonempty by giving one of its elements.
2. Give two elements $\mathbf{a}$ and $\mathbf{b}$ in $\mathcal{S}_2$ such that $a_2 = b_2$ and $a_3 \neq b_3$.
3. Prove that the projections $\pi_k : \mathcal{S}_2 \to \mathbb{T}$, $\mathbf{x} \mapsto x_k$ are onto.
4. Prove that $\mathcal{S}_2$ is a compact abelian group.

Let us study now the minimality of translations on $\mathcal{S}_2$. Let $\mathbf{a} \in \mathcal{S}_2$ such that the addition of a_1 on $\mathbb{T}$ is minimal, that is, any real number representing a_1 is irrational.

1. Prove that the addition of any coordinate a_k on $\mathbb{T}$ is also minimal.
2. Prove that the addition of $\mathbf{a}$ on $\mathcal{S}_2$ is onto.
3. Prove that the addition of $\mathbf{a}$ on $\mathcal{S}_2$ is minimal.

Let us finally study some spectral properties of the minimal addition of **a** on the 2-adic solenoid.

1. Prove that the spectrum of the addition of **a** on $\mathcal{S}_2$ contains all the numbers $e^{2i\pi a_k}$, $k \in\geq 1$, and that it is exactly equal to the group generated by
$$\cup_{k\geq 1} \exp(2i\pi\mathbb{Z}[\alpha/2^k]) = \{e^{2i\pi \frac{m\alpha}{2^k}},\ k \geq 1, m \in \mathbb{Z}\}.$$
(Hint: a description of the characters of $\mathcal{S}_2$ is given in [247].)
2. Let α be an irrational number. Let $\mathbf{a} = (\alpha, \alpha/2, \ldots, \alpha/2^k, \ldots)$. Prove that the addition of **a** on $\mathcal{S}_2$, that we denote $(\mathcal{S}_2, R_{\mathbf{a}})$, is a measure-theoretic factor of the ergodic dynamical system (X, T, μ) if and only if for every $k \geq 1$, $e^{2i\pi \frac{\alpha}{2^k}}$ is an eigenvalue of (X, T, μ).

Example. The spectrum of the addition of $(\sqrt{2}/2^k)_{k\geq 1}$ on $\mathcal{S}_2$ is the group generated by $\cup_{k\geq 1} \exp(2i\pi\mathbb{Z}[\sqrt{2}/2^k])$.

1.6.3 Application to the spectrum.

Let (X, T, μ) be an invertible ergodic dynamical system with discrete spectrum. From Theorem 1.5.4, this system is a compact group rotation. Let us collect the results we have obtained on the existence of translation factors in order to describe this rotation:

- if $\exp(2i\pi\alpha)$, with α irrational, is an eigenvalue of (X, T, μ), then the system $(\mathbb{T}, R_\alpha)$ is a factor of (X, T, μ); its spectrum contains $\exp(2i\pi\mathbb{Z}\alpha)$;
- let R_α be a minimal rotation on the torus $\mathbb{T}^d$, with $\alpha = (\alpha_1, \ldots, \alpha_d)$; if for every $1 \leq i \leq d$, $\exp(2i\pi\alpha_i)$ is an eigenvalue of (X, T, μ), then the system $(\mathbb{T}^d, R_\alpha)$ is a factor of (X, T, μ); its spectrum contains $\exp(2i\pi \sum \mathbb{Z}\alpha_i)$;
- if $\exp(2i\pi/r)$, with $r \in \mathbb{N}^+$ is an eigenvalue of (X, T, μ), then the system $(\mathbb{Z}/r\mathbb{Z}, 1)$ is a factor of (X, T, μ); its spectrum contains $\exp(2i\pi\frac{\mathbb{Z}}{r})$;
- if all the elements of $\cup_{k\geq 1} \exp(2i\pi\mathbb{Z}[\alpha/p^k])$, with α irrational and p prime, are eigenvalues of (X, T, μ), then the system $(\mathcal{S}_p, R_{\mathbf{a}})$, where $\mathbf{a} = (\alpha, \alpha/p, \ldots, \alpha/p^k, \ldots) \in \mathcal{S}_p$, is a factor of (X, T, μ);
- if all the elements of $\cup_{k\geq 1} \exp(2i\pi\frac{\mathbb{Z}}{p^k})$, with p prime, are eigenvalues of (X, T, μ), then the system $(\mathbb{Z}_p, 1)$ is a factor of (X, T, μ).

Let us partition now the spectrum into disjoint components which are rationally independent; note that we have lost no information in restricting ourselves in the above enumeration to prime integers, since we require rational independence between the components. By using the results stated in this enumeration, one can construct a minimal compact group rotation which admits exactly the same spectrum: this group will be built as a direct product of tori, finite groups, p-adic groups or solenoids.

1.6.4 Maximal equicontinuous factor

Theorem 1.5.6 has numerous consequences in topological dynamics. Indeed, for any topological dynamical system, one can look at a specific set of topological factors, which contains all the rotation factors, that is, the equicontinuous factors.

Equicontinuity. A topological dynamical system (X, T) is said to be *equicontinuous* if the set of maps $\{T^n, n \in \mathbb{Z}\}$ is uniformly equicontinuous, that is:

$$(\forall \varepsilon > 0)\,(\exists \alpha > 0)\,(\forall x, y \in X)\,\big(d(x,y) < \alpha \implies (\forall n \in \mathbb{Z}), d(T^n(x), T^n(y)) < \varepsilon\,\big).$$

In particular, any rotation on a compact abelian metric group is equicontinuous. For more details on this notion, see [50, 160, 161, 134].

One interest of this notion is that equicontinuous factors can be compared:

Maximal equicontinuous factor. Every topological dynamical system (X, T) admits a largest equicontinuous factor, (Y, T_1), in the sense that any topological equicontinuous factor of (X, T) is a topological factor of (Y, T_1). This largest equicontinuous factor is called *maximal equicontinuous factor* [161].

Application to the spectrum of substitutive dynamical systems. One interesting point in the spectral study of substitutive dynamical systems is that we do not need to distinguish between topological and measure-theoretic eigenfunctions, and thus between topological and measure-theoretic factors (see also Exercise 7.3.12 in Chap. 7).

Theorem 1.6.9 (B. Host [215]). *Let σ be a primitive and not shift-periodic substitution, and let (X_σ, S) denote the uniquely ergodic topological dynamical system associated with it. Then, any class (in $\mathcal{L}^2$) of eigenfunctions contains a continuous eigenfunction.*

An invertible and uniquely ergodic equicontinuous topological dynamical system has a discrete spectrum [183]. Furthermore every rotation factor of (X_σ, S) is a factor of its maximal equicontinuous factor. As a consequence, the maximal equicontinuous factor of the dynamical system (X_σ, S) is nothing else but a rotation on a compact abelian group, determined by the discrete part of its spectrum.

An important literature is devoted to the determination of the maximal equicontinuous factor of a substitutive dynamical system, with a special emphasis on the discrete spectrum case. A review of these results is given in Chap. 7. The description of the spectrum of some representative examples of substitutions is detailed in Chap. 5.

Part I

Arithmetics and combinatorics of substitutions

2. Substitutions, arithmetic and finite automata: an introduction

The aim of this chapter is to introduce substitutions by showing some typical and important examples of situations in number theory where they appear. Special stress will be given to the statistical properties of these sequences. We first recall some properties of the Morse sequence, then we introduce the Rudin-Shapiro sequence and focus on its spectral properties. We also evoke the Baum-Sweet sequence, the Cantor sequence and the Fibonacci sequence.

For all the definitions related to words, substitutions and automata, we refer the reader to Chap. 1 (see also [158] and [339]).

2.1 The Morse sequence

As mentioned in Chap. 1, the *Morse sequence* u is defined as the fixed point beginning by a of the *Morse substitution* σ defined over the alphabet $\{a, b\}$ by $\sigma(a) = ab$ and $\sigma(b) = ba$:

$$u = abbabaabbaababbabaababbaabbabaab \dots$$

Let us insist on the importance of the order of the letters in the definition of a substitution. For example the substitution over the alphabet $\{a, b\}$ defined by $\sigma(a) = ab$ and $\sigma(b) = ab$ has only one (non-empty) fixed point, that is, the periodic sequence $abababababab \dots$.

We can deduce from the equality

$$\sigma^{r+1}(a) = \sigma(\sigma^r(a)) = \sigma^r(\sigma(a)),$$

the following combinatorial properties of the Morse sequence u:

- $\forall n \in \mathbb{N}$, $u_{2n} = u_n$ and $u_{2n+1} = 1 - u_n$;
- at position $k2^n$ of the sequence occurs $\sigma^n(a)$, if $u_k = 0$, and $\sigma^n(b)$ if $u_k = 1$.

Let $U_r = \sigma^r(a)$ and $U'_r = \sigma^r(b)$. These sequences of words over the alphabet $\{a, b\}$ are uniquely defined by the following relation:

[1] This chapter has been written by C. Mauduit

$$\begin{cases} U_0 = a \\ U'_0 = b \end{cases} \quad \text{and} \quad \forall r \geq 0 \begin{cases} U_{r+1} = U_r U'_r \\ U'_{r+1} = U'_r U_r \end{cases} .$$

By construction, we have $u = \lim_{r \to +\infty} U_r$. Since the word U_r is equal to the word U'_r modulo the exchange of the letters a and b, the other fixed point of the Morse substitution is equal to $\lim_{r \to +\infty} U'_r$. This provides an alternative construction of the Morse sequence by catenative rules of the form $U_{r+1} = U_r U'_r$.

2.1.1 Arithmetic definition of the Morse sequence

The Morse sequence can be defined by a property of the dyadic development of the integers. As shown in Chap. 1, this is a special case of a general property of sequences obtained as a fixed point of a constant length substitution: if the sequence v is a fixed point of a substitution of constant length q, then v can be defined by a property of the q-adic development of the integers (Proposition 1.3.1). Moreover this property is simple enough to be recognizable by a finite automaton.

Consider the Morse sequence. Let us denote by $\mathbb{N}_a$ (respectively $\mathbb{N}_b$) the set of integers n such that the $(n+1)$-th letter of the Morse sequence is a (respectively b):

$$\mathbb{N}_a = \{0, 3, 5, 6, 9, 10, 12, 15, \ldots\},$$

$$\mathbb{N}_b = \{1, 2, 4, 7, 8, 11, 13, 14, \ldots\}.$$

For any integer n, we denote by $S_2(n)$ the sum of the dyadic digits of n, i.e.,

$$S_2(n) = \sum_{i \geq 0} n_i, \text{ if } n = \sum_{i \geq 0} n_i 2^i, \; n_i \in \{0, 1\},$$

is the dyadic development of n.

Proposition 2.1.1. *We have*

$$\mathbb{N}_a = \{n \in \mathbb{N}, \quad S_2(n) \textit{ is even } \},$$

$$\mathbb{N}_b = \{n \in \mathbb{N}, \quad S_2(n) \textit{ is odd } \}.$$

Proof. Obviously $\mathbb{N}_a$ and $\mathbb{N}_b$ form a partition of $\mathbb{N}$. It is thus enough to show by induction over r the following property:

$$(\mathcal{P}_r) \begin{cases} n \in \mathbb{N}_a \text{ and } n < 2^r \Longrightarrow S_2(n) \text{ is even }, \\ n \in \mathbb{N}_b \text{ and } n < 2^r \Longrightarrow S_2(n) \text{ is odd}. \end{cases}$$

Note that $(\mathcal{P}_0)$ is obvious. Suppose that $(\mathcal{P}_r)$ is true and consider $n \in \mathbb{N}_a$ (respectively $\mathbb{N}_b$) such that $2^r \leq n < 2^{r+1}$. Then $n = 2^r + n'$ with $n' \in \mathbb{N}_b$

(respectively $\mathbb{N}_a$) and $n' < 2^r$. As $S_2(n) = S_2(2^r + n') = S_2(n') + 1$, $(\mathcal{P}_{r+1})$ follows. ■

Let us recall that if we consider integers as words over the alphabet $\{0,1\}$ (via their dyadic development), then it is clear that the 2-automaton shown in Fig. 2.1.1 with initial state a and exit map $Id_{\{a,b\}}$ recognizes the Morse sequence (either in direct reading or in reverse reading).

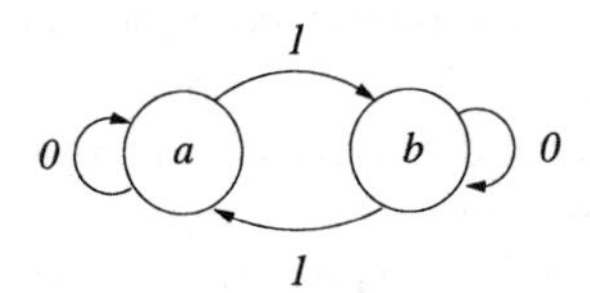

Fig. 2.1. Automaton associated with the Morse sequence.

2.1.2 The problem of Prouhet

In fact, it seems that the first mathematician who introduced the Morse sequence was E. Prouhet. In 1851 he solved in [338] the following problem (also known as the Tarry-Escott problem): given the positive integers q and r, find an infinite number of sequences of q^r numbers that can be cut in q sets of q^{r-1} elements such that, for any $k < r$, the sum of all the k-th powers of the elements of each set is the same.

If $q = 2$, the solution to Prouhet's problem is given by the Morse sequence in the following sense:

Proposition 2.1.2. *For any nonnegative integers k, r and n_0 such that $k < r$, we have*

$$\sum_{\substack{n \in n_0+\mathbb{N}_a \\ n < n_0+2^r}} n^k = \sum_{\substack{n \in n_0+\mathbb{N}_b \\ n < n_0+2^r}} n^k .$$

Proof. As for $\alpha \in \{a,b\}$

$$\sum_{\substack{n \in n_0+\mathbb{N}_\alpha \\ n < n_0+2^r}} n^k = \sum_{\substack{n \in \mathbb{N}_\alpha \\ n < 2^r}} (n_0+n)^k = \sum_{i \le k} \binom{k}{i} n_0^{k-i} \sum_{\substack{n \in \mathbb{N}_\alpha \\ n < 2^r}} n^i,$$

it is enough to prove the property in the case $n_0 = 0$. This is equivalent to saying that for any nonnegative integers k and r such that $k < r$, we have $\sum_{n<2^r} (-1)^{S_2(n)} n^k = 0$.

Let us introduce the polynomial $F_r \in \mathbb{Z}[X]$ defined by

$$F_r(X) = \sum_{n<2^r} (-1)^{S_2(n)} X^n = \prod_{k<r} (1 - X^{2^k}).$$

One checks that $F_r(X) = (1-X)^r G_r(X)$ with $G_r \in \mathbb{Z}[X]$, so that $F_r^{(k)}(1) = \sum_{n<2^r}(-1)^{S_2(n)} n(n-1)\ldots(n-k+1) = 0$ for $k < r$, from which we deduce that $\sum_{n<2^r}(-1)^{S_2(n)} n^k = 0$ for $k < r$. ∎

For more details, see for instance [461, 462] and the survey [84].

2.1.3 A statistical property of the Morse sequence

By a theorem of Fréchet, any monotone function f can be decomposed as $f = f_1 + f_2 + f_3$, where f_1 is a monotone step-function, f_2 a monotone function which is the integral of its derivative, and f_3 a monotone continuous function which has almost everywhere a derivative equal to zero.

In [457] Wiener extended the spectrum theory to the harmonic analysis of functions defined for a countable set of arguments (that he called arrays).

As an application of some theorems proved in [457], Mahler gives in [278] a construction based on the array $(-1)^{S_2(n)}$ for which $f_3 \neq 0$ in the Fréchet decomposition. The crucial point is the following property:

Proposition 2.1.3. *For any positive integer k and any positive integer N, if*

$$\gamma_N(k) = \frac{1}{N} \sum_{n<N} (-1)^{S_2(n)}(-1)^{S_2(n+k)},$$

then for any k the sequence $(\gamma_N(k))_{N>0}$ converges and its limit is non-zero for infinitely many k.

The convergence of the sequence $(\gamma_N(k))_{N\geq 1}$ can be understood as a consequence of the unique ergodicity of the symbolic dynamical system associated with the infinite word u (see [339, 341] for this approach). It follows also from [235] that $\lim_{N\to+\infty} \gamma_N(k)$ is equal to the k-th Fourier coefficient of the *correlation* measure associated with the Morse sequence, which is the Riesz product $\prod_{n\geq 0}(1 - \cos 2^n t)$. See also Sec. 1.5.3 in Chap. 1.

Let us present here a proof of Proposition 2.1.3 independent of these results.

Proof.

- For $k = 0$, we have for any positive integer $N, \gamma_N(0) = 1$.
- For $k = 1$, we have for any positive integer N,

$$\begin{aligned}\gamma_{2N}(1) &= \frac{1}{2N}\sum_{n<2N}(-1)^{S_2(n)}(-1)^{S_2(n+1)}\\ &= \frac{1}{2N}\sum_{2n<2N}(-1)^{S_2(2n)}(-1)^{S_2(2n+1)}\\ &\qquad\qquad + \frac{1}{2N}\sum_{2n+1<2N}(-1)^{S_2(2n+1)}(-1)^{S_2(2n+2)}\\ &= \frac{1}{2N}\sum_{n<N}-(-1)^{S_2(n)}(-1)^{S_2(n)} + \frac{1}{2N}\sum_{n<N}-(-1)^{S_2(n)}(-1)^{S_2(n+1)}\\ &= -\frac{1}{2}-\frac{1}{2}\gamma_N(1),\end{aligned}$$

and for any nonnegative integer N

$$\begin{aligned}\gamma_{2N+1}(1) &= \frac{1}{2N+1}\sum_{n<2N+1}(-1)^{S_2(n)}(-1)^{S_2(n+1)}\\ &= \frac{1}{2N+1}\sum_{2n<2N+1}(-1)^{S_2(2n)}(-1)^{S_2(2n+1)}\\ &\qquad\qquad + \frac{1}{2N+1}\sum_{2n+1<2N+1}(-1)^{S_2(2n+1)}(-1)^{S_2(2n+2)}\end{aligned}$$

$$\begin{aligned}\gamma_{2N+1}(1) &= \frac{1}{2N+1}\sum_{n<N+1}-(-1)^{S_2(n)}(-1)^{S_2(n)}\\ &\qquad\qquad + \frac{1}{2N+1}\sum_{n<N}-(-1)^{S_2(n)}(-1)^{S_2(n+1)}\\ &= -\frac{N+1}{2N+1}-\frac{N}{2N+1}\gamma_N(1).\end{aligned}$$

If we put $\delta_N = \gamma_N(1)+\frac{1}{3}$ for any strictly positive integer N, these relations become

$$\begin{cases}\delta_1 &= \gamma_1(1)+\dfrac{1}{3} = -\dfrac{2}{3}\\ \delta_{2N} &= -\dfrac{1}{2}\delta_N\\ \delta_{2N+1} &= -\dfrac{N}{2N+1}\delta_N - \dfrac{2}{3(2N+1)}.\end{cases}$$

It is easy to check by induction over N that

$$\forall N \geq 2 \qquad |\,\delta_N\,| \leq \frac{2}{3}\frac{\log_2 N}{N}.$$

Indeed, this is true for $N = 2$ and $N = 3$ ($\delta_2 = \frac{1}{3}$ and $\delta_3 = 0$); for $2N \geq 4$ we have

$$| \delta_{2N} | = \frac{1}{2} | \delta_N | \leq \frac{1}{2}\frac{2}{3}\frac{\log_2 N}{N} \leq \frac{2}{3}\frac{\log_2 2N}{2N},$$

and for $2N+1 \geq 5$ we have

$$\begin{aligned} | \delta_{2N+1} | &\leq \frac{N}{2N+1}\frac{2}{3}\frac{\log_2 N}{N} + \frac{2}{3(2N+1)} \\ &\leq \frac{2}{3(2N+1)}(\log_2 N + 1) \leq \frac{2}{3}\frac{\log_2(2N+1)}{2N+1}. \end{aligned}$$

This computation shows that $\gamma_N(1) = -\frac{1}{3} + O(\log N/N)$ so that the sequence $(\gamma_N(1))_{N>0}$ converges to $-\frac{1}{3}$.

It is now easy to deduce by induction over k the convergence of the sequences $(\gamma_N(k))_{N>0}$ for $k \geq 2$ from the convergence of the sequences $(\gamma_N(0))_{N>0}$ and $(\gamma_N(1))_{N>0}$:

- If $2k \geq 2$, we have for any positive integer N,

$$\begin{aligned} \gamma_{2N}(2k) &= \frac{1}{2N}\sum_{n<2N}(-1)^{S_2(n)}(-1)^{S_2(n+2k)} \\ &= \frac{1}{2N}\sum_{2n<2N}(-1)^{S_2(2n)}(-1)^{S_2(2n+2k)} \\ &\qquad + \frac{1}{2N}\sum_{2n+1<2N}(-1)^{S_2(2n+1)}(-1)^{S_2(2n+2k+1)} \\ &= \gamma_N(k), \end{aligned}$$

and for any nonnegative integer N,

$$\begin{aligned} \gamma_{2N+1}(2k) &= \frac{1}{2N+1}\sum_{n<2N+1}(-1)^{S_2(n)}(-1)^{S_2(n+2k)} \\ &= \sum_{2n<2N+1}(-1)^{S_2(2n)}(-1)^{S_2(2n+2k)} \\ &\qquad + \frac{1}{2N+1}\sum_{2n+1<2N+1}(-1)^{S_2(2n+1)}(-1)^{S_2(2n+2k+1)} \\ &= \frac{N+1}{2N+1}\gamma_{N+1}(k) + \frac{N}{2N+1}\gamma_N(k). \end{aligned}$$

It follows that if $\gamma(k) = \lim_{N\to+\infty}\gamma_N(k)$, both sequences $(\gamma_{2N}(2k))_{N>0}$ and $(\gamma_{2N+1}(2k))_{N\geq 0}$ converge to $\gamma(k)$, so that the sequence $(\gamma_N(2k))_{N>0}$ converges and $\gamma(2k) = \lim_{N\to+\infty}\gamma_N(2k) = \gamma(k)$.

- If $2k+1 \geq 3$, we have for any positive integer N,

$$
\begin{aligned}
\gamma_{2N}(2k+1) &= \frac{1}{2N}\sum_{n<2N}(-1)^{S_2(n)}(-1)^{S_2(n+2k+1)} \\
&= \frac{1}{2N}\sum_{2n<2N}(-1)^{S_2(2n)}(-1)^{S_2(2n+2k+1)} \\
&\qquad + \frac{1}{2N}\sum_{2n+1<2N}(-1)^{S_2(2n+1)}(-1)^{S_2(2n+2k+2)} \\
&= -\frac{1}{2}\gamma_N(k) - \frac{1}{2}\gamma_N(k+1),
\end{aligned}
$$

and for any nonnegative integer N,

$$
\begin{aligned}
\gamma_{2N+1}(2k+1) &= \frac{1}{2N+1}\sum_{n<2N+1}(-1)^{S_2(n)}(-1)^{S_2(n+2k+1)} \\
&= \frac{1}{2N+1}\sum_{2n<2N+1}(-1)^{S_2(2n)}(-1)^{S_2(2n+2k+1)} \\
&\qquad + \frac{1}{2N+1}\sum_{2n+1<2N+1}(-1)^{S_2(2n+1)}(-1)^{S_2(2n+2k+2)} \\
&= \frac{N+1}{2N+1}\gamma_{N+1}(k) - \frac{N}{2N+1}\gamma_N(k+1).
\end{aligned}
$$

It follows that if $\gamma(k) = \lim_{N\to\infty}\gamma_N(k)$ and $\gamma(k+1) = \lim_{N\to\infty}\gamma_N(k+1)$, both sequences $(\gamma_{2N}(2k+1))_{N>0}$ and $(\gamma_{2N+1}(2k+1))_{N\geq 0}$ converge to $-\frac{1}{2}\gamma(k) - \frac{1}{2}\gamma(k+1)$, so that the sequence $(\gamma_N(2k+1))_{N>0}$ converges and $\gamma(2k+1) = \lim_{N\to\infty}\gamma_N(2k+1) = -\frac{1}{2}\gamma(k) - \frac{1}{2}\gamma(k+1)$.

• It is now easy to deduce from the relations

$$
\begin{cases}
\gamma(0) &= 1 \\
\gamma(2k) &= \gamma(k) \\
\gamma(2k+1) &= -\frac{1}{2}\gamma(k) - \frac{1}{2}\gamma(k+1)
\end{cases}
$$

the values of $\gamma(k)$ for any nonnegative integer k.

In particular for any integer i we have $\gamma(2^i) = -\frac{1}{3}$, which ends the proof of Proposition 2.1.3. ∎

2.2 The Rudin-Shapiro sequence

2.2.1 Definition

For any sequence $\varepsilon = (\varepsilon_n)_{n\in\mathbb{N}} \in \{-1,+1\}^{\mathbb{N}}$ we have

$$
\int_0^1 \left|\sum_{n<N}\varepsilon_n e(n\theta)\right|^2 d\theta = \int_0^1 \sum_{\substack{n<N \\ n'<N}} \varepsilon_n\varepsilon_{n'} e((n-n')\theta)\, d\theta = N,
$$

because

$$\int_0^1 e(k\theta)\, d\theta = \begin{cases} 1 \text{ if } k = 0 \\ 0 \text{ if } k \in \mathbb{Z} \setminus \{0\} \end{cases}$$

(with the notation $e(x) = e^{2i\pi x}$).

We hence have

$$\sup_{\theta\in[0,1]} \left| \sum_{n<N} \varepsilon_n e(n\theta) \right| \geq \left(\int_0^1 \left| \sum_{n<N} \varepsilon_n e(n\theta) \right|^2 d\theta \right)^{\frac{1}{2}} = \sqrt{N}.$$

R. Salem asked in 1950 the following question, linked to several problems in harmonic analysis (see Chap. X from [227]) : is it possible to find a sequence $\varepsilon \in \{-1,+1\}^{\mathbb{N}}$ such that there is a constant $c > 0$ for which

$$\sqrt{N} \leq \sup_{\theta\in[0,1]} \left| \sum_{n<N} \varepsilon_n e(n\theta) \right| \leq c\sqrt{N}$$

holds for any positive integer N?

H.S. Shapiro in 1951 and then W. Rudin in 1959 gave a positive answer to this question in [393] and [365].

Definition 2.2.1. *The Rudin-Shapiro sequence ε over the alphabet $\{-1,+1\}$ is defined by the relation $\varepsilon_0 = 1$ and for any nonnegative integer n*

$$\begin{cases} \varepsilon_{2n} &= \varepsilon_n \\ \varepsilon_{2n+1} &= (-1)^n \varepsilon_n. \end{cases}$$

Remark. The relations $u_0 = 1$, $u_{2n} = u_n$, and $u_{2n+1} = -u_n$, for any nonnegative integer n define the Morse sequence over the alphabet $\{+1,-1\}$.

By analogy with the fact that u_n gives the parity of the sum of digits of n, i.e., the parity of the number of one's in the dyadic development of n, it is easy to verify that ε_n gives the parity of the number of words 11 in the dyadic development of n.

Proposition 2.2.2. *For any nonnegative integer n with dyadic development $n = \sum_{i\geq 0} n_i 2^i$ $(n_i \in \{0,1\})$ we have*

$$\varepsilon_n = (-1)^{\sum_{i\geq 0} n_i n_{i+1}}.$$

Furthermore for any nonnegative integers a, b and n such that $b < 2^n$, we have

$$\varepsilon_{2^{n+1}a+b} = \varepsilon_a \varepsilon_b.$$

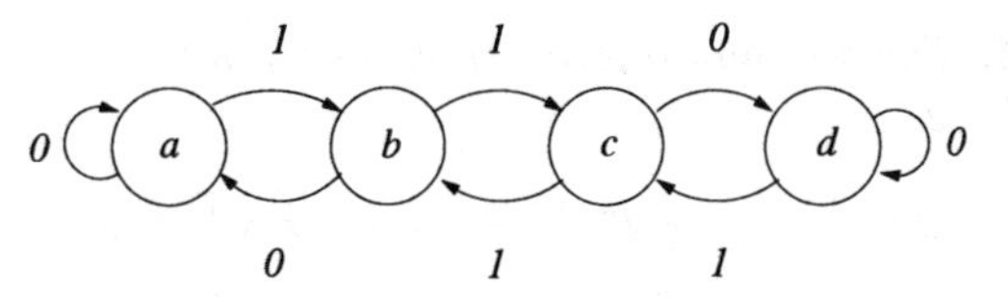

Fig. 2.2. Automaton associated with the the Rudin-Shapiro substitution.

Proof. The proof can be deduced by induction from the relations $\varepsilon_{2n} = \varepsilon_n$ and $\varepsilon_{2n+1} = (-1)^n \varepsilon_n$. ∎

If we consider again integers as words over the alphabet $\{0,1\}$, it is clear that the 2-automaton on Fig. 2.2 (in reverse reading or direct reading) with initial state a and exit map φ defined by $\varphi(a) = \varphi(b) = +1$ and $\varphi(c) = \varphi(d) = -1$ recognizes the Rudin-Shapiro sequence.

Proposition 2.2.3. *For any nonnegative integer N*

$$\sup_{\theta \in [0,1]} \left| \sum_{n<N} \varepsilon_n e(n\theta) \right| \le (2+\sqrt{2})\sqrt{N}\,.$$

Proof. For any nonnegative integer n, put

$$S_n(\theta) = \sum_{k<2^n} \varepsilon_k\, e(k\theta), \quad S'_n(\theta) = \sum_{k<2^n} (-1)^k \varepsilon_k\, e(k\theta), \quad \mathbf{M}(\theta) = \begin{pmatrix} 1 & e(\theta) \\ 1 & -e(\theta) \end{pmatrix}.$$

As $S_{n+1}(\theta) = S_n(2\theta) + e(\theta) S'_n(2\theta)$ and $S'_{n+1}(\theta) = S_n(2\theta) - e(\theta) S'_n(2\theta)$, we have

$$\begin{pmatrix} S_n(\theta) \\ S'_n(\theta) \end{pmatrix} = \prod_{k<n} \mathbf{M}\left(2^k\theta\right) \begin{pmatrix} 1 \\ 1 \end{pmatrix}$$

and

$$|S_n(\theta)| \le \sqrt{|S_n(\theta)|^2 + |S'_n(\theta)|^2} \le \sqrt{2} \prod_{k<n} \left\|\mathbf{M}\left(2^k\theta\right)\right\|_2 = \sqrt{2}\, 2^{n/2}\,.$$

Now, if $N = 2^{n_1} + 2^{n_2} + \cdots + 2^{n_k}$, with $n_1 > n_2 > \cdots > n_k$, we have

$$\sum_{k<N} \varepsilon_k\, e(k\theta) = S_{n_1}(\theta) + e(2^{n_1}\theta) S_{n_2}(\theta) + \cdots + e\left(\left(2^{n_1} + \cdots + 2^{n_{k-1}}\right)\theta\right) S_{n_k}(\theta),$$

since it follows from Proposition 2.2.2 that for any nonnegative integers k, n, we have $\varepsilon_{2^n+k} = \varepsilon_k$ as soon as $k < 2^{n-1}$.

We deduce from this equality that

$$\left| \sum_{k<N} \varepsilon_k e(k\theta) \right| \le \sqrt{2} \left(2^{n_1/2} + 2^{n_2/2} + \cdots + 2^{n_k/2} \right)$$

and Proposition 2.2.3 follows from the following lemma:

Lemma 2.2.4. *For any nonnegative integers $n_1, n_2, \cdots, n_k$ with $n_1 > n_2 > \cdots > n_k$, we have*

$$2^{n_1/2} + 2^{n_2/2} + \cdots + 2^{n_k/2} \leq (1+\sqrt{2})\sqrt{2^{n_1} + \cdots + 2^{n_k}}\,.$$

Proof. The proof works by induction over k. The property is obvious in the case $k = 1$. Suppose that for any nonnegative integers $n_1, n_2, \cdots, n_k$ with $n_1 > n_2 > \cdots > n_k$, we have

$$2^{n_1/2} + 2^{n_2/2} + \cdots + 2^{n_k/2} \leq (1+\sqrt{2})\sqrt{2^{n_1} + \cdots + 2^{n_k}}\,.$$

Then for any nonnegative integers $n_1, n_2, \cdots, n_{k+1}$ with $n_1 > n_2 > \cdots > n_{k+1}$, we get

$$2^{n_1/2} + 2^{n_2/2} + \cdots + 2^{n_{k+1}/2} \leq 2^{n_1/2} + (1+\sqrt{2})\sqrt{2^{n_2} + \cdots + 2^{n_{k+1}}}$$

by induction.

But as $2^{n_2} + \cdots + 2^{n_{k+1}} \leq 2^{n_1}$, we have $2^{n_1/2}\sqrt{2^{n_2} + \cdots + 2^{n_{k+1}}} \leq 2^{n_1}$ and

$$\begin{aligned} 2^{n_1} + 2(1+\sqrt{2})\, 2^{n_1/2}\sqrt{2^{n_2} + \cdots + 2^{n_{k+1}}} + (3+2\sqrt{2})\,(2^{n_2} + \cdots + 2^{n_{k+1}}) \\ \leq (3+2\sqrt{2})\,(2^{n_1} + 2^{n_2} + \cdots + 2^{n_{k+1}}) \end{aligned}$$

i.e.,

$$2^{n_1/2} + (1+\sqrt{2})\sqrt{2^{n_2} + \cdots + 2^{n_{k+1}}} \leq (1+\sqrt{2})\sqrt{2^{n_1} + 2^{n_2} + \cdots + 2^{n_{k+1}}},$$

which ends the proof. ■

2.2.2 Statistical properties of the Rudin-Shapiro sequence

As the Rudin-Shapiro sequence is defined by a very simple algorithm, one should expect its behavior to be very far from that of a "random" sequence. In particular, one should expect, as in the case of the Morse sequence (see Proposition 2.1.3) that the Rudin-Shapiro sequence has positive correlations. See also Sec. 1.5.3 in Chap. 1.

Nevertheless the following proposition shows a result in the opposite direction:

Proposition 2.2.5. *For any nonnegative integers k and N, define*

$$\gamma_N(k) = \frac{1}{N}\sum_{n<N} \varepsilon_n\, \varepsilon_{n+k}.$$

Then for any nonnegative integer k the sequence $(\gamma_N(k))_{N>0}$ converges and

$$\lim_{N\to\infty} \gamma_N(k) = 0 \quad \textit{for every} \quad k \geq 1\,.$$

Proposition 2.2.5 is a corollary of the more precise following result, due to C. Mauduit and A. Sárközy [284]:

Proposition 2.2.6. *For any positive integers k and N, we have*

$$|\sum_{n<N} \varepsilon_n \varepsilon_{n+k}| < 2k + 4k \log_2 \frac{2N}{k}.$$

Proof.

• Let us first consider the case where N is a power of 2, $N = 2^M$ with M a nonnegative integer.

If we put for any positive integers k and M

$$S_M(k) = \sum_{n<2^M} \varepsilon_n \varepsilon_{n+k} \text{ and } S'_M(k) = \sum_{n<2^M} = (-1)^n \varepsilon_n \varepsilon_{n+k},$$

then we have

$$\begin{aligned} S_{M+1}(2k) &= \sum_{n<2^{M+1}} \varepsilon_n \varepsilon_{n+2k} = \sum_{n<2^M} \varepsilon_{2n} \varepsilon_{2n+2k} + \sum_{n<2^M} \varepsilon_{2n+1} \varepsilon_{2n+2k+1} \\ &= \sum_{n<2^M} \varepsilon_n \varepsilon_{n+k} + (-1)^k \sum_{n<2^M} \varepsilon_n \varepsilon_{n+k} \\ &= (1 + (-1)^k) S_M(k), \end{aligned}$$

and

$$\begin{aligned} S_{M+1}(2k+1) &= \sum_{n<2^{M+1}} \varepsilon_n \varepsilon_{n+2k+1} \\ &= \sum_{n<2^M} \varepsilon_{2n} \varepsilon_{2n+2k+1} + \sum_{n<2^M} \varepsilon_{2n+1} \varepsilon_{2n+2k+2} \\ &= \sum_{n<2^M} (-1)^{n+k} \varepsilon_n \varepsilon_{n+k} + \sum_{n<2^M} (-1)^n \varepsilon_n \varepsilon_{n+k+1} \\ &= (-1)^k) S'_M(k) + S'_M(k+1). \end{aligned}$$

By a similar computation, we get

$$S'_{M+1}(2k) = (1 - (-1)^k) S_M(k)$$

and

$$S'_{M+1}(2k+1) = (-1)^k S'_M(k) - S'_M(k+1).$$

In particular we have $S'_M(0) = S'_M(1) = \delta_0(M)$ and $S_M(1) = \delta_0(M) + 2\delta_1(M)$, where δ_0 and δ_1 are Dirac measures.

It is now easy to deduce by induction over k from these four relations that for any positive integer k and any nonnegative integer M, we have

$$|S_M(k)| \leq 2k \text{ and } |S'_M(k)| \leq 2k.$$

Now, if $N = \sum_{i=1}^{l} 2^{M_i}$ is the dyadic expansion of the positive integer N, it is easy to check that

$$\sum_{n<N} \varepsilon_n \varepsilon_{n+k} = \sum_{i=1}^{l} \sum_{n<2^{M_i}} \varepsilon_{\left(n+\sum_{1\leq j<i} 2^{M_j}\right)} \varepsilon_{\left(n+k+\sum_{1\leq j<i} 2^{M_j}\right)}.$$

Let us define the integer ν by

$$2^\nu \leq k < 2^{\nu+1}.$$

If $i \in I^+ = \{j \in \{1,\ldots,l\}, M_j > \nu\}$ we have

$$\sum_{n<2^{M_i}} \varepsilon_{\left(n+\sum_{1\leq j<i} 2^{M_j}\right)} \varepsilon_{\left(n+k+\sum_{1\leq j<i} 2^{M_j}\right)}$$

$$= S_{M_i}(k) + \sum_{n<2^{M_i}} \left(\varepsilon_{\left(n+\sum_{1\leq j<i} 2^{M_j}\right)} \varepsilon_{\left(n+k+\sum_{1\leq j<i} 2^{M_j}\right)} - \varepsilon_n \varepsilon_{n+k} \right),$$

because it follows from Proposition 2.2.3 that

$$\sum_{n<2^{M_i}-k} \varepsilon_{\left(n+\sum_{1\leq j<i} 2^{M_j}\right)} \varepsilon_{\left(n+k+\sum_{1\leq j<i} 2^{M_j}\right)}$$

$$= \sum_{n<2^{M_i}-k} \left(\varepsilon_{\left(\sum_{1\leq j<i} 2^{M_j}\right)} \right)^2 \varepsilon_n \, \varepsilon_{n+k} = \sum_{n<2^{M_i}-k} \varepsilon_n \varepsilon_{n+k}.$$

This shows that

$$\left| \sum_{i\in I^+} \sum_{n<2^{M_i}} \varepsilon_{\left(n+\sum_{1\leq j<i} M_j\right)} \varepsilon_{\left(n+k+\sum_{1\leq j<i} 2^{M_j}\right)} \right| \leq \sum_{i\in I^+} (|S_{M_i}(k)| + 2k)$$

$$\leq 4k \text{ Card } I^+.$$

Now

$$\text{Card } I^+ = M_1 - \nu = [\log_2 N] - [\log_2 k] \leq \log_2 N - \log_2 k + 1,$$

which gives

$$\left| \sum_{i\in I^+} \sum_{n<2^{M_i}} \varepsilon_{\left(n+\sum_{1\leq j<i} 2^{M_j}\right)} \varepsilon_{\left(+k+\sum_{1\leq j<i} 2^{M_j}\right)} \right| \leq 4k \log_2 \frac{2N}{k}.$$

If we put $I^- = \{i \in \{1,\ldots,l\}, M_i \leq \nu\}$, then we have

$$\left| \sum_{i \in I^-} \sum_{n < 2^{M_i}} \varepsilon_{\left(n + \sum_{1 \le j < i} 2^{M_j}\right)} \varepsilon_{\left(n+k+\sum_{1 \le j < i} 2^{M_j}\right)} \right| \le \sum_{i \in \dot{I}^-} 2^{M_i}$$
$$\le 2^{\nu+1} - 1 \ < \ 2k.$$

Finally we proved that

$$\left| \sum_{n<N} \varepsilon_n \varepsilon_{n+k} \right| \le 2k + 4k \log_2 \frac{2N}{k}.$$

■

Exercise 2.2.7. What can be said about the multiple correlations of the Rudin-Shapiro sequence defined for nonnegative integers $k_1, k_2, \cdots, k_p$ and N by

$$\gamma_N(k_1, k_2, \ldots, k_p) \ = \ \frac{1}{N} \sum_{n<N} \varepsilon_n \varepsilon_{n+k_1} \varepsilon_{n+k_2} \cdots \varepsilon_{n+k_p} \ ?$$

2.3 The Baum-Sweet sequence

It is well known that the continued fraction expansion of an irrational algebraic number x is ultimately periodic if and only if x is quadratic (this result is known as *Lagrange's theorem*). But we know almost nothing about the expansion of nonquadratic irrational algebraic numbers. In particular, the following question seems to be of the highest difficulty: is there any algebraic number of degree at least 3 with a bounded continued fraction expansion (i.e., such that the partial quotients in the expansion are bounded)? If we replace $\mathbb{R}$ by the field $\mathbb{F}_2((X^{-1}))$ of formal power series in X^{-1} on $\mathbb{F}_2$ (where $\mathbb{F}_2 = \mathbb{Z}/2\mathbb{Z}$ denotes the finite field with two elements), the analogous problem was solved in 1976 by L. E. Baum and M. M. Sweet. Indeed they give in [57] an example of an algebraic element of degree 3 with a bounded continued fraction expansion (i.e., with partial quotients in $\mathbb{F}_2[X]$ of bounded degree).

Definition 2.3.1. *The* Baum-Sweet sequence $(f_n)_{n \in \mathbb{N}}$ *with values in the alphabet* $\mathbb{F}_2$ *is defined by*

$$\begin{aligned} f_n &= 0 \ \textit{ if the dyadic development of } n \textit{ contains} \\ &\qquad \textit{at least one odd string of } 0's, \\ &= 1 \ \textit{ if not}. \end{aligned}$$

It is clear from this definition that the Baum-Sweet sequence is recognized by the 2-automaton (in direct reading) shown in Fig. 2.3, with initial state a and exit map φ defined by $\varphi(a) = \varphi(b) = 1$ and $\varphi(c) = \varphi(d) = 0$.

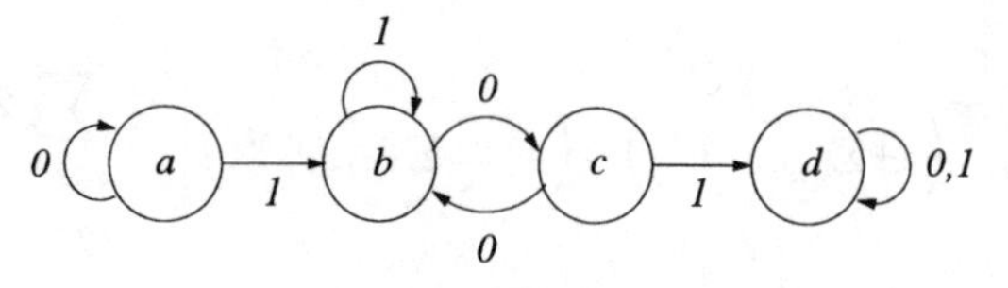

Fig. 2.3. Automaton associated with the the Baum-Sweet sequence.

The following proposition illustrates a typical property of sequences generated by an automaton. It is an illustration of a classical theorem on q-automata (for q power of a prime) due to G. Christol, T. Kamae, M. Mendès France and G. Rauzy that will be discussed in Chap. 3 (see also [118]).

A formal power series f is said to be *algebraic* over $\mathbb{F}_2(X)$ if there exists a nontrivial polynomial P with coefficients in $\mathbb{F}_2(X)$ such that $P(f) = 0$.

Proposition 2.3.2. *Let $f = \sum_{n\geq 0} f_n X^{-n}$ be the formal power series with coefficients given by the Baum-Sweet sequence. Then*

1. *f is an algebraic element of degree 3 over $\mathbb{F}_2(X)$;*
2. *furthermore, the continued fraction expansion of f is bounded and consists of elements of the set $\{1, X, X+1, X^2, X^2+1\}$.*

Proof. Let us give a proof of the first assertion. We do not give here a proof of the second one which is rather a special and surprising property of the Baum-Sweet sequence than a general property of automata. Note that there is no characterization of algebraic elements over $\mathbb{F}_q(X)$ (q power of a prime) the continued fraction expansion of which is bounded, and this problem seems to be hopelessly difficult.

It follows from the definition that for any nonnegative integer n we have

$$\begin{cases} f_{2n+1} = f_n \\ f_{4n} \quad = f_n \\ f_{4n+2} = 0 \end{cases}, \text{ with } f_0 = 1.$$

Now if we write

$$f = \sum_{n\geq 0} f_n X^{-n} = \sum_{n\geq 0} f_{2n+1} X^{-(2n+1)} + \sum_{n\geq 0} f_{4n+2} X^{-(4n+2)} + \sum_{n\geq 0} f_{4n} X^{-4n},$$

we see that

$$f = X^{-1} \sum_{n\geq 0} f_n X^{-2n} + \sum_{n\geq 0} f_n X^{-4n} = X^{-1} f^2 + f^4$$

and f is solution of the algebraic equation of degree 3

$$X f^3 + f + X = 0 .$$

■

For more details on continued fraction expansions of formal power series, see the surveys [76, 258, 377]. For transcendence results for Laurent formal power series with coefficients in a finite field, see Chap. 3.

2.4 The Cantor sequence

The *Cantor substitution* τ is defined over the alphabet $\{a, b\}$ by

$$\tau(a) = aba$$
$$\tau(b) = bbb\,.$$

Definition 2.4.1. *The sequence of words* $(\tau^n(a))_{n\in\mathbb{N}}$ *converges to the fixed point* C *of the Cantor substitution beginning with* a. *This sequence is called the* Cantor sequence. *Its first terms are:*

$$aba\, bbb\, aba\, bbb\, bbb\, bbb\, aba\, bbb\, aba \ldots$$

Let us denote by $\mathbb{C}_a$ *the set of integers* n *such that the* $(n+1)$*-th letter of the Cantor sequence is* a.

It follows from Proposition 1.3.1 that the Cantor sequence v is recognized by the 3-automaton in Fig. 2.4 in direct reading with initial state a and exit map $Id_{\{a,b\}}$:

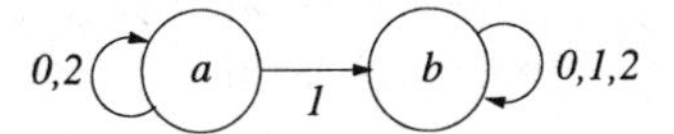

Fig. 2.4. Automaton associated with the Cantor sequence.

This proves the following proposition:

Proposition 2.4.2.

$$\mathbb{C}_a = \left\{ n \in \mathbb{N};\ n = \sum_{i\geq 0} n_i 3^i,\ \textit{with}\ \forall i \geq 0\ \ n_i \in \{0, 2\} \right\}.$$

2.5 An application of substitutions to criteria of divisibility

The four examples of sequences we have seen (Morse, Rudin-Shapiro, Baum-Sweet and Cantor) are fixed points of substitutions of constant ngth 2 or

3. As we pointed out several times, this means that they can be defined by an automatic property of the q-adic development of the integers.

We want to give now an application of Proposition 1.3.1 to the resolution of the following problem: given integers d and q greater or equal to 2, is it possible to decide only from its q-adic development whether a given positive integer is divisible by d or not? From an algebraic point of view, this corresponds to consider the powers of q in $\mathbb{Z}/d\mathbb{Z}$. Let us give an interpretation of this, in terms of automata.

Let v be the periodic sequence with values in the alphabet $\{0,1,\ldots,d-1\}$:

$$01\ldots(d-1)01\ldots(d-1)01\ldots(d-1)\ldots.$$

To solve our problem, we need to construct a q-automaton which recognizes the sequence v.

To obtain this q-automaton, it is enough to find a substitution ϱ of constant length q such that v is a fixed point of ϱ.

This can be easily done by cutting v into words of length q and rewriting v as $v = \varrho(0)\varrho(1)\ldots\varrho(d-1)\,\varrho(0)\varrho(1)\ldots\varrho(d-1)\ldots$.

To avoid heavy notations we will give explicitly the final automaton only in the case $q = 2$ and $d = 5$:

$$v = 01234012340123401234012340123401234012340123401234\ldots.$$

$$\begin{aligned}\varrho(0) &= 01\\ \varrho(1) &= 23\\ \varrho(2) &= 40\\ \varrho(3) &= 12\\ \varrho(4) &= 34\end{aligned}$$

Clearly v is fixed point of ϱ, and it follows from Proposition 1.3.1 that v is recognized by the 2-automaton shown in Fig. 2.5 (in direct reading) with initial state 0 and exit map $Id_{\{0,1,2,3,4\}}$.

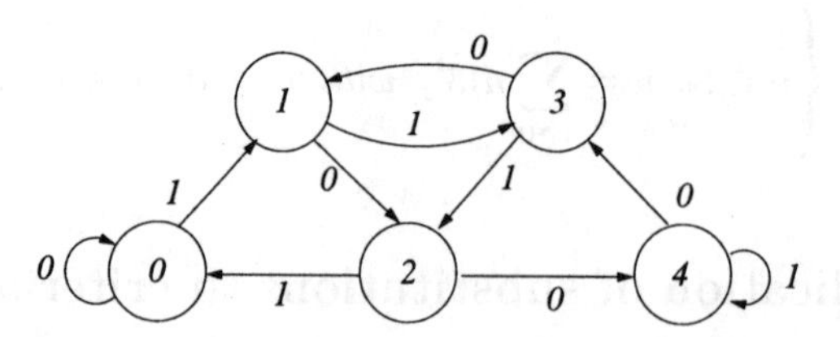

Fig. 2.5. Automaton associated with the substitution ϱ.

2.6 The Fibonacci sequence

2.6.1 Definition

After all these examples of sequences obtained as a fixed point of a constant length substitution and their arithmetic description, it is natural to look for non-constant length substitutions and to ask whether their fixed point can be defined by some simple arithmetic property of the integers.

This leads us to associate some generalized systems of numeration with substitutions. In the most general context, this problem is still linked to several open questions : we do not yet understand all the connections between the arithmetics and the geometry of the substitution dynamical systems (see [339]). A detailed survey of the literature on this subject is given in Chap. 7.

This fifth and last introductory example shows what happens in a case where both the arithmetics and the geometry are well understood. It is the "simplest" sequence among Sturmian sequences introduced in the 1940's by G. A. Hedlund and M. Morse in [302] and [303]. More details on the Fibonacci sequence will be found in Chaps. 5, 6, and 9.

Definition 2.6.1. *The* Fibonacci sequence w *is the unique non-empty fixed point* w *of the* Fibonacci substitution φ *defined over the alphabet* $\{a,b\}$ *by*

$$\begin{cases} \varphi(a) = ab \\ \varphi(b) = a. \end{cases}$$

The first terms of w *are*

$$abaababaabaababaababaabaababaabaab\ldots.$$

2.6.2 The Fibonacci system of numeration

Let $(F_n)_{n\in\mathbb{N}}$ be the sequence of integers defined by $F_0 = 1$, $F_1 = 2$ and for any integer $n \geq 1$, $F_{n+1} = F_n + F_{n-1}$.

Proposition 2.6.2. *Every nonnegative integer* n *can be written in a unique way as* $n = \sum_{i\geq 0} n_i F_i$ *with* $n_i \in \{0,1\}$ *and* $n_i n_{i+1} = 0$ *for any* $i \geq 0$. *This numeration system is called the* Zeckendorff numeration system *[468].*

Proof.

- Let us first prove the existence of the decomposition. It is true for $n = 0$, $n = 1$ and $n = 2$. Let us suppose that the property is true for any integer $n < F_k$ (with $k \geq 2$).
  Then if $F_k \leq n < F_{k+1} = F_k + F_{k-1}$, we have $n - F_k < F_{k-1}$ and by hypothesis, $n - F_k = \sum_{i=0}^{k-2} n_i F_i$, hence $n = F_k + \sum_{i=0}^{k-2} n_i F_i$, and the property is true by induction over k.

- Let us prove now the unicity of the expansion. Let us remark that if $(n_0, n_1, \ldots, n_k) \in \{0,1\}^{k+1}$ and $n_i n_{i+1} = 0$ for any $i < k$, then $\sum_{i=0}^{k} n_i F_i < F_{k+1}$.
This can be easily be proved by induction. Indeed if $n_k = 1$, then $n_{k-1} = 0$ and $\sum_{i=0}^{k} n_i F_i = F_k + \sum_{i=0}^{k-2} n_i F_i$. Hence if we suppose that $\sum_{i=0}^{k-2} n_i F_i < F_{k-1}$ we get that $\sum_{i=0}^{k} n_i F_i < F_k + F_{k-1} = F_{k+1}$.
It is now easy to prove the unicity of the decomposition : let us suppose that $\sum_{i \geq 0} n_i F_i = \sum_{i \geq 0} n'_i F_i$ with $n_i, n'_i \in \{0,1\}$ and $n_i n_{i+1} = n'_i n'_{i+1} = 0$ for any $i \geq 0$.
Only to simplify the terms of higher indices, we can suppose that $n_k = 1$, $n'_k = 0$ and $n_i = n'_i = 0$ for $i > k$. This would imply that $\sum_{i \geq 0} n_i F_i \geq F_k$ and, because of our previous remark, that $\sum_{i \geq 0} n'_i F_i < F_k$, which is a contradiction.

■

Definition 2.6.3. *If* $n = \sum_{i=0}^{k} n_i F_i$ *with* $n_k = 1$, $n_i \in \{0,1\}$ *and* $n_i n_{i+1} = 0$ *for any* $i < k$, *we say that* $Fib(n) = n_k n_{k-1} \cdots n_0 \in \{0,1\}^{k+1}$ *is the* Fibonacci expansion *of the positive integer* n.

Let us denote by $\mathbb{F}_a$ (respectively $\mathbb{F}_b$) the sets of integers n such that the $(n+1)$-th letter of the Fibonacci sequence is a (respectively b):

$$\mathbb{F}_a = \{0, 2, 3, 5, 7, 8, 10, 11, 13, 15, \ldots\},$$

$$\mathbb{F}_b = \{1, 4, 6, 9, 12, 14, 17, 19, 22, 25, \ldots\}.$$

Proposition 2.6.4. *We have*

$$\mathbb{F}_a = \{n \in \mathbb{N},\ Fib(n) \in \{0,1\}^* 0\},$$

$$\mathbb{F}_b = \{n \in \mathbb{N},\ Fib(n) \in \{0,1\}^* 1\}.$$

Proof. It follows from Definition 2.6.1 that $F = \lim_{n \to \infty} \varphi^r(a)$, $|\varphi^r(a)| = F_r$, and $\varphi^{r+1}(a) = \varphi^r(a)\varphi^{r-1}(a)$, for any $r \geq 1$.

This last relation shows that if $\text{Fib}(n) = n_k n_{k-1} \cdots n_0$, then $n \in \mathbb{F}_a$ if and only if $n - F_k \in \mathbb{F}_a$. As $0 \in \mathbb{F}_a$ and $1 \in \mathbb{F}_b$ we deduce from this that $n \in \mathbb{F}_a$ (respectively $n \in \mathbb{F}_b$) if and only if $n_0 = 0$ (respectively $n_0 = 1$). ■

Remark. For more results on numeration systems, see the nice survey [179].

Chapter 4 develops this approach in a more systematic way by associating partitions of the set of positive integers first with Sturmian sequences (these are the so-called *Beatty partitions*) and second, to fixed points of substitutions.

3. Automatic sequences and transcendence

The aim of this chapter is to investigate the connections between automatic sequences and transcendence in fields of positive characteristic. In the real case, it is well known that the expansion in a given integer basis of a rational number is ultimately periodic, which implies that its complexity function is bounded. More generally, the expansion of an algebraic irrational number in a given integer basis is supposed to be normal, whereas real numbers having as expansion a sequence with a low complexity function are conjectured to be either rational, or transcendental. In particular, real numbers having as binary expansion a Sturmian sequence, or a fixed point of a substitution over a two-letter alphabet which is either of constant length or primitive, are transcendental. The situation is much simpler in the case of formal power series with coefficients in a finite field. Indeed it is possible to characterize algebraicity in a simple way: a formal power series is algebraic if and only if the sequence of its coefficients is automatic; this criterion is known as Christol, Kamae, Mendès France, and Rauzy's theorem. In a similar vein, the continued fraction expansion of an algebraic number is supposed to be unbounded if this number is neither quadratic nor irrational; here again much more is known in the case of formal power series with coefficients in a finite field. For instance, examples of algebraic series with unbounded partial quotients can be produced. For more details, see Sec. 3.3.

This chapter is organized as follows. We first recall in Sec. 3.1 some definitions concerning the transcendence of Laurent formal power series with coefficients in a finite field and then state in Sec. 3.2 the main theorem of transcendence of this chapter, that is, the Christol, Kamae, Mendès France, and Rauzy theorem. We also consider some generalizations of this theorem, to the multidimensional case and to general fields of positive characteristic. Section 3.3 alludes to transcendence results for real numbers with expansion in a given integer basis having a low complexity sequence. We introduce in Sec. 3.4 some functions defined by Carlitz (exponential, logarithm, zeta) which are analogous to the corresponding real functions and which give us examples of application of the Christol, Kamae, Mendès France, and Rauzy theorem. For this purpose, we will recall the notions of absolute value and valuation. We then briefly review transcendence results for these functions as

[1] This chapter has been written by V. Berthé

well as proof methods, and emphasize the results involving automata. We end this chapter by going into the details of the most representative automaton proofs of transcendence.

3.1 Introduction

The first results of classical transcendence theory concerned special values of the exponential and logarithm functions. Analogous results in the field of p-adic numbers or in fields of positive characteristic have been introduced later, as for instance the Carlitz functions, that we define in Sec. 3.4.

Let us first recall some basic definitions concerning transcendence. A complex number α is called *algebraic* over $\mathbb{Q}$ if there exists a nontrivial polynomial P with coefficients in $\mathbb{Q}$ such that $P(\alpha) = 0$. Otherwise, α is called *transcendental* over $\mathbb{Q}$. For instance the real number π is transcendental over $\mathbb{Q}$ whereas $\sqrt{2}$ is algebraic.

We can similarly extend these notions to Laurent formal power series with values in a finite field. For this purpose, let us define the sets which play respectively the role of $\mathbb{Z}$, $\mathbb{Q}$, $\mathbb{R}$ and $\mathbb{C}$. Let $\mathbb{F}_q$ be the finite field with q elements, and let p denote its characteristic. We thus have p prime and $q = p^s$, where s is a positive integer. If the reader is not familiar with these notions, he can suppose in all that follows that $s = 1$, i.e., that $\mathbb{F}_q$ is the field $\mathbb{Z}/p\mathbb{Z}$ of the integers reduced modulo p. The analogue of $\mathbb{Z}$ will then be the ring $\mathbb{F}_q[X]$ of polynomials with coefficients in $\mathbb{F}_q$, the analogue of $\mathbb{Q}$ will be its fraction field $\mathbb{F}_q(X)$ and the analogue of $\mathbb{R}$ will be the field $\mathbb{F}_q((1/X))$ of formal power series defined below. Note that the coefficients in $\mathbb{F}_q$ play the role of "digits" in the numeration basis given by the powers of the indeterminate X.

Definition 3.1.1. *The field* $\mathbb{F}_q((1/X))$ *of* Laurent formal power series *with coefficients in* $\mathbb{F}_q$ *is the field of series of the form*

$$u_{-d}X^d + \cdots + u_0 + u_1X^{-1} + \cdots,$$

where u_i *belongs to* $\mathbb{F}_q$. *We furthermore denote by* $\mathbb{F}_q[[1/X]]$ *the ring of series of the form*

$$u_0 + u_1X^{-1} + u_2X^{-2} + \cdots,$$

where u_i *belongs to* $\mathbb{F}_q$.

We similarly denote by $\mathbb{F}_q((X))$ *the field of formal power series of the form* $u_{-d}X^{-d} + \cdots + u_0 + u_1X + \cdots$, *and by* $\mathbb{F}_q[[X]]$ *the ring of formal power series of the form* $u_0 + u_1X + u_2X^2 + \cdots$, *where the coefficients* u_i *belongs to* $\mathbb{F}_q$.

We work in Secs. 3.1 and 3.2 with series of $\mathbb{F}_q((X))$, for simplicity of notations, whereas Carlitz functions are defined in Sec. 3.4 with respect to series of $\mathbb{F}_q((1/X))$.

We can now define the notion of transcendence on $\mathbb{F}_q(X)$ for formal power series of $\mathbb{F}_q((1/X))$ or $\mathbb{F}_q((X))$.

Definition 3.1.2. *A formal power series F is* algebraic over $\mathbb{F}_q(X)$ *if there exists a nontrivial polynomial P with coefficients in $\mathbb{F}_q(X)$ such that $P(F) = 0$. Otherwise, F is called* transcendental *over $\mathbb{F}_q(X)$.*

This definition holds for series of $\mathbb{F}_q((X))$ as well as for series belonging to $\mathbb{F}_q((1/X))$.

In fact, we can restrict ourselves in the definition of algebraicity to *linear polynomials*, i.e., to polynomials with coefficients in $\mathbb{F}_q(X)$ of the form $\sum_{j=0}^{k-1} a_j T^{q^j}$, as stated in the following lemma, called *Ore's lemma.* Note that the q-power of a series with coefficients in $\mathbb{F}_q$ has a very simple expression. Recall that the Frobenius map $x \mapsto x^p$ is a morphism over any field of characteristic p, i.e., $(u+v)^p = u^p + v^p$, for any two elements u, v of such a field. We have furthermore $a^q = a$, for any $a \in \mathbb{F}_q$. Hence $(\sum u_i X^i)^q = \sum u_i X^{qi}$, where the coefficients u_i belong to $\mathbb{F}_q$.

Lemma 3.1.3 (Ore). *Every nontrivial polynomial P of $\mathbb{F}_q[X][T]$ divides a nontrivial linear polynomial of $\mathbb{F}_q[X][T]$ of the form $\sum_{j=0}^{k} a_j T^{q^j}$, where the coefficients a_j belong to $\mathbb{F}_q[X]$. In particular, if a series F is algebraic over $\mathbb{F}_q(X)$ then there exists $a_0, \ldots, a_k$ in $\mathbb{F}_q[X]$ such that $a_0 F + a_2 F^q + \ldots + a_k F^{q^k} = 0$, and $a_k \neq 0$.*

Proof. Consider the Euclidean division of T^{q^i} by P, for $0 \leq i \leq d = \deg P$. There exist Q_i and R_i in $\mathbb{F}_q(X)[T]$, with $\deg_T R_i < d$ such that $T^{q^i} = PQ_i + R_i$. We can eliminate the powers T^{q^i}, for $0 \leq i \leq d-1$, in these $d+1$ equations. Indeed, there exists a nontrivial choice of $a_0, a_1, \ldots, a_d$ in $\mathbb{F}_q[X]$ (after multiplication by a suitable polynomial in $\mathbb{F}_q[X]$) such that $a_0 R_0 + a_1 R_1 + \ldots + a_d R_d = 0$. We thus have $\sum_{i=0}^{d} a_i T^{q^i} = P \times (\sum_{i=0}^{d} a_i Q_i)$. ∎

Example 3.1.4. Consider the Morse sequence $u = (u_n)_{n \in \mathbb{N}}$: recall that u is defined over $\mathbb{F}_2$ as $u_n = S_2(n)$, where $S_2(n)$ is the sum modulo 2 of the digits in the base 2 expansion of the integer n (for more details, see Chap. 2). We thus have $u_0 = 0$, $u_{2n} = u_n$ and $u_{2n+1} = 1 + u_n$, for every integer n. Recall that in $\mathbb{F}_2[[X]]$, we have $(\sum_{n \geq 0} u_n X^n)^2 = \sum_{n \geq 0} u_n X^{2n}$. Let $F(X) = \sum_{n \geq 0} u_n X^n$. We have

$$
\begin{aligned}
F(X) = \textstyle\sum_{n \geq 0} u_n X^n &= \textstyle\sum_{n \geq 0} u_{2n} X^{2n} + \sum u_{2n+1} X^{2n+1} \\
&= \textstyle\sum_{n \geq 0} u_n X^{2n} + \sum (1 + u_n) X^{2n+1} \\
&= \textstyle(\sum_{n \geq 0} u_n X^n)^2 + X(\sum u_n X^n)^2 + \frac{X}{1+X^2} \\
&= F(X)^2 (1 + X) + \frac{X}{1+X^2}.
\end{aligned}
$$

Hence the series $F(X)$ is algebraic over $\mathbb{F}_2(X)$ and satisfies

$$(1+X)^3 F(X)^2 + F(X)(1+X)^2 + X = 0,$$

by using $(1+X)^2 = 1 + X^2$.

Exercise 3.1.5. Prove that if $v = (v_n)_{n\in\mathbb{N}}$ denotes the Rudin-Shapiro sequence (see Chap. 2), then $G(X) = \sum_{n\geq 0} v_n X^n$ satisfies in $\mathbb{F}_2((X))$

$$(1+X)^5 G^2(X) + (1+X)^4 G(X) + X^3 = 0.$$

3.2 The Christol, Kamae, Mendès France, and Rauzy theorem

The automaton proofs of transcendence are based on the following theorem due to Christol, Kamae, Mendès France, and Rauzy (see [117] and also [118]) which gives a necessary and sufficient condition of algebraicity for a formal power series with coefficients in a finite field. For more references on automatic sequences, see [11] and the impressive [28].

Theorem 3.2.1 (Christol, Kamae, Mendès France, and Rauzy).

Let $u = (u(n))_{n\in\mathbb{N}}$ be a sequence with values in $\mathbb{F}_q$. The following conditions are equivalent:

1. *the formal power series $\sum_{n\geq 0} u(n)X^n$ is algebraic over the field $\mathbb{F}_q(X)$,*
2. *the* q-kernel *$N_q(u)$ of the sequence u is finite, where $N_q(u)$ is the set of subsequences of the sequence $(u(n))_{n\in\mathbb{N}}$ defined by*

$$N_q(u) = \{(u(q^k n + r))_{n\in\mathbb{N}};\ k \geq 0;\ 0 \leq r \leq q^k - 1\},$$

3. *the sequence u is q-automatic,*
4. *the sequence u is the image by a letter-to-letter projection of a fixed point of a substitution of constant length q.*

Remarks.

- The last equivalence is actually due to Cobham (see [120]) and the equivalence between statements 2. and 3. dates back to Eilenberg in [158] (for a proof of both equivalences, see Chap. 1). This theorem is also called Christol's theorem since the equivalence between statements 1. and 2. can be found in [117].
- We consider here, except otherwise stated, automata in reverse reading since we will focus on the notion of q-kernel. Let us recall that there is equivalence between reverse and direct reading automaticity for a sequence (see Chap. 1, Proposition 1.3.4).
- We easily deduce from the Christol, Kamae, Mendès France, and Rauzy theorem the following facts. The notion of k-automaticity is stable by finite modification (as for instance changing a finite number of terms, see [120]) or by applying the shift $S((u_n)_{n\in\mathbb{N}}) = (u_{n+1})_{n\in\mathbb{N}}$. Furthermore a ultimately periodic sequence is k-automatic.

- The notion of p-automaticity can also be expressed as follows in terms of first-order logic: a sequence is generated by a p-substitution if and only if it is p-definable (it can be defined in the theory $(\mathbb{N}, +, V_p)$, where V_p is the function "valuation" that associates with x the highest power of p that divides x (or 1 if $x = 0$)). For more details, the reader is referred to the survey [98].
- The same theorem holds by considering a series $\sum u_n X^{-n}$ in $\mathbb{F}_q((1/X))$ with the same definition for the q-kernel. Indeed, $\sum u_n X^{-n}$ is transcendental over $\mathbb{F}_q(X)$ if and only if $\sum u_n X^n$ is transcendental over $\mathbb{F}_q(X)$.

Exercise 3.2.2. 1. Show that a sequence is p-automatic if and only if it is p^k-automatic for any positive power of the prime p.
2. Build a d-automaton generating the characteristic sequence of the set of powers of a fixed positive integer d.
3. Build a d-automaton generating the characteristic sequence of the set of integers divisible by d.
4. Build a 2-automaton generating the characteristic sequence of the set of integers with base 2-expansion of the form $1^n 0^m 1$, for $n, m > 0$ and $n+m$ odd.
5. Prove that the characteristic sequence of the set of integers with base 2-expansion of the form $1^n 0^{n+1} 1$, for $n > 0$, is not 2-automatic.

3.2.1 Proof of the theorem

The proof of Theorem 3.2.1 we give here is exactly the proof of [11]. We also refer the reader to the original proof in [118].

- Let us first prove that the q-kernel of an algebraic series is finite.
The idea of the proof is the following. Suppose we have a finite set $\mathcal{N}$ of subsequences of a given sequence u such that u belongs to this set and such that $\mathcal{N}$ is stable under the maps $\mathcal{A}_r$ $(0 \leq r \leq q-1)$, where $\mathcal{A}_r$ associates with a subsequence $(v(n))_{n \in \mathbb{N}}$ the subsequence $(v(qn + r))_{n \in \mathbb{N}}$. As this set contains the q-kernel of u, hence the q-kernel is finite.
Let us work here on the (algebraic) formal power series $\sum u(n)X^n$ and let us introduce the corresponding maps A_r. Let r be an integer in $[0, q-1]$. Let $A_r : \mathbb{F}_q[[X]] \to \mathbb{F}_q[[X]]$ defined by:

$$A_r(\sum_{n \geq 0} v(n)X^n) = \sum_{n \geq 0} v(qn + r)X^n,$$

where $v(n)$ belongs to $\mathbb{F}_q$, for all n. Let $F = \sum_{n \geq 0} v(n)X^n$; we have $F = \sum_{r=0}^{q-1} A_r(F)^q X^r$. Moreover, if $F = \sum_{r=0}^{q-1} B_r^q X^r$, with $B_r \in \mathbb{F}_q[[X]]$ for $r \in [0, q-1]$, then $A_r(F) = B_r$, for every $0 \leq r \leq q-1$. We deduce from this that for any polynomial P in $\mathbb{F}_q[X]$ and for any formal power series F: $A_r(PF^q) = A_r(P)F$. Namely, $P = \sum_{r=0}^{q-1} A_r(P)^q X^r$ and $PF^q =$

$\sum_{r=0}^{q-1}(A_r(P)F)^q X^r$. Note furthermore that $A_r(P)$ is a polynomial and that $\deg A_r(P) \leq \deg P/q$.
Let $F = \sum u(n)X^n$ be an algebraic formal power series. From Ore's lemma (Lemma 3.1.3), there exists an integer j and polynomials a_0, $a_1, \cdots, a_k$ in $\mathbb{F}_q[X]$, not all equal to 0, such that

$$\sum_{i=j}^{k} a_i F^{q^i} = 0. \tag{3.1}$$

We can furthermore suppose $a_0 \neq 0$. Indeed, let j be the smallest index such that a relation of type (3.1) holds with $a_j \neq 0$. Let us show that $j = 0$. We have $a_j = \sum_{r=0}^{q-1} A_r(a_j)^q X^r$. As $a_j \neq 0$, there exists an integer r such that $A_r(a_j) \neq 0$. Suppose $j \neq 0$. We have

$$0 = A_r(\sum_{i=j}^{k} a_i F^{q^i}) = \sum_{i=j}^{k} A_r(a_i) F^{q^{i-1}},$$

which is of type (3.1) and which contradicts the minimality of j.
Suppose now that F satisfies equality (3.1) with $a_0 \neq 0$. Let $G = F/a_0$. We thus have:

$$G = \sum_{i=1}^{k} b_i G^{q^i},$$

with $b_i = -a_i a_0^{q^i - 2}$. We want to show that there exists a finite set of formal power series $\mathcal{H}$ stable under the maps A_r and containing the series F. Let $\mathcal{H}$ denote the set of formal power series of the form

$$\sum_{i=0}^{k} c_i G^{q^i}, \tag{3.2}$$

with $c_i \in \mathbb{F}_q[X]$ and $\deg c_i \leq \sup(\deg a_0, \deg b_1, \cdots, \deg b_k)$.

Let H be a series of type (3.2). For any integer r in $[0, q-1]$, we have

$$\begin{aligned} A_r(H) &= A_r(c_0 G + \textstyle\sum_{i=1}^{k} c_i G^{q^i}) \\ &= A_r(\textstyle\sum_{i=1}^{k} (c_0 b_i + c_i) G^{q^i}) \\ &= \textstyle\sum_{i=1}^{k} A_r(c_0 b_i + c_i) G^{q^{i-1}}. \end{aligned}$$

As

$$\deg(c_0 b_i + c_i) \leq 2 \sup(\deg a_0, \deg b_1, \cdots, \deg b_k),$$

which implies that

$$\deg A_r(c_0 b_i + c_i) \leq 2 \sup(\deg a_0, \deg b_1, \cdots, \deg b_k)/q,$$

the set $\mathcal{H}$ is stable under the action of the maps A_r. We conclude the proof by noticing that $\mathcal{H}$ is finite and that $\mathcal{H}$ contains $F = a_0 G$.

- Let us prove now the converse implication. Suppose that the q-kernel $N_q(u)$ of the sequence $u = (u(n))_{n\in\mathbb{N}}$ with values in $\mathbb{F}_q$ is finite. Let $F = \sum u(n)X^n$. Note $u = u_1, u_2, \cdots, u_d$ the sequences of $N_q(u)$ and for any integer j in $[1, d]$, put $F_j = \sum u_j(n)X^n$. We have, for any integer r in $[0, q-1]$:

$$F_j = \sum_{r=0}^{q-1} A_r(F_j)^q X^r,$$

and

$$A_r(F_j)^q = \sum u_j(qn+r)X^{qn} = \sum (u_j(qn+r)X^n)^q.$$

Let i $(i = i(j,r))$ be such that $(u_j(qn+r))_{n\in\mathbb{N}} = u_i$. We thus have $A_r(F_j)^q = F_i^q$. Hence, for any j, the series F_j belongs to the vector space spanned over $\mathbb{F}_q(X)$ by $F_1(X)^q, \cdots, F_d(X)^q$.
Similarly, for any j, the series F_j^q, and hence the series F_j, belong to the vector space spanned over $\mathbb{F}_q(X)$ by $F_1(X)^{q^2}, \cdots, F_d(X)^{q^2}$.
We thus prove by induction that the $(d+1)$ series $F, F^q, \cdots, F^{q^d}$ belong to the vector space spanned over $\mathbb{F}_q(X)$ by $F_1(X)^{q^{d+1}}, \cdots, F_d(X)^{q^{d+1}}$. This vector space has dimension less than or equal to d. We deduce from this the algebraicity of F.

∎

Remarks.

- The equivalence between statements 2., 3., and 4. is true whatever the cardinality of the alphabet: we have seen in Chap. 1 that the field structure is not required in the proof of this equivalence.
- Several results of k-automaticity (see for instance [32]) or more generally of k-regularity [25] appear in the literature, with proofs in a similar vein to that of the proof of equivalence between statements 1. and 2..

3.2.2 Applications

The aim of this section is to give some applications of Theorem 3.2.1. The first two are easy consequences of the theorem.

- Let $\sum_{n\geq 0} u(n)X^n$ be an algebraic formal power series. Let a and b be two natural integers. The series $\sum_{n\geq 0} u(an+b)X^n$ is algebraic.
- Let $p \geq 2$ be a prime. Let $S_p(n)$ be the sum modulo p of the digits of n in base p. The series $\sum_{n\geq 0} S_p(n)X^n$ is algebraic.

Remark. It can be proved that the series $\sum_{n\geq 0} S_p(n^2)X^n$ is transcendental. More generally, let R be a polynomial with coefficients in $\mathbb{Q}$ such that $R(\mathbb{N}) \subset \mathbb{N}$. The formal power series $\sum_{n\geq 0} S_p(R(n))X^n$ is algebraic over $\mathbb{F}_p$ if and only if the degree of R is less than or equal to 1 (see [9]).

Note that the following theorem, due to Cobham (see [119]), produces more examples of transcendental series. We do not give here the proof of this theorem, which is rather difficult.

Theorem 3.2.3 (Cobham). *Let u be a sequence which is both k-automatic and k'-automatic. If k and k' are multiplicatively independent (i.e., if $\frac{\log(k)}{\log(k')}$ is irrational), then the sequence u is ultimately periodic.*

Remark. This theorem has received numerous generalizations and has given rise to much literature. See for instance the survey [98], see also [97, 333].

Exercise 3.2.4. Give an example of a 6-automatic sequence which is neither 3-automatic nor 2-automatic.

We deduce from Cobham's theorem the following result of transcendence, which resembles an open question of Mahler: let $(u_n)_{n\in\mathbb{N}}$ be a binary sequence such that the real numbers $\sum_{n\geq 0} u_n 2^{-n}$ and $\sum_{n\geq 0} u_n 3^{-n}$ are algebraic over $\mathbb{Q}$; is this sequence ultimately periodic?

Proposition 3.2.5. *Let $(u_n)_{n\in\mathbb{N}}$ be a binary sequence such that $\sum u_n X^n$ considered as an element of $\mathbb{F}_2[[X]]$ and $\sum u_n X^n$ considered as an element of $\mathbb{F}_3[[X]]$ are algebraic. Then, this sequence is ultimately periodic, i.e., both series are rational.*

The *Hadamard product* of two series $\sum u_n X^n$ and $\sum v_n X^n$ is defined as the series $\sum u_n v_n X^n$. In particular, the Hadamard product of the characteristic series of a subset A of $\mathbb{N}$ (defined as $\sum_{n\in A} X^n$) by the characteristic series of a subset B is equal to the characteristic series of $A \cap B$. By considering the notion of q-kernel, we easily deduce the following.

Theorem 3.2.6. *The Hadamard product of two algebraic formal power series with coefficients in a finite field is algebraic.*

Another application of Cobham's theorem (Theorem 3.2.3) is the following. We give here the proof of [14].

Theorem 3.2.7. *Let r be an integer greater than or equal to 2. The series $\sum_{k=0}^{+\infty} X^{r^k}$ is algebraic over $\mathbb{F}_q(X)$ if and only if r is a power of p.*

Proof. Write

$$\sum_{k=0}^{+\infty} X^{r^k} = \sum_{n\geq 1} u_n X^n,$$

where $u = (u_n)_{n\in\mathbb{N}}$ is the characteristic sequence of the set of powers of r. The series $\sum_{k=0}^{+\infty} X^{r^k}$ is algebraic over $\mathbb{F}_q(X)$ if and only if the sequence u is p-automatic. But it is easily seen that the sequence u is r-automatic (Exercise 3.2.2) and not ultimately periodic. Hence the series $\sum_{k=0}^{+\infty} X^{r^k}$ is algebraic over $\mathbb{F}_q(X)$ if and only if r is a power of p. ∎

Remark. The formal power series $\sum u_n X^n$ belongs to $\mathbb{F}_q(X)$ if and only if the sequence $(u_n)_{n\in\mathbb{N}}$ is ultimately periodic. Note that in the real case we just have the following implication: if the sequence $(u_n)_{n\in\mathbb{N}}$ is ultimately periodic, then the series $\sum u_n X^n$ belongs to $\mathbb{Q}(X)$. The rational series $\sum nX^n$ shows that the converse is not true.

3.2.3 The multidimensional case

Christol, Kamae, Mendès France, and Rauzy's theorem was generalized by Salon to the multidimensional case, i.e., to the case of a formal power series with a finite number of indeterminates and with coordinates in a finite field, say $\sum_{n_i\geq 0} u(n_1, n_2, \cdots, n_d) X_1^{n_1} \cdots X_d^{n_d}$ (see for instance [368] and [369]).

The generalized q-kernel is given in this case by:

$$N_q(u(n_1, n_2, \cdots, n_d)) = \{u(q^k n_1 + r_1, q^k n_2 + r_2, \cdots, q^k n_d + r_d),$$

$$k \geq 0,\ 0 \leq r_i \leq q^k - 1, \text{ for } i = 1 \text{ to } d\}.$$

Recall that a formal power series $F = \sum_{n_i\geq 0} u(n_1, n_2, \cdots, n_d) X_1^{n_1} \cdots X_d^{n_d}$ is called *algebraic* over $\mathbb{F}_q(X_1, X_2, \cdots, X_d)$ if there exists a nontrivial polynomial P with coefficients in $\mathbb{F}_q(X_1, X_2, \cdots, X_d)$ such that $P(F) = 0$. We thus have the following theorem due to Salon (see [368] and [369]).

Theorem 3.2.8. *The power series $\sum u(n_1, n_2, \cdots, n_d) X_1^{n_1} \cdots X_d^{n_d}$ is algebraic over $\mathbb{F}_q(X_1, X_2, \cdots, X_d)$ if and only if the q-kernel of the sequence u is finite.*

The following results are easy applications of this theorem.

- Let $\sum u(n)X^n$ be an algebraic formal power series. The double formal power series $\sum u(n+m)X^n Y^m$ is algebraic.
- Let $\sum u(n,m)X^n Y^m$ be algebraic. Let a, b, c, d be four integers. The series $\sum u(an+b, cm+d)X^n Y^m$ is algebraic.

The notions of automaton and substitution can also be generalized into two dimensions. A *two-dimensional substitution* of constant length k associates with each letter a square array of letters of size (k,k). A *two-dimensional k-automaton* is defined similarly as a one-dimensional k-automaton but in this case the edges are labelled by pairs of integers in $[0, k-1]^2$. A sequence $(u(n,m))$ is generated by the automaton by reading simultaneously the digits of the base k expansions of n and m, the shortest expansion being completed with leading zeroes to get two strings of symbols of the length of the longest expansion (without leading zeroes).

Exercise 3.2.9. Consider the substitution $\sigma : \{0,1\} \to \{0,1\}^{2\times 2}$ defined by

$$\sigma(0) = \begin{matrix} 00 \\ 00 \end{matrix},$$

$$\sigma(1) = \frac{11}{10}.$$

Prove that the two-dimensional sequence defined over $\mathbb{N}^2$ as the fixed point of this substitution generated by the successive images of 1 is equal to Pascal's triangle reduced modulo 2. Find the substitution generating Pascal's triangle modulo p, where p is a prime integer.

Remarks.

- More generally, the two-dimensional sequence corresponding to Pascal's triangle modulo an integer d is automatic if and only if d is a power of a prime (for more details, see [30, 31, 32]). See also [19] for an expression of the rectangular complexity of the two-dimensional sequence associated with Pascal's triangle modulo 2.
- Note that it seems more difficult to generalize to the multidimensional case the notion of substitution of non-constant length. For a discussion of such generalized substitutions, see Chaps. 8 and 12.

3.2.4 Diagonals

Another interesting consequence of this generalization to the multidimensional case is given by the following results. The *diagonal* of a double formal power series $\sum u_{m,n}X^nY^m$ is defined as the series $\sum u_{n,n}X^n$.

Theorem 3.2.10. *The diagonal of an algebraic formal power series with coefficients in a finite field is algebraic.*

Proof. Consider either the notion of q-kernel or the one-dimensional substitution defined by associating with each letter the "diagonal" of the square array of letters associated through the initial substitution. ∎

Theorem 3.2.10 was first proved by Furstenberg in [183] and can be compared to the following theorem, also due to Furstenberg.

Theorem 3.2.11. *A series with coefficients in a finite field is algebraic if and only if there exists a rational double formal power series such that the initial series is the diagonal of this double series.*

For more details, the reader is referred to [13].

Exercise 3.2.12. 1. Let $(u_n)_{n\in\mathbb{N}}$ be the Morse sequence. Prove that the series $\sum u_nX^n$ is the diagonal of the rational function in $\mathbb{F}_2(X,Y)$ defined by $Y(1+Y(1+XY)+X(1+XY)^{-2})^{-1}$ (for details, see [11]).

2. Let $(v_n)_{n\in\mathbb{N}}$ be a sequence with values in the finite field $\mathbb{F}_q$. Prove that if $\sum v_nX^n$ is algebraic, then, for any $a\in\mathbb{F}_q$, $\sum_{v_n=a} X^n$ is algebraic.
3. Let $(v_n)_{n\in\mathbb{N}}$ be the characteristic sequence of the set of powers of the prime p. Prove that the series $\sum v_nX^n$ is the diagonal of the rational function $G(X) = X/(1-(X^{p-1}+Y))$ in $\mathbb{F}_p(X,Y)$.

4. Let $(w_n)_{n\in\mathbb{N}}$ be the characteristic sequence of the set of integers with base 2-expansion of the form 1^i0^j1, for $i,j>0$ and $i+j$ odd. Let $(x_n)_{n\in\mathbb{N}}$ be the characteristic sequence of the set of integers with base 2-expansion of the form $1^i0^{i+1}1$, for $i>0$. Let $(y_n)_{n\in\mathbb{N}}$ be the characteristic sequence of the set of squares. Prove that the Hadamard product of the series $\sum_{n\geq 0} w_nX^n$ and $\sum_{n\geq 0} y_nX^n$ is equal to $\sum_{n\geq 0} x_nX^n$. Deduce from Exercise 3.2.2 that the sequence $(y_n)_{n\in\mathbb{N}}$ is not 2-automatic (for more details, see [358] and the survey [427]).

3.2.5 Fields of positive characteristic

Christol, Kamae, Mendès France, and Rauzy's theorem can also be extended to a general field of positive characteristic, which is not necessarily finite. Such a generalization is due to Sharif and Woodcock (see [13]) and Harase (see [196]).

Theorem 3.2.13. *Let u be a sequence with values in a field $\mathbb{K}$ of positive characteristic p. Let s be any positive integer, $q=p^s$, and let $\overline{\mathbb{K}}$ be a perfect field (i.e., a field in which the map $x\mapsto x^p$ is onto) containing $\mathbb{K}$. The series $\sum u_nX^n$ is algebraic over $\mathbb{K}(X)$ if and only if the vector space spanned over $\overline{\mathbb{K}}$ by the following set of subsequences*

$$N_q(u)=\{(u^{1/q^k}(q^kn+r))_{n\in\mathbb{N}}, k\geq 0,\ 0\leq r\leq q^k-1\}$$

has a finite dimension.

Remarks.

- The set $N_q(u)$, which plays here the role of a generalized q-kernel, is exactly the q-kernel of the sequence u when the field $\mathbb{K}$ is finite, and this theorem reduces in this case to the theorem of Christol, Kamae, Mendès France, and Rauzy.
- We have seen that the notions of substitution and finite automaton can also be extended to the multidimensional case when the ground field is finite. In the general case of a field with nonzero characteristic, only the notion of substitution can be extended (for more details, see the survey [13]).
- The results for the Hadamard product and for the diagonal still hold in this context as consequences of Theorem 3.2.13. In fact we can deduce the following corollary first proved by Deligne in [142].

Corollary 3.2.14. *The Hadamard product of two algebraic formal power series with coefficients in a field of positive characteristic is algebraic. The diagonal of an algebraic formal power series with coefficients in a field of positive characteristic is algebraic.*

- Fresnel, Koskas, and de Mathan give a quantitative version of Christol, Kamae, Mendès France, and Rauzy's theorem in the case of an infinite ground field (see [177]), by giving an effective upper bound for the rank of the q-kernel of an algebraic element with respect to the height and to the degree of this element.

3.3 Transcendence in the real case and continued fractions

3.3.1 Some transcendence results in the real case

It is natural to look for connections between transcendence in the real case and transcendence for formal power series with coefficients in a finite field. Indeed, a formal power series is algebraic in positive characteristic if the sequence of its coefficients has some kind of order, whereas irrational algebraic real numbers cannot have an expansion which is too regular. Loxton and van der Poorten [272] have partially proved the following conjecture (this conjecture is often quoted as a theorem, but there seems to be a gap in the proof):

Conjecture 3.3.1. If the sequence of the coefficients in the base q-expansion of a real number is automatic, then this number is either rational or transcendental.

This conjecture illustrates, like Cobham's theorem, the fact that transcendence deeply depends on the frame in which it is considered. Some partial results in the direction of this conjecture have been obtained in [33, 172] (see also Chap. 4). The proof of the first assertion in the next theorem is given in [33], and the proof of the second one is given in [172].

Theorem 3.3.2. *Let α be a positive real number whose base k digit expansion is a fixed point of a substitution over a two-letter alphabet. If the morphism is either of constant length greater than or equal to 2, or primitive, then α is either rational or transcendental [33].*

If there exists k such that the expansion in base k of a positive real number α is a Sturmian sequence, then α is a transcendental number.

These proofs involve a clever use of Ridout's theorem [356] which is an improvement of Roth's theorem [363]. The underlying idea is to exhibit infinitely many $(2+\varepsilon)$-powers of blocks (that is, VVV' for a nonempty word V and a prefix V' of V with $|V'| \geq \varepsilon|V|$ $(\varepsilon > 0)$) that occur at ranks which are not too much larger than $|V|$.

Theorem 3.3.3. *Let $\alpha \in [0,1]$ such that its base k-expansion ($k \geq 2$) satisfies the following: there exist blocks $U_n, V_n, V_n' \in \{0, 1, \ldots, k-1\}^*$ such that α has base k-expansion*

$$\alpha = 0, U_n V_n V_n V_n' \dots,$$

V_n' is a prefix of V_n, $\liminf \frac{|V_n'|}{|V_n|} > 0$, $\limsup \frac{|U_n|}{|V_n|} < \infty$. *Then the number α is either rational or transcendental.*

In the particular case of a primitive substitution, it is sufficient to get the following:

Proposition 3.3.4 (([172], Proposition 5)). *If the expansion of α in some base k is a non-ultimately periodic fixed point of a primitive substitution, and does contain at least one word of the form $V^{2+\varepsilon}$, then α is transcendental.*

3.3.2 Continued fraction expansions and automaticity

In a similar vein, transcendence results concerning those real numbers whose sequence of partial quotients in their continued fraction expansion is automatic are obtained in [128, 342, 343, 21]. The theorem corresponding to Ridout's theorem in this framework is due to Schmidt [376]: if an irrational positive number is too well approximated by quadratic numbers, then it is either quadratic or transcendental. For a nice survey of these results, see [18].

Let us end this section by surveying related results of automaticity concerning the "digits" in the continued fraction expansion of formal power series with coefficients in a finite field.

In the real case very few explicit examples of continued fraction expansions are known; one can mostly expand power series or roots of some particular equations. Furthermore, it is a still open problem to determine if the set of partial quotients in the expansion of an algebraic number of degree greater than 2 can be bounded. Only a few examples of such expansions are known (for more references see the survey [390]). The situation is drastically different for formal power series with coefficients in a finite field. Indeed Baum and Sweet [57] have produced a cubic series the continued fraction expansion of which has partial quotients with bounded degree (see also Chap. 2 and [58]). Mills and Robbins have extended Baum and Sweet's approach to produce in [297] explicit expansions of algebraic elements in characteristic $p > 2$ for which the degrees of the partial quotients are bounded.

The algebraic series with bounded partial quotients produced respectively by Baum and Sweet [57], and by Mills and Robbins [297], have raised many interrogations, in particular concerning the automaticity of their coefficients. In the case where the partial quotients in the continued fraction expansion of an algebraic Laurent formal power series take finitely many values, Mendès France asked whether this sequence is itself p-automatic. A positive answer to this question has been given in [12, 20] in the case of Mills and Robbins examples [297] in characteristic > 2. But Mkaouar [298] (see also [463]) showed that the Baum and Sweet series [57] provides a negative answer to the question asked by Mendès France. Note also Thakur's results on the continued fraction expansion of the Carlitz analogue of the exponential [422, 423, 424].

3.4 Some functions defined by Carlitz

In the preceding pages we used the following notion of transcendence: a formal power series F is algebraic over $\mathbb{F}_q(X)$ if there exists a nontrivial polynomial P with coefficients in $\mathbb{F}_q(X)$ such that $P(F) = 0$. By analogy with the real case, we have seen that the set which plays here the role of $\mathbb{Z}$ is the ring $\mathbb{F}_q[X]$ of polynomials with coefficients in $\mathbb{F}_q$ and the analogue of $\mathbb{Q}$ is the fraction field of $\mathbb{F}_q[X]$, i.e., the field $\mathbb{F}_q(X)$ of fractions with coefficients in $\mathbb{F}_q$. We will use in all that follows the notation $\mathbf{Z} := \mathbb{F}_q[X]$.

Let us see in what respect the field $\mathbb{F}_q((1/X))$ is an analogue of $\mathbb{R}$. Let us recall that the field $\mathbb{R}$ is the *completion* of $\mathbb{Q}$ with respect to the usual absolute value. Hence our next step will be to define an *absolute value* over $\mathbb{F}_q(X)$. For more details on valuations, see [34, 467].

3.4.1 Absolute value and valuation

Definition 3.4.1. *An absolute value over a field* $\mathbb{K}$ *is a map (that we denote by* $|\ |$*) from* $\mathbb{K}$ *to* $\mathbb{R}^+$ *which satisfies the following properties:*

- $|x| = 0$ *if and only if* $x = 0$*;*
- $|xy| = |x||y|$*;*
- $|x + y| \leq |x| + |y|$.

The classical absolute value over $\mathbb{R}$ or the function which associates with a complex number its modulus are examples of absolute values.

It is easily seen that the function which associates with a fraction R in $\mathbb{F}_q(X)$ the number $a^{\deg R}$ (denoted by $|R|$), where $a > 1$ and where $\deg R$ denotes the degree of R (with the convention that the degree of the zero fraction is equal to $-\infty$), is an absolute value over $\mathbb{F}_q(X)$. But this absolute value does not have the same properties as the classical real absolute value. In particular, this absolute value satisfies for any two fractions R and S:

$$|R + S| \leq \max(|R|, |S|),$$

$$|R + S| = \max(|R|, |S|), \text{ if } |R| \neq |S|.$$

The degree obviously satisfies the same property.

Definition 3.4.2. *An absolute value over a field* $\mathbb{K}$ *is called* non-Archimedean *if for any two elements* x, y *of* $\mathbb{K}$*, we have:*

$$|x + y| \leq \max(|x|, |y|).$$

Lemma 3.4.3. *A non-Archimedean absolute value satisfies for any two elements* x, y *of* $\mathbb{K}$ *with* $|x| \neq |y|$*:*

$$|x + y| = \max(|x|, |y|).$$

Proof. Let us suppose $|x| < |y|$. We have $|x+y| \leq \max(|x|, |y|) = |y|$. But $|y| \leq \max(|-x|, |x+y|)$. Since $|-1|^2 = |1|^2 = |1|$, we have $|-1| = |1|$ and $|-x| = |x|$. As $|x| < |y|$, we thus have $\max(|-x|, |x+y|) = |x+y|$, from which the lemma follows. ∎

Lemma 3.4.4. *An absolute value is non-Archimedean if and only if the set of values*

$$\{|n.1_{\mathbb{K}}|,\ n \in \mathbb{Z}\},$$

where $1_{\mathbb{K}}$ denotes the unit of the field $\mathbb{K}$, is bounded.

Proof. Let M be such that $|n.1_{\mathbb{K}}| \leq M$, for any $n \in \mathbb{Z}$. Let x, y be in $\mathbb{K}$ with $|x| \leq |y|$. We have:

$$|x+y|^n = |(x+y)^n| = |x^n + \binom{n}{1} x^{n-1}y + \cdots + \binom{n}{n-1} xy^{n-1} + y^n|.$$

Thus

$$|x+y|^n \leq (n+1)M|y|^n,$$

i.e.,

$$|x+y| \leq |y|((n+1)M)^{1/n}.$$

Letting n tend towards $+\infty$, we obtain the desired property:

$$|x+y| \leq |y| = \max(|x|, |y|).$$

The converse is immediate. ∎

Remark. An absolute value defined over a field of positive characteristic is non-Archimedean. Namely, the set of values $\{|n.1_{\mathbb{K}}|,\ n \in \mathbb{Z}\}$ is finite in this case.

Definition 3.4.5. *A valuation over a field $\mathbb{K}$ is a map (that we denote by v) from $\mathbb{K}$ to $\mathbb{R} \cup \{+\infty\}$ which satisfies the following properties:*

- $v(x) = +\infty$ *if and only if* $x = 0$*;*
- $v(xy) = v(x) + v(y)$*;*
- $v(x+y) \geq \min(v(x), v(y))$.

For instance, the function $v : \mathbb{F}_q(X) \to \mathbb{F}_q(X)$, $R \mapsto -\deg R$, is a valuation over $\mathbb{F}_q(X)$. This valuation is called the $1/X$-adic valuation, by analogy with the p-adic valuation v_p which is defined as follows over $\mathbb{Z}$: $v_p(x)$ is the exponent of the greatest power of p which divides x. The completion of $\mathbb{Z}$ with respect to this valuation is called the ring of *p-adic integers*.

Proposition 3.4.6. *Let v be a valuation defined over the field $\mathbb{K}$. Let $0 < a < 1$. The function a^v is a non-Archimedean absolute value over $\mathbb{K}$. Conversely, if $|\ |$ is an non-Archimedean absolute value defined over the field $\mathbb{K}$ and if $b > 0$, then the function v defined over $\mathbb{K}$ by $v(0) = +\infty$ and $v(x) = -\log_b(|x|)$, if $x \neq 0$, is a valuation.*

Recall that the field $\mathbb{F}_q((1/X))$ of Laurent formal power series with coefficients in $\mathbb{F}_q$ is the set of series of the form

$$a_{-d}X^d + \cdots + a_0 + a_1 X^{-1} + \cdots ,$$

where a_i belongs to $\mathbb{F}_q(X)$. The notion of degree can be easily extended to this field: we define the degree of $a_{-d}X^d + \cdots + a_0 + a_1X^{-1} + \cdots$, with $a_{-d} \neq 0$, as d. The function v, which associates the quantity $v(z) = -\deg z$ with a formal power series z, is still a valuation over $\mathbb{F}_q((1/X))$.

In fact, the field $\mathbb{F}_q((1/X))$ is the completion of $\mathbb{F}_q(X)$ with respect to this valuation. For more details, the reader is referred to [34, 467].

The question now is to define an analogue of $\mathbb{C}$. The field $\mathbb{C}$ is algebraically closed and complete. Let $\mathbf{R}$ be the field $\mathbb{F}_q((1/X)$; this field is complete but not algebraically closed; let $\mathbf{C}$ be the completion of an algebraic closure of $\mathbf{R}$; a fundamental property here is that $\mathbf{C}$ is still algebraically closed (see [34]). This field will be an analogue of $\mathbb{C}$. The valuation v extends to $\mathbf{C}$, and we extend the definition of the degree by $\deg = -v$.

3.4.2 Convergence

One particular point of the non-Archimedean nature of the absolute value in $\mathbf{C}$ is the following:

Proposition 3.4.7. *A series $\sum_{n>0} u_n$ with coefficients in $\mathbb{F}_q((1/X))$ (respectively $\mathbf{C}$) converges in $\mathbb{F}_q((1/X))$ (respectively $\mathbf{C}$) if and only if its general term u_n tends to 0, i.e., if and only if the degree of u_n tends to $-\infty$.*

Proof. The field $\mathbb{F}_q((1/X))$ (respectively $\mathbf{C}$) is complete. We thus have to check that if u_n tends to 0, then the series $\sum u_n$ is a Cauchy sequence, which is an easy consequence of the non-Archimedean nature of the valuation. ∎

Example 3.4.8. For instance, the series

$$\sum_{k=1}^{+\infty} \frac{1}{X^{q^k} - X}$$

converges in $\mathbb{F}_q((1/X))$. Indeed, the degree of its general term is equal to $-q^k$. We will see in the following that this series, called the *bracket series*, is transcendental over $\mathbb{F}_q(X)$.

Proposition 3.4.9. *A product $\prod_{n\geq 0}(1+u_n)$ is convergent if and only if u_n tends to 0, i.e., if and only if the degree of u_n tends to $-\infty$.*

Proof. It is easily seen that the condition is necessary. Suppose now that u_n tends to 0. Let $P_N = \prod_{n=0}^{N}(1+u_n)$. We have $P_{N+1} - P_N = P_N u_{N+1}$. There exists an integer n_0 such that for $n \geq n_0$, we have $|u_n| \leq 1$, $|1+u_n| \leq 1$ (since the absolute value is non-Archimedean) and thus $|P_n| \leq |P_{n_0}|$. We deduce from this that $|P_{N+1} - P_N| \leq |u_{N+1}||P_{n_0}|$, for $N \geq n_0$, which implies that the sequence $(P_{N+1} - P_N)_{N\in\mathbb{N}}$ tends towards 0 and thus the convergence of P_N, from Proposition 3.4.7. ∎

Application. The series

$$\prod_{j=1}^{+\infty}(1 - \frac{X^{q^j} - X}{X^{q^{j+1}} - X})$$

is convergent. This product is called Π and was introduced by Carlitz in [104]. We will see in the sequel in what respect this product is the analogue of the real number π.

3.4.3 The Carlitz functions

In 1935 Carlitz defined two functions ψ and λ analogous respectively to the exponential and to the logarithm (see [104]). These functions are defined as follows:

$$\psi(t) = \sum_{k=0}^{+\infty} \frac{(-1)^k t^{q^k}}{F_k} \quad \text{for all } t \text{ in } \mathbf{C},$$

$$\lambda(t) = \sum_{k=0}^{+\infty} \frac{t^{q^k}}{L_k} \quad \text{for every } t \text{ such that } \deg t < \frac{q}{q-1},$$

$$\text{with } [k] = X^{q^k} - X,$$
$$F_k = [k][k-1]^q ... [1]^{q^{k-1}} \text{ and } F_0 = 1,$$
$$L_k = [k][k-1]...[1] \text{ and } L_0 = 1.$$

It is easily seen that $\psi(t)$ is convergent for all t. Indeed $\deg(\frac{1}{F_j} t^{q^j}) = q^j(\deg t - j)$, which tends to $-\infty$, when j tends to $+\infty$. But $\lambda(t)$ is convergent if and only if $\deg t < \frac{q}{q-1}$. Indeed, $\deg(\frac{1}{L_j} t^{q^j}) = q^j(\deg t - \frac{q}{q-1}) + \frac{q}{q-1}$.

Theorem 3.4.10 ([104]). *The function ψ satisfies the following properties.*

- *It is $\mathbb{F}_q$-linear, i.e., $\psi(t+u) = \psi(t) + \psi(u)$ and $\psi(ct) = c\psi(t)$, for any t, u in $\mathbf{C}$ and c in $\mathbb{F}_q$.*
- *For any t in $\mathbf{C}$, $\psi(t) = 0$ if and only if $t = E\xi$, with E in $\mathbb{F}_q[X]$.*
- *We deduce by linearity that it is periodic, i.e.,*

$$\forall t \in \mathbf{C},\ \forall E \in \mathbf{Z}(= \mathbb{F}_q[X]),\ \psi(t + E\xi) = \psi(t),$$

$$\textit{with } \xi = (X^q - X)^{1/(q-1)}\Pi \ \textit{ and } \ \Pi = \prod_{j=0}^{+\infty}(1 - \frac{X^{q^j} - X}{X^{q^{j+1}} - X}).$$

- *For any t in $\mathbf{C}$, $\psi(t) = t\prod_{E \in \mathbb{F}_q[X],\ E \neq 0}(1 - \frac{t}{E\xi})$.*
- *The function ψ satisfies the following functional equation for any t in $\mathbf{C}$:*

$$\psi(Xt) = X\psi(t) - \psi(t)^q.$$

The formal power series Π is of course the analogue of the real number π and ξ (which is defined up to multiplication by an element of $\mathbb{F}_q^*$ as a $(q-1)$-th root) is the analogue of $2i\pi$.

Theorem 3.4.11 ([104]). *The function λ satisfies the following properties.*

- *The function λ is $\mathbb{F}_q$-linear.*
- *For any t in $\mathbf{C}$ such that $\deg t < \frac{q}{q-1}$, $\psi\lambda(t) = t$.*
- *The function λ satisfies the following functional equation for any t in $\mathbf{C}$ such that $\deg t < \frac{1}{q-1}$:*

$$\lambda(Xt) - X\lambda(t) = \lambda(t^q).$$

Carlitz showed in [104] that the definition of λ can be extended to $\mathbf{C}$, by defining modulo $\mathbf{Z}\xi$ a function $\overline{\lambda}$ with values in $\mathbf{C}$, which is an inverse of the function ψ.

Carlitz also introduces in [104] an analogue of the Riemann zeta function. The Carlitz zeta function is defined as follows:

$$\zeta(m) = \sum_{G \in \mathbb{F}_q[x] \text{ and } G \text{ unitary}} 1/G^m \ , \quad m \in \mathbb{N}, \ m \geq 1.$$

It is easily seen that $\zeta(m)$ is convergent for any $m \geq 1$.

3.4.4 Some results of transcendence

Most of the "classical" transcendence properties analogous to the real case as, for instance, the Hermite-Lindemann theorem and the Gelfond-Schneider theorem, were stated in the 40's by Wade (see [441] and [442]).

Theorem 3.4.12. *If α is a nonzero element of $\mathbb{F}_q((1/X))$ algebraic over $\mathbb{F}_q(X)$, then $\psi(\alpha)$ and $\lambda(\alpha)$ are transcendental.*

We deduce from this theorem the transcendence of Π.

Theorem 3.4.13. *Suppose that α is nonzero and β is an irrational element of $\mathbb{F}_q((1/X))$. Then one of the three numbers α, β, $\psi(\beta\lambda(\alpha))$ is transcendental.*

Carlitz showed in [104] that $\zeta(s)/\Pi^s \in \mathbb{F}_q(X)$, for any s multiple of $(q-1)$. This property is analogous to Euler's result on the even values of the Riemann ζ function, i.e., $\zeta(2n)/\pi^{2n}$ is rational for every nonzero integer n. Note that the group of units of $\mathbb{F}_q[X]$ is $\mathbb{F}_q^*$ and has cardinality $q-1$, whereas the multiplicative group of $\mathbb{Z}$ is $\{+1, -1\}$ and has cardinality 2. The congruences modulo 2 in the complex case are replaced here by congruences modulo $q-1$. For more details, see the survey [443].

There are essentially four methods to prove transcendence results for these functions: the "classical" method which imitates the real case (see for instance

[442, 127] and Sec. 3.4.5), the use of Drinfeld modules [147, 465, 425, 443, 189], the Diophantine approximation method [132] and the finite-automata method [14, 15, 17, 29, 71, 70, 254, 177, 287, 426, 427].

The first automaton result states the transcendence of the period $\xi = (X^q - X)^{\frac{1}{q-1}} \Pi$ of the exponential. More precisely, Allouche shows in [14] the transcendence of $\frac{\alpha}{\Pi}$, where α is an algebraic series. Indeed the formal power series expansion of $\frac{\alpha}{\Pi}$ is particularly interesting, in the sense that the terms of this expansion are computable in an explicit way. Furthermore most of its coefficients are zero. We expose the proof of [14] of this result in Sec. 3.5.2.

Allouche and Shallit introduce in [25] the notion of regular sequences which generalizes the concept of automatic sequences. Becker shows in [60] an analogue of the theorem of Christol, Kamae, Mendès France, and Rauzy: namely, regular power series satisfy Mahler-type functional equations; the reciprocal is true under extra hypotheses. For regular sequences with values in a finite field, these Mahler-type functional equations lead naturally to transcendence results. In particular, Becker shows in [60] the transcendence of the series $\sum_{k\geq 0} \alpha^{r^k}$, where α is a power series in $\mathbb{F}_q[[X_1^{-1}, \ldots, X_d^{-1}]]$ with nonzero constant term, which is algebraic over $\mathbb{F}_q(X_1, \ldots, X_d)$, and where r is not a power of the characteristic p. This result generalizes Theorem 3.2.7. See also [149, 344].

Results of transcendence concerning the Carlitz-Goss gamma function (see for instance [17] and [287]), the transcendence of the period of the Tate elliptic curve (see [426] and [29]) were also recently proved via automata (see also the surveys [15, 427]).

3.4.5 Proof of the transcendence of $\psi(1)$ by Wade's method

The aim of this section is to prove the transcendence of $\psi(1)$ by Wade's method [442]. This proof can be considered as classical in the sense that it is close to the real-case method.

Theorem 3.4.14. *The series $\psi(1)$ is transcendental over $\mathbb{F}_q(X)$.*

Proof. Suppose that $\psi(1)$ is algebraic over $\mathbb{F}_q(X)$. From Ore's lemma, there exist $a_l, \cdots, a_d$ in $\mathbb{F}_q[X]$, with $a_d \neq 0$ and $a_l \neq 0$, such that

$$\sum_{i=l}^{d} a_i \psi(1)^{q^i} = 0.$$

We thus have

$$\sum_{k=0}^{+\infty} \sum_{i=l}^{d} a_i (-1)^k \frac{1}{F_k^{q^i}} = 0.$$

Multiply this equality by F_j, where j will be chosen large enough. Let $I = \sum_{k+i\leq j} \frac{F_j a_i (-1)^k}{F_k^{q^i}}$ and $Q = \sum_{k+i\geq j+1} \frac{F_j a_i (-1)^k}{F_k^{q^i}}$. We have for j large enough

$$I + Q = 0,\ I \in \mathbb{F}_q[X] \text{ and } \deg(Q) < 0.$$

Let us prove that $I \neq 0$, in order to get the desired contradiction.

Consider modulo $[j-l]$ the quantity

$$\frac{F_j}{F_k^{q^i}} = \frac{[j][j-1]^q \dots [k]^{q^{j-k}} \dots [1]^{q^{j-1}}}{[k]^{q^i} \dots [1]^{q^{i+k-1}}},$$

for $k+i \leq j$. Recall that $a_i = 0$, if $i < l$. If $k+i < j$ or if $k+i = j$, with $i \neq l$, then $k < j-l$ and $[j-l]$ divides $\frac{F_j}{F_k^{q^i}}$. If $i = l$ and $k = j-l$, then

$$\frac{F_j}{F_k^{q^i}} = [j][j-1]^q \dots [j-l+1]^{q^{l-1}} \equiv [l][l-1]^q \dots [1]^{q^{l-1}} = F_l \bmod [j-l].$$

For j large enough, $\deg a_l F_l < \deg[j-l]$. Thus

$$I \equiv a_l \frac{F_j}{F_{j-l}^{q^l}} \not\equiv 0 \mod [j-l],$$

hence the theorem. ∎

3.5 Some examples of automaton proofs

3.5.1 Transcendence and q-kernel

Let us first illustrate how to use Christol, Kamae, Mendès France, and Rauzy's theorem by showing that the q-kernel is infinite. Indeed one can distinguish in some cases the subsequences of the q-kernel by considering their first terms.

Suppose we have the following situation. Let $(v_k)_{k\in\mathbb{N}} = (v_k(n))_{k\in\mathbb{N}}$ be subsequences in the q-kernel of the sequence v satisfying the following three conditions:

1. there exists an integer $m(k)$ such that, for any $n < m(k)$: $v_k(n) = 0$;
2. the sequence $(m(k))_{k\in\mathbb{N}}$ tends to $+\infty$;
3. the sequences v_k are not identically equal to 0 for infinitely many k.

Then the set $\{v_k;\ k \in \mathbf{N}\}$ (and hence the q-kernel of v) is infinite.

Let us consider an example of formal power series for which one can exhibit subsequences of the q-kernel fulfilling Conditions 1, 2 and 3.

3.5.2 Transcendence of Π

Let us give here an automaton proof of the following theorem:

Theorem 3.5.1. *The series Π is transcendental over $\mathbb{F}_q(X)$.*

This proof is strongly inspired by the proof of [14]. Consider the series

$$\alpha = \prod_{j=0}^{+\infty}(1 - \frac{X^{q^j}}{X^{q^{j+1}}}).$$

It is easily seen that α is algebraic over $\mathbb{F}_q(X)$ (consider α^q). Hence the transcendence of $\frac{\alpha}{\Pi}$ will imply the transcendence of Π.

We have

$$\Pi = \prod_{j=0}^{+\infty}(1 - \frac{X^{q^j} - X}{X^{q^{j+1}} - X}).$$

We thus obtain

$$\frac{\alpha}{\Pi} = \prod_{j=1}^{+\infty}(1 - (\frac{1}{X})^{q^j - 1}).$$

Note that if

$$n = \sum_{k=1}^{+\infty} \varepsilon_k(q^k - 1) \text{ with } \varepsilon_k = 0 \text{ or } 1, \varepsilon_k = 0 \text{ for } k \text{ large enough,}$$

then such a decomposition is unique.

Let $(v(n))_{n\in\mathbb{N}}$ be the sequence defined by $\sum_{n\geq 0} v(n)X^{-n} = \frac{\alpha}{\Pi}$. If

$$n = \sum_{k=1}^{+\infty} \varepsilon_k(q^k - 1) \text{ with } \varepsilon_k = 0 \text{ or } 1, \varepsilon_k = 0 \text{ for } k \text{ large enough,}$$

then

$$v(n) = (-1)^{\sum_{k=1}^{+\infty} \varepsilon_k}, \quad \text{otherwise } v(n) = 0.$$

Let $(v_k(n))_{n\in\mathbb{N}}$ be the sequence defined by:

$$\forall n \in \mathbb{N},\ v_k(n) = v(q^k n + q^k - k).$$

Let $k \geq 2$. Let $n \in \mathbb{N}$ such that $v_k(n) \neq 0$. Thus

$$\exists\ (\varepsilon_j)_{j\in\mathbb{N}} \text{ with } \varepsilon_j = 0 \text{ or } 1, \varepsilon_j = 0 \text{ for } j \text{ large enough, such that}$$

$$q^k n + q^k - k = \sum_{j=1}^{+\infty} \varepsilon_j(q^j - 1).$$

We set $\sigma = \sum_{j\geq k} \varepsilon_j$. We have

$$q^k n + q^k - k = \sum_{1\leq j\leq k-1} \varepsilon_j(q^j - 1) + \sum_{j\geq k} \varepsilon_j q^j - \sigma.$$

Hence

$$\sigma \equiv \sum_{1 \le j \le k-1} \varepsilon_j (q^j - 1) + k \mod q^k.$$

Since

$$2 \le \sum_{1 \le j \le k-1} \varepsilon_j (q^j - 1) + k < q^k,$$

necessarily

$$\sigma \ge \sum_{1 \le j \le k-1} \varepsilon_j (q^j - 1) + k \ge k.$$

On the other hand,

$$q^k n + q^k - k \ge \sum_{j \ge k} \varepsilon_j (q^j - 1) \ge \sum_{j=k}^{k+\sigma-1} (q^j - 1) \ge \sum_{j=k}^{2k-1} (q^j - 1) = \frac{q^k - 1}{q - 1} q^k - k.$$

Hence $n \ge \frac{q^k - q}{q-1}$.

Conversely, if $n = \frac{q^k - q}{q-1}$, $q^k n + q^k - k$ is indeed of the form $\sum_{j \ge 1} \varepsilon_j (q^j - 1)$, with $\varepsilon_j = 1$, for $k \le j \le 2k-1$, $\varepsilon_j = 0$ otherwise.

Setting $m(k) = \frac{q^k - q}{q-1}$, the sequences v_k satisfy Conditions 1, 2, and 3, which ends the proof. ∎

Remark. Hellegouarch has defined in [202] an exponential function associated with a periodic sequence of endomorphisms generalizing the Carlitz exponential. Inspired by the previous proof, Recher has shown in [354] the transcendence of the period of this generalized exponential for a certain choice of endomorphisms.

3.5.3 General case

Let us come back to Conditions 1, 2, and 3 of Sec. 3.5.1. In the examples we consider here (as illustrated by the previous example), the sequence v is equal (modulo p) to the number $u(n)$ of expansions of the integer n under a given form. But it is often a difficult combinatorial problem to give an expression of $u(n)$. It is easier to work with zero elements of the sequence $(u(n))$ (i.e., with integers n which cannot be expanded in the desired form). This explains why we consider zero elements in Conditions 1, 2, and 3. The method can be applied when the sequence u, and thus the sequence v, has long ranges of consecutive zero elements. But this method cannot be used in the case where the sequence $u = (u_n)_{n \in \mathbb{N}}$ takes nonzero values for n large enough, as illustrated by the following example.

Consider $\lambda(1)$. We have

$$\lambda(1) = \sum_{k=0}^{+\infty} \frac{1}{L_k}, \quad \text{with } [k] = X^{q^k} - X, \quad L_k = [k][k-1]...[1].$$

We thus get

$$\begin{aligned}\lambda(1) &= 1+\sum_{k\geq 1}\prod_{j=1}^{k}(\frac{1}{X^{q^j}-X})\\ &= 1+\sum_{k\geq 1}(1/X)^{(q+\ldots+q^k)}\prod_{j=1}^{k}\frac{1}{(1-(\frac{1}{X})^{q^j-1})}\\ &= 1+\sum_{k\geq 1}(1/X)^{(q+\ldots+q^k)}\prod_{j=1}^{k}\sum_{n_j\geq 0}(1/X)^{n_j(q^j-1)}\\ &= 1+\sum_{n\geq 0}v(n)X^{-n},\end{aligned}$$

where $v(n)$ equals the value modulo p of the cardinality $u(n)$ of the set

$$\{(k,n_1,...,n_k) \text{ such that } k\geq 1,\ n_j\geq 0 \text{ for } 1\leq j\leq k \text{ and}$$

$$n = q+...+q^k+\sum_{j=1}^{k}n_j(q^j-1)\}. \tag{3.3}$$

It seems difficult to give an evaluation of this cardinality. Furthermore, every integer large enough is easily seen to have a decomposition under the form (3.3); namely, there is no limitation on the size of the coefficients n_j. The sequence u eventually takes only non-zero values. Let us introduce now a limitation on the size of the coefficients n_j. Let us multiply $\lambda(1)-1$ by the product $\mathcal{P}=\prod_{j=1}^{+\infty}(1-(\frac{1}{X})^{q^j-1})$. We thus obtain

$$\begin{aligned}\mathcal{P}(\lambda(1)-1) &= \sum_{k\geq 1}(1/X)^{(q+\ldots+q^k)}\prod_{j=k+1}^{+\infty}(1-(\frac{1}{X})^{q^j-1})\\ &= \sum_{n\geq 0}w(n)X^{-n},\end{aligned}$$

with $(w(n))_{n\in\mathbb{N}}$ defined as follows: if $n = q+...+q^k+\sum_{j=k+1}^{+\infty}\varepsilon_j(q^j-1)$, with $k\geq 1$, $\varepsilon_j = 0$ or $1, \varepsilon_j = 0$ for k large enough, then $w(n) = (-1)^{\sum_{j=k+1}^{+\infty}\varepsilon_j}$, otherwise $w(n)=0$.

One can check that the family of subsequences of the q-kernel $(w(q^kn+1+q+\cdots+q^{k-1}))_{n\in\mathbb{N}}$ satisfies Conditions 1, 2, and 3 (see for instance [70, 71]). The quantity $\mathcal{P}$ is furthermore equal to the quotient $\frac{\alpha}{\Pi}$. We thus deduce the transcendence of $\frac{\lambda(1)-1}{\Pi}$.

Remarks.

- Using the same methods, one can also obtain the transcendence of $\frac{\alpha}{\Pi}\lambda(1)$. More generally, automaton methods seem to suit particularly sums and products of transcendental formal power series.

- The transcendence of every linear combination over $\mathbb{F}_q(X)$ of $\frac{\zeta(s)}{\Pi^s}$, for $1 \leq s \leq q-2$, can be shown in a similar way (see [71]), as well as the transcendence of the quotients $\frac{\lambda(P)}{\Pi^s}$, with $1 \leq s \leq q-2$ and $P = XQ$, with $Q \in \mathbb{F}_q[1/X]$ (see [70]). Here again, the idea is to exhibit an infinite family of subsequences of the q-kernel and also to work on these quantities multiplied by a suitable power of the algebraic series α. Let us note that the restriction $s \leq q-2$ is a combinatorial limitation due to the fact that the properties of uniqueness in the expansions used (when explicitly expanding in formal power series) do not hold for $s > q$. More generally the transcendence of $\frac{\zeta(s)}{\Pi^s}$ for $s \not\equiv 0$ modulo $q-2$ was proved by Yu in [465], by using Drinfeld modules.

3.5.4 Transcendence of the bracket series

The purpose of this section is to prove the following result, which gives us an example of an application of the Christol, Kamae, Mendès France, and Rauzy theorem. This result was first proved by Wade in [441]; the proof below is due to Allouche (see [14] and also [287]).

Theorem 3.5.2. *The series $\sum_{k=1}^{+\infty} \frac{1}{[k]}$ is transcendental over $\mathbb{F}_q(X)$.*

This proof makes use of the following consequence of the Christol, Kamae, Mendès France, and Rauzy theorem.

Proposition 3.5.3. *Let $(u(n))_{n\in\mathbb{N}}$ be a sequence with values in $\mathbb{F}_q$. If the series $\sum_{n\geq 0} u(n)X^{-n}$ is algebraic over $\mathbb{F}_q(X)$, then the sequence $(u(q^n-1))_{n\in\mathbb{N}}$ is ultimately periodic, that is, $\sum_{n\geq 0} u(q^n-1)X^{-n}$ is rational.*

Proof. Suppose that the series $\sum_{n\geq 0} u_n X^{-n}$ is algebraic; the sequence $u = (u(n))_{n\in\mathbb{N}}$ is thus q-automatic. Let A denote a finite q-automaton which generates the sequence u. The subsequence $(u(q^n-1))_{n\in\mathbb{N}}$ is obtained by reading in the automaton A strings of digits all equal to $q-1$. As the number of states of A is finite, a long enough string of identical digits meets the same state twice. The sequence of states met is thus eventually periodic, which implies that $(u(q^n-1))_{n\in\mathbb{N}}$ is also eventually periodic. ∎

Remark. Let $\overline{UT^nV}$ denote, for any natural integer n, the integer of base-q expansion UT^nV, where U, T, V are words defined over $\{0, 1, \ldots, q-1\}$. We can similarly prove that if the series $\sum_{n\geq 0} u_n X^{-n}$ is algebraic, then the sequence $(u(\overline{UT^nV}))_{n\in\mathbb{N}}$ is eventually periodic. This result corresponds to the classical *pumping lemma* in automata theory.

Proof of Theorem 3.5.2. We have:

$$\sum_{k\geq 1}\frac{1}{[k]}=\sum_{k\geq 1}\frac{1}{(X^{q^k}-X)}=\sum_{k\geq 1}\frac{1}{X^{q^k}(1-(\frac{1}{X})^{q^k-1})}$$

$$=\sum_{k\geq 1}\frac{1}{X^{q^k}}\sum_{j\geq 0}(\frac{1}{X})^{j(q^k-1)}=\frac{1}{X}\sum_{k\geq 1,\ j\geq 0}(\frac{1}{X})^{(j+1)(q^k-1)}$$

$$=\frac{1}{X}\sum_{k\geq 1,\ j\geq 1}\frac{1}{X^{j(q^k-1)}}=\frac{1}{X}\sum_{n\geq 1}a(n)X^{-n},$$

where $a(n)$ is the number (modulo the characteristic p) of decompositions of the integer n as $n=j(q^k-1)$, with $k\geq 1$ and $j\geq 1$, i.e.,

$$a(n)=\sum_{k\geq 1,\ (q^k-1)|n}1.$$

Clearly the series $\sum_{k\geq 1}\frac{1}{[k]}$ is transcendental over $\mathbb{F}_q(X)$ if and only if the series $X\sum_{k\geq 1}\frac{1}{[k]}$ is transcendental. Suppose that the series $\sum_{n\geq 1}a(n)X^{-n}$ is algebraic over $\mathbb{F}_q(X)$. This implies that the sequence $(a(n))_{n\in\mathbb{N}}$ is q-automatic and in particular that the subsequence $a((q^n-1))_{n\in\mathbb{N}}$ is ultimately periodic. This assertion leads to a contradiction.

Indeed, it is easily seen that q^k-1 divides q^n-1 if and only if k divides n. We thus have

$$a(q^n-1)=\sum_{k\geq 1,\ (q^k-1)|(q^n-1)}1=\sum_{k\geq 1,\ k|n}1.$$

The subsequence $a((q^n-1))_{n\in\mathbb{N}}$ is supposed to be ultimately periodic. Thus there exist $n_0\geq 1$ and $T\geq 1$ such that:

$$\forall n\geq n_0,\ \sum_{k\geq 1,\ k|n}1=\sum_{k\geq 1,\ k|n+T}1\mod p.$$

Then let $N\geq n_0$ such that T divides N; we have

$$\sum_{k\geq 1,\ k|N^p}1=\sum_{k\geq 1,\ k|(2N^p)}1\mod p.$$

By considering the decomposition of N^p into a product of prime factors, it is easily seen that $\sum_{k\geq 1,\ k|N^p}1=1\mod p$, whereas $\sum_{k\geq 1,\ k|(2N^p)}1=2\mod p$, which is the desired contradiction.

Remark. By using the same kind of arguments it can be proved that the series $\frac{1}{\Pi}\sum_{k\geq 1}\frac{1}{[k][k+i]}$, where $i\geq 1$, is transcendental over $\mathbb{F}_q(X)$.

Exercise 3.5.4. The aim of this exercise is first, to give another proof of this result, and second, to prove Theorem 3.5.5 below. This very satisfactory proof is due to Mendès France and Yao [287].

1. Prove that for any positive integers u, v, w, the number $q^w - 1$ divides $q^u(q^v - 2) + 1$ if and only if w divides the greatest common divisor of u and v.
2. Define the sequence $a_u = (a(q^u n + 1))_{n \in \mathbb{N}}$, for a fixed positive integer u. Let u, v be two distinct positive integers. Let h be the smallest integer such that h divides u and h does not divide v. Prove that $a_u(q^h - 2) - a_v(q^h - 2) \equiv 1$.
3. Deduce from this the transcendence of $\sum_{n \geq 1} a(n) X^{-n}$.
4. Prove in the same way the following theorem ([287]).

Theorem 3.5.5. *Let $(n_k)_{k \in \mathbb{N}}$ be a sequence of elements of $\mathbb{F}_q$ which is not eventually equal to zero. Then the formal power series*

$$\sum_{k \geq 1} \frac{n_k}{[k]}$$

is transcendental over $\mathbb{F}_q(X)$.

3.5.5 Derivation and transcendence of Π

We will see in this section how to infer from the transcendence of the bracket series the transcendence of the series Π, by using the derivation of formal power series following the method of [15]; see also [425] or [143].

Definition 3.5.6. *The* derivative *of the series $\sum_{n \geq 0} u_n X^{-n}$, where u_n belongs to $\mathbb{F}_q$, is the series $(\sum_{n \geq 0} u_n X^{-n})' = -\sum_{n \geq 0} n u_n X^{-n-1}$.*

Proposition 3.5.7. *The derivative of an algebraic series is algebraic.*

The proof is immediate and left as an exercise.

Remark. The converse is generally not true. For instance, if $\sum u_n X^{-n}$ is transcendental then $(\sum u_n X^{-n})^q = \sum u_n X^{-qn}$ is also transcendental but its derivative is equal to 0 (in $\mathbb{F}_q((1/X))$).

Let us prove that the transcendence of the bracket series implies the transcendence of Π. We have

$$\Pi = \prod_{j=1}^{+\infty} \left(1 - \frac{[j]}{[j+1]}\right),$$

i.e.,

$$\Pi = \prod_{j=1}^{+\infty}(1 - \frac{X^{q^j} - X}{X^{q^{j+1}} - X}) = \prod_{j=1}^{+\infty} \frac{X^{q^{j+1}} - X^{q^j}}{X^{q^{j+1}} - X}.$$

The derivative of $\frac{X^{q^{j+1}}-X^{q^j}}{X^{q^{j+1}}-X}$ is equal to $\frac{X^{q^{j+1}}-X^{q^j}}{(X^{q^{j+1}}-X)^2}$. The logarithmic derivative of Π, $\frac{\Pi'}{\Pi}$, is thus equal to

$$\frac{\Pi'}{\Pi} = \sum_{j\geq 1} \frac{1}{X^{q^{j+1}} - X},$$

i.e.,

$$\frac{\Pi'}{\Pi} = \sum_{j\geq 2} \frac{1}{[j]}.$$

The series Π is hence transcendental. Otherwise the series $\frac{\Pi'}{\Pi}$ would be algebraic and so would be the bracket series. ■

3.6 Conclusion

How can one recognize that a sequence is not k-automatic, or in other words, that a formal power series with coefficients in a finite field is transcendental? Let us review some more techniques to disprove the automaticity of a sequence.

- An automatic sequence has a low complexity function, due to its strong underlying structure: the complexity of a fixed point of a substitution of constant length satisfies [120]

$$\exists C,\ \forall n,\ p(n) \leq Cn.$$

  This disproves the automaticity of a high complexity sequence. More generally, the complexity of a fixed point of a substitution satisfies (see [156, 319])

$$\exists C,\ \forall n,\ p(n) \leq Cn^2,$$

  but if the substitution is primitive, then we get (see Proposition 5.4.6)

$$\exists C,\ \forall n,\ p(n) \leq Cn.$$

- If the frequencies of the factors of an automatic sequence exist, then they are rational numbers (for more precise results on frequencies of automatic sequences, see [120]). Let us recall that if the corresponding substitution is primitive, then the frequencies exist (see Chap. 1). In particular, a Sturmian sequence cannot be automatic.

- Shallit introduces in [391] a *measure of automaticity* of a sequence $u = (u(n))_{n \in \mathbb{N}}$ over a finite alphabet: the k-automaticity of a sequence is defined as the smallest possible number of states in any deterministic finite automaton which generates the prefix of size n of this sequence. This measure tells quantitatively how "close" a sequence is to being k-automatic.
- Yao gives in [463] non-automaticity criteria motivated by a transcendence criterion due to de Mathan (see [132]). Originally the work of de Mathan used Diophantine approximation but Koskas gave a proof using automata of this criterion in [254]. In particular, Yao gives a simple proof of a result due to Mkaouar: the sequence of partial quotients in the continued fraction expansion of the Baum-Sweet series (see Chap. 2) is not p-automatic, for any integer $k \geq 2$.
- Durand has generalized in [152, 151] Christol, Kamae, Mendès France, and Rauzy's theorem to uniformly recurrent sequences that are letter-to-letter projections of fixed points of primitive substitutions. This characterization is based on the notion of *return words* and *derived sequence.* A return word over the factor W is a word separating two successive occurrences of W; a derived sequence of a given minimal sequence u is a sequence obtained by coding u by return words over a non-empty prefix of the sequence u (see also Definition 7.3.21 in Chap. 7). The characterization is the following: a uniformly recurrent sequence is a letter-to-letter projection of a fixed point of a primitive substitution if and only if the set of its derived sequences is finite. This result was proved independently by Holton and Zamboni [208]. It has been extended to substitutive tilings by Priebe [336]. Note that the notion of return words has many applications in arithmetics with generalizations of Cobham's theorem [153], or in dynamics with a description of minimal stationary Bratteli-Vershik diagrams [155]. See also Chap. 12.

4. Substitutions and partitions of the set of positive integers

There is a natural duality between the symbolic sequences, as they are studied throughout this book, and the *partitions of the set of positive integers* $\mathbb{N}^+$. Any right infinite word $\omega = \omega_1\omega_2\omega_3\ldots$ strictly with values in the alphabet

$$\mathcal{A}_s := \{a_0, a_1, \ldots, a_s\},$$

(i.e., every symbol a_i eventually occurs in ω), gives rise to a partition of $\mathbb{N}^+$ into $s+1$ parts:

$$\mathop{\overset{\circ}{\bigcup}}_{0\leq i\leq s} \chi(\omega; a_i) = \mathbb{N}^+, \tag{4.1}$$

where $\chi(\omega; a)$ is the characteristic set of ω with respect to $a \in \mathcal{A}_s$

$$\chi(\omega; a) := \{n \in \mathbb{N}^+; \omega_n = a\}.$$

This partition will be referred to as the *partition corresponding* to ω (throughout this chapter $\overset{\circ}{\cup}$ indicates a disjoint union). Conversely, any partition of $\mathbb{N}^+$ into nonempty sets $\chi_1, \ldots, \chi_s$ gives rise to a word $w = w_1w_2w_3\ldots$ strictly over $\mathcal{A}_s$ defined by $\omega_n = i$ whenever $n \in \chi_i$.

We say that the partition (4.1) is *nonperiodic* (respectively *totally nonperiodic*) if ω (respectively $\partial\chi(\omega; a_i)$ for all i) is not an ultimately periodic word (or sequence), and vice versa, where we mean by ∂C the sequence $(c_{n+1} - c_n)_{n\geq 1}$ for a given sequence $C := (c_n)_{n\geq 1}$.

Throughout this chapter, we also identify a set $\{s_n; n \in \mathbb{N}^+\} \subset \mathbb{N}^+$ such that $s_1 < s_2 < s_3 < \ldots$ with a sequence $(s_n)_{n\geq 1}$.

In what follows, we shall give some examples of classes of nonperiodic partitions of $\mathbb{N}^+$ (or some classes of nonperiodic infinite words). We also state some results and problems related to transcendence and complexity.

[1] This chapter has been written by J. -I. Tamura

4.1 Beatty and Sturmian sequences, and associated partitions

4.1.1 Partitions associated with Beatty sequences

We denote by $[x]$ ($x \in \mathbb{R}$) the largest integer not exceeding x, by $\{x\}$ the fractional part of $x \in \mathbb{R}$, i.e., $\{x\} := x - [x]$, and by $\mathbb{R}$ (respectively $\mathbb{Z}, \mathbb{Q}, \mathbb{R}^+$) the set of real numbers (respectively integers, rational numbers, positive numbers).

Let us recall that a *Beatty sequence* is a sequence of the form $([\alpha n+\beta])_{n\geq 1}$. It is well-known that for positive numbers α and β, two Beatty sequences (or sets) $([\alpha n])_{n\geq 1}$ and $([\beta n])_{n\geq 1}$ make a partition of the set $\mathbb{N}^+$ into two parts if and only if α, β are irrationals numbers satisfying $1/\alpha + 1/\beta = 1$. This fact can be written in an equivalent form as follows:

$$\mathop{\mathring{\bigcup}}_{(\gamma_0,\gamma_1)\in A} \{[\gamma_0 n] + [\gamma_1 n]; n \in \mathbb{N}^+\} = \mathbb{N}^+ \tag{4.2}$$

holds if and only if $\alpha > 0$ is an irrational number, where

$$A := \{(1,\alpha), \alpha^{-1}(1,\alpha)\}.$$

We thus get a partition of $\mathbb{N}^+$ into two parts.

Proposition 4.1.1 below is a generalization of (4.2), which gives a partition of $\mathbb{N}$ into $s+1$ parts by specific sums of Beatty sequences (see [418], Theorem 1).

We denote by $\alpha S, S+\alpha, S+T$, and ST the sets $\{\alpha s; s \in S\}, \{s+\alpha; s \in S\}, \{s+t; s \in S, t \in T\}$, and $\{st; s \in S, t \in T\}$, respectively, for given sets $S, T \subset \mathbb{R}$, and for a given number $\alpha \in \mathbb{R}$.

Proposition 4.1.1. *Let s be a positive integer, and $\alpha_i > 0, \beta_i$ $(0 \leq i \leq s)$ be real numbers. Then the condition*

$$(\alpha_i^{-1}\mathbb{Z} - \alpha_i^{-1}\beta_i) \cap (\alpha_j^{-1}\mathbb{Z} - \alpha_j^{-1}\beta_j) \cap \mathbb{R}^+ = \emptyset \quad \textit{for all } i \neq j \tag{4.3}$$

is necessary and sufficient to have a partition

$$\mathop{\mathring{\bigcup}}_{(\underline{\gamma},\underline{\delta})\in B} \left\{ \sum_{(0\leq j\leq s)} [\gamma_j n + \delta_j]\,; n \in \mathbb{N}^+ \right\} = \mathbb{N}^+ \tag{4.4}$$

where $(\underline{\gamma}, \underline{\delta}) = \gamma_0, \gamma_1, \ldots, \gamma_s, \delta_0, \delta_1, \ldots, \delta_s)$, *and*

$$\begin{aligned} B := \{\, (\underline{\alpha}, \underline{\beta});\ & \underline{\alpha} = \alpha_i^{-1}(\alpha_0, \alpha_1, \ldots, \alpha_s), \\ & \underline{\beta} = -\alpha_i^{-1}\{\beta_i\}(\alpha_0, \alpha_1, \ldots, \alpha_s) \\ & \qquad + (\{\beta_0\}, \{\beta_1\}, \ldots, \{\beta_s\}), 0 \leq i \leq s \,\}. \end{aligned}$$

We denote by $\omega(B)$ the word corresponding to the partition. Setting $\alpha_0 = 1, \beta_i = 0$ for all i, we have

Corollary 4.1.2. *Let $s \in \mathbb{N}^+$, $\alpha_0 = 1$, $\alpha_i \in \mathbb{R}^+$ $(1 \le i \le s)$. Then the condition $\alpha_i \notin \mathbb{Q}$, and $\alpha_i/\alpha_j \notin \mathbb{Q}$ for all $1 \le i < j \le s$ is necessary and sufficient to have a partition*

$$\mathop{\dot{\bigcup}}_{(\gamma_0,\ldots,\gamma_s)\in C} \left\{ \sum_{0\le j\le s} [\gamma_j n]; n \in \mathbb{N}^+ \right\} = \mathbb{N}^+,$$

where $C := \{\alpha_i^{-1}(\alpha_0, \ldots, \alpha_s), 0 \le i \le s\}$.

Remark. If we take $s = 1$ in Corollary 4.1.2, we obtain (4.2). We remark that we can choose $\alpha_0 = 1$, $\beta_0 = 0$ in Proposition 4.1.1 without changing the form of the components of the partition. If $\alpha_0 = 1$, the partition (4.4) is nonperiodic if one of the α_i $(1 \le i \le s)$ is irrational, since the irrationality of f_{a_i}/f_{a_0} implies the nonperiodicity of $\omega(B)$, where f_{a_i} is the frequency, in the asymptotic sense (see Chap. 1), of a symbol a_i appearing in the word $\omega(B)$ corresponding to the partition (4.4).

Related to the condition (4.3), we can show the implications (4.8) $\Rightarrow$ (4.7) $\Rightarrow$ (4.6) $\Rightarrow$ (4.3) $\Rightarrow$ (4.5) in the case of $\alpha_0 = 1, \beta_0 = 0$, where (4.5)–(4.8) are the following conditions:

- $-\beta_i \notin \alpha_i\mathbb{N}^+ + \mathbb{Z}$ for all $1 \le i \le s$; (4.5)
- $\alpha_i\beta_j - \alpha_j\beta_i \notin \alpha_i\mathbb{Z} + \alpha_j\mathbb{Z}$ for all $1 \le i < j \le s$; (4.6)
- $1, \alpha_i, \beta_i$ are linearly independent over $\mathbb{Q}$ for each $1 \le i \le s$, and $(\alpha_i(\mathbb{Z} + \beta_i)\mathbb{Q}) \cap (\alpha_j(\mathbb{Z} + \beta_j)\mathbb{Q}) = \{0\}$ for all $1 \le i < j \le s$; (4.7)
- the $2s + 1$ numbers 1, and $\alpha_i, \alpha_i\beta_i (1 \le i \le s)$ are linearly independent over $\mathbb{Q}$. (4.8)

We remark that a result obtained by J. V. Uspensky [435] says the impossibility of having a partition into t parts by Beatty sequences for $t \ge 3$. Proposition 4.1.3 is a generalization of Proposition 4.1.1 (see [418], Theorem 3).

Proposition 4.1.3. *Let $f_i : \mathbb{R}^+ \cup \{0\} \to \mathbb{R}$ $(0 \le i \le s, 1 \le s \in \mathbb{N}^+)$ be a continuous, strictly monotone increasing function with $\lim_{x\to\infty} f_i(x) = \infty$ for all i. Then the condition*

$$f_i^{-1}(\mathbb{Z}) \cap f_j^{-1}(\mathbb{Z}) \cap \mathbb{R}_+ = \emptyset \text{ for all } i \ne j \tag{4.9}$$

is necessary and sufficient to have a partition

$$\mathop{\dot{\bigcup}}_{0\le i\le s} \left\{ \sum_{0\le j\le s} ([f_j(f_j^{-1}(n + [f_i(0)]))] - [f_j(0)]); n \in \mathbb{N}^+ \right\} = \mathbb{N}^+. \tag{4.10}$$

The condition $\lim_{x\to\infty} f_i(x) = \infty$ can be omitted from Proposition 4.1.3, at most, for s indices i. In that case, some of the components of the partition (4.10) turn out to be a finite set. We remark that in this sense, any partition of $\mathbb{N}^+$ into $s+1$ parts can be given by (4.10) under a suitable choice of the functions f_i (without loss of generality, we may assume that all the f_i are of $\mathcal{C}^\infty$ class with $f_0(x) = x$), that will be clear from the following argument.

Proof. The idea of the proof of the proposition is very simple. We denote by Π_i, $\Pi_i \subset \mathbb{R}^{s+1}$ (respectively $K \subset \mathbb{R}^{s+1}$) the set of hyperplanes (respectively the curve) defined by

$$\Pi_i := \{(x_0, \ldots, x_i, \ldots, x_s);\ x_j \in \mathbb{R}\ (j \neq i),\ x_i \in \mathbb{Z}\}\ (0 \leq i \leq s),$$

$$\Pi := \bigcup_{0 \leq i \leq s} \Pi_i\,;\ K := \{\underline{f}(x) = (f_0(x), \ldots, f_s(x)); x \in \mathbb{R}_+\}.$$

We consider an infinite word $\omega = \omega(K) = \omega_1\omega_2\omega_3\ldots$ given by

$$\omega_n = a_i \text{ if } \underline{f}(x_n) \in \Pi_i\,,$$

where the sequence $(x_n)_{n\geq 1}$ is defined by

$$\{\underline{f}(x_n)\,;\, 0 < x_1 < x_2 < \cdots < x_n < \ldots\} := K \cap \Pi\,.$$

Note that the sequence $(x_n)_{n\geq 1}$ is well-defined, since the set $K \cap \Pi$ is a discrete one in $\mathbb{R}^{s+1}$ if the functions f_i are continuous, and strictly monotone increasing; and the word ω is well-defined by (4.9).

Under the assumption that the functions f_i are continuous, strictly monotone increasing, we can calculate the n-th term of the sequence (or the set) $\chi(\omega; a_i)$ by using the intermediate value theorem, and we can obtain Proposition 4.1.3. Taking K to be a half-line $L := \{\underline{\alpha}t + \underline{\beta};\ t \in \mathbb{R}\}$, and considering the word $w = w(L)$, we get Proposition 4.1.1. For further details of the proof, see [418]. ∎

4.1.2 Billiards and Sturmian sequences

Proposition 4.1.1 has some connection with higher dimensional ***billiards*** (see also Chap. 6). Let $I^{s+1} (I := [0,1])$ be the unit cube of dimension $s+1$ with the faces

$$\{(x_0, \ldots, x_i, \ldots, x_s); x_j \in I\ (\forall j \neq i),\ x_i = 0, \text{ or } 1\},\ 0 \leq i \leq s$$

labelled by a_i. Let a particle start at a point $\underline{\beta} \in [0,1)^{s+1}$ along a vector $\underline{\alpha} \in \mathbb{R}^{s+1}$ with the condition for α_i, β_j stated in Proposition 4.1.1, and be reflected at each face of I^{s+1} specularly. We note L the set called B in the proposition. Then the word $\omega(L)$ coincides with a word obtained by writing down the label a_i of the faces which the particle hits in order of collision.

Let us recall that the complexity $p = p_w$ of an infinite word w is the function $p_w : \mathbb{N} \to \mathbb{N}$ defined as the number of factors of length n of w (see Chap. 1). An infinite word w (or a sequence) is called *Sturmian*(over $s+1$ letters) if there exists a positive integer s such that $p_w(n) = n + s$ for every n.

If w is a word defined strictly over the alphabet $\mathcal{A}_s$ which is not ultimately periodic, then $p_w(n) \geq n + s$, see [172, 303] and Proposition 1.1.1. For $s = 1$, $\omega(L)$ is Sturmian over 2 letters provided that ω is not periodic.

G. Rauzy has conjectured that $p_{\omega(L)}(n) = n^2 + n + 1$ for $s = 2$, when $\alpha_1, \alpha_2, \alpha_3$ are linearly independent over $\mathbb{Q}$, which was proved affirmatively in [46], see also [350, 351, 47]. An exact formula for $p_{\omega(L)}(n) = p_{\omega(L)}(n, s)$ as a function of n and s in the case where $\alpha_0, \dots, \alpha_s$ are linearly independent over $\mathbb{Q}$ was conjectured in [46]:

$$p_{\omega(L)}(n, s) = \sum_{0 \leq i \leq min\{n,s\}} \frac{n!\, s!}{(n-i)!\, i!\, (s-i)!}$$

and proved affirmatively by Y. Baryshnikov [56]. Consequently, we have for every n, s:

$$p_{\omega(L)}(n, s) = p_{\omega(L)}(s, n) \text{ and } p_{\omega(L)}(n; 3) = n^3 + 2n + 1.$$

This astonishing symmetry property was a major conjecture made by J.-I. Tamura; P. Arnoux and C. Mauduit derived from it the exact formula by adding some minor hypotheses. It is an interesting question (posed by C. Mauduit), to ask for a direct (or combinatorial) proof of the symmetry; it still remains mysterious why $p(n, s)$ is a symmetric function.

Quite recently, S. Ferenczi and C. Mauduit [172] have obtained the following remarkable result: the numbers having a Sturmian sequence (with values in the set $\{0, \dots, h\}$) as base $g(\geq h+1)$ expansion are transcendental. They gave further results on transcendence of numbers having an infinite word with low complexity as base g expansion.

4.2 Partitions given by substitutions

4.2.1 Existence theorem

By $\mathcal{A} = \mathcal{A}_s$ we mean the alphabet $\{a_0, a_1, \dots, a_s,\}(s \geq 1)$ as in Sec. 4.1.

We denote by $|W|$ the length of a finite word w, and by $|W|_a$ the number of occurrences of a symbol $a \in \mathcal{A}$ appearing in a word $W \in \mathcal{A}^*$. For a given sequence $C = (c_n)_{n \geq 1}$, $\int_i C$ indicates the sequence

$$\int_i C := \left(i + \sum_{1 \leq m \leq n-1} c_m \right)_{n \geq 1}.$$

Then we can show the following

Proposition 4.2.1. *Let σ be the substitution over the alphabet $\mathcal{A}$ defined by*

$$\sigma(a_j) := {a_0}^{k_s - j} a_{j+1} \ (1 \le j \le s-1), \ \sigma(a_s) := a_0,$$

where k_i $(1 \le i \le s)$ are integers satisfying $k_s \ge k_{s-1} \ge \cdots \ge k_0 = 1$.

Let L_j be the set $\{|\sigma^j(a_0)|, |\sigma^j(a_0)| + |\sigma^j(a_1)|, \ldots, |\sigma^j(a_0)| + |\sigma^j(a_s)|\}(1 \le j \le s)$, and let $\tau_j : \mathcal{A}^* \to {L_j}^*$ be the monoid morphism defined by

$$\tau_j(a_i) := (|\sigma^j(a_0)|)^{k_s - i - 1}(|\sigma^j(a_0)| + |\sigma^j(a_{i+1})|) \ (0 \le i \le s-1),$$

$$\tau_j(a_s) := |\sigma^j(a_0)|, \ 0 \le j \le s.$$

Then

$$\mathop{\mathring{\bigcup}}_{0 \le j \le s} \int_{|\sigma^j(a_0)|} \tau_j(\omega) = \mathbb{N}^+ \tag{4.11}$$

where ω is the fixed point of σ.

It is clear that (4.11) follows from $\chi(\omega; a_j) = \int_{|\sigma^j(a_0)|} \tau_j(\omega)$, which is Theorem 4 of [417].

Note that the partition (4.11) is a totally nonperiodic one for all $s \ge 1$, and all $k_i \in \mathbb{Z}$ satisfying $k_s \ge k_{s-1} \ge \cdots \ge k_0 = 1$; this follows from [417], Lemma 11:

$$\lim_{n \to \infty} |\sigma^n(a_0)|_{a_i} / |\sigma^n(a_0)| = \alpha^{s-i}/(\alpha^s + \alpha^{s-1} + \cdots + \alpha + 1), \tag{4.12}$$

where $\alpha > 1$ is an algebraic number with minimal polynomial $f(x) := x^{s+1} - \sum_{0 \le i < s} k_i x^i$; the minimality follows from [417], Lemma 10.

We remark that in general, the partition (4.11) cannot be a partition of the form (4.4). For instance, suppose that (4.11) with $s = 2$, $k_1 = k_2 = 1$ coincides with (4.4) corresponding to some infinite word $\omega = \omega(L)$, where $L = \{t(1, \alpha_1, \alpha_2) + \underline{\beta}; t \in \mathbb{R}_+\}$. Then (4.12) implies that $\alpha_i = \alpha^{-i} (i = 1, 2)$. The minimality of $f(x)$ implies that $1, \alpha_1, \alpha_2$ are linearly independent over $\mathbb{Q}$. Hence, $p_\omega(n) = n^2 + n + 1$, which contradicts that $p_\omega(n) = 2n + 1$ is the complexity of the fixed point of the substitution σ with $s = 2$, $k_1 = k_2 = 1$ (the fixed point is an *Arnoux-Rauzy sequence*, see [49], [172]).

On the other hand, in the case $s = 1$, the partition (4.11) turns out to be the partition (4.2); that will be seen by the following argument : Proposition 4.2.1 with $s = 1$ implies $f(x) = x^2 - kx - 1$ $(k := k_1)$, so that $\alpha = (k + (k^2 + 4))^{1/2}/2$. Setting $\chi(\omega; a_i) = \{{t_1}^{(i)} < {t_2}^{(i)} < \cdots < {t_n}^{(i)} < \ldots\}$, for $i = 0, 1$, we get by Proposition 4.2.1

$${t_n}^{(1)} = kn + {t_n}^{(0)}, \qquad \chi(\omega; a_0) \mathring{\bigcup} \chi(\omega; a_1) = \mathbb{N}^+. \tag{4.13}$$

Noting that the sets $\chi(\omega; a_i)$ are uniquely determined by (4.13), and

$$[\eta_1 n] = kn + [\eta_0 n], 1/\eta_0 + 1/\eta_1 = 1 \quad (\eta_0 := 1 + 1/\alpha, \quad \eta_1 := 1 + \alpha),$$

we obtain $\chi(\omega; a_i) = \{[\eta_i n]; \ n \in \mathbb{N}^+\}$, for $i = 0, 1$.

4.2.2 Some transcendence results

Definition 4.2.2. *Let* $G = G_g := \{0, 1, \ldots, g-1\}$, *where* $2 \le g \in \mathbb{N}^+$. *Let* $\tau : \mathcal{A} \to G^\star$ *be a monoid morphism such that* $\tau(a) \ne \varepsilon$ *for all* $a \in \mathcal{A}$. *We denote by* $(0.\tau(w))_g$ *the number defined by* $\sum_{i \ge 1} w_i/g^i$, *where* $\tau(w) = w_1 w_2 w_3 \cdots \in G^\infty$, $w_i \in G$, $w \in \mathcal{A}^{\mathbb{N}}$.
We say w *is* transcendental *if* $(0.\tau(w))_g$ *is transcendental for an integer* g *and a morphism* τ.

The fixed point ω of the substitution σ given in Proposition 4.2.1 is not only totally nonperiodic, but also transcendental:

Proposition 4.2.3. *([417], Theorem 3) Let* ω *be as in Proposition 4.2.1,* $g \ge 2$ *an integer,* τ *a monoid morphism such that* $\tau(a) \ne \varepsilon$ *for all* $a \in \mathcal{A}$, *satisfying*

$$\text{rank } (|\tau(a_i)|_j)_{0 \le i \le s,\, 0 \le j \le g-1} > 1.$$

Then the number $(0.\tau(w))_g$ *is transcendental.*

The key for the proof of Proposition 4.2.3 is to show that the infinite word ω has a prefix which is a $(2+\varepsilon)$-power of w for infinitely many prefixes w (see Proposition 3.3.4, and [417], Lemma 13); that can be connected with Roth's theorem (see also Sec. 3.3). A stronger argument works in [172], where S. Ferenczi and C. Mauduit make use of a theorem of Ridout ([279], pp. 147-148) instead of Roth's theorem. We shall mention their results in the following section.

4.3 Similis partitions

4.3.1 A linguistic problem

Let D be a subset of $\mathbb{N}^+$, with $1 \in D$. In some cases, we can show that there exists a subset Γ such that

$$\mathop{\dot{\bigcup}}_{d \in D} d\Gamma = \mathbb{N}^+ \qquad (D \ne \emptyset, \{1\}). \tag{4.14}$$

Such a partition will be referred to as a *similis partition* (of $\mathbb{N}^+$ with respect to D). Some results on similis partitions are given in [416]. A higher-dimensional version of similis partitions are considered in [420, 419]. Let us mention that a simple example of similis partitions comes from a linguistic phenomenon in Hungarian and Japanese language that is probably well-known to linguists, see [421, 431]: Numerals one, two, three, four, ... in Hungarian (respectively Japanese) are egy, kettő, három, négy, ... (hi, fu, mi, yo, ...). We thus can make the following diagram, where in each language, underlined consonants of two numerals in each row are common, or they have a resemblance (e.g., n and ny= palatalized n in the 3rd stage of the diagram); and simultaneously, in each row, the number corresponding to the right group is exactly twice the left:

	Γ		2Γ	
1):	1	egy (hi←fi←pi)	2	kettő (fu←pu)
2):	3	három (mi)	6	hat (mu)
3):	4	négy (yo)	8	nyolc (ya)
4):	5	öt (itsu←itu)	10	tíz (to)

Here, among the numerals in Japanese language that are written in parentheses, for instance, itsu←itu indicates that the contemporary Japanese word itsu comes from the old Japanese word itu, that is, a kind of palatalization. If we look at the numerals of older Japanese in parentheses, the consonants correspondence turns out to be an exact one. (Related to vowels, see, e.g., [199]; vowel harmony is also common in Hungarian and old Japanese.)

Considering what will happen, apart from numerals in natural language, when we formally extend the diagram downwards, we get a similis partition (4.14) with $D = \{1, 2\}$, which is uniquely determined. In fact, it is clear that $\gamma_1 := 1 \in \Gamma$, so that $2\gamma_1 \in 2\Gamma$, which gives the first stage 1) of the diagram. Now, consider the smallest positive integer γ_2 among the numbers that have not appeared in the stage 1). Then the minimality of γ_2 implies $\gamma_2 \in \Gamma$, otherwise $\gamma_2 \in 2\Gamma$, so that $\gamma_2 > \gamma_2/2 \in 2\Gamma$, i.e., the second stage is of the form $\gamma_2/2 \in \Gamma$, $\gamma_2 \in \Gamma$, which contradicts the minimality of γ_2. (Forget that $\gamma_2 \in \Gamma$ follows from $\gamma_2 = 3$ is odd; we shall see that $\gamma_3 (= 4)$ is even in the following argument.) Suppose that we have obtained a diagram with stages 1)-n). Consider the number γ_{n+1} defined to be the smallest positive integer that differs from all the numbers appearing in the stages 1)-n). Then $\gamma_{n+1} \in \Gamma$ follows from its minimality. We can continue the process in general, and we must have $\Gamma = \{\gamma_1, \gamma_2, \gamma_3, \ldots\}$ as far as all the numbers $d\gamma_{n+1}$ ($d \in D$) are different from the numbers $d\gamma_m$ ($d \in D, 1 \leq m \leq n$). Hence, noting that the argument given above is valid for any nonempty, finite or infinite, subset D of $\mathbb{N}$, we obtain

Proposition 4.3.1. *If there exists a similis partition (4.14) for a given nonempty subset D of $\mathbb{N}^+$, then the partition is uniquely determined by the set D.*

4.3.2 The Hungarian-Japanese partition

On the other hand, it is clear that a similis partition (4.14) for $D = \{1, 2\}$ exists, since $\Gamma = \{2^{2j}m;\ j \geq 0, m \geq 1, m \text{ is odd}\}$ satisfies (4.14). This partition will be referred to as the *H.-J. (Hungarian-Japanese) partition.* The H.-J. partition can be easily generalized as

Proposition 4.3.2. *Let $D = D(k; q_1, \dots, q_k; e_1, \dots, e_k)$ be a set defined by*

$$D := \{\prod_{1 \le i \le k} q_i^{j_i} \; ; \; 0 \le j_i \le e_i \quad (1 \le i \le k)\},$$
$$k \ge 1, q_i \ge 2, e_i \ge 1 \, (1 \le i \le k), G.C.D.(q_i, q_j) = 1 \text{ for all } i \ne j. \tag{4.15}$$

Then

$$\Gamma := \{ \prod_{1 \le i \le k} q_i^{(e_i+1)j_i} m; \; j_i \ge 0, m \ge 1, \; m \not\equiv 0 \text{ modulo } q_j^{e_j+1}, \; 1 \le j \le k\}$$

satisfies (4.14).

We conjectured that if a similis partition (4.14) is a partition of $\mathbb{N}^+$ into finite components, then there exist numbers k, and $q_1, \dots, q_k, e_1, \dots, e_k$ satisfying (4.15); that is probably still open. It is easily seen that there are no partitions (4.14) for some explicitly given D which are not of the form (4.15), see [416], Theorem 12. For example, if we take $D = \{1, 2, 3\}$, and trace the uniqueness proof of the uniqueness of (4.14) above, we see $2\gamma_4 = 12 = 3\gamma_2$, which contradicts the fact that (4.14) is a disjoint union. Proposition 4.3.2 can be extended to partitions into infinite parts with respect to D given by (4.15) with $0 \le j_i$ for some indices i instead of $0 \le j_i \le e_i$ $(1 \le i \le k)$:

$$\begin{aligned} D := \{ &\prod_{1 \le i \le k-h} q_i^{j_i} \cdot \prod_{k-h+1 \le i \le k} q_i^{j_i} \; ; \\ &0 \le j_i \le e_i \, (1 \le i \le k-h), \; 0 \le j_i \; (k-h+1 \le i \le k) \; \} \\ &k \ge 1, k \ge h \ge 1, \\ &q_i \ge 2, e_i \ge 1 \, (1 \le i \le k-h), \; G.C.D.(q_i, q_j) = 1 \text{ for all } i \ne j. \end{aligned}$$

For D given as above, we can show

$$\begin{aligned} \Gamma = \{ &\prod_{1 \le i \le k-h} q_i^{(e_i+1)j_i} m \; ; \\ &j_i \ge 0, \; m \ge 1, \; m \not\equiv 0 \mod q_i^{e_i+1}, \; 1 \le i \le k-h, \\ &G.C.D.(m, q_{k-h+1} \cdots q_k) = 1 \; \} \; (k \ge 2). \end{aligned}$$

If $k = h = 1$, then $\Gamma = \mathbb{N}^+ \backslash q_1 \mathbb{N}^+$, and the partition (4.14) is periodic (this case is not interesting). We remark that for some partitions (4.14) into infinite parts, D is not always of the form above. For instance, if we take $D = \{p_i^{j_i}; \; j_i \ge 0, (0 \le i \le s)\}$ with prime numbers p_i $(p_0 > p_1 > \cdots > p_s, \; s \le 1)$, then

$$\Gamma = \{(p_0 \dots p_s)^i m\}; \; i \ge 0, \; m \ge 1, \; G.C.D.(m, p_0 \dots p_s) = 1 \; \}$$

satisfies (4.14). By the way, we remark that a sequence $\omega = \omega_1 \omega_2 \dots \omega_n \dots$ over $\mathcal{A}_s$ defined by

$$\omega_n := a_i \text{ if } m_n \in \{p_i^j;\ j \geq 0\}$$
$$(\{1 < m_1 < m_2 < \cdots < m_n < \ldots\} := D = \{p_i^{j_i};\ j_i \geq 0,\ 1 \leq i \leq s\})$$

coincides with a word $\omega(L)$ defined by the billiards in I^{s+1} with $\underline{\alpha} = (\alpha_0 \ldots \alpha_s), \underline{\beta} = \underline{0}, \alpha_i = \log p_i / \log p_0$, see [351].

4.3.3 Some properties of the H.-J. partition

We return to to the first example of (4.14), the H.-J. partition. Let us show the following:

Proposition 4.3.3. *The word $\omega = \omega_1\omega_2\omega_3\ldots (\omega_n \in \mathcal{A}_1)$ corresponding to the H.-J. partition is a totally nonperiodic word, which is a fixed point of a substitution.*

Proof. We mean by UV the set $\{uv; u \in U, v \in V\}$, by U^* the set $\{u_1 \ldots u_n; u_i \in U\ (1 \leq i \leq n), n \geq 0\}$ for subsets U, V of a monoid, and by Γ (with $1 \in \Gamma$) the component of the H.-J. partition. Let E_g denote the base-g expansion of $\gamma \in \mathbb{N}$ ($E_g(0) := \varepsilon$), and W^n ($W \in G_g^*$) be the word obtained by concatenating n copies of W.

We have $\gamma \in \Gamma$ if and only if $E_2(\gamma) = u0^{2n}$, where $n \geq 0$, and $u \in \{0,1\}^*$ is a word having 1 as a prefix and a suffix. Hence $\gamma \in \Gamma$ if and only if $E_2(\gamma - 1) = v1^{2n}$ $(n \geq 0)$, where $v \in G_2^* = \{0,1\}^*$ is a word such that 1 is not a suffix of v. Consequently the set $\{0^*E_2(\gamma - 1);\ \gamma \in \Gamma\}$ coincides with the language accepted by the automaton M defined by

$$M := (\mathcal{A}_1, G_1, \delta, a_0, \{a_0\})$$

with the transition function δ

$$\delta(a_0, i) := a_i,\ \delta(a_1, i) := a_0,\ (i = 0, 1).$$

For the definitions and notation related to automata, see [211]. Therefore, noting that $\omega = \delta(a_0, E_2(0)) \ldots \delta(a_0, E_2(n-1)) \ldots$, we see that ω is the fixed point of the substitution over $\mathcal{A}_1$ defined by

$$\sigma(a_0) := a_0a_1,\ \sigma(a_1) := a_0a_0. \tag{4.16}$$

Using this fact shown above, we can prove that ω is a totally nonperiodic word in the following manner. We remark that so far as similis partition are concerned, nonperiodicity implies total nonperiodicity. Hence it suffices to show the nonperiodicity of ω.

Suppose that ω is an ultimately periodic word, then $\theta := (0.\tau(\omega))_2 \in \mathbb{Q}$ $(\tau(a_i) := i)$. We put $a = a_0$, $b = a_1$, $\theta_n = (0.\tau(u_n)^*)_2$ $(u_n = \sigma^n(a))$, where u^* denotes the periodic word $uuu\ldots$ for a nonempty word u.

We write $u \neg v$ if v is a prefix of u. The binary relation $\neg$ is transitive. In view of (4.16), we get $u_2 = abaa = u_1 u_0^2$, so that $u_{n+2} = u_{n+1} u_n^2$ for all $n \geq 0$, $|u_n| = 2^n$, and $\omega \neg u_{n+1} \neg u_n u_{n-1}$. Hence, we obtain $|\theta - \theta_n| \leq 2^{-3 \cdot 2^{n-1}}$.

For any $n \geq 1$, we can put $\theta_n = E_2^{-1}(u)/(2^n - 1)$ with a certain $u \in 1G_2^*$. Let θ_n equal P_n/Q_n, $G.C.D.(P_n, Q_n) = 1$. Then $|\theta - P_n/Q_n| \leq Q_n^{-3/2}$, which together with $\theta \in \mathbb{Q}$ implies that $\{P_n/Q_n; n \geq 0\}$ is a finite set. Therefore $\theta_i = \theta_{i+j}$ for some $i \geq 0$ and $j \geq 1$, so that $u_{i+j} = u_i^{2^j}$. Since $u_{i+j} = u_{i+j-1} u_{i+j-2}^2$, we get $u_{i+j-1} = u_i^{2^{j-1}}$, and inductively, $u_{i+1} = u_i^2$. By $u_{i+1} = u_i u_{i-1}^2$, we get $u_i = u_{i-1}^2$. Repeating the argument, we obtain $u_1 = u_0^2$ which contradicts $u_1 = ab \neq aa = u_0^2$. ∎

Remark. By direct computation, we see that $\partial \Gamma = 2112221121121122\ldots$ for the H.-J. partition. We can show that the sequence (or word) $\partial \Gamma$ is the fixed point of a substitution over $\{1, 2\}$ in the following manner. Let ω be the fixed point of the substitution σ (4.16). Noting that bb does not occur in ω, we can factorize ω into two words $A := ab$ and $B := a$, and we get a new word $\tilde{\omega}$ over $\{A, B\}$:

$$\begin{array}{lccccccccccccc} \omega = & ab & a & a & ab & ab & ab & a & a & ab & a & a & ab & \ldots, \\ \tilde{\omega} = & A & B & B & A & A & A & B & B & A & B & B & A & \ldots, \end{array}$$

and noting that ω is the fixed point of σ, we see that $\tilde{\omega}$ is the fixed point of a substitution τ over $\{A, B\}$:

$$\tau: \begin{array}{ll} A = ab & \mapsto \sigma(ab) = abaa = ABB \\ B = a & \mapsto \sigma(a) = ab = A. \end{array}$$

Since $\partial \Gamma = \zeta(\tilde{\omega})$ with $\zeta(A) = 2$, $\zeta(B) = 1$, $\partial \Gamma$ becomes the fixed point of a substitution $2 \mapsto 211$, $1 \mapsto 2$.

4.3.4 General case

We can generalize all the statements given above for the H.-J. partition to those for the partition with $D = \{q^i; 0 \leq i \leq e\}$ $(q \geq 2)$ as in Propositions 4.3.4-4.3.5, by considering the automaton

$$M_{e,q} := (\mathcal{A}_e, G_q, \delta, a_0, \{a_0\})$$

with a transition function $\delta = \delta_{e,q}$ defined by

$$\begin{array}{l} \delta(a_i, j) := a_0, \;\; \delta(a_i, q-1) := a_{i+1} \;\; (0 \leq i \leq e-1, \;\; 0 \leq j \leq q-2), \\ \delta(a_e, j) := a_0 \;\; (0 \leq j \leq q-1). \end{array}$$

Proposition 4.3.4. *Let ω be the word corresponding to a similis partition (4.14) with respect to $D = \{q^i; 0 \leq i \leq e\}$ $(e \geq 1, q \geq 2)$. Then ω is a totally nonperiodic word over $\mathcal{A}_e$, which is the fixed point of a substitution over $\mathcal{A}_e$ defined by*

$$\sigma(a_i) := a_0^{q-1} a_{i+1} \;\; (0 \leq i \leq e-1), \;\; \sigma(a_e) := a_0^q. \tag{4.17}$$

Proposition 4.3.5. *Let (4.14) be a similis partition with respect to D as in Proposition 4.3.4. Let* $\tau : \mathcal{A}_e^* \to \mathcal{A}_e^*, \kappa : \mathcal{A}_e^* \to \{1,2\}^*$ *be the morphisms defined by*

$$\text{for } q=2, \begin{cases} \tau(a_i) := a_0 a_{i+1} \ (0 \le i \le e-2), \\ \tau(a_{e-1}) := a_0 a_e^2, \\ \tau(a_e) := a_0, \end{cases}, \begin{cases} \kappa(a_i) = 2 \ (0 \le i < e), \\ \kappa(a_e) = 1, \end{cases},$$

$$\text{for } q>2, \begin{cases} \tau(a_0) := a_0^{q-2} a_1, \\ \tau(a_i) := a_0^{q-2} a_1 a_0^{q-2} a_{i+1} \ (0 \le i < e), \\ \tau(a_e) := a_0^{q-2} a_1 a_0^q, \end{cases}, \begin{cases} \kappa(a_i) = 1 \ (0 \le i < e), \\ \kappa(a_e) = 2. \end{cases}$$

Then $\partial\Gamma$ *is a nonperiodic word over* $\{1,2\}$*, which is given by*

$$\partial\Gamma = \kappa(\omega'),$$

where ω' *is the fixed point of* τ.

Remark. Let ω be as in Proposition 4.3.4. Then, in view of the locally catenative formula $\sigma^{n+e}(a_0) = (\sigma^{n+e-1}(a_0))^{q-1} \dots (\sigma^{n+1}(a_0))^{q-1} (\sigma^n(a_0))^q$, we can easily find that the frequency of a_i appearing in ω is rational for all i. This fact together with the nonperiodicity of ω implies that a similis partition with respect to D given by (4.15) can neither be a partition (4.4) nor a partition (4.11).

Using Proposition 4.3.4, we can show the total nonperiodicity of the partition given by Proposition 4.3.2. For instance, let us consider a similis partition (4.14) with respect to $D = \{1,2,3,6\}$. Then $\Gamma = \{2^{2i}3^{2j}k;\ i \ge 0,\ j \ge 0,\ k \ge 1,\ G.C.D.(2 \cdot 3, k) = 1\}$ which equals $\{2^{2i}k;\ i \ge 0,\ (2,k) = 1\} \cap \{3^{2i}k;\ i \ge 0,\ (3,k) = 1\}$. It is clear that

$$\Gamma = \chi(\omega(1,2); a_0) \cap \chi(\omega(1,3); a_0),$$

where $\omega(e,q)$ is the fixed point of a substitution over $\mathcal{A}_e$ defined by (4.17). In general, we can show, considering the languages accepted by the automata $M_{e_i,q_i,j_i} := (S_{e_i}, G_{q_i}, \delta_{e_i,q_i}, a_0, \{a_{j_i}\})$ $(0 \le j_i \le e_i \ (1 \le i \le k))$, that

$$q_1^{j_1} \dots q_k^{j_k} \Gamma = \bigcap_{1 \le i \le k} \chi(\omega(e_i, q_i); a_{j_i}) \ \ (0 \le j_i \le e_i \ (1 \le i \le k)) \qquad (4.18)$$

holds for any finite similis partition with respect to D given by (4.15). We denote by Ω the word (strictly over $\prod_{1 \le i \le k} (e_i + 1)$ letters) corresponding to a partition given by Proposition 4.3.2. Then, it follows from (4.18) that Ω is an interpretation of $\omega(e_i, q_i)$ (i.e., whenever the i-th symbol counted from the beginning differs from the j-th symbol in ω, then the same happens in Ω, see [367]). Hence, Ω is not an ultimately periodic word by Proposition 4.3.4. Therefore, any similis partition given according to Proposition 4.3.2 is totally nonperiodic.

Related to the transcendence of the word $\partial\Gamma$ and the word corresponding to a finite similis partition, Proposition 3.3.4 due to S. Ferenczi and C. Mauduit [172] is useful. Indeed, if we apply Proposition 3.3.4 to the word ω in Proposition 4.3.4, and note that the transcendence of ω implies the transcendence of Ω, we obtain Proposition 4.3.6 (respectively Proposition 4.3.7) by Proposition 4.3.4 (respectively Proposition 4.3.5) as follows:

Proposition 4.3.6. *Let (4.14) be a similis partition into finite parts with respect to D given by (4.15) with the word Ω corresponding to (4.14). Then Ω is transcendental.*

Proposition 4.3.7. *Let Γ be as in Proposition 4.3.5. Then $\partial\Gamma$ is transcendental.*

4.4 Log-fixed points and Kolakoski words

4.4.1 Definition

A word (or a sequence) ξ' over $\{1,2\}$ is referred to as a *Kolakoski word* if the word defined by its run-lengths is equal to ξ' itself:

$$\begin{aligned}\xi' &= 22\ 11\ 2\ 1\ 22\ 1\ 22\ 11\ 2\ 11\ 22\ 1\ 2\ 11\ 2\ 1\ 22\ \ldots\\ &\quad\ 2\ \ 2\ \ 1\,1\,2\ \ 1\,2\ \ 2\ 1\ 2\ \ 2\ \ 1\,1\,2\ \ 1\,1\,2\cdots = \xi',\end{aligned}$$

where we mean by a *run* a maximal subword consisting of identical letters (for more details, see [249], and [137] with the references therein). The word $\xi := 1\xi'$ is the only other word having this property. It can be easily seen that ξ is not an ultimately periodic word, see [434]. Related to the complexity $p(n) := p_\xi(n)$, F. M. Dekking has shown that

$$n+1 \le p(n) \le n^{7.2}\ (\forall n \ge 0).$$

Furthermore, he has conjectured (see [135, 137]) that

$$p(n) \asymp n^{\log 2/\log(3/2)}.$$

Let $\mathcal{A}$ be an alphabet with Card $\mathcal{A} \ge 2$. We denote by $\mathcal{A}^\times$ the set

$$\mathcal{A}^\times := (\mathcal{A}^* \cup \mathcal{A}^\infty)\backslash(\bigcup_{a\in\mathcal{A}} \mathcal{A}^*\{a^*\}),$$

i.e., $\mathcal{A}^\times$ is the set of all finite or infinite words that are different from the words of the form ua^* ($u \in \mathcal{A}^*, a \in \mathcal{A}$). We shall write $\mathcal{A}^* w$ instead of $\mathcal{A}^*\{w\}$. For any word $\omega = \omega_1\omega_2\ldots\omega_n\cdots \in \mathcal{A}^\times$, we can define two words $\log\omega$, and base ω by

$$\log \omega := e_1 e_2 e_3 \ldots, \qquad \text{base } \omega := b_1 b_2 b_3 \ldots,$$
$$\text{if } \omega = b_1^{e_1} b_2^{e_2} b_3^{e_3} \ldots \ (e_i \geq 1,\ b_i \in \mathcal{A},\ b_i \neq b_{i+1}, \text{ for all } i \geq 1).$$

In what follows, we take $\mathcal{A} \subset \mathbb{N}^+$.

Definition 4.4.1. *A word $\omega \in \mathcal{A}^\times$ satisfying $\omega = \log \omega$ will be referred to as a* log-fixed point. *The* Kolakoski word *ξ is defined to be a log-fixed point with base $\xi = (12)^*$.*

If Card $\mathcal{A} = 2$, then Card $\{\omega \in \mathcal{A}^\infty; \omega = \log \omega\} = 2$; if Card $\mathcal{A} \geq 3$, then the set $\{\omega \in \mathcal{A}^\infty; \omega = \log \omega\}$ has continuum cardinality, since so does the set $\{\text{base } \omega\}$. It can be easily seen by a similar manner to that given by [434] that all the log-fixed points are not ultimately periodic.

Consider, for instance, a log-fixed point ω with base $\omega = (26)^*$, and factorize it into the words of length 2:

$$\begin{array}{llllllllllll} \omega & = 22 & 66 & 22 & 22 & 22 & 66 & 66 & 66 & 22 & 66 \ldots \\ & = \mathrm{A} & \mathrm{B} & \mathrm{A} & \mathrm{A} & \mathrm{A} & \mathrm{B} & \mathrm{B} & \mathrm{B} & \mathrm{A} & \mathrm{B} \ldots \\ & = W_1 & W_2 & W_3 & W_4 & W_5 & W_6 & W_7 & W_8 & W_9 & W_{10} \ldots. \end{array}$$

Then it is clear that W_i is $A := 22$ or $B := 66$ (since the length of the period of base ω is 2, which divides $2, 6 \in \mathcal{A}$), and ω as a word over $\{A, B\}$ is invariant under the morphism

$$A = 22 \mapsto 2266 = AB$$
$$B = 66 \mapsto 222222666666 = AAABBB.$$

Note that such an argument does not work at all for the Kolakoski words, but it can be applied to some general cases:

Proposition 4.4.2. *Let $s \geq 1$ be an integer, $\mathcal{A} = \{a_0, \ldots, a_s\} \subset \mathbb{N}^+$ such that s divides a_i for all $1 \leq i \leq s$. Let σ be a substitution over $\{A_0, \ldots, A_s\}$ defined by $\sigma(A_i) = A_0^{a_i/s} \ldots A_s^{a_i/s}$ $(0 \leq i \leq s)$, and let Ω be its fixed point. Then the log-fixed point ω with base $\omega = (a_0 \ldots a_s)^*$ can be given by $\omega = \tau(\Omega)$, where τ is the morphism defined by $\tau(A_i) = a_i^{s+1}$ $(0 \leq i \leq s)$.*

4.4.2 Generalized substitutions

Return now to the Kolakoski word ξ'. Consider what could be the substitution σ in Proposition 4.4.2 for the infinite word ξ' in a formal sense. It would become a "substitution" defined by

$$\sigma(A_0) = A_0 A_1,\ \sigma(A_1) = A_0^{1/2} A_1^{1/2},$$

where we mean by $A^{1/2}$ a half of a symbol. Define $(W_1 W_2 \ldots W_n)^{1/2}$ (each W_i is a symbol, or a half-symbol) to be a "word" $W_1 \ldots W_{[n/2]}$ (respectively, $W_1 \ldots W_{[n/2]} W_{[n/2]+1}^{1/2}$, which is possibly a "word" containing a fourth of a symbol) for even n (respectively, odd n), and define $\sigma(W^{1/2})$ to be a "word" $\sigma(W)^{1/2}$. Consider an infinite word $\Omega = \lim \sigma^n(A_0)$, then

$$A_0 \xrightarrow{\sigma} A_0A_1 \xrightarrow{\sigma} A_0A_1A_0^{1/2}A_1^{1/2} \xrightarrow{\sigma} A_0A_1A_0^{1/2}A_1^{1/2}A_0A_0^{1/2} \xrightarrow{\sigma} \dots$$

$$\longrightarrow \Omega = A_0A_1A_0^{1/2}A_1^{1/2}A_0A_0^{1/2}A_0A_1A_0^{1/2}A_0A_1A_0^{1/2}A_1^{1/2}A_0 \dots .$$

We can define the sequence Ω to be the fixed point of a substitution over an alphabet $\{a,\ b,\ c,\ d\}$ in the usual sense, where we identify $a = A_0$, $b = A_1$, $c = A_0^{1/2}$, $d = A_1^{1/2}$. The following question is thus natural: can we find any relation between ξ' and Ω ? (Probably, not!; and then, we must find a better treatment for half-symbols.) It will be remarkable that the word ξ is a fixed point of the map

$$\Psi : (\mathbb{N}^+)^\infty \cup ((\mathbb{N}^+)^* \backslash \mathbb{N}^* 1) \to \{1,\ 2,\ 3\}^\infty,$$
$$\Psi(\omega) := B_3(B_{2\to 3}(\varphi(c(\omega))) + 1/2),$$

where c, φ, $B_{2\to 3}$, B_3 are maps defined as follows:

1. $c : (\mathbb{N}^+)^\infty \cup (\mathbb{N}^{+*} \backslash (\mathbb{N}^+)^* 1) \to I = [0,1]$, $c(a_1a_2a_3\dots) := [0; a_1, a_2, a_3, \dots]$ for $a_1a_2a_3 \dots \in (\mathbb{N}^+)^\infty \cup ((\mathbb{N}^+)^* \backslash (\mathbb{N}^+)^* 1)$, where the right-hand side denotes the usual continued fraction expansion;
2. $\varphi : I \to I$ is the so called *question-mark-function* introduced by Minkowski determined by the following conditions:
   a) the map φ is continuous with $\varphi(0) = 0$, $\varphi(1) = 1$,
   b) $\varphi((p+p')/(q+q')) = (\varphi(p/q) + \varphi(p'/q'))/2$ for all $p, q, p', q' \in \mathbb{N}^+ \cup \{0\}$ such that p/q, $p'/q' \in I$, $p'q - pq' = \pm 1$;
3. $B_{2\to 3} : I \to [0, 1/2], B_{2\to 3}((0.b_1b_2b_3\dots)_2) := (0.b_1b_2b_3\dots)_3$ for $b_1b_2b_3 \dots \in \{0,1\}^\infty \backslash \{0,1\}^* 0^*$ $(B_{2\to 3}(0) := 0)$;
4. $B_3 : I \to \{0,1,2\}^\infty$, $B_3(x) = c_1c_2c_3\dots$ for $x = (0.c_1c_2c_3\dots)_3$ with $c_1c_2c_3 \dots \in \{0,1,2\}^\infty \backslash \{0,1,2\}^* 0^*$ $(B_3(0) := 0^*)$.

We can see that ξ is uniquely determined by $\Psi(\xi) = \xi$ and by the fact that

$$\varphi([0; a_1, a_2, a_3, \dots]) = (0.\ 0^{a_1}\ 1^{a_2}\ 0^{a_3}\ 1^{a_4}\ 0^{a_5}\ \dots)_2, \tag{4.19}$$

see [321].

The problem of the existence of frequencies in the Kolakoski word is known as Keane's problem: the question is whether the frequency of 1 in ξ exists, and whether it equals 1/2 [238]. This is still open. If the frequency does not exist, or if it equals 1/2 (it probably does!), then it is easy to see that the words ω corresponding to the one of the partitions defined at (4.4), (4.12) and (4.15) cannot be the word ξ.

Remark. Note that this notion of generalized substitution has nothing to do with the generalized substitutions discussed in Chaps. 8 and 12. This later generalization is a higher-dimensional one.

4.4.3 Some connections with continued fraction expansions

Instead of (4.19), we may ask about the existence of a number $x \in I$ satisfying

$$x = [0; a_1, a_2, a_3, \ldots] = (0.a_1 a_2 a_3 \ldots)_g, \tag{4.20}$$

where

$$a_n \in \mathbb{Z},\ 1 \leq a_n \leq g-1\ (i \geq 1). \tag{4.21}$$

Such a number x exists for a square number $g = h^2$, for $2 \leq h \in \mathbb{Z}$, since $[0; h] = 1/h = h/g = (0.h)_g$; this is not interesting. Now, we ask for an irrational number $x \in I$ satisfying (4.20) with (4.21). If we take $g = 10$, then by simple computations, we can show that such a number does not exist. If we consider (4.20) with

$$a_1 a_2 a_3 \cdots \in (\{0, \ldots, g-1\}^\infty \backslash \{0, \ldots, g-1\}^* 0^* \tag{4.22}$$

instead of (4.21), then it seems very likely that a number $x \in I$ satisfying (4.20) exists; a computation says that

$$[0; 3, 3, 5, 8, 3, 4, 7, \ldots] = (0.3358347\ldots)_{11},$$

where we mean, for example,

$$[0; 3, 3, 5, \ldots, 1, 1, 9, 10, \underbrace{0, \ldots, 0}_{\text{odd number of 0s}}, 2, 9, \ldots] = [0; 3, 3, 5, \ldots, 1, 1, 9, 10+2, 9, \ldots],$$

$$[0; 3, 3, 5, \ldots, 1, 1, 9, 10, \underbrace{0, \ldots, 0}_{\text{even number of 0s}}, 2, 9, \ldots] = [0; 3, 3, 5, \ldots, 1, 1, 9, 10, 2, 9, \ldots].$$

The difficulty in showing the existence of a number x satisfying (4.20) for $g = 11$ comes from the possibility of a long run of 0's. Probably, the length of a run of 0's which begins with the n-th symbol counted from the beginning is bounded by a function of n taking sufficiently small values; and probably, such an irrational number x, satisfying (4.20) with (4.22) exists for infinitely many g. It is clear that if an irrational number x, satisfying (4.20) with (4.21) exists, then x is an irrational number being different from all the quadratic irrationals. Note that a periodic, or a nonperiodic infinite continued fraction with (4.22) can be a rational number, for instance $[0; 3, 1, 0, 3, 0, 0, 0, 5, 0, 0, 0, 0, 0, 7, \ldots] = [0; 3, \infty] = 1/3$, $[0; 3, 1, 0, 7, 0, 7, 0, 7, \ldots] = 1/3$.

4.5 Problems

Let us end this chapter by evoking some open problems.

1. We denote by $\psi_i(z)$ the analytic function on the unit disc defined by

$$\psi_i(z) = \psi_i(z;\omega) := \sum_{n\in\chi(\omega;a_i)} z^n \quad (0 \le i \le s)$$

for $\omega \in \mathcal{A}^\infty$, where $\mathcal{A} = \{a_0, \dots, a_s\}$, and we take ω to be the word $\omega(L)$ defined by the billiard as in Sec. 4.1 with $\{t\underline{\alpha} + \underline{\beta}; t \in \mathbb{R}^+\}$. Then

$$\psi_i(z) = \sum_{1\le n<\infty} z^{\sum_{0\le j\le s}[\alpha_i^{-1}\alpha_j n - \alpha_i^{-1}\{\beta_i\}\alpha_j + \{\beta_j\}]}$$

follows from Proposition 4.1.1. We suppose that $\alpha_0, \dots, \alpha_s$ are linearly independent over $\mathbb{Q}$.
Question. Is the number $\sum_{0\le i\le s} c_i \cdot \psi_i(g^{-1})$ $(c_i, g \in \mathbb{Z},\ g \ge 2)$ always transcendental except for the case where $c_i = c$ for all i?

It follows from Proposition 2 in [172], that for $s = 1$, $\sum_{0\le i\le s} c_i \cdot \psi_i(g^{-1})$ $(c_0 \ne c_1)$ is transcendental since ω is Sturmian for $s = 1$, as we have mentioned in Sec. 4.1. For a proof of the transcendence of $\omega = \omega(L)$, it suffices to show the transcendence of $\psi_i(g^{-1}; s)$ for some i. Taking $s = 2$, $\alpha_0 = 1$, $\beta_0 = 0$, $0 < \beta_i < 1$, we have

$$\psi_0(z) = \sum_{1\le n<\infty} z^{[(\alpha_1+1)n+\beta_1]+[\alpha_2 n+\beta_2]}. \tag{4.23}$$

Question. Can we show the transcendence of the value $\psi_0(g^{-1})$? (Probably, yes; (4.23) is a simple expression similar to that in the case $s = 1$.)
For problems related to linear independence and transcendence for $\psi_i(z)$, see [418], (i)-(v), pp.213.

2. It is difficult to show that there is no number x satisfying (4.20) with (4.22) for $g = 10$. The difficulty comes from that, for example,

$$[0; 2, 0, 2, 1, 0, 0, 9, 0, 8, \dots] = 0.202100908\dots \text{ (in base 10)}$$

may be a solution for (4.20).
We may ask about the existence of a number x satisfying (4.20) for irrational g; for example, such a number may be:

$$[0; 3, 2, 4, 6, 9, 8, 2, \dots] = 0.3246982\dots \text{ (in base } \beta = ((1+5^{1/2})/2)^5).$$

It is easy to show that there exists a number $\beta = \beta(\omega)$ satisfying

$$[0; a_1, a_2, a_3, \dots] = 0.a_1a_2a_3\dots \text{ (in base } \beta) \tag{4.24}$$

for any given $\omega = a_1a_2a_3\cdots \in \{0, 1, \dots, h\}^\infty \backslash \{0, 1, \dots, h\}^*0^*$. For instance, for the Kolakoski sequence ξ,

$$[0; 1, 2, 2, 1, 1, 2, 1, 2, 2, \dots] = (0.122112122\dots)_\beta,\ \beta = 2.837559\dots;$$

for the fixed point of a substitution $1 \mapsto 10$, $0 \mapsto 1$,

$$[0; 1, 0, 1, 1, 0, 1, 0, 1, 1, \dots] = (0.101101011\dots)_\beta,\ \beta = 2.729451\dots.$$

Question. Can we show the transcendence of such a number $\beta(\omega)$ for a nonperiodic fixed point ω of a substitution?

Let us mention two conjectures.

Conjecture 4.5.1. For any integer $g = h^2 + h + i$ ($i = 0, 1,\ h \geq 3$), there exists an irrational number satisfying (4.20) with (4.22); such an irrational number is always transcendental.

Conjecture 4.5.2. The number $\beta(\omega)$ defined by (4.24) is transcendental for any nonperiodic word ω.

3. Let ω be the word corresponding to the partition (4.10), i.e., $\omega = \omega(K)$ for a curve $K = \{f_0(x), \ldots, f_s(x);\ x \in \mathbb{R}^+\}$ for f_i as in Proposition 4.1.3. Suppose that $f_i(x) \in \mathbb{Q}(x)$ for all i.

   Question. Can we show that $p_{\omega(K)}(n)$ is bounded by a polynomial in n (respectively s) for fixed s (respectively n) (see [418], (vi,vii), p.214)?

Part II

Dynamics of substitutions

5. Substitutions and symbolic dynamical systems

The aim of this chapter is to introduce the fundamental notions of ergodic theory, through the study of a few examples of symbolic dynamical systems.

To improve lisibility, some of the definitions and propositions stated in Chap. 1 are repeated in the present chapter.

In the course of this chapter, we shall use at some points notions of measure theory and spectral theory; when there is no explanation in the text, the reader is referred to any standard book, such as [121] and [339]; however, we have tried to isolate these places, and the reader can skip them without damage to his general understanding.

5.1 The Morse sequence: elementary properties

5.1.1 A geometrical problem

In 1920, M. Morse was studying *geodesics*, that is, the curves realizing the minimum distance between two points, on connected surfaces with constant negative curvature; he was looking at infinite geodesics which stay in a small part of the space; more precisely:

Definition 5.1.1. *A geodesic G is* uniformly recurrent *if for every $\varepsilon > 0$ there exists L such that for every segment S in G of length bigger than L, every point in G is at a distance smaller than ε from some point in S.*

Of course, a *closed* geodesic is a recurrent geodesic; but are there non-closed recurrent geodesics?

To answer this question, in [301], using a method initiated by Hadamard, Morse did a *coding* of geodesics, by infinite sequences of 0 and 1, according to which boundary of the surface they meet: thus, we arrive in the space $\{0,1\}^{\mathbb{N}}$ of infinite symbolic sequences (or, more properly, in the space $\{0,1\}^{\mathbb{Z}}$ of biinfinite symbolic sequences, but we shall consider only their positive coordinates), and to advance along a geodesic translates into looking at the next element of the sequence. The coding sends under suitable conditions the topology of the surface onto the product topology in $\{0,1\}^{\mathbb{N}}$.

[1] This chapter has been written by S. Ferenczi

Under these conditions, a closed geodesic corresponds to a periodic sequence. In the same way, by replacing points by elementary segments, the reader shall be able to check that a recurrent geodesic corresponds to what is now called a *minimal* sequence: as defined in Chap. 1, a sequence $u = (u_n)_{n\in\mathbb{N}}$ is said to be minimal if for every $i < j$, there exists s such that for every n, there exists $n < m < n + s$ such that $u_m = u_i$, ..., $u_{m-i+j} = u_j$.

It was stated in Chap. 1 that this definition can also be read by using language terminology: the sequence u is minimal if every word occurring in u occurs in an infinite number of positions with bounded gaps.

Hence what we need is a minimal non-periodic sequence, and the example Morse gave is the sequence that bears his name, though it had already been discovered by Prouhet in 1851 [338] and Thue [429] in 1907: we consider the map σ from $\{0, 1\}$ to $\{0, 1\}^\star$ defined by $\sigma(0) = 01$ and $\sigma(1) = 10$. We extend σ into a *morphism* of $\{0, 1\}^\star$ for the concatenation by $\sigma(abcd...) = \sigma(a)\sigma(b)\sigma(c)\sigma(d)\ldots$ We can then iterate σ: $\sigma^2(0) = \sigma(0)\sigma(1) = 0110$, $\sigma^n(0) = \sigma^{n-1}(0)\sigma^{n-1}(1)$, and the nested words $\sigma^n(0)$ (called *n-words*) converge (in $\{0,1\}^{\mathbb{N}} \cup \{0,1\}^\star$) to the only infinite sequence which begins with $\sigma^n(0)$ for every n, the *Morse sequence* (also called the *Prouhet-Thue-Morse sequence*):

$$u = u_0 u_1 ... = 0110100110010110...$$

This sequence satisfies the fundamental equality

$$u = \sigma(u).$$

Some properties of this sequence were stated in Chap. 2. A consequence is the following:

Proposition 5.1.2. *The Morse sequence u is minimal and neither periodic nor ultimately periodic.*

Proof. Any word W occurring in u must occur at a position between 0 and $2^m - |W|$ for some m, hence must occur in some $\sigma^m(0)$; but $\sigma^m(0)$ occurs in every $m+1$-word, hence infinitely often in u with gaps bounded by 2.2^{m+1}, and so does W. Hence u is minimal.

As u_{2n+1} is always different from u_n, $n + 1$ cannot be a period; and for any p and n_0, there exists $n \geq n_0$ such that $n + 1$ is a multiple of p, hence u is not ultimately periodic. ∎

5.1.2 Combinatorial properties

For the study of the dynamics of the Morse sequence, the following lemma, due to Del Junco ([139]), will be useful. It is a simple example of the notion of *recognizability*: for the Morse sequence, we know where a given word should occur, or even occur approximately.

Definition 5.1.3. *The* Hamming distance *between two words of equal length is*

$$\overline{d}((v_1 \dots v_n),(w_1 \dots w_n)) = \frac{1}{n} Card\ \{1 \le i \le n;\ v_i \ne w_i\}.$$

Proposition 5.1.4. *If U and V are two n-words, if W is a word of length 2^{n+1} occurring in u at position i, and such that $\overline{d}(UV,W) < \frac{1}{4}$, then $W = UV$ and i is a multiple of 2^n.*

Proof. The result is empty for $n = 0$; we suppose it is true at stage $n-1$; if U, V, W, i are as above, $W = W_1W_2$, with either $\overline{d}(U, W_1) < \frac{1}{4}$ or $\overline{d}(V, W_2) < \frac{1}{4}$; as U and V are each made with two $(n-1)$-words, the induction hypothesis implies that there exists an integer k with $i = k2^{n-1}$; hence, writing $W_1 = W_{11}W_{12}$, $W_2 = W_{21}W_{22}$, $U = U_1U_2$, $V = V_1V_2$, the words W_{11}, W_{12}, W_{21}, W_{22} must be $(n-1)$-words, and are at a distance $\overline{d} < \frac{1}{2}$ of the corresponding $(n-1)$-words U_1, U_2, V_1, V_2; as the distance between two $(n-1)$-words can only be 0 and 1, we must have $W_1 = U$ and $W_2 = V$.

It remains to be proved that k is even; if it is odd, let $W = abb'c$, $U = U_1U_1'$, $V = V_1V_1'$ their expressions as concatenation of $(n-1)$-words, where, for every word $M = m_1 \dots m_q$, M' denotes the word made with the letters $(1-m_1)\dots(1-m_q)$. Then we must have $a' = b = c$, and $a'abb'cc' = bb'bb'bb'$ must occur, which is impossible as 000 and 111 do not occur. ∎

Recognizability will be used several times in the sequel (see Lemmas 5.1.23, 5.2.4, 5.2.5 below), and it is an important notion for the general study of substitutions [305]. For more details on recognizability, see Chap. 7.

The same techniques allow us to prove an interesting combinatorial property which has opened a wide field of investigation in word combinatorics.

Indeed, the Morse sequence has been also studied at the beginning of this century by A. Thue. In 1906 and 1912 he wrote two papers that he considered as an attempt to open new ways to number theory the purpose of which was to construct sequences over finite alphabets containing no square or no cube (see [428, 429]). In 1938, G. A. Hedlund and M. Morse used the same property to link the Morse sequence to the problem of unending chess [304].

Definition 5.1.5. *A sequence is called* square-free *if no word of the form UU occurs in it, for any nonempty word U.*

It is not difficult to see that there is no infinite square-free sequence on two letters; but, as was shown in [428] (see also Corollary 2.2.4 from [270]), the Morse sequence can be said to be *free of powers* $2+\varepsilon$ for any $\varepsilon > 0$, in the following sense:

Proposition 5.1.6. *In the Morse sequence no word of the form UUv occurs, where U is any nonempty word and v is the first letter of U.*

Proof. Let l be the length of U; suppose for example that l is odd, and that UUv occurs at an odd position. Let $U = va_1b_1 \ldots a_pb_p$. If we put bars to indicate the separations between the 1-words occurring at positions $2k$ in u, the word $UUv = v|a_1b_1| \ldots |a_pb_p|va_1| \ldots |b_pv|$ must occur in the Morse sequence, with the indicated position of the bars. Hence $a_0 + v = a_1 + b_1 = \cdots = a_p + b_p = v + a_1 = b_1 + a_2 = \cdots = b_p + v = 1$. We thus get $a_i + b_i = b_i + a_{i+1} = a_{i+1} + b_{i+1}$, for $i = 1, 2, \ldots, p-1$. Hence, $a_1 = a_2 = \cdots = a_p$, $a_0 = a_1 = b_p$, $b_1 = b_2 = \cdots = b_p$, and $1 - v = a_0 = \cdots = a_p = 1 - v$, which is impossible as $v = 0$ or 1. The same reasoning applies if l is odd and UUv appears in an even position. In particular, if U is of length one, we check immediately that 000 and 111 do not occur.

If the length l is even and UUv occurs at the even position $2k$, then $U = |vb_1|a_2b_2| \ldots |a_pb_p|$, and if $U' = va_2..a_p$, $\sigma(U'U'v)$ occurs in the Morse sequence at position $2k$, and then $U'U'v$ must occur at position k, v is the first letter of U', and the length of U' is strictly smaller than l. The same reasoning applies to the case where l is even and UUv occurs in an odd position. ■

An immediate consequence of this result is that two occurrences of the same word in the Morse sequence have to be disjoint; another one is that the Morse sequence does indeed provide a square-free sequence on a three-letter alphabet (see also Sec. 2.3 from [270]):

Proposition 5.1.7. *If u is the Morse sequence, and v is the sequence over the alphabet $\{-1, 0, 1\}$ defined by: $\forall n \in \mathbb{N}$, $v_n = u_{n+1} - u_n$, v is an infinite square-free sequence.*

Proof. If the word $a_1 \ldots a_ka_1 \ldots a_k$ occurs at position i in v, then $a_1 + \cdots + a_k = 0$, as $u_{2k+i} - u_i = 2(a_1 + \cdots + a_k)$ is in $\{-1, 0, 1\}$; but this implies that for some $e = 0$ or 1, the word $e(e + a_1) \ldots (e + a_1 + .. + a_{k-1})e(e + a_1) \ldots (e + a_1 + .. + a_{k-1})e$ occurs in u, which is impossible. ■

In 1902 W. Burnside submitted the following problem: if G is a group of finite type such that there is an integer n such that $g^n = 1$ for any g in G, is G finite? W. Burnside gave a positive answer to this question in the case $n \leq 3$, Sanov in the case $n = 4$ in 1940, followed by M. Hall for the case $n = 6$ in 1957. In 1968, S. I. Adian and P. S. Novikov showed that the general answer to Burnside's problem is negative: there are counter-examples at least for any odd integer greater than or equal to 665. It is interesting to point out that their method uses fundamentally the construction of square-free sequences (see [5]).

5.1.3 Complexity

Let us recall the definition of the complexity function (for more details, see Chap. 1): we call *complexity function* of a sequence u, and denote by $p_u(n)$, the function which with each integer $n \geq 1$ associates Card $\mathcal{L}_n(u)$, that is, the number of different words of length n occurring in u.

Let us recall that an n-word is $\sigma^n(a)$ for any letter a.

Lemma 5.1.8. *In the Morse sequence, every word of length at least five has a unique decomposition into 1-words, possibly beginning with the last letter of a 1-word and possibly ending with the first letter of a 1-word.*

Proof. In other words, there is only one way to put the bars; it is true if our word contains 00 or 11, as they cannot occur between bars, so there must be a bar in the middle; if not, the word must contain 01010 or 10101; but 01010 is either 0|10|10 or 01|01|0, which do not occur as 010101 and 101010 do not occur, and 10101 does not occur for the same reason. ■

Proposition 5.1.9 ([91, 131]). *For the Morse sequence,* $p_u(1) = 2$, $p_u(2) = 4$, *and, for* $n \geq 3$, *if* $n = 2^r+q+1$, $r \geq 0$, $0 < q \leq 2^r$, *then* $p_u(n) = 6.2^{r-1}+4q$ *if* $0 < q \leq 2^{r-1}$ *and* $p_u(n) = 8.2^{r-1} + 2q$ *if* $2^{r-1} < q \leq 2^r$.

Proof. If $n \geq 4$, our lemma implies that $p_u(n) = p_0(n) + p_1(n)$, where $p_0(n)$ is the number of words of length n beginning just after a bar, and $p_1(n)$ just before. Let $n = 2k + 1$ and W a word of length n; if W is in the first category, $W = |a_1b_1| \ldots |a_kb_k|a_{k+1}$, $a_i + b_i = 1$, and there are as many such words as words $a_1..a_{k+1}$, hence $p_0(2k+1) = p_u(k+1)$. In the same way $p_1(2k+1) = p_u(k+1)$, $p_0(2k) = p_u(k)$, $p_1(2k) = p_u(k+1)$. The first values can be computed by hand, and we check that the proposed expression $p_u(n)$ has the same initial values and satisfies the induction relations $p_u(2k) = p_u(k) + p_u(k+1)$, $p_u(2k+1) = 2p_u(k+1)$. ■

Note that $p_u(n)$ is smaller than $4n$ and that the differences $p_u(n+1) - p_u(n)$ take only two values.

5.1.4 The associated topological dynamical system

Recall that the topological dynamical system associated with a sequence u with values in the finite alphabet $\mathcal{A}$ is the system (X_u, S), where

- $\mathcal{A}^{\mathbb{N}}$ equipped with the product topology of the discrete topology on each copy of $\mathcal{A}$, or equivalently with the distance

$$d(w, w') = 2^{-\min\{n \in \mathbb{N}; w_n \neq w'_n\}},$$

- $S(w_0, w_1, w_2, \ldots) = (w_1, w_2, \ldots)$,
- $X_u \subset \mathcal{A}^{\mathbb{N}}$ is the closure of the set $\{S^n u, n \in \mathbb{N}\}$.

We have seen in Chap. 1 (Lemma 1.1.2) that $w \in X_u$ if and only if there exists a sequence k_n such that $w_0 \ldots w_n = u_{k_n} \ldots u_{k_n+n}$ for every $n \geq 0$, which is also equivalent to $\mathcal{L}_n(w) \subset \mathcal{L}_n(u)$ for all n.

Proposition 5.1.10. *If u is minimal, $w \in X_u$ implies $u \in X_w$ and X_u is the set of all sequences having the same language as u.*

Proof. If u is minimal and $w \in X_u$, $u_0 \ldots u_p$ occurs in every long enough word occurring in u, hence in every possible $u_{k_n} \ldots u_{k_n+n}$ for n large enough, and it occurs in $w_0 \ldots w_n$; hence $u_0 \ldots u_p = w_{l_p} \ldots w_{l_p+p}$ and $u \in X_w$; and consequently $\mathcal{L}_n(u) = \mathcal{L}_n(w)$ for all n; and if $\mathcal{L}_n(u) = \mathcal{L}_n(w)$ for all n, then $w \in X_u$ because of Lemma 1.1.2. ∎

Lemma 5.1.11. *If u is recurrent, then the shift S is surjective onto X_u.*

Proof. If $w = w_0 w_1 \ldots$ is in X_u, then $w_0 \ldots w_n$ occurs in u, and not only in position 0; hence for each n there exists $a \in \mathcal{A}$ such that $a w_0 \ldots w_n$ occurs in u; and there exists at least one a such that $a w_0 \ldots w_n$ occurs in u for infinitely many values of n; but that implies that $a w_0 w_1 \ldots$ is in X_u, and it is an antecedent of w. ∎

Most sequences we shall consider have an at most linear complexity function ($\forall n \in \mathbb{N}$, $p_u(n) \leq Cn$). In this case, even if the shift is not injective, the following proposition shows that it can be made invertible up to a set which is at most countable.

Proposition 5.1.12. *Let u be a recurrent sequence the complexity of which satisfies $p_u(n) \leq Cn$ for all n and some constant C; then there exists a finite set F such that, if D is the (at most countable, and S-invariant) set $\bigcup_{n\in\mathbb{Z}} S^n F$, S and all its iterates S^n, $n \geq 0$, are one-to-one from $X_u \setminus D$ to $X_u \setminus D$.*

Proof. We suppose first $p_u(n+1) - p_u(n) \leq C$ for all n; since u is recurrent, every word w of length n has at least one *left extension* (that is, a word aw occurring in u for some $a \in \mathcal{A}$); hence there can be no more than C words of length n which have two or more left extensions. Let F be the set of sequences w in X_u such that $S^{-1}w$ has at least two elements; if the sequence $w = (w_n)_{n\in\mathbb{N}} \in F$, then there exists $a \neq b$ such that the sequences $a w_0 w_1 \ldots$ and $b w_0 w_1 \ldots$ belongs to X_u, and hence the word $w_0 \ldots w_n$ has at least two left extensions for every n; so F has at most C elements.
This is still true if $p_u(n) \leq Cn$ as then $p_u(n+1) - p_u(n) \leq C$ on a subsequence tending to infinity. ∎

Suppose that u is recurrent and not eventually periodic. The set X_u is not countable, hence $X_u \setminus D$ is not empty. In the sequel, we may use the map S^{-1}, tacitly assuming that we use it only on $X_u \setminus D$.

Let us come back now to the notion of minimality. The minimality of a sequence is equivalent to that of the associated dynamical system: a topological dynamical system (X, S) is said to be *minimal* if the only closed sets $E \subset X$ such that $S(E) \subset E$ are $\emptyset$ and E, or, equivalently, if for every point $x \in X$, the orbit of x, that is, $\{S^n x, x \in \mathbb{N}\}$, is dense in X.

Proposition 5.1.13. *The system (X_u, S) is minimal if and only if the sequence u is minimal.*

Proof. If u is minimal, we saw that for every $w \in X_u$, u, and hence its closed orbit, must be in X_w, the closed orbit of w, hence $X_w = X_u$.
If (X_u, S) is minimal; then, for any $w \in X_u$, the closed orbit X_w is equal to X_u, hence u is in X_w and $\mathcal{L}_n(u) \subset \mathcal{L}_n(w)$ for all n. Now, if W occurs in u, then W must occur in every $w \in X_u$, which implies $X_u = \cup_{n=0}^{+\infty} S^{-n}[W]$; hence, by compacity, $X_u = \cup_{i=1}^{p} S^{-n_i}[W]$ and hence for every $k \geq 0$, $S^k u \in \cup_{i=1}^{p} S^{-n_i}[W]$; this means that W occurs in u infinitely often with gaps bounded by $\max_i n_i$. ∎

Let us recall that a *topological dynamical system* (X, T) is usually defined as a compact set X together with a continuous map T acting on the set X. Two topological dynamical systems (X, T) and (Y, T') are *topologically conjugate* if there exists a bicontinuous bijection ϕ from X to Y such that $\phi T = T' \circ \phi$.

Lemma 5.1.14. *If u and v are sequences on finite alphabets and if (X_u, S) is topologically conjugate to (X_v, S), then there exists a finite integer q such that for every $i \in \mathbb{N}$ $(\phi(w))_i$ depends only on $(w_i, \ldots w_{i+q})$.*

Proof. The idea of the proof is the following : ϕ associates with a sequence (w_n) a sequence (w'_n); for $p \in \mathbb{Z}$, the map $\phi_p : X_u \to \mathcal{A}$, $(w_n)_{n\in\mathbb{N}} \mapsto w'_p$ is continuous because of the product topology; we deduce that $\phi_p^{-1}[a]$ is open and closed; hence by compacity it must be a finite union of cylinders; therefore there exists a finite integer q such that $\phi_0(w)$ depends only on $w_0 \ldots w_q$, and the ϕ_p, for $p \neq 0$, depend also only on $w_0 \ldots w_q$, as ϕ commutes with the shift S. ∎

Corollary 5.1.15. *Under the same hypothesis, if $p_u(n+1) - p_u(n)$ is bounded, so is $p_v(n+1) - p_v(n)$; and if $p_u(n) \leq Cn + C'$, then $p_v(n) \leq Cn + C''$.*

Proof. We have $p_u(n - q') \leq p_v(n) \leq p_u(n+q)$ for constants q and q'. ∎

Exercise 5.1.16. Prove that the topological entropy (see Chap. 1) is invariant under topological conjugacy.

5.1.5 Unique ergodicity

A topological dynamical system (X, T) always has an *invariant probability measure* that is, a measure μ such that

$$\mu(X) = 1 \text{ and } \int f(Tx)d\mu(x) = \int f(x)d\mu(x)$$

for any integrable Borel function f. We can take for μ any cluster point for the weak-star topology of the sequence of probability measures

$$\mu_n = \frac{1}{n}\sum_{k=0}^{n-1} \delta_{T^k x}$$

for any point $x \in X$.

Definition 5.1.17. *A topological dynamical system is* uniquely ergodic *if and only if it has only one invariant probability measure.*

Proposition 5.1.18. *Topological conjugacy preserves unique ergodicity.*

Proof. The proof is left as an exercise.

Lemma 5.1.19. *Let σ be the Morse substitution, and $N(W,V)$ be the number of occurrences of the word W in the word V; then, for any factor W, when n tends to infinity, $\frac{N(W,\sigma^n(e))}{2^n}$ tends to a limit f_W, independent of* e $= 0$ *or* 1.

Proof. As $\sigma^{n+1}(0) = \sigma^n(0)\sigma^n(1)$ and $\sigma^{n+1}(1) = \sigma^n(1)\sigma^n(0)$, we have

$$N(W,\sigma^{n+1}(0)) + N(W,\sigma^{n+1}(1)) \geq 2N(W,\sigma^n(0)) + 2N(W,\sigma^n(1))$$

as we may have forgotten occurrences at the junction of two n-words. So the quantity

$$\frac{N(W,\sigma^n(0)) + N(W,\sigma^n(1))}{2^{n+1}},$$

which is smaller than 1, has a limit, denoted by f_W. But the quantity $2^{-n-1}N(W,\sigma^{n+1}(0))$ differs from the preceding one only because of the occurrences at the junction between n-words, and their contribution is smaller than $2^{-(n+1)}|W|$, so it has the same limit; and the same is true if we replace 0 by 1. ■

Lemma 5.1.20. *Let u be the Morse sequence. For any factor W,*

$$\frac{N(W,u_k\dots u_{k+n})}{n+1} \to f_W, \textit{ uniformly in } k.$$

Proof. Let $V = u_k \dots u_{k+n}$. We write $V = A\sigma^p(u_j \dots u_{j+l-1})B$, for some p much smaller than n, with $|A| < 2^p$ and $|B| < 2^p$. We have $n+1 = |A| + |B| + l2^p$, and, because of the previous result, $|N(W,\sigma^p(u_i)) - 2^p f_W| < \varepsilon 2^p$ for n and p large enough. And

$$N(W,V) \leq N(W,A) + N(W,B) + \sum_{i=j}^{j+l-1} N(W,\sigma^p(u_i)) + (l+1)|W|,$$

the last term, which is necessary because of occurrences at the junctions between p-words, satisfying $(l+1)|W| \leq \frac{n+1}{2^p}|W|$, while

$$N(W,V) \geq \sum_{i=j}^{j+l-1} N(W,\sigma^p(u_i)),$$

which yields the conclusion by dividing the inequalities by $n+1$ and taking first p then n large enough; note that p and n can be chosen independently of k, hence the uniform convergence. ∎

Proposition 5.1.21. *Let v be a sequence such that, for any factor W, the sequence $\frac{N(W,v_k\dots v_{k+n})}{n+1}$ tends to a limit f_W, uniformly in k. Then, the system (X_v,S) is uniquely ergodic.*

Proof. For every word W, we define the measure of the associated cylinder by $\mu([W]) = \mu(S^{-n}[W]) = f_W$, for every $n \in \mathbb{N}$. This measure extends to all Borel sets and is an S-invariant probability measure. By hypothesis, when N tends to infinity,

$$\frac{1}{N}\sum_{n=0}^{N-1} 1_{[W]}(S^{n+j}u) \to \mu(W) = \int 1_{[W]} d\mu$$

uniformly in j for every cylinder $[W]$ (because $N(W,v_j\dots v_{j+n}) = \sum_{n=0}^{N-|W|} 1_{[W]}(S^{n+j}u)$, and the limit does not change if we replace $N-|W|$ by $N-1$); hence

$$\frac{1}{N}\sum_{n=0}^{N-1} g(S^{n+j}u) \to \int g\,d\mu$$

uniformly in j for every continuous function g; hence for any sequence of integers n_k,

$$\frac{1}{N}\sum_{n=0}^{N-1} g(S^{n+n_k}u) \to \int g\,d\mu$$

uniformly in k for every continuous function g; hence

$$\frac{1}{N}\sum_{n=0}^{N-1} g(S^{n}w) \to \int g\,d\mu \tag{5.1}$$

for every $w \in X_u$ and for every continuous function g.

But if there exists another S-invariant probability measure ν, then the last result and the dominated convergence theorem applied to $\frac{1}{N}\int \sum_{n=0}^{N-1} g(S^n w)$ implies that $\int f\,d\mu = \int f\,d\nu$ for every continuous function, hence $\mu = \nu$. ∎

Proposition 5.1.22. *If the topological dynamical system (X,T) is uniquely ergodic, and μ is its unique invariant probability measure, then the measure-theoretic dynamical system (X,T,μ) is ergodic.*

Proof. Otherwise, the measure ν defined over the Borel sets by: $\nu(H) = \frac{\mu(H\cap E)}{\mu(E)}$ is another T-invariant probability. ■

Lemma 5.1.23. *If W is a word of length l occurring in the Morse sequence, $\mu([W]) \leq \frac{6}{l}$.*

Proof. Because of Proposition 5.1.4, $N(\sigma^n(e)\sigma^n(f), \sigma^p(0)) \leq 2^{-n}|\sigma^p(0)|$ if $p > n$ and hence $\mu([\sigma^n(e)\sigma^n(f)]) \leq 2^{-n}$ for any $n \in \mathbb{N}$, $e \in \mathcal{A}$, $f \in \mathcal{A}$. If $3.2^n \leq l < 6.2^n$, then W must contain some $\sigma^n(e)\sigma^n(f)$, hence $\mu([W]) \leq \mu([\sigma^n(e)\sigma^n(f)]) \leq 2^{-n} < \frac{6}{l}$. ■

Corollary 5.1.24. *For the system associated with the Morse sequence, the measure μ is a* non-atomic measure*: for any point w, $\mu(\{w\}) = 0$.*

Proof. For every point w, $\mu(\{w\}) \leq \mu([w_0 \ldots w_n])$. ■

Hence, because of Proposition 5.1.9, Proposition 5.1.12 and Corollary 5.1.24, the shift S is bijective on a subset of X_u of measure one.

5.1.6 Digression: the ergodic theorem

Proposition 5.1.25. *If (X, T) is uniquely ergodic, μ being its invariant probability measure, if g is a continuous function on X, then*

$$(1/N) \sum_{n=0}^{N-1} g \circ T^n \to \int g \, d\mu$$

uniformly.

Proof. Otherwise, there exists $\delta > 0$, a continuous function f, a sequence of integers $n_k \to +\infty$ and a sequence of points x_k such that

$$\left| (1/n_k) \sum_{n=0}^{n_k-1} g(T^n x_k) - \mu(g) \right| > \delta.$$

By compacity, we may extract a subsequence m_k of n_k such that for every continuous function g, $\lim_{k\to+\infty}(1/m_k)\sum_{n=0}^{m_k-1} g \circ T^n$ exists, and defines a measure ν. Then this measure must be a T-invariant probability, hence is μ; this contradicts the hypothesis. ■

Formula (5.1) is also the conclusion of the primordial *ergodic theorem*, known as Birkhoff's theorem (1931) though the Russian school prefers to attribute it to Khinchin. Indeed, what we have already proved is that the result holds for every x provided g is continuous. We shall now see that it holds for almost every x under the (much weaker) hypothesis that f is integrable. The proof is slightly outside the scope of this chapter (though we do use the Theorem in Sec. 5.5.3), as it uses techniques of measure-theoretic ergodic theory; but we thought convenient to give here what is believed to be the shortest proof at this date of the ergodic theorem, due to Petersen (private communication, 1998). To skip the remainder of this section would not prevent the reader to understand the sequel.

In the remainder of this section, we take a measure-theoretic system (X, T, μ), where μ is a T-invariant probability; we do not require the transformation T to be invertible. For $f \in \mathcal{L}^1(X, \mu)$, we define

$$A_k f = \frac{1}{k}\sum_{j=0}^{k-1} f \circ T^j, \quad \overline{A}(f) = \limsup_{k\to+\infty} A_k f, \quad f_N^\star = \sup_{1\le k\le N} A_k f, \quad f^\star = \sup_N f_N^\star.$$

The following proposition is generally known as *Hopf's maximal lemma.*

Proposition 5.1.26. *Let h be an invariant ($h(Tx) = x$ for almost all x) function on X such that $h^+ = h \vee 0 \in \mathcal{L}^1(X, \mu)$. Then*

$$\int_{\{f^\star > h\}} (f - h) \geq 0.$$

Proof. We may suppose

$$\int_{\{f^\star > h\}} |h| < +\infty,$$

as otherwise, because $h^+ \in \mathcal{L}^1$, we would have

$$\int_{\{f^\star > h\}} (f - h) = +\infty \geq 0.$$

And hence $h \in \mathcal{L}^1(X, \mu)$, as on the set $\{f^\star \leq h\}$ we have also $f \leq h$, and hence on this set $h^- \leq -f + h^+$, an integrable function.

We suppose first that $f \in \mathcal{L}^\infty(X, \mu)$, and we fix N. Let

$$E_N = \{f_N^\star > h\}.$$

We remark that if $x \notin E_N$, $(f - h)(x) \leq 0$, and hence

$$(f - h)1_{E_N} \geq (f - h).$$

For an m much bigger than N, we consider

$$\sum_{k=0}^{m-1} (f - h)1_{E_N}(T^k x).$$

The term $(f - h)1_{E_N}(T^k x)$ is zero if $T^k x \notin E_N$; if $T^k x \in E_N$, by definition there exists $1 \leq l \leq N$ such that

$$A_l f(T^k x) = \frac{1}{l} \sum_{j=k}^{k+l-1} f \circ T^j x \geq h(x).$$

As h is invariant, we deduce

$$\frac{1}{l} \sum_{j=k}^{k+l-1} (f-h) \circ T^j(x) \geq 0$$

and then

$$\sum_{j=k}^{k+l-1} (f-h) 1_{E_N} \circ T^j(x) \geq \sum_{j=k}^{k+l-1} (f-h) \circ T^j(x) \geq 0.$$

Thus we can cut this sum into two kinds of parts: some of these parts are made with terms all equal to zero, while the others have at most N terms and the sum of these terms is positive. The last part of the sum is itself of one of these two types, hence either $\sum_{k=0}^{m-1} (f-h)1_{E_N}(T^k x) \geq 0$, or for one $m - N + 1 \leq j \leq m$,

$$\sum_{k=0}^{m-1} (f-h) 1_{E_N}(T^k x) \geq \sum_{k=j}^{m-1} (f-h) 1_{E_N}(T^k x).$$

And then

$$\sum_{k=0}^{m-1} (f-h) 1_{E_N}(T^k x) \geq -N(||f||_\infty + h^+(x)).$$

We take the integral and divide by m, thence

$$\int_{E_N} (f-h) \geq \frac{-N}{m}(||f||_\infty + ||h^+||_1).$$

Hence, by taking the limit when m goes to infinity,

$$\int_{E_N} (f-h) \geq 0.$$

By making N go to infinity, the dominated convergence theorem implies our proposition.

In the case $f \in \mathcal{L}^1(X, \mu)$, we approximate f by $g_s = f1_{|f| \leq s}$; $g_s \in \mathcal{L}^\infty(X, \mu)$, and $g_s \to f$ in $\mathcal{L}^1(X, \mu)$ and almost everywhere. Hence for fixed N, $(g_s)^\star_N \to f^\star_N$ in $\mathcal{L}^1$ and almost everywhere, and $\mu(\{(g_s)^\star_N > h\} \delta \{f^\star_N > h\}) \to 0$. Hence

$$0 \leq \int_{\{(g_s)^\star_N > h\}} (g_s - h) \to \int_{\{f^\star_N > h\}} (f-h)$$

by the dominated convergence theorem. Then N goes to infinity as in the previous case. ■

Corollary 5.1.27. *If $f \in \mathcal{L}^1$,*

$$\int \overline{A}(f) \leq \int f.$$

Proof. We consider first f^+ ; for fixed n, $h = -\frac{1}{n} + (\overline{A}(f^+) \wedge n)$ is an invariant function such that $\{(f^+)^\star > h\} = X$, hence by last proposition

$$\int f^+ \geq \int h$$

and, n going to infinity

$$\int f^+ \geq \int \overline{A}(f^+).$$

Hence $(\overline{A}(f))^+ \leq \overline{A}(f^+)$ is integrable.

Now we take an arbitrary $\epsilon > 0$ and we apply the previous proposition to $h = \overline{A}(f) - \epsilon$, which gives

$$\int f \geq \int h$$

and, with ϵ going to 0, the claimed results. ■

Theorem 5.1.28 (Birkhoff, 1931). *The sequence $A_k f$ converges almost everywhere.*

Proof. We define

$$\underline{A}(f) = \liminf_{k \to +\infty} A_k f.$$

The above corollary gives

$$\int \overline{A}(f) \leq \int f$$

and, by taking $-f$,

$$\int -\underline{A}(f) \leq \int -f.$$

We get

$$\int \overline{A}(f) \leq \int f \leq \int \underline{A}(f) \leq \int \overline{A}(f)$$

and hence $\underline{A}(f) = \overline{A}(f)$ almost everywhere. ■

This theorem is the foundation of ergodic theory, and is called the *ergodic theorem*. The physical meaning of the ergodic theorem is that, under the (relatively mild) condition of ergodicity, the time and space averages coincide, as is stated in the following corollary

Corollary 5.1.29. *Let (X, T, μ) be a measure-theoretic system. If T is ergodic, and if $f \in \mathcal{L}^1(X, \mu)$, then*

$$A_k f = \frac{1}{k} \sum_{j=0}^{k-1} f \circ T^j \to \mu(f), \quad \mu \text{ almost everywhere.}$$

Proof. If T is ergodic, every invariant function h is a constant almost everywhere as the sets $\{h \geq c\}$ are invariant. This is true in particular for the function $h = \lim_{k \to +\infty} A_k f$. And this constant is $\int h d\mu = \int f d\mu$. ■

We give now a spectacular application of the ergodic theorem to normal numbers.

Example 5.1.30. The system $\Omega = (\{0,1\}^{\mathbb{N}}, S)$, where S is the shift, is not uniquely ergodic.

Proof. Among the many measures we can define (including for example the one defined by the Morse sequence) there is the measure δ_0 giving measure 1 to the constant sequence $000\ldots$, and 0 to its complement, and also the measure ν giving to each cylinder $[w_1 \ldots w_n]$ measure 2^{-n}. ■

Lemma 5.1.31. *The system (Ω, S, ν) defined above is ergodic.*

Proof. Let A be the cylinder $[w_0, \ldots, w_r]$; if $n > r$, we have

$$\nu(A \cap S^{-n} A) = \nu(A)^2.$$

By a standard approximation argument, we deduce that for every Borel set A,

$$\lim_{n \to +\infty} \nu(A \cap S^{-n} A) = \nu(A)^2.$$

Hence an invariant set has to satisfy $\nu(A) = \nu(A)^2$. ■

Definition 5.1.32. *A sequence u is the expansion in base* 2 *of a* normal number *if for every word W of length n on $\{0,1\}$, we have*

$$\lim_{m \to +\infty} \frac{1}{m} N(W, u_0 \ldots u_{m-1}) \to 2^{-n}.$$

Proposition 5.1.33. *Almost every number $0 < x < 1$ (for the Lebesgue measure) is normal.*

Proof. We apply the ergodic theorem in the system (Ω, S, ν) to the indicator function of each cylinder $[w_0 \ldots w_{n-1}]$. We check that the mapping associating with u the number $\sum_{j=O}^{+\infty} \frac{u_j}{2^{j+1}}$ sends the measure ν onto the Lebesgue measure on $[0,1]$, as a cylinder of length n maps onto an interval of length 2^{-n}. ■

5.2 The Morse sequence: advanced properties

5.2.1 Representation by Rokhlin stacks

The aim of this section is to give a geometric representation by Rokhlin stacks of the Morse system (X_u, S). We give first a general, though slightly unpalatable, definition:

Definition 5.2.1. *A sequence of partitions* $P^n = \{P_1^n, \dots, P_{k_n}^n\}$ generates *a measure-theoretic dynamical system* (X, T, μ) *if there exists a set* E *with* $\mu(E) = 0$ *such that for every pair* $(x, x') \in (X/E)^2$*, if* x *and* x' *are in the same set of the partition* P^n *for every* $n \geq 0$*, then* $x = x'$*.*

Definition 5.2.2. *A system* (X, T, μ) *is of* rank one *if*

1. *there exist sequences of positive integers* $(q_n)_{n\in\mathbb{N}}$*, and* $(a_{n,i})_{n\in\mathbb{N}, 1\leq i\leq q_n-1}$*, such that the sequence of integers* h_n *defined by the recurrence* $h_0 = 1$*,* $h_{n+1} = q_n h_n + \sum_{j=1}^{q_n-1} a_{n,i}$ *satisfies*

$$\sum_{n=0}^{+\infty} \frac{h_{n+1} - q_n h_n}{h_{n+1}} < +\infty;$$

2. *there exist subsets of* X*, denoted by* $(F_n)_{n\in\mathbb{N}}$*, by* $(F_{n,i})_{n\in\mathbb{N}, 1\leq i\leq q_n}$*, and by* $(C_{n,i,j})_{n\in\mathbb{N}, 1\leq i\leq q_n-1, 1\leq j\leq a_{n,i}}$ *such that for every fixed* n
   a) $(F_{n,i})_{1\leq i\leq q_n}$ *is a partition of* F_n*,*
   b) the sets $(T^k F_n)_{1\leq k\leq h_n-1}$ *are disjoint,*
   c) $T^{h_n} F_{n,i} = C_{n,i,1}$ *if* $a_{n,i} \neq 0$ *and* $i < q_n$*,*
   d) $T^{h_n} F_{n,i} = F_{n,i+1}$ *if* $a_{n,i} = 0$ *and* $i < q_n$*,*
   e) $TC_{n,i,j} = C_{n,i,j+1}$ *if* $j < a_{n,i}$*,*
   f) $TC_{n,i,a_{n,i}} = F_{n,i+1}$ *if* $i < q_n$*,*
   g) $F_{n+1} = F_{n,1}$*;*
3. *the sequence of partitions* $\{F_n, TF_n, ..., T^{h_n-1}F_n, X \setminus \cup_{k=0}^{h_n-1} T^k F_n\}$ *generates the system* (X, T, μ)*.*

The union of the disjoint $(T^k F_n)_{1\leq k\leq h_n-1}$ *is called a* Rokhlin stack *of* base F_n*. We say also that the system is* generated *by the sequence of Rokhlin stacks with bases* F_n*.*

More generally, if we replace the sequence of partitions $\{F_n, TF_n, ..., T^{h_n-1}F_n, X \setminus \cup_{k=0}^{h_n-1} T^k F_n\}$ by partitions of the type $\{F_n^1, TF_n^1, ..., T^{h_n^1-1}F_n^1, \dots, F_n^r, TF_n^r, ..., T^{h_n^r-1}F_n^r, X \setminus \cup_{p=1}^{r} \cup_{k=0}^{h_n^p-1} T^k F_n^p\}$, built in a similar way, we say the system is of *rank at most* r. The general formulas are quite tedious, but happily in the cases of the systems we study they are much simpler, and the reader should not feel obliged to remember the general definitions. In particular, the sets $C_{n,i,j}$ will not appear before Sec. 5.5.1, where they correspond to letters called *spacers* in the associated symbolic sequences. Other definitions of the rank, together with a survey of many examples and properties of finite rank systems can be found in [167].

Lemma 5.2.3. *Every substitution* σ *defines a continuous map from* X_u *to* X_u*.*

Proof. The proof is left as an exercise.

We are now back to the particular case of the Morse sequence.

Lemma 5.2.4. *Let u be the Morse sequence and $n \in \mathbb{N}$. We have*

$$X_u = \bigcup_{e=0,1} \bigcup_{k=0}^{2^n-1} S^k \sigma^n[e],$$

and this union is disjoint.

Proof. The set on the right is closed, so we have only to prove that $S^m u$ is in it, for every $m \in \mathbb{N}$. But if $m = 2^n a + b$, with $0 \leq b < 2^n$, $S^m u = S^b S^{a2^n} u$ and $S^{a2^n} u = u_{a2^n} u_{a2^n+1} \cdots = \sigma^n(u_a u_{a+1} \ldots .) \in \sigma^n([u_a])$.

Suppose that $S^p \sigma^n[a] \cap S^q \sigma^n[b] \neq \emptyset$, with $q \geq p$; then $\sigma^n(w) = S^{q-p}\sigma^n(w')$ for a point w in $[a]$ and a point w' in $[b]$; we have $w' = \lim S^{m_i} u$ and $w = \lim S^{p_i} u$; hence for a fixed i large enough, $S^{q-p}(\sigma^n(u_{m_i} \ldots u_{m_i+i}))$ is a prefix of $\sigma^n(u_{p_i} \ldots u_{p_i+i})$. Hence, for i large enough, a word of the form $\sigma^n(u_r)\sigma^n(u_s)$ occurs in u at a position $j2^n + q - p$ and, because of the recognizability (Proposition 5.1.4), $q - p$ is a multiple of 2^n, and so must be 0; from which we deduce $a = b$. ■

Lemma 5.2.5. *We have*

$$\sigma^n[e] = [\sigma^n(e)\sigma^n(0)] \cup [\sigma^n(e)\sigma^n(1)]$$

for $e = 0$ or 1. For every $n \geq 0$,

$$\sigma^n[e] = \sigma^{n+1}[e] \cup S^{2^n} \sigma^{n+1}[e'],$$

where $e + e' = 1$. The measure of the cylinder $[0]$ is $\mu([0]) = 1/2$, and $\mu(\sigma^n[e']) = \mu(\sigma^n[e])$, if $e + e' = 1$.

Proof. If $w \in \sigma^n[e]$, we have $w = \sigma^n(w')$, with $w'_0 = e$, and so $w \in [\sigma^n(e)\sigma^n(w'_1)]$. Conversely, if $w \in [\sigma^n(e)\sigma^n(f)]$ and $w = \lim_{p \to +\infty} S^{k_p} u$, then for every large enough element M of the sequence k_p, $w_0 \ldots w_{2^{n+1}-1} = \sigma^n(e)\sigma^n(f) = u_M \ldots u_{M+2^{n+1}-1}$, and, because of Proposition 5.1.4, M is a multiple of 2^n. Hence $k_p = l_p 2^n$, and $w = \sigma^n(w')$, where $w' = \lim S^{l_p}(u)$, the sequence converging because $S^{k_p} u$ converges, and $w' \in [e]$.

The second assertion comes easily from the first, while the third and fourth ones are immediate by symmetry. ■

We may build the sets $F_{n,e} = \sigma^n[e]$ and their images $(S^k F_{n,e})_{0 \leq k \leq 2^n - 1}$ in the following way:

- conventionally, for fixed e and n, we denote the set $S^k F_{n,e}$ by dots drawn one above the other as k increases.

- At stage n, we have two *Rokhlin stacks* (each n-stack being $\cup_{k=0}^{2^n-1} S^k F_{n,e}$, $e = 0$ or 1), with *bases* $F_{n,0}$ and $F_{n,1}$, with *height* 2^n, whose *levels* $S^k F_{n,e}$ are disjoint sets of measure 2^{-n-1}.
- The shift map S sends each level of each stack, except the top ones, onto the level immediately above; S is not explicit on the top levels.
- In the beginning, $F_{0,0}$ and $F_{0,1}$ are two disjoint sets of measure $1/2$.
- At stage n, we cut $F_{n,e}$ into two subsets of equal measure $F_{n+1,e}$ and $H_{n+1,e}$. The shift map S becomes explicit on part of the levels where it was not yet so, as it sends $S^{2^n-1}F_{n+1,0}$ onto $H_{n+1,1}$ and $S^{2^n-1}F_{n+1,1}$ onto $H_{n+1,0}$. This defines the $(n+1)$-stacks, which will have height 2^{n+1}.

An illustration is given on Fig. 5.1.

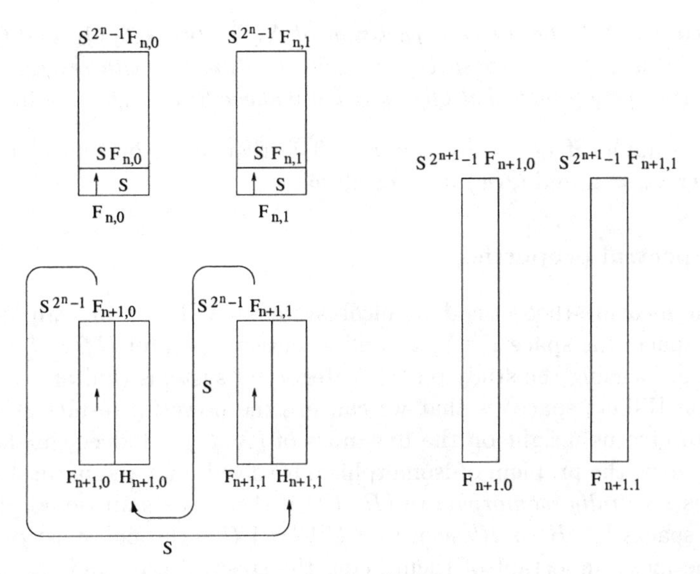

Fig. 5.1. Rokhlin stacks representation for the Morse sequence: steps n and $n+1$.

Proposition 5.2.6. *There exists a countable set E such that for every pair $(w, w') \in (X_u/E)^2$, if w and w' are in the same stack and the same level $S^k F_{n,e}$, for all $n \geq 0$, $0 \leq k \leq 2^n - 1$ and $e = 0, 1$, then $w = w'$.*

Proof. If w and w' are always in the same level of the same stack, they are in the same $S^{k_n}\sigma^n[e_n]$ for all n; and hence the sequences w and w' coincide between the indices 0 and $2^n - 1 - k_n$; this implies $w = w'$ if $2^n - 1 - k_n \to +\infty$. If not, there exists p such that $(S^m w)_0 = (S^m w')_0$ for $0 \leq m \leq 2^n - p$ for every n, hence for every positive m. Now, the Morse sequence has an at most linear complexity (Proposition 5.1.9), and the reasoning of Proposition 5.1.12, applied to right instead of left extensions, implies that there exists a countable set E on the complement of which $(S^m w)_0 = (S^m w')_0$ for every positive m implies $w = w'$. ■

Remark. As an exercise, the reader can check that E is made with two (non-closed, of course) orbits under S: the orbit of u and the orbit of the sequence $u' = (1 - u_n)_{n \in \mathbb{N}}$.

Hence, we have proved that the stacks of bases $F_{n,e}$ generate the system (X_u, T, μ) in the sense of Definition 5.2.2 (we have even proved more, as E is countable, and the stacks fill the whole space) and we have generated S by Rokhlin stacks. We need two stacks at each stage, so S of rank at most two. In fact, it is shown in [139] that S is not of rank one, and thus we say that S is of *rank two*.

Definition 5.2.7. *Let Q be a partition of X_u in two sets Q_1 and Q_2. For every point $w \in X_u$, its (positive) Q-*name *(or itinerary with respect to Q) is the sequence $Q(w)$ such that $Q(w)_n = i$ whenever $S^n w \in Q_i$, $n \geq 0$.*

For example, if $Q_i = F_{0,i}$ for $i = 0, 1$, then by definition $Q(w)_n = i$ whenever $w_n = i$, and $Q(w) = w$ for all w.

5.2.2 Spectral properties

Given a measure-theoretic dynamical system (X, T, μ), we can define a Hilbert space: the space $\mathcal{L}^2(X, \mu)$, and a unitary operator $Uf = f \circ T$. The interest of focusing the study on this category of systems (unitary operators acting on Hilbert space) is that we can use the powerful results of *spectral theory* to give us insight on the dynamics of (X, T, μ). Indeed, in the spectral category, the problem of isomorphism is solved – for the natural notion, (H, U) is *spectrally isomorphic* to (H', U') if there exists an isomorphism of Hilbert spaces $V : H \to H'$ such that $U'V = VU$ – through a set of invariants, the most important of them being the spectral type and the spectral multiplicity (see [234], Chap. 1 and definitions below). Hence, we compute these invariants for the present system, that is, the Morse system.

Lemma 5.2.8. *The eigenvalues of the unitary operator U are of modulus one; S is ergodic if and only if the eigenvalue one is a simple eigenvalue for U; in that case, every eigenvalue is simple and every eigenfunction is of constant modulus.*

Proof. The eigenvalues have modulus one because U is unitary.

If E is an invariant set of nontrivial measure, 1_E is a nonconstant invariant function; if f is a nonconstant invariant function, we can cut its image into two disjoint sets, whose inverse images have a nontrivial measure and are invariant sets.

If f is an eigenfunction for the eigenvalue β, $|f|$ is an eigenfunction for the eigenvalue $|\beta| = 1$ and hence is a constant. If f_1 and f_2 are eigenfunctions for

β, $|f_2|$ is a nonzero constant, and f_1/f_2 is an eigenfunction for 1 and hence a constant. ■

Definition 5.2.9. *Given a unitary operator U on a Hilbert space H, a* cyclic space *is the closure of the subspace generated by the set $\{U^n(f), n \in \mathbb{Z}\}$ for an element f in H.*

The spectral type *of f, or of the generated cyclic space, is the finite positive measure ϱ on the torus $\mathbb{T}^1$ defined by $\hat{\varrho}(n) = (U^n f, f)$. Its total mass is $||f||_H^2$.*

We say that U has a spectrum of multiplicity *at most k if H is the direct sum of k cyclic spaces; if $k = 1$, we say the spectrum is* simple.

We say that U has a discrete spectrum, *if the vector space generated by its eigenfunctions is dense in $\mathcal{L}^2(X, \mu)$.*

A useful criterion to show that a space is cyclic is the following:

Lemma 5.2.10. *If L is a separable Hilbert space approximated by an increasing sequence of cyclic spaces, L itself is a cyclic space.*

Proof. We denote by $H(f)$ the cyclic space generated by the function f. Then we have $L = \cup(H(f_n))$, and we want to find g such that $L = H(g)$. It suffices to show that for a dense sequence of functions g_n in L, $\cap_{n,p}\{g; d(g_n, H(g)) < 1/p\}$ is nonempty. Because of Baire's theorem, it is enough to show that for given f and ε, $\{g; d(f, H(g)) < \varepsilon\}$ is dense. So we take a function h, which we may choose of norm one, as well as f. For some m, $d(f, H(f_m)) < \varepsilon$ and $d(h, H(f_m)) < \varepsilon$. Hence $d(h, P(U)f_m) < \varepsilon$ for some polynomial P; hence we can find a polynomial Q, nonzero on the unit circle, such that $d(h, Q(U)f_m) < 2\varepsilon$. Then $Q(U)$ is invertible and its inverse is approximated by polynomials in U, hence $g = Q(U)f_m$ satisfies $d(h, g) < 2\varepsilon$ and $d(f, H(g)) < 2\varepsilon$. ■

We have to admit the following result of spectral theory, as its proof would need too many prerequisites:

Lemma 5.2.11. *If H and H' are cyclic spaces whose spectral types ϱ and ϱ' are* mutually singular, *that is, $\mathbb{T}^1 = E \cup E', \varrho(E) = \varrho'(E') = 1, \varrho'(E) = \varrho(E') = 0$, then $H \perp H'$ and $H + H'$ is a cyclic space, of spectral type $\frac{\varrho+\varrho'}{2}$. This lemma is still valid for a countable sum of subspaces.*

Corollary 5.2.12. *If U defined on a Hilbert space H has a discrete spectrum and all its eigenvalues β_i, $i \in I$, are simple, then U has simple spectrum and the spectral type of the cyclic space H is equivalent to the weighted sum over i of the Dirac masses δ_{β_i}*

$$\sum_i \alpha_i \delta_{\beta_i},$$

where $\alpha_i > 0$, for all i, and $\sum_i \alpha_i = 1$.

Proof. We write $\mathcal{L}^2(X,\mu)$ as the sum of the cyclic spaces generated by the eigenfunctions f_β, whose spectral types δ_β are mutually singular. ∎

5.2.3 The dyadic rotation

The aim of the remainder of our study of the Morse system is to compare it with other "classical" systems, and particularly to geometric systems, living on manifolds, or, in the most elementary cases, on intervals.

First, we need to define new notions of isomorphism, as the topological conjugacy of Sec. 5.1.4 is too strong to be really useful.

Definition 5.2.13. *Two measure-theoretic dynamical systems (X,T,μ) and (Z,R,ϱ) are* measure-theoretically isomorphic *if there exist $X_1 \subset X$, $Z_1 \subset Z$, and ϕ a (bimeasurable) bijection from X_1 to Z_1 such that $\mu(X_1) = \varrho(Z_1) = 1$, $\phi\mu = \varrho$ and $R\phi = \phi T$.*

Note that if two uniquely ergodic topological dynamical systems (X,T) and (Z,R) are topologically conjugate, then the corresponding measure-theoretic dynamical systems (X,T,μ) and (Z,R,ϱ) are measure-theoretically isomorphic. But in this case, the latter notion is in fact much weaker than the topological one, as the sets X_1 and Z_1 may be very wild. An intermediate notion is the following:

Definition 5.2.14. *Two systems (X,T) and (Z,R) are* semi-topologically conjugate *if if there exist $X_1 \subset X$, $Z_1 \subset Z$, and ϕ a bicontinuous bijection from X_1 to Z_1 such that $X \setminus X_1$ and $Z \setminus Z_1$ are countable, and $R \circ \phi = \phi \circ T$.*

In that case, if (X,T) is a symbolic system, we say that it is a coding *of (Z,R).*

Proposition 5.2.15. *Semi-topological conjugacy preserves unique ergodicity, and, for uniquely ergodic systems, the semi-topological conjugacy of (X,T) and (Z,R) implies the measure-theoretic isomorphism of (X,T,μ) and (Z,R,ϱ).*

Proof. The proof is left as an exercise.

The first classical systems we know are the rotations.

Definition 5.2.16. *We call* rotation *the dynamical system made with a compact group G, a translation R of G and the Haar measure λ on G.*

The most famous case is when G is the torus $\mathbb{T}^1$, and $Rx = x+\alpha$ modulo 1, for an irrational α. We now introduce another important kind of rotation.

Let R_0 be the map from $Y_0 = [0,1]$ to itself defined by

$$R_0(1 - \frac{1}{2^n} + x) = \frac{1}{2^{n+1}} + x, \quad \text{for} \quad 0 \le x < \frac{1}{2^{n+1}}, \quad n \in \mathbb{N},$$

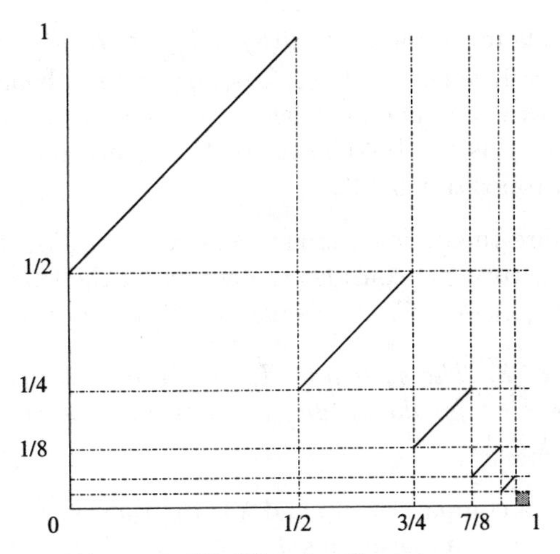

Fig. 5.2. The Van der Corput map.

and $R_0(1) = 0$; under this form it is called the *Van der Corput map.* It preserves the Lebesgue measure λ_0. A representation is given on Fig. 5.2.

We now send $[0,1]$ onto $Y = \{0,1\}^{\mathbb{N}}$ by the mapping $\chi(x) = (\omega_0\omega_1\ldots)$ whenever $x = \sum \omega_j 2^{-j-1}$ (there may be ambiguities at dyadic points, but still the coding is well-defined outside a countable set). We see how the intervals of continuity of R_0 are coded, and notice that

$$R := \chi R_0 \chi^{-1}$$

sends points of the form $0abc\ldots$ to $1abc\ldots$, $10abc\ldots$ to $01abc\ldots$, $110abc\ldots$ to $001abc\ldots$, and hence S is the *addition of* 1 *in base two* of $100\ldots0\ldots$ with infinite carry. The Lebesgue measure is sent to the infinite product of measures $1/2(\delta_0 + \delta_1)$, denoted by λ (δ_0 and δ_1 denote here the Dirac measures).

The set Y, equipped with the addition in base two, is a compact group. It contains a copy of $\mathbb{N}$ by $\phi(1) = 10\ldots0\ldots$, $\phi(2) = \phi(1) + \phi(1), \ldots$ and a copy of $\mathbb{Z}$ as we check that $10\ldots0\ldots. + 11111\ldots. = 0000\ldots.$, hence we can define $\phi(-1) = 111\ldots.$ And we see that the integers are dense in Y equipped with the usual topology; we call $(Y,+)$ the *group of* 2*-adic integers.* Then, R is the rotation $w \mapsto w+1$. It is immediate that (Y_0, R_0) and (Y, R) are semi-topologically conjugate, and we treat them as the same system: we call it the *dyadic rotation* or the *dyadic odometer.*

Proposition 5.2.17. *The system (Y,R) is uniquely ergodic, and hence the system (Y,R,λ) is ergodic.*

Proof. The *characters*, that is, the continuous homomorphisms of Y to the torus $\mathbb{T}^1$, define for each measure ν on Y, a Fourier transform $\hat{\nu}(\gamma) =$

$\int_{g\in Y}\gamma(g)d\nu$. If μ is invariant by R, $\hat{\mu}(\gamma)=\int_{g\in Y}\gamma(Rg)d\mu=\gamma(1)\hat{\mu}(\gamma)$, hence if $\hat{\mu}(\gamma)\neq 0$, $\gamma(1)=1$, hence $\gamma(n)=1$ for every integer, hence γ is identically 1. So $\hat{\mu}$ is zero on the characters which are not identically 1, and one on the character 1, hence μ is the Haar measure λ. The ergodicity of (Y,R,λ) is a consequence of Proposition 5.1.22. ■

For the measure-theoretic dynamical system (Y,R,λ), there is an underlying *spectral* structure: we consider the action on the space $\mathcal{L}^2(Y,\lambda)$ of the operator $U:\mathcal{L}^2(Y,\lambda)\to\mathcal{L}^2(Y,\lambda)$, defined by $Uf(w)=f(Rw)$.

Proposition 5.2.18. *The system (Y,R) has a discrete spectrum. Its eigenvalues are all the $e^{2i\pi\alpha}$ for the dyadic rationals, i.e., the rationals of the form $\alpha=p2^{-k}$, for $p,k\in\mathbb{Z}$.*

Proof. As Y is a compact group and λ is the Haar measure, we know that the characters generate a dense subspace of $\mathcal{L}^2(Y,\lambda)$. But if γ is a character, $\gamma(Rg)=\gamma(1)\gamma(g)$. Hence all the characters are eigenfunctions, we get thus all the eigenfunctions, and the eigenvalues are all the $\gamma(1)$.

Hence we have to find all the characters of Y. Such a character γ must be also a character of $\mathbb{Z}$, hence $\gamma(n)=e^{2\pi in\alpha}$ for an α in $[0,1]$; if γ can be extended in a continuous way to Y, it will remain a character. But $\omega=\omega_0\omega_1\cdots=\lim_{k\to+\infty}\omega_0\ldots\omega_k00\ldots0\cdots=\lim_{k\to+\infty}n_k(\omega)$, where $n_k(\omega)=\sum_{j=0}^{k}\omega_j2^j$. Hence we have to find all the α such that $\forall\omega, e^{2\pi i\alpha n_k(\omega)}$ converges when $k\to+\infty$.

We write $\alpha=\sum_{k=0}^{+\infty}\alpha_k2^{-k-1}$; if the α_i are ultimately equal to 1, the expansion is improper; otherwise, either they are ultimately equal to 0, or for infinitely many k, $\alpha_k=1,\alpha_{k+1}=0$. We choose an ω such that $\omega_k=1$ for all these values of k. Then we have, for this ω, $\gamma(n_k(\omega))=\gamma(n_{k-1}(\omega))e^{2\pi i\omega_k2^k\alpha}=\gamma(n_{k-1}(\omega))e^{2\pi i2^k\alpha}$, and $2^k\alpha=m+1/2+1/8+\ldots$, where m is an integer, and hence $2^k\alpha$ falls between 1/2 and 3/4, modulo 1 and the sequence $\gamma(n_k(\omega))$ cannot converge.

Hence α must have a dyadic expansion consisting ultimately of zeros, hence it must be a dyadic rational number; conversely, it is clear that every dyadic rational number α yields an eigenvalue $e^{2\pi i\alpha}$. ■

Proposition 5.2.19. *The system (Y,R) is of rank one.*

Proof. Let G_0 be the interval $[0,1[$, G_n the interval $[0,2^{-n}[$. We see that S sends in an affine way R^iG_n onto $R^{i+1}G_n$ for $i=0,1,\ldots 2^n-1$. We can write these intervals one above the other, and this makes a stack filling all the space; the action of R is to climb up one level in the n-stack and is not defined on the top level. The $(n+1)$-stack is built by cutting the n-stack vertically into two equal halves and stacking the right half on the left half; this operation defines R on a greater part of the space, and allows us to know eventually R on the whole space. As the R^kG_n are arbitrarily small intervals, outside a countable set (the set of all the endpoints of the R^kG_n), a point w

is determined by the sequence $0 \leq k_n(w) \leq 2^n - 1$ such that $w \in R^{k_n(w)}G_n$, $n \in \mathbb{N}$. Hence R is generated by the stacks of bases G_n. ∎

Under that form, R is known as the *Von Neumann–Kakutani map.*

5.2.4 Geometric representation of the Morse system

In the Morse system (X_u, S) we introduce the following equivalence relation: $w \sim \overline{w}$ if either $w = \overline{w}$ or $w_n + \overline{w}_n = 1$ for every $n \in \mathbb{N}$. Each equivalence class has two elements, and we check that on the quotient space $\overline{X}_u$, the shift and the measure define naturally a measure $\overline{\mu}$ and a shift $\overline{S}$; we say that $(\overline{X}_u, \overline{S}, \overline{\mu})$ is a *factor with fiber two* of (X_u, S, μ).

To come back from $\overline{X}_u$ to X_u, we associate with $\overline{w} \in \overline{X}$ the two elements of the class $\overline{w}$: we denote by $(\overline{w}, 0)$ the one whose first coordinate is 0, by $(\overline{w}, 1)$ the other one. So we have $X_u = \overline{X}_u \times \{0, 1\}$, as topological spaces too, and μ is the tensor product of $\overline{\mu}$ by $1/2(\delta_0 + \delta_1)$. And $S(w, e) = (\overline{S}w, z(e, w))$, where $z(0, w) + z(1, w) = 1$; hence we can write, in $\mathbb{Z}/2\mathbb{Z}$, that $z(e, w) = e + \psi_0(w)$, where ψ_0 is a measurable map from $\overline{X}_u$ to $\mathbb{Z}/2\mathbb{Z}$. We say that (X_u, S, μ) is a *skew product* of $(\overline{X}_u, \overline{S}, \overline{\mu})$ by $\mathbb{Z}/2\mathbb{Z}$ built with the map ϕ, or a *two-point extension* of $(\overline{X}_u, \overline{S}, \overline{\mu})$.

For fixed k, n, both levels $S^k F_{n,e}$ project into $\overline{X_u}$ on the same level $\overline{S}^k F_n$; because of Proposition 5.2.6, the stacks F_n generate the system $(\overline{X}_u, \overline{S})$, which is of rank one. And, from the construction of the stacks with bases $F_{n,e}$, we deduce that the stacks with base F_n are built in the same way as the stacks for the dyadic rotation (Y, R): the $n+1$-stack is built by cutting the n-stack vertically into two equal halves and stacking the right half on the left half, starting from $F_0 = \overline{X}$.

Proposition 5.2.20. *The Morse system is a coding of a two-point extension of the dyadic rotation.*

Proof. Let (Y', S', λ) be the following system: Y' is the interval $[0, 1[$, λ the Lebesgue measure; for $n \geq 0$ and $0 \leq k \leq 2^n - 1$, we define inductively intervals (closed on the left, open on the right) $G_{n,0,k}$ and $G_{n,1,k}$ of length 2^{-n-1} such that for $n \geq 0$, $0 \leq k \leq 2^n - 1$, $e \in \{0, 1\{$, $G_{n,0,k} \cup G_{n,1,k} = R^k G_n$ and $G_{n,e,k} = G_{n+1,e,k} \cup G_{n+1,1-e,2^n+k}$. The intervals $G_{n,e,k}$, $e \in \{0, 1\}$, $0 \leq k \leq 2^n - 1$, constitute for fixed n a partition of Y' into sets of measure 2^{-n-1}. We define S' by sending by a translation each $G_{n,e,k}$ onto $G_{n,e,k+1}$, if $k \leq 2^n - 1$; S' is not explicitly defined on the $G_{n,e,2^n-1}$. We check that these definitions are compatible, and that S' becomes explicit on part of the levels where it was not yet so, in the same way as for the stacks of the Morse system (Sec. 5.2.1), $G_{n,e,k}$ playing the role of $S^k F_{n,e}$ and $G_{n+1,1-e,2^n}$ the role of the auxiliary $H_{n+1,e}$. The system (Y', S') has rank at most two, and is generated by stacks whose levels are the sets $G_{n,e,k} = S'^k G_{n,e,0}$: except on a countable set E' (which is just the set of all endpoints of the $G_{n,e,k}$), if x and x' are in the same $G_{n,e(n),k(n)}$ for all n, then $x = x'$.

We define now a bijection ϕ between $X_u \setminus E$ (of Proposition 5.2.6) and the corresponding $Y' \setminus E'$ by associating with a point w such that $w \in S^{k(n)} F_{e(n)}$ for all n the unique point x such that $x \in S'^{k(n)} G_{n,e(n),0}$ for all n, and conversely from Y' to X_u; that is possible as, the stacks being defined by the same recursion formulas, the possible sequences $k(n), e(n)$ are the same, and we check from the structure of E and E' that by this process all points in $X \setminus E$ are indeed sent to points in $Y' \setminus E'$, and conversely. The construction of S and definition of S' ensure also that $\phi(Sw) = S'(\phi w)$; and ϕ is bicontinuous because all the $S^k F_{n,e} \cap X_u \setminus E$ and the $S'^k G_{n,e,0} \cap X' \setminus E'$ are open sets. Hence ϕ is a coding (in the sense of Definition 5.2.14).

And, by construction, ϕ associates the factor of S generated by the stacks with base F_n with the factor of S' generated by the stacks with base G_n, and provides a semi-topological conjugacy between the dyadic rotation (Y, R) and the factor $(\overline{X}_u, \overline{S})$ in (X_u, S). Hence we can carry the map ψ_0 to Y: (Y', S', λ) is a skew product of (Y, R, λ) by $\mathbb{Z}/2\mathbb{Z}$ built with the map $\psi = \psi_0 \circ \phi^{-1}$, or a two-point extension of the dyadic rotation. ∎

Note that for a point $x \in Y' \setminus E'$, $\phi^{-1}(x)$ is simply its Q'-name, where $Q'_e = G_{0,e,0}$, $e = 0, 1$.

Thus we have a geometric model for the Morse system. Note that (X_u, S) and (Y', S') cannot be topologically conjugate, as S' is not continuous while S is.

Note also that for the dyadic rotation R, we do not know any simple coding, for example using one sequence of words corresponding to the sequence of stacks (partitions into levels give periodic sequences, while R is ergodic and hence aperiodic); it can be shown that a coding of R is given by the system associated with the *period-doubling sequence*, the fixed point of the substitution $0 \mapsto 11$, $1 \mapsto 10$, sometimes called the *Toeplitz substitution.*

Proposition 5.2.21 ([139]). *The Morse system has a nondiscrete simple spectrum; its eigenvalues are all the $e^{2\pi i\alpha}$ where α is a dyadic rational number.*

Proof. Let τ be the map on X_u which associates with $(w_n)_{n\in\mathbb{N}}$ the point $(1-w_n)_{n\in\mathbb{N}}$. The map τ sends each $S^j F_{i,n}$ onto $S^j F_{i',n}$ where $i + i' = 1$, and commutes with S (τ is called the *flip*). Let V be the operator on $\mathcal{L}^2(\overline{X}_u, \overline{\mu})$ defined by $Vf = f \circ \tau$, H the space of functions f such that $Vf = f$, K the space of functions such that $Vf = -f$. The sets H and K are U-invariant. The action of U on H is also the action of the spectral operator associated with $\overline{S}$ on $\mathcal{L}^2(\overline{X}, \overline{\mu})$; as measure-theoretic isomorphism implies spectral isomorphism, it is isomorphic to the action of the spectral operator associated with R on $\mathcal{L}^2(Y)$, hence the restriction V_H of U to H has a discrete spectrum, and simple spectrum, as S is ergodic; the eigenvalues are the the $e^{2\pi i\alpha}$ where α is a dyadic rational number.

Let $f_n = 1_{F_{1,n}} - 1_{F_{0,n}}$, $K(f_n)$ the cyclic space it generates under U; these increase to K, hence K is cyclic.

Let g be an eigenfunction for the restriction V_K of U to H; then $g \in K, V_K g = \beta g$, hence $g^2 \in H, \beta^2 g^2 = V_H g^2$. But then the argument of β^2 is a dyadic rational, and so is the argument of β, but then U has two orthogonal eigenfunctions for the eigenvalue β, which contradicts ergodicity. Hence V_K has a continuous spectrum, and the spectral type of K is singular with every discrete measure.

But $\mathcal{L}^2(X,\mu) = H + K$, by writing $2f = (f + Vf) + (f - Vf)$, hence U has a simple spectrum; its eigenvalues and eigenvectors are those of V_H, and the eigenvectors are not dense. ∎

For a precise description of the spectrum of the Morse system in terms of Riesz products, see [339].

5.3 The Rudin-Shapiro sequence

For all the definitions concerning automata and automatic sequences, see Chap. 1. Recall that if σ is a substitution of length q over the alphabet $\mathcal{A}$ and u a fixed point of σ, then u is recognized by a q-automaton (Proposition 1.3.1).

5.3.1 The Rudin-Shapiro substitution

The Rudin-Shapiro sequence was first introduced in [393] and then in [365] for some estimations in harmonic analysis (see Chap. 2 and Sec. 5.3.2 below); it is defined over the four-letter alphabet $\{a, b, c, d\}$ as the fixed point u beginning with a of the substitution

$$a \mapsto ab \qquad b \mapsto ac \qquad c \mapsto db \qquad d \mapsto dc.$$

The associated 2-automaton is given on Fig. 5.3.

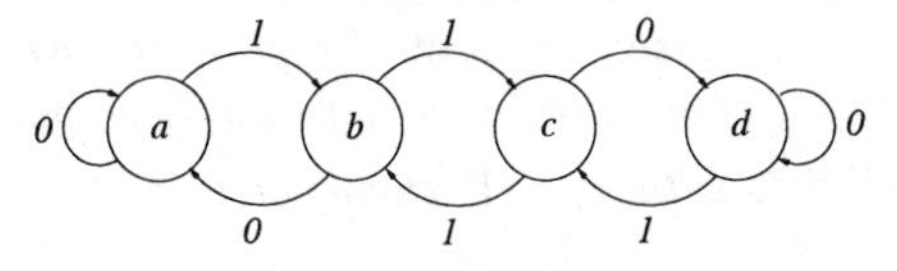

Fig. 5.3. Automaton associated with the Rudin-Shapiro substitution.

Lemma 5.3.1. *We have $u_n = a$ or b if and only if there is an even number of words* 11 *in the expansion of n in base* 2.

This lemma was already proved in Chap. 2 (see Proposition 2.1.1).

Let v be the sequence deduced from u by the letter-to-letter projection ϕ, which replaces a and b by 1, c and d by -1. The sequence v is then the limit in the usual sense of the words $\phi(\sigma^n(a))$. We check by induction that $\phi(\sigma^n(c)) = -\phi(\sigma^n b)$; hence if $A_n = \phi(\sigma^n(a))$ and $B_n = \phi(\sigma^n(b))$, the sequence v is the limit A_n given by the recursion formula $A_{n+1} = A_n B_n$, $B_{n+1} = A_n(-B_n)$. We remark that v is a 2-automatic sequence (i.e., recognized by a 2-automaton) though it is not a priori a fixed point of a substitution.

Like the Morse sequence, u is minimal. Its complexity is $8n - 8$ for every $n \geq 2$, and is easy to compute by the same method as for Morse. The complexity of v is also $8n - 8$ ultimately, and X_u and X_v are topologically conjugate ([26]). The same methods as for Morse (see Sec. 5.2.4, but see also Sec. 5.3.2 below for the general setting) allow us also to prove that the dynamical system (X, S) generated by the Rudin-Shapiro sequence u (or v) is uniquely ergodic, with a non-atomic invariant probability measure μ.

A similar construction also shows that the associated shift transformation S has rank at most four, and is an (at most) four-point extension of the dyadic rotation: after the same kind of isomorphism as in Sec. 5.2.4, X is the product of Y by the set (a, b, c, d), and $S(w, e) = (Sw, \phi(w)e)$ for a map ϕ of Y into the group of permutations on four points. Note that we do not prove that this extension is non-trivial; (X, S) might be isomorphic to an extension of (Y, R) with less than four points, and actually is a two-point extension of the dyadic rotation (this is a particular case of a result in [262]).

5.3.2 Spectral properties

Though we do not want to give a complete description of the spectral structure of (X, S, μ), it has a remarkable property: the question whether the Lebesgue spectrum (that is: some function f has a spectral type equivalent to the Lebesgue measure) and finite multiplicity (see Definition 5.2.9), could co-exist stayed open for a long time, and this system solved it.

Lemma 5.3.2. *Let f be the function defined over X_v taking the value 1 whenever $w_0 = 1$, 0 otherwise; its spectral type is the weak-star limit of the measures with density $\frac{1}{N}|f_N|^2$ with respect to the Lebesgue measure λ, where $f_N(x) = \sum_{n=0}^{N-1} z_n e^{2\pi i n x}$, with $z_n = 1$ whenever $v_n = 1$, 0 otherwise.*

Proof. We use a standard method to compute this spectral type ϱ_f. The measure ϱ_f is defined by

$$\hat{\varrho}_f(p) = (f, U^p f) = \mu(\{w; w_0 = 1,\ w_p = 1\}).$$

By unique ergodicity, we read this measure on the sequence v as a correlation measure (see Chap. 1), that is, as the sum of the frequencies of the words $a_0 \dots a_p$, where $a_0 = 1$, $a_p = 1$; we deduce that it is exactly

$$\lim_{N \to +\infty} \frac{1}{N} \sum_{n=0}^{N-1} 1_{v_n = v_{n+p} = 1}.$$

Let then f_N be the function defined on the torus $\mathbb{T}^1$ by

$$f_N(x) = \sum_{n=0}^{N-1} z_n \, e^{2\pi i n x},$$

where $z_n = (v_n + 1)/2$; and let $d\varrho_N = \frac{1}{N}|f_N|^2 d\lambda$. We check immediately that $\hat{\varrho}_N(p) = \frac{1}{N}\sum_{n=0}^{N-p-1} z_n z_{n+p}$; but $z_n z_{n+p} = 1$ if $v_n = v_{n+p} = 1$, 0 otherwise. The convergence of Fourier coefficients is by definition equivalent to the weak-star topology convergence of measures, hence we have identified ϱ_f as the weak-star topology limit of the ϱ_N. ■

Proposition 5.3.3. *The associated operator U has a spectrum with multiplicity at most four; all the dyadic rationals are eigenvalues; there exist functions the spectral type of which is equivalent to the Lebesgue measure.*

Proof. Because (X, S) is a four-point extension of a system with a simple spectrum, the Hilbert space $\mathcal{L}^2(X, \mu)$ is a limit of $H_n^1 + H_n^2 + H_n^3 + H_n^4$, where the H_n^i are (not necessarily nonempty or orthogonal) cyclic space, and a generalization of Lemma 5.2.10 shows that $\mathcal{L}^2(X, \mu)$ is generated by at most four cyclic spaces. One of them, as for the Morse system, contains the eigenvectors of the dyadic rotation, hence the dyadic rationals are eigenvalues.

We normalize f to get a function which is orthogonal to the constants: it is immediate that $\int f d\mu = 1/2$, hence we take $g = 2f - 1$. We deduce from the previous lemma that ϱ_g is the vague limit of $\varrho'_n = 2^{-n}|g_n|^2\lambda$, where $g_n(x) = \sum_{k=0}^{2^n-1} v_k e^{2\pi i k x}$. If $B_n = w_{n,0} \ldots w_{n,2^n-1}$, we take $h_n(x) = \sum_{k=0}^{2^n-1} w_{n,k} e^{2\pi i k x}$. From the structure of v, we deduce that $g_{n+1}(x) = g_n(x) + e^{2i\pi 2^n x} h_n(x)$ and $h_{n+1}(x) = g_n(x) - e^{2i\pi 2^n x} h_n(x)$, hence $|g_{n+1}^2(x)| + |h_{n+1}^2(x)| = 2(|g_n^2(x)| + |h_n^2(x)|)$, and, going back to $n = 0$, this quantity is equal to 2^{n+1} for all x. So $2^{-n-1}|g_n|^2 \leq 1$, for all n; by dominated convergence, the densities of the ϱ'_n must converge in $\mathcal{L}^2(\mathbb{T}^1, \lambda)$; hence ϱ_g has a density with respect to Lebesgue measure. Furthermore, if $2^{-n}|g_n|^2(x) \to 0$, then $2^{-n-1}|h_n|^2(x) \to 1$, which is incompatible with the recursion formula, so the density of ϱ_g is strictly positive, and ϱ_g is equivalent to Lebesgue measure. ■

In particular, this implies that there exists a function f', of norm one, such that $(U^n f', f') = 0$ if $n \neq 0$; on some (but not all!) functions S is strongly mixing (see Definition 5.5.3 below), and even more. The complete analysis [340] shows that the system is generated by two cyclic spaces with spectral types λ and $\lambda + \delta$, λ being the Lebesgue measure and δ the sum of the atomic measures on the dyadic rationals. By the way, let us recall (see Chap. 2) that the original Rudin-Shapiro sequence, v, was precisely introduced to minimize $\sup_{x \in [0,1]} |\sum_{n=0}^{N-1} \varepsilon_n e^{2\pi i n x}|$, when ε_n is a sequence taking values $+1$ and -1.

For the sequence v_n, there is a bound $(2+\sqrt{2})\sqrt{N}$, and v is the sequence which gives the lowest bound.

5.4 The Fibonacci sequence

5.4.1 Unique ergodicity of primitive substitutions

Let us recall some properties of primitive substitutions; see also Chap. 1.

Definition 5.4.1. *The substitution σ over the alphabet $\mathcal{A}$ is* primitive *if there exists k such that, for every a and b in $\mathcal{A}$, the letter a occurs in $\sigma^k(b)$.*

Let us recall the statement of Proposition 1.2.3: *if σ is primitive, any of its fixed points is a minimal sequence.*

Lemma 5.4.2. *With the notations of Sec. 5.1.5, if σ is primitive, then if $N(e,\sigma^n(a)) = |\sigma^n(a)|_e$ denotes the number of occurrences of the letter e in the word $\sigma^n(a)$, $\frac{N(e,\sigma^n(a))}{|\sigma^n(a)|}$ tends to a positive limit f_e independent of a when n tends to infinity, for every $a \in \mathcal{A}$ and every letter $e \in \mathcal{A}$.*

Proof. Let $\mathbf{l}(V)$ be the vector $(|V|_e, e \in \mathcal{A})$ for any word V, and let $\mathbf{M}_\sigma$ be the *matrix of the substitution σ* (also called *incidence matrix*) defined by $\mathbf{M}_\sigma = ((m_{i,j}))_{i\in\mathcal{A},j\in\mathcal{A}}$, where $m_{i,j} = N(j,\sigma(i))$. For any V we have $\mathbf{l}(\sigma(V)) = \mathbf{M}_\sigma \mathbf{l}(V)$.

Primitivity implies that all coefficients of the matrix $\mathbf{M}_\sigma^k$ are positive, and by *Perron-Frobenius theorem* (see Chap. 1 and [339] for example) the matrix $\mathbf{M}_\sigma$ has a real positive eigenvalue α, which is bigger than 1 as $\mathbf{M}_\sigma$ has integer coefficients, simple, corresponding to a positive eigenvector $\mathbf{u}$, and such that $\alpha > |\lambda|$ for any other eigenvalue λ; we thus deduce that $\frac{\mathbf{M}_\sigma^n \mathbf{l}(a)}{\alpha^n}$ converges to a positive multiple $\mathbf{u^a}$ of $\mathbf{u}$ for every $a \in \mathcal{A}$.

Thus, $|\sigma^n(a)|$ is the scalar product of $\mathbf{l}(\sigma^n(a))$ with the vector $\mathbf{e} = (1,\dots,1)$, hence $\frac{N(\sigma^n(a))}{|\sigma^n(a)|} \to \frac{\mathbf{u^a}}{<\mathbf{u^a},\mathbf{e}>}$, which is the multiple of $\mathbf{u}$ whose sum of coordinates equals 1, and hence is independent of a. ■

Lemma 5.4.3. *If σ is primitive, then $\frac{N(W,\sigma^n(a))}{|\sigma^n(a)|}$ tends to a positive limit f_W independent of a when n tends to infinity, for every $a \in \mathcal{A}$ and every word W.*

Proof. We write each $V = v_1 \dots v_l$ as a one-letter word for a new primitive substitution ζ_l over the alphabet $\mathcal{A}_l$ whose letters are the words of length l occurring in u: if $W = w_1 \dots w_l$, $\sigma(W) = w'_1 \dots w'_m$ and $q = |\sigma(w_1)|$, we define $\zeta_l(W) = (w'_1 \dots w'_l)(w'_2 \dots w'_{l+1}) \dots (w'_q \dots w'_{q+l-1})$ which is well defined as $q+l-1 \le m$. We check that $U_l = (u_0 \dots u_{l-1})(u_1 \dots u_l) \dots (u_n \dots u_{n+l-1}) \dots$ is a fixed point for ζ_l.

The substitution ζ_l is primitive: by definition of u and finiteness of $\mathcal{A}_l$, there exists a positive integer p such that every W in $\mathcal{A}_l$ is a factor of $\sigma^p(a)$, and, by primitivity of σ, there exists a positive integer m such for any letter b, a occurs in $\sigma^m b$; then if V and W are in $\mathcal{A}_l$, W is a letter of $\zeta_l^{m+p}(V)$. Then we check that $|\sigma^n(v_1)| = |\zeta_l^n(V)|$ and that $N(W, \zeta_l^n(V))$, which tends to infinity, is close to $N(W, \sigma^n(v_1))$ when n is large. Hence the previous lemma applies. ∎

Proposition 5.4.4. *If u is a fixed point of a primitive substitution, then the system (X_u, T) is uniquely ergodic.*

Proof. We can prove again Lemma 5.1.20 in the same way, except that $\sigma^p(u_i)$ is no longer equal to 2^p; but, by Perron–Frobenius, if p is large enough,

$$c\,\alpha^p < \inf_{a\in\mathcal{A}} |\sigma^p(a)| < \sup_{a\in\mathcal{A}} |\sigma^p(a)| < d\,\alpha^p,$$

which is enough to reach the same conclusion. Then we apply Proposition 5.1.21. ∎

Proposition 5.4.5. *If u is a non-periodic fixed point of a primitive substitution, then the unique invariant probability measure of the system (X_u, T) is non-atomic.*

Proof. If $\mu(\{w\}) = \nu > 0$, then $\mu([W_n]) \geq \nu$ for all n, where $W_n = w_0 \dots w_n$. Hence, because $\mu([W_n]) = f_{W_n}$, for every n there must exist $j_n > i_n$ such that $j_n - i_n < \frac{1}{\nu}$ and W_n occurs in u at positions i_n and j_n; if $n > \frac{2}{\nu}$, this implies that $w_0 \dots w_{[\frac{n}{2}]} = w_{j_n - i_n} \dots w_{[\frac{n}{2}]+j_n-i_n}$; as $j_n - i_n$ takes only a finite number of values, there exists k such that $w_0 \dots w_{[\frac{n}{2}]} = w_k \dots w_{[\frac{n}{2}]+k}$ for infinitely many n, hence for every n. Hence w is periodic, and so is u which has the same language. ∎

The following result is proved in [319], using the results in [156], or [120] in the particular case of constant length:

Proposition 5.4.6. *If σ is primitive, then its fixed points have an at most linear complexity.*

Proof. For every n, we find p such that

$$\inf_{a\in\mathcal{A}} |\sigma^{p-1}(a)| \leq n \leq \inf_{a\in\mathcal{A}} |\sigma^p(a)|.$$

Every word of length n has to be included either in a $\sigma^p(a)$ or in a $\sigma^p(ab)$; for fixed a and b, there are at most $|\sigma^p(ab)|$ such words, according to the position of the first letter; and there are at most $K = (\mathrm{Card}\mathcal{A})^2$ possible choices for a and b. Hence for n large enough

$$p_u(n) < 2Kd\,\alpha^p < 2K\,\frac{d}{c}\,\alpha\, n$$

with c and d as in Proposition 5.4.4. ∎

5.4.2 A trajectory of a rotation

We are interested here in a problem which has, at first sight, no connection with substitutions: given an irrational rotation of the torus $\mathbb{T}^1$, of angle α, and a particular partition (corresponding to the point of discontinuity of the rotation when it is considered as acting on the fundamental domain $[0, 1[$), we want to find explicitly the name of a point x under that partition. This problem is solved for any α and x in [252], using formal power series, and in [41], using the dynamical notion of *Rauzy induction* ([348] and see also Chap. 6). We shall make here this computation for a particular value of α and x. About the general case, and its link with the continued fraction approximation of α, more information can be found in Chap. 6.

We will define first what is the induction:

Definition 5.4.7. *For a transformation T on X, a set $A \subset X$, and a point $x \in A$, we call* first return time *of x in A and denote by $n_A(x)$ the (possibly infinite) smallest integer $m > 0$ such that $T^m x \in A$. The* induced map *of T on A is the map $T^{n_A(x)}x$ defined on $A \cap \{x; n_A(x) < +\infty\}$.*

Each time we use these notions in this course, $n_A(x)$ will be finite and the induced map defined everywhere.

We look at the irrational rotation of angle $\alpha = \frac{1}{2}(\sqrt{5} - 1)$, defined on the torus $\mathbb{T}^1$ by $\overline{R}x = x + \alpha \bmod 1$, and, after an immediate semi-topological conjugacy, on the interval $[0, 1[$ by

$$Rx = x + \alpha \quad \text{if} \quad x \in [0, 1 - \alpha[,$$
$$Rx = x + \alpha - 1 \quad \text{if} \quad x \in [1 - \alpha, 1[.$$

Let P_0 be the set $[0, 1 - \alpha[$ and P_1 the set $[1 - \alpha, 1[$; let v be the P-name of the point α under R (see Definition 5.2.7 above). We shall now prove the following result:

Proposition 5.4.8. *The sequence v is the image by $S\tau_0$ of the fixed point of the substitution τ, where S is the shift, $\tau_0(0) = 10, \tau_0(1) = 0$, $\tau(0) = 001, \tau(1) = 01$.*

Proof. We first write $v = Sv''$, where $v''_n = 0$ if $R^n 0 \in [1 - \alpha, 1[$, $v''_n = 1$ if $R^n 0 \in [0, 1 - \alpha[$, S being the shift. We make now an induction on a particular interval, which is the so-called Rauzy induction; let I be the interval $[0, \alpha[$ and R' the induced map of R on I, $n(x)$ being the first return time of x in I; we compute easily that:

$$\text{if } x \in [0, 1 - \alpha[, \ n(x) = 2, \quad R'x = x + 2\alpha - 1$$
$$\text{if } x \in [1 - \alpha, \alpha[, \ n(x) = 1, \quad R'x = x + \alpha - 1.$$

Let v' be the sequence defined by $v'_n = 0$ if $R'^n 0 \in [0, 1 - \alpha[$, $v'_n = 1$ if $R'^n 0 \in [1 - \alpha, \alpha[$.

We compare v' with v''. If $n(x)$ is the first return time of x in I, $n_2(x) = n(x) + n(R'x)$ is the second return time, $n_i(x) = n_{i-1}(x) + n(R'^{i-1}x)$ is the i-th return time of x in I. Suppose $v'_i = 0$; then $R'^i 0$ is in $[0, 1-\alpha[$; hence $R^{n_i(0)}0$, which is the same point, is in $[0, 1-\alpha[$, $R^{n_i(0)+1}0$ is in $[\alpha, 1[\subset [1-\alpha, 1[$, and $R^{n_i(0)+2}$ is again in $[0, \alpha[$, hence must be $R'^{i+1}0$; hence we can write that $n_{i+1}(0) - n_i(0) = 2$, $v''_{n_i(0)} = 1$, and $v''_{n_i(0)+1} = 0$. Suppose on the contrary that $v'_i = 1$; then $R'^i 0$ is in $[1-\alpha, \alpha[$, hence $R^{n_i(0)}0$ is in $[1-\alpha, \alpha[\subset [1-\alpha, 1[$ and $R^{n_i(0)+1}$ is still in $[0, \alpha[$. Hence we can write that $n_{i+1}(0) - n_i(0) = 1$ and $v''_{n_i(0)} = 0$. The previous analysis allows us, from the knowledge of v', to determine v'' between two consecutive $n_i(0)$: we have proved that $v'' = \tau_0(v')$, where τ_0 is the substitution $\tau_0(0) = 10, \tau_0(1) = 0$. Hence the initial v is $S\tau_0(v')$.

We do not change v' if we make a homothety of ratio $1/\alpha$; R' becomes a rotation of $2 - 1/\alpha = 1 - \alpha$ on the interval $[0, 1[$. And $v'_n = 1$ if $R'^n 0 \in [\alpha, 1[$, $v'_n = 0$ if $R'^n 0 \in [0, \alpha[$.

Let Q be the induced transformation of R' on $[0, \alpha[$, $m(x)$ being the first return time; we have:

$$\text{if } x \in x \in [0, 2\alpha - 1[,\ m(x) = 1, \quad Qx = x + 1 - \alpha$$
$$\text{if } x \in [2\alpha - 1, \alpha[,\ m(x) = 2, \quad Qx = x + 1 - 2\alpha.$$

Let w be the sequence defined by $w_n = 0$ if $Q^n 0 \in [0, 2\alpha - 1[$, $w_n = 1$ if $Q^n 0 \in [2\alpha - 1, \alpha[$. By the same method as above, we check that $v' = \tau_1(w)$, where $\tau_1(0) = 0, \tau_1(1) = 01$.

We normalize again, dividing by α. We see that Q becomes the rotation R itself, and that $w_n = 1$ if $R^n 0 \in [1-\alpha, 1[$, $w_n = 0$ if $R^n 0 \in [0, 1-\alpha[$.

By a new Rauzy induction, we come back to R' and v', and show that $w = \tau_2(v')$, where $\tau_2(0) = 01, \tau_2(1) = 1$.

Hence $v' = \tau(v')$, where $\tau(0) = 001, \tau(1) = 01$, i.e. $\tau = \tau_1\tau_2$; and $v = S\tau_0(v')$, which proves our proposition. ∎

5.4.3 The Fibonacci substitution: geometric representation

In this section, σ is the *Fibonacci substitution*

$$0 \mapsto 01 \quad 1 \mapsto 0.$$

Its unique fixed point u is the *Fibonacci sequence* 0100101... Note that the length of $\sigma^n(0)$ is the n-th *Fibonacci number* f_n, given by the recursion formulas $f_0 = 1$, $f_1 = 2$, $f_{n+1} = f_n + f_{n-1}$.

The substitution σ is primitive, hence the system (X_u, S) is uniquely ergodic. The dominant eigenvalue of the matrix is the *golden ratio number* $\alpha_0 = \frac{1+\sqrt{5}}{2} = 1 + \alpha$.

Let R and $\overline{R}$ be the irrational rotation of angle α defined respectively on the interval $[0,1[$ and on the torus $\mathbb{T}^1$ as in Sec. 5.4.2. The irrational rotation $\overline{R}$ is a translation on a compact group; hence, by the same proof as in Sec. 5.2.3, it is uniquely ergodic; as the invariant measure, which is the Lebesgue measure, gives a strictly positive measure to every open set, unique ergodicity implies by a standard argument [339] that $\overline{R}$ is minimal (which is also a consequence of Kronecker's theorem). We check that these properties are shared by R.

Proposition 5.4.9. *We have* $u_n = 0$ *whenever* $R^n\alpha \in [1-\alpha, 1[$, $u_n = 1$ *whenever* $R^n\alpha \in [0, 1-\alpha[$.

Proof. We just have to identify the sequence u with the sequence v in the previous proposition. But $\tau_0\tau(0) = 10100$, $\tau_0\tau(1) = 100$; we check that $\tau_0\tau^n(0)$ is made with a 1 followed by $\sigma^{2n+1}(0)$ (minus its last letter), and that $\tau_0\tau^n(1)$ is made by the f_{2n+1} last letters of $\tau_0\tau^n(1)$; this is achieved by using the reverse decomposition of u: $\sigma^n(0) = \sigma^{n-1}(0)\sigma^{n-2}(0) = \sigma^{n-2}(0)(\sigma^{n-3}(0)\sigma^{n-2}(0))$; hence v begins with $\sigma^{2n+1}(0)$, and $v = u$. ■

Corollary 5.4.10. *The complexity function of the Fibonacci sequence is*

$$p_u(n) = n + 1 \textit{ for every } n.$$

Proof. A word $w_0 \dots w_{n-1}$ occurs in u if and only if $\cap_{i=0}^{n-1} R^{-i} P_{w_i} \neq \emptyset$ (by minimality). The sets $\cap_{i=0}^{n-1} R^{-i} P_{w_i}$, when w ranges over $\mathcal{L}_n(u)$, are intervals, and the partition of the interval $[0,1[$ by them is the partition of the interval by the points $R^{-i}0, 1 \leq i \leq n$; hence there are $n+1$ nonempty intervals. ■

As an exercise, we propose a direct and combinatorial proof of this result.

Exercise 5.4.11. 1. Prove that every factor W of the Fibonacci sequence can be uniquely written as follows:

$$W = A\sigma(V)B,$$

where V is a factor of the Fibonacci sequence, $A \in \{\varepsilon, 1\}$, and $B = 0$, if the last letter of W is 0, and $B = \varepsilon$, otherwise.

2. Prove that if W is a left special factor distinct from the empty word, then there exists a unique left special factor V such that $W = \sigma(V)B$, where $B = 0$, if the last letter of W is 0, and $B = \varepsilon$, otherwise. Deduce the general form of the left special factors.
3. Prove that the Fibonacci sequence is not ultimately periodic.
4. Prove that the complexity function of the Fibonacci sequence is $p_u(n) = n + 1$ for every n.

We say that the Fibonacci sequence is a *Sturmian sequence*; it has the lowest possible complexity for a nonperiodic sequence.

Proposition 5.4.12. *The system* (X_u, S, μ) *associated with the Fibonacci sequence is a coding of the rotation* R *on the interval, or* $\overline{R}$ *on the torus, preserving the Lebesgue measure* λ.

Proof. Let P be the partition $(P_1 = [0, 1-\alpha[, P_0 = [1-\alpha, 1[)$ and $P(x)$ be the P-name of x. We have $P(\alpha) = u, P(R^n\alpha) = S^n u$; we check that when $R^{n_k}\alpha \to x$, we have $P(R^{n_k}\alpha) \to P(x)$, for the product topology on $\{0,1\}^{\mathbb{N}}$, except if x is in the orbit of α (because P_0 and P_1 are semi-closed); and $P(x) = \lim S^{n_k}u$ is then a point of X_u; as the set $\{n\alpha, n \in \mathbb{N}\}$ is dense in $[0,1[$, we see that $P([0,1[/D) \subset X$, where D is a countable set.

Conversely, two points with the same P-name are not separated by arbitrarily small intervals, hence are identical; and, after deleting a countable number of points, every point in X_u, written under the form $\lim S^{n_k}u$, may be written, after taking a subsequence such that $R^{n'_k}\alpha$ converges, as $P(\lim R^{n'_k}\alpha)$; hence $P(x)$ is a bicontinuous bijection, except on a countable set, and $PR = SP$, while P sends the only invariant measure λ for R to the only invariant measure for S.

As for $\overline{R}$, it is semi-topologically conjugate to R. ■

Remark 5.4.13. The systems (X_u, S), $([0,1[, R)$ and $(\mathbb{T}^1, \overline{R})$ are not mutually topologically conjugate: $\overline{R}$ and S are continuous while R is not; and between S and $\overline{R}$, the topology of X_u is generated by clopen sets while the topology of $\mathbb{T}^1$ is not.

The system $(\mathbb{T}^1, \overline{R})$ has a discrete spectrum and the eigenvalues are all the $e^{2i\pi n\alpha}$, $n \in \mathbb{Z}$ (see Chap. 1, Lemma 1.6.2). Also, $\overline{R}$ has rank one (see for example [167]), though we do not know any explicit sequence of stacks generating the system (of course, the Fibonacci substitution can be used to produce a sequence of stacks, but this would only give R a rank at most 2).

5.5 The Chacon sequence

5.5.1 Elementary properties

The *Chacon substitution*

$$0 \mapsto 0012 \qquad 1 \mapsto 12 \qquad 2 \mapsto 012$$

defines a primitive substitution, hence the system (X_0, S) associated with the fixed point v beginning with 0 is uniquely ergodic.

This system is a new form [165] of an historical system, which we shall study in its traditional form: we remark that v begins with the word w_n defined by

- $w_0 = 0$,
- $w_{n+1} = w_n w_n 1 w'_n$,
- w'_n is deduced from w_n by changing the first letter from 0 to 2.

Let now u be the sequence beginning with b_n, where

$$b_0 = 0, \qquad b_{n+1} = b_n b_n 1 b_n, \quad \forall n \in \mathbb{N},$$

and (X, S) the dynamical system associated with u; it is called *Chacon's map* ([114]). Note that u is also a fixed point of a substitution $(0 \mapsto 0010, 1 \mapsto 1)$ but a non-primitive one.

We send X_0 onto X by replacing every 2 by 0 in every point; to come back from a point in X to its pre-image in X_0, we replace 0 by 2 whenever there is a 1 just before. So the two systems are topologically conjugate, and in particular (X, S) also is uniquely ergodic. Let μ be the invariant measure given by the frequencies of words in u.

The system (X, S) is a rank one system, with the following construction, illustrated in Fig. 5.4:

- we cut the interval $[0, 1[$ into two intervals P_0 and P_1, of respective lengths 2/3 and 1/3; we take $F_0 = P_0$;
- we cut F_0 into three intervals of equal length, and send by a translation defining S the first one onto the second one, the second one onto a sub-interval a_1 of P_1, beginning at the left end of P_1 and of corresponding length, namely 2/9, and this sub-interval of P_1 is sent onto the third piece of P_0;
- at stage n, we cut the n-stack vertically into three equal *columns*, and send (by a translation) the top of the first one onto the bottom of the second one, the top of the second one onto a sub-interval a_{n+1} of P_1 of corresponding length, beginning at the left end of the yet unused part of P_1, and this sub-interval onto the bottom of the first column.

We check that by associating with a point its P-name, we send F_n, the basis of the n-stack, onto the cylinder $[u_0, \ldots, u_{h_n - 1}]$, where $h_0 = 1$, $h_{n+1} = 3h_n + 1$, which gives

$$h_n = \frac{3^{n+1} - 1}{2}.$$

Note that the parameters we have chosen ensure that $\cup_1^\infty a_n = P_1$, and that the Lebesgue measure of the n-th stack $\cup_{i=0}^{h_n - 1} S^i F_n$ tends to 1.

The word $b_n = u_0 \ldots u_{h_n - 1}$ is called the *n-block*; every point in X is a concatenation of n-blocks (that is, copies of b_n) and letters 1; the 1's are called *spacers*, or n-spacers if they are between two n-blocks.

Lemma 5.5.1. *Chacon's system is* weakly mixing *: there are no eigenfunctions in $\mathcal{L}^2(X, \mu)$ except the constants, which are simple.*

Proof. Let f be an eigenfunction for the eigenvalue λ, of norm 1; we approximate f in $\mathcal{L}^2(X, \mu)$ by a sequence of functions f_n of norm 1 constant on the levels of the n-stack. Let C_1 and C_2 be the first and second column of the n-stack, their measure is close to 1/3 and bigger than 1/4, and

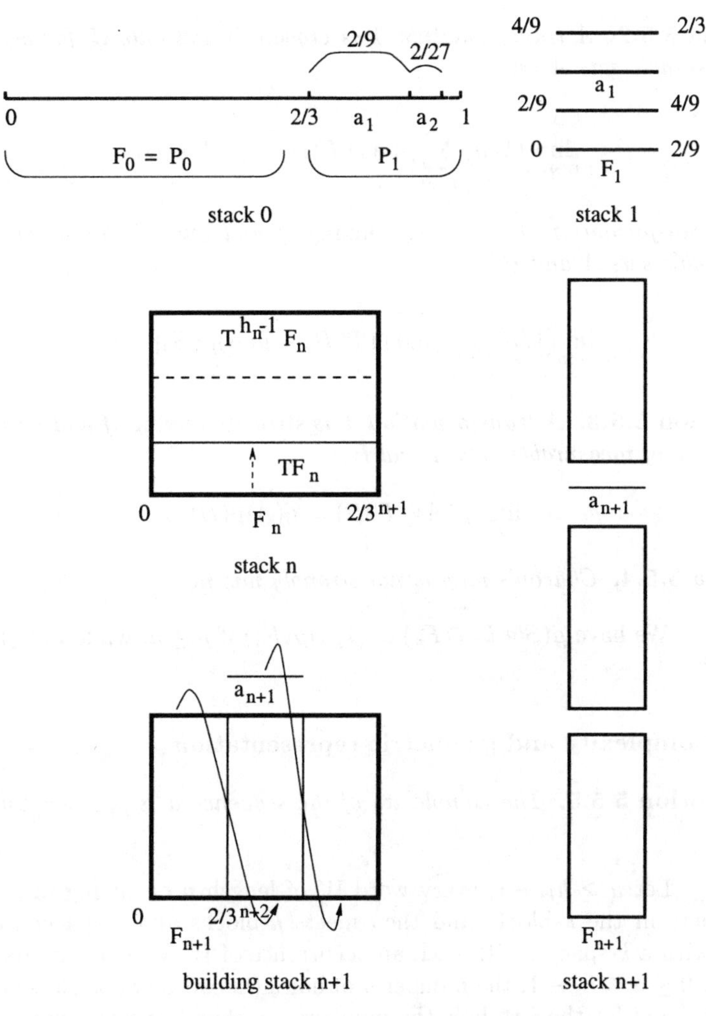

Fig. 5.4. Chacon's map is a rank one system.

$$\int_{C_1} |f_n(S^{h_n}) - \lambda^{h_n} f_n|^2 \leq \int_{C_1} |f_n(S^{h_n}) - f(S^{h_n})|^2 + \int_{C_1} |\lambda^{h_n} f - \lambda^{h_n} f_n|^2 < 2\varepsilon,$$

if n is large enough. But, as f_n is constant on the levels of the n-stack, $S^{h_n} f_n = f_n$ on C_1 and so $|\lambda^{h_n} - 1|^2 < 8\varepsilon$ if n is large enough.

In the same way $S^{h_n+1} f_n = f_n$ on C_2 implies that $|\lambda^{h_n+1} - 1|^2 < 8\varepsilon$ if n is large enough. We conclude that $\lambda = 1$, and it is a simple eigenvalue because unique ergodicity implies ergodicity. ■

Weak mixing and ergodicity can be expressed also as convergences of Cesaro averages; the following lemma is proved for example in [121] or [339]:

Lemma 5.5.2. *A transformation T is ergodic if and only if, for every pair of measurable sets A and B,*

$$\lim_{n\to\infty}(1/n)\sum_{k=0}^{n-1}\mu(A\cap T^kB)=\mu(A)\mu(B).$$

A transformation T is weakly mixing *if and only if, for every pair of measurable sets A and B*

$$\lim_{n\to\infty}(1/n)\sum_{k=0}^{n-1}|\mu(A\cap T^kB)-\mu(A)\mu(B)|=0.$$

Definition 5.5.3. *A transformation T is* strongly mixing *if and only if, for every pair of measurable sets A and B,*

$$\lim_{n\to\infty}\mu(A\cap T^nB)=\mu(A)\mu(B).$$

Lemma 5.5.4. *Chacon's map is not strongly mixing.*

Proof. We have $\mu(S^{h_p}F_n\cap F_n)>(1/4)\mu(F_n)$ if $p\geq n$, while $\mu(F_n)<1/4$ if $n\geq 2$. ■

5.5.2 Complexity and geometric representation

Proposition 5.5.5. *The complexity of the sequence u is $p_u(n)=2n-1$ if $n\geq 2$.*

Proof. Let $n\geq h_k+1$; every word W of length n occurring in u begins somewhere in the k-block, and then meets k-blocks and k-spacers, or else begins with a k-spacer; with such an occurrence of W we associate its initial position $0\leq a\leq h_k-1$, the number d of holes (between two k-blocks) which it crosses, and for the i-th hole the number $s_i=0$ or 1 of k-spacers inside it, or else, if the occurrence of W begins with a k-spacer, we write that $a=-1$ and there is a hole at the beginning, with $s_1=1$ (in the same way, if the occurrence ends with a spacer, we write that there is a hole at the end). We can rebuild W from n, d, a, $s_1,\ldots,s_d$; these parameters, which we call the *k-configuration*, depend a priori on the particular occurrence and not only on the word W; we define the hypothesis $H(k,n)$ by: *each word of length n occurring in u has only one k-configuration.*

For $n=h_k+1$, for each initial position between 0 and h_k-1 there are two possible k-configurations, according to whether after the k-block we see another k-block or a spacer; there is also a configuration beginning with a k-spacer. This makes $2h_k+1$ possible configurations. Hence the hypothesis $H(k,h_k+1)$ is equivalent to $p_u(h_k+1)=2h_k+1$.

Let $h_k + 1 \leq n \leq h_{k+1} = 3h_k + 1$, and suppose $H(k,n)$ is satisfied; a word of length n corresponds to one configuration, with $1 \leq d \leq 3$, we call $(s_1, \ldots, s_d)$ its *spacer configuration*; we check that every spacer configuration is allowed, with zero or one spacer in each hole, except that there cannot be three consecutive holes without a spacer nor three consecutive holes with one spacer.

We say a spacer configuration with c holes is *bi-extendable* if, by adding at its end another hole with either zero or one spacer, we get two allowed spacer configurations with $c+1$ holes. For $c = 1, 2$, there are two bi-extendable spacer configurations with c holes.

Because of $H(k,n)$, a word W of length n is bi-extendable (that is: has two right extensions $W0$ and $W1$ occurring in u) if and only if it ends with a k-block, and its spacer configuration is bi-extendable. So, if n is such that all the words of length n ending with a k-block have the same number of holes, this number is 1 or 2 (not 3) and there are two bi-extendable words of length n. Otherwise, there remain the case $n = 2h_k + 1$, where the only words ending with a k-block have two holes if they do not contain any spacer and one hole if they contain one spacer, and there are still two bi-extendable words, and the case $n = 3h_k + 1$ where $1b_kb_kb_k$ has 3 holes and is not bi-extendable, and $b_k1b_kb_k$ and $b_kb_k1b_k$ have 2 holes and are bi-extendable. The two right extensions of a bi-extendable word are two different words by definition; hence $H(k,n)$ implies simultaneously $H(k,n+1)$ and $p_u(n+1) - p_u(n) = 2$ as long as $n \leq h_{k+1}$. And so, starting from $H(k, h_k+1)$, we deduce that $p_u(h_{k+1}+1) = 2h_{k+1}+1$, which implies $H(k+1, h_{k+1}+1)$ and we may continue the recursion; we check that for $n = 2 = h_0 + 1$, $p_u(n) = 2n - 1$. ■

We check that the complexity of the sequence v is $2n+1$ (for each n, two words of length n of u correspond each to two words of v, and every other one corresponds to one word); this provides an example of a complexity function changed by a topological conjugacy.

Proposition 5.5.6. *The induced map of Chacon's transformation on the cylinder $w_0 = 0$ is semi-topologically conjugate to the triadic rotation.*

Proof. We consider the substitution τ defined by $a \mapsto aab$, $b \mapsto bab$, and u', its fixed point beginning with a;

$$u' = aabaabbabaabaabbabbabaabbab\ldots,$$

is deduced from u by $0 \mapsto a$, $10 \mapsto b$. Let T be the induced map (see Definition 5.4.7 above) of S on the cylinder $[0] \subset X$; we see that $Tw = Sw$ if $w_0 = w_1 = 0$, $Tw = S^2w$ if $w_0 = 0, w_1 = 1$. Let (Y,T) denote this dynamical system. We call Q_a and Q_b these cylinders, and we code the trajectory of a point w' under T by $w'_n = a$ if $T^nw' \in Q_a$, and $w'_n = b$ if $T^nw' \in Q_b$. We check that if w is a point in the cylinder $\{w_0 = 0\}$ and the sequence $v(w)$ its Q-name (under T), we deduce Sw from $v(w)$ by $a \mapsto 0$, $b \mapsto 10$. By the same reasoning as in Proposition 5.4.12, we check that the dynamical system associated with τ is topologically conjugate with the system (Y, S).

The structure of the system associated with τ can be analyzed in the same way as for the Morse system: we get two stacks, with bases $F_{n,0}$ and $F_{n,1}$, and, if we identify corresponding levels of the two stacks, we get the sequence of stacks with bases F_n associated with the triadic rotation (which is a straightforward generalization of the dyadic one). But here $\tau^n(0)$ and $\tau^n(1)$ differ only by the first letter; the reasoning of Proposition 5.2.6 shows then that, except on a countable set, a point is determined by the sequence of levels of F_n to which it belongs (the knowledge that w is in the 0 or 1 stack does not provide extra information). Hence the system associated with τ is a coding of the triadic rotation. ∎

Thus we have a geometric representation of Chacon's map:

Definition 5.5.7. *A system (X, T) is a* Rokhlin-Kakutani exduction *of the system (X', T') if there exists a finite partition $(X'_1, \ldots, X'_p)$ of X' and positive integers $n_1, \ldots, n_p$ such that $X = \cup_{i=1}^{p} (X'_i \times \{0, \ldots, n_i - 1\})$, and for $w \in X'_i$ $T(w, j) = (w, j + 1)$ if $0 \le j < n_i - 1$, $T(w, n_i - 1) = (T'w, 0)$.*

We have proved that, up to topological conjugacy, Chacon's map is a Rokhlin-Kakutani exduction of the triadic rotation. The set X'_1 is the cylinder $[a]$ in the system associated with τ, $p = 2$, $n_1 = 1$, $n_2 = 2$.

However, being an exduction is a much weaker property than being a two (or more)-point extension, as many properties are lost; for example, the triadic rotation has a discrete spectrum, while Chacon's transformation has no eigenvalue. If the only condition we ask from the sets X'_i is to be measurable, the theory of *Kakutani equivalence* ([316]) says we can get, from the triadic rotation, many different systems, such as (up to measure-theoretic isomorphism) any ergodic substitution or even the *horocycle flow* (see [345]); the interest of the above construction is that the exduction is on an explicit and, after a coding, clopen set.

The study of Morse, Rudin-Shapiro and τ show us particular cases of the following theorem [134], which we state without proof. For more details, see Chap. 7.

Proposition 5.5.8. *If ζ is a primitive substitution of constant length q, that is, if $\forall a \in \mathcal{A}$, $|\zeta(a)| = q$, then the associated system has a discrete spectrum if and only if there exist k and i such that the i-th letter of $\zeta^k(a)$ is the same letter for every a in $\mathcal{A}$.*

The Fibonacci substitution, on the other side, is a *Pisot substitution*, as the dominant eigenvalue is a Pisot number (it is strictly larger than 1 and all the other eigenvalues have nonzero modulus strictly smaller than 1); the existence of an analogous theorem for this family of substitutions is an open question, that we discuss in Chap. 7. The general study of eigenvalues for any substitution can be found in [173], see also Chap. 7.

5.5.3 Joinings

Our aim is now to find every *joining* of S with itself.

Definition 5.5.9. *For a system (X,T,μ), a* self-joining *is an ergodic probability measure on $X \times X$, invariant by $T \times T$, and whose marginals on X are μ*

Even if T has only one invariant probability measure, μ, $T \times T$ preserves already the measure $\mu \times \mu\,(A \times B) = \mu(A)\mu(B)$ (it is ergodic whenever T is weakly mixing - this is a nontrivial result), but also $\nu(A \times B) = \mu(A \cap B)$ and others.

Definition 5.5.10. *A system (X,T,μ) has* minimal self-joinings of order two *if its joinings with itself are the product measure $\mu \times \mu$ and the* diagonal measures *$\nu(A \times B) = \mu(A \cap S^i B)$ for an integer i.*

The joinings of Chacon's map have been studied in [141], and we shall now expose their beautiful results and proofs; we need first to recall a classical ergodic result:

Definition 5.5.11. *For a transformation S, defined as the shift over a space of sequences, and an invariant measure ν, a point w is* generic *if for every cylinder A, $(1/N)\sum_{n=1}^{N} 1_A(S^n w) \to \nu(A)$ if $N \to +\infty$.*

Lemma 5.5.12. *If (T,ν) is ergodic, ν-almost every point is generic.*

Proof. Apply the ergodic theorem (Theorem 5.1.28) to the indicator function of each cylinder. Thus for every cylinder A, there exists a set X_A of measure one such that

$$\frac{1}{N}\sum_{1}^{N} 1_A(T^n w) \to \nu(A)$$

for every $w \in X_A$.

Hence every point in $\bigcap_{A \text{ cylinder}} X_A$ is generic, and this intersection has measure one as there are only countably many cylinders. ■

Note that in the same way Propositions 5.4.4 and 5.1.25 imply that if σ is a primitive substitution, every point of the associated X_u is generic for the invariant probability measure.

We shall use Lemma 5.5.12 for the Cartesian powers of Chacon's map; in fact, there is a deep result ([140]) showing that for $S \times S$, all but a countable set of points are generic.

In Chacon's system (X,S), let $R(k,n)$ be the set of w such that w_0 is in the k-th n-block inside its $n+1$-block, $k = 1,2,3$.

Lemma 5.5.13. *The sets $R(k,n)$ and $R(k',n+1)$ are, for fixed k and k', independent sets for all n. Almost every point w of X is in $R(k,n) \cap R(k',n+1)$ for n in a set of integers of density at least 1/10. Every point w with these properties is called* admissible.

Proof. The first assertion is immediate. The second comes from the strong law of large numbers applied to the variables

$$X_n(k,k') = 1_{R(k,2n)} 1_{R(k',2n+1)} / \mu(\cup_{i=0}^{2n-1} S^i F_{2n}) \mu(\cup_{i=0}^{2n} S^i F_{2n+1}),$$

which are independent and equidistributed, and the same after exchanging $2n$ and $2n+1$. ■

Lemma 5.5.14. *If w and w' are admissible points belonging to different orbits, then, for infinitely many values of n, w_0 and w'_0 are in different n-blocks inside their $n+1$-block, and, if $w_{-a} \dots w_b$ and $w'_{-c} \dots w'_d$ are these n-blocks, their* overlap length *$(a \wedge c) + (b \wedge d)$ is at least $h_n/10$.*

Proof. We take m such that w_0 is in the second m-block. If, for every $n > m$, w_0 and w'_0 are in the same n-block, they are always in the same column of the n-stack, and their difference of level $i(w,w')$ is constant: then w and $S^i w'$ are not separated by the partition into levels of the stacks and w and w' are on the same orbit. We take the first $n > m$ such that w_0 and w'_0 are in different n-blocks. Then , either w_0 is in the second $n-1$-block (if $n = m+1$) or w_0 and w'_0 are in the same $n-1$-block, which guarantees an overlap length of at least h_{n-1} to the left or to the right of (w_0, w'_0). ■

Lemma 5.5.15. *If (Y, T, ϱ) is an ergodic transformation, ν an ergodic measure on $Y \times Y$, with marginals ϱ on each copy of Y, invariant under $T \times I$, where I is the identity, then $\nu = \varrho \times \varrho$.*

Proof. Let $B \subset Y$ measurable, and $P_B(x)$ a *conditional probability* of $Y \times B$ relatively to the product of the Borel σ-algebra Y by the σ-algebra $\{Y, \emptyset\}$). We have $\nu(A \times B) = \int_A P_B(x) d\varrho(x)$. But $\nu(TA \times B) = \nu(A \times B)$, hence $\int_A P_B(x) d\varrho(x) = \int_{TA} P_B(x) d\varrho(x) = \int_A P_B(T^{-1}x) d\varrho(x)$,and, by unicity of the conditional expectation, P_B is T-invariant, and hence constant by ergodicity. Its value can only be $\varrho(B)$ and hence $\varrho(A \times B) = \varrho(A)\varrho(B)$. ■

Proposition 5.5.16. *Chacon's system (X, S, μ) has minimal self-joinings of order 2.*

Proof. Let ν be a joining; we choose a point (w, w') generic for ν, such that w and w' are admissible (it is possible because the marginals are μ). If w and w' are on the same orbit under S, we check that ν is diagonal. Henceforth we suppose w and w' are not on the same orbit.

Let $P(k) = (P_1(k), \dots, P_{h_k}(k), P_s(k))$ be the partition into levels of the k-stack and the complement of the whole k-stack. We shall show that for all k , and all $1 < i \leq h_k, 1 \leq j \leq h_k$, $\nu(P_i \times P_j) = \nu(P_{i-1} \times P_j)$. As unions of these atoms approximate every measurable set, we shall deduce that ν is $S \times I$-invariant, and so is the product measure.

We fix a k, $P = P(k)$, and an ε.

We choose n such that w_0 and w'_0 are in different n-blocks, with overlap on at least $h_n/10$ indices. If n is large enough, genericity ensures that on every segment $(w,w')_0, \ldots, (w,w')_l$ or $(w,w')_{-l}, \ldots, (w,w')_0$, where $l > \varepsilon h_n/1000$, the pairs (i,j) appear with a frequency $\varepsilon/100$-close to $\nu(P_i \times P_j)$. We deduce that these pairs appear with a frequency $\varepsilon/50$-close to $\nu(P_i \times P_j)$ in every segment of (w,w') of length at least $h_n/10$ containing the origin.

Suppose for example that w is in the second and w' in the first n-block, and that the overlap between the n-blocks of w and w' goes from $-q$ to r, with $q + r + 1 \geq h_n/10$. Let $Q_{i,j}$ be the density of l in $(-q,r)$ for which $S^l w \in P_i, S^l w' \in P_j$; we have $|Q_{i,j} - \nu(P_i \times P_j)| < \varepsilon/50$. Let $R_{i,j}$ be the density of l in $(-q + h_n, r + h_n)$ for which $S^l w \in P_i, S^l w' \in P_j$: we have $|R_{i,j} - \nu(P_i \times P_j)| < \varepsilon/2$, for otherwise there would be a proportion bigger than $\varepsilon/(2 \times 11) - \varepsilon/50$ of errors on the segment $(-q, r + h_n)$. But for $l \in (-q,r)$, because of the position of w and w' in their $n+1$-block, if $S^l w \in P_i$ and $S^l w' \in P_j$, then $S^{l+h_n} w \in P_{i-1}$ and $S^{l+h_n} w' \in P_j$. Hence, if $i \neq 1$, $R_{i-1,j} = Q_{i,j}$ and $|\nu(P_i \times P_j) - \nu(P_{i-1} \times P_j)| < \varepsilon$. We conclude in the same way for other positions of w and w', possibly replacing the translation of h_n by $-h_n$ or $2h_n$. ∎

It is worth mentioning that, while the above proof uses the presence of isolated spacers between n-blocks, a similar property, called the *R-property*, was used in [345] to compute the joinings of the *horocycle flows*, and was the basis of the famous papers of Ratner which culminate in the proof of Ragunathan's conjecture [346].

The interest of minimal self-joinings, as well as the notion itself, appear in [366]:

Proposition 5.5.17. *Let (X,T,μ) be a system with minimal self-joinings of order two and such that T^n is ergodic for all n. Then every transformation T' preserving μ and commuting with T is a power T^n, and every Borel sub-σ-algebra invariant by T is trivial.*

Proof. Suppose that T' is measure-preserving and commutes with T; then $\mu(T'^{-1}A) = \mu(A)$ for every Borel set. If T' is invertible, we have also $\mu(T'A) = \mu(A)$ and $\nu(A \times B) = \mu(A \cap T'B)$ is a joining, and so $T' = T^n$; if T' is not invertible, and if ξ is the full Borel σ-algebra, $T'^{-1}\xi$ is a non-trivial invariant sub-σ-algebra and the second assertion will imply the first.

Let H be an invariant sub-σ-algebra for T; we write $x \equiv y$ if x and y are not separated by H; the set Z of equivalence classes is equipped with the trace measure of μ, and the dynamical system (Z,T,μ) is called a *factor* of (X,T,μ) (as in Sec. 5.2.4, but here not necessarily with a finite fiber); by abuse of language, we speak of the factor H. We define $\nu(A \times B) = \int_X E_H(1_A)(x) E_H(1_B)(x) d\mu(x)$. We check that a set A is measurable for H if and only if $A \times X = X \times A$, ν-almost everywhere: $\nu(A \times X \Delta X \times A) = \nu(A \times A^c) + \nu(A^c \times A) = 0$ if and only if $E_H(1_A)$ takes only values 0 or 1, that is, $A \in H$.

The measure ν is an invariant measure for $T \times T$, not necessarily ergodic, but it admits a decomposition into *ergodic components*, and these are joinings. Hence $\nu = a\mu \times \mu + (1-a)\sum_{j\in Z} b_j \Delta_j$, where $\Delta_j(A' \times B) = \mu(A \cap T^j B)$.

Suppose there exists a set $A_0 \subset Z$ of non-trivial measure, and let $A \subset X$ be the union of all elements in A_0, which has the same measure. We have $\mu(A) =' \nu(A \times A) = a(\mu(A)^2) + (1-a)\sum b_j \mu(A \cap T^j A)$. Because of the ergodicity of every T^j, we have a convex combination of terms $< \mu(A)$, except the term for $j = 0$; hence only this last one can have a nonzero coefficient, and $\nu = \Delta_0$. Hence $\nu(A \times B) = \mu(A \cap B)$, and by the above criterion, every set is H-measurable. ∎

Corollary 5.5.18. *Such a transformation has no measure-preserving roots.*

Proof. Suppose $T'^n = T$; then $T'T = TT' = T'^{n+1}$ and we conclude. Even if we know only that T'^n is measure-theoretically isomorphic to T, T'^n still has minimal self-joinings and we can conclude. ∎

Furthermore, it can be shown also [141] that Chacon's map S has *minimal self-joinings of all orders*: for every p-uple of nonzero integers $(k_1, \ldots, k_p)$, the ergodic measures invariant by $S^{k_1} \times \cdots \times S^{k_p}$, and with marginals μ, are all the products of diagonal measures, defined here by $\nu(A_1 \times \ldots A_n) = \mu(S^{l_1} A_1 \cap \ldots S^{l_n} A_n)$; these include the product measure $\mu \times \cdots \times \mu$.

A transformation having minimal self-joinings of all orders can be used in the so-called *counter-example machine* ([366]): by using Cartesian products of the S^{k_n}, we can build a large family of transformations with surprising properties, for example a transformation with a continuum of non-isomorphic square roots, or two non-isomorphic systems where each one is a factor of the other.

6. Sturmian Sequences

In this chapter, we will study symbolic sequences generated by an irrational rotation. Such sequences appear each time a dynamical system has two rationally independent periods; this is a very typical situation, arising for example in astronomy (with the rotation of the moon around the earth, and of the earth around the sun), or in music (with the building of musical scales, related to the properties of $\log 3/\log 2$), and such sequences have been studied for a long time. These sequences, or related objects, appear in the mathematical literature under many different names: rotation sequences, cutting sequences, Christoffel words, Beatty sequences, characteristic sequences, balanced sequences, Sturmian sequences, and so on.

It is easy to obtain such sequences; the most intuitive way is to consider a line with irrational slope in the plane, and to build a sequence by considering its intersections with an integer grid, counting **0** when the line intersects an horizontal, and **1** when it intersects a vertical, see Fig. 6.1. All Sturmian sequences can in fact be obtained in this way (there is a problem of definition if the line meets an integral point: there are then 2 possible sequences). We will see along the chapter several other ways to obtain these sequences; however, this figure often shows immediately the reason of some properties.

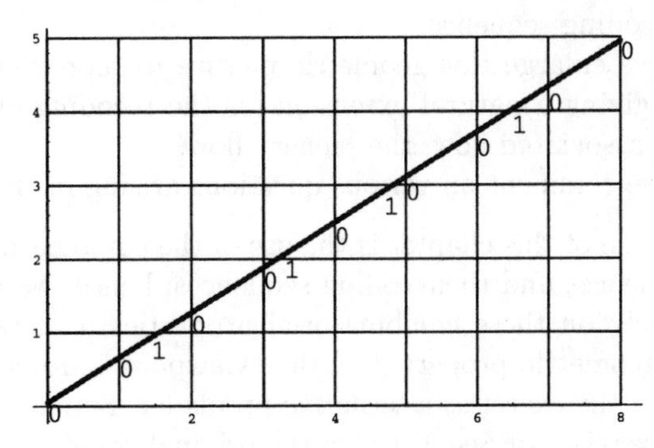

Fig. 6.1. A typical Sturmian sequence.

[1] This chapter has been written by P. Arnoux

The term "Sturmian sequence" was coined by Hedlund and Morse in 1943; they showed that, in a sense, they are the simplest non-trivial sequences, and gave two simple combinatorial characterizations of these sequences. The name comes from a relation with the Sturm comparison theorem (see for example [414], p. 104). In elementary terms, for an irrational number $\alpha \in [0,1]$, the position of the sequence of zeroes of $\sin(\alpha\pi x+\beta\pi)$ with respect to the integers (zeroes of $\sin(\pi x)$) gives a sequence u, defining u_n as the number of zeroes in the interval $[n, n+1[$; every Sturmian sequence is of this type.

In Sec. 6.1, we define three types of sequences: Sturmian sequences, balanced sequences and rotation sequences. We prove that Sturmian sequences are exactly the non-eventually periodic balanced sequences, and that all rotations sequences are Sturmian sequences. We also discuss dynamical systems generated by Sturmian sequences, in the sense of Chap. 5, and prove that each such system contains a sequence with peculiar properties, the so-called "special sequence".

In. Sec. 6.2, we generalize these notions to biinfinite sequences; this is technically easier for the following discussion.

In Sec. 6.3, we discuss another way to present Sturmian sequences, using a special encoding process related to an expanding map on the set of all Sturmian sequences, with a simple Markov partition, and derive various consequences; we give several variations of the encoding process, and establish their properties.

In Sec. 6.4, we show how these coding sequences translate to arithmetic; in particular, we use this to prove that all Sturmian sequences are rotation sequences, and give explicit expression for the angle and initial point of the rotation sequence in term of the coding sequence.

In Sec. 6.5, we study the case of periodic coding sequences. We obtain various consequences, algebraic and geometric, among them an explicit expression of the Markov coding for the toral automorphism canonically associated with a periodic coding sequence.

In Sec. 6.6, we enlarge this geometric picture to represent all Sturmian sequences, thus giving a natural extension for the recoding of Sturmian sequences, and an associated flow, the scenery flow.

In Sec. 6.7, we comment on various questions arising in this chapter.

The main theme of this chapter is the use of the combinatorial properties of Sturmian sequences and their coding sequences; hence, we tried, as much as possible, to rely on these combinatorial properties to derive from them arithmetic and geometric properties. Other viewpoints are clearly possible, and we provide in the exercises alternative proofs for a number of results. In particular, the exercises of Sec. 6.1 give the original proof that all Sturmian sequences are rotation sequences, a result which is obtained in Sec. 6.4. The combinatorial properties of Sturmian sequences are often easier to understand if one remembers that these are rotation sequences, and especially the link with continued fractions which arises in the recoding process.

The present chapter is deeply linked with Chap. 9, which studies the properties of the substitutions that preserve Sturmian words and the combinatorial properties of their fixed words, and in particular the simplest of these substitutions, the Fibonacci substitution. It is also linked in a less obvious way with Chap. 10, which gives, among other things, a non-trivial characterization of invertible substitutions in terms of matrices.

For more references on the subject, the reader is referred to [69], in particular the very complete chapter on Sturmian sequences, and to its impressive bibliography; see also [8, 42, 65, 72, 95, 292, 293].

6.1 Sturmian sequences. Basic properties

We begin this section with giving two equivalent definitions of Sturmian sequences: as sequences of minimal unbounded complexity, and as balanced non-ultimately periodic sequences. In the second part, we give a typical example of Sturmian sequence: the so-called rotation sequences; we will later prove that all Sturmian sequences are of this type (an alternative proof is found in the exercises of the present section). In the last part, we investigate properties of the dynamical system defined by a Sturmian sequence.

6.1.1 Two definitions of Sturmian sequences

Let u be a sequence, with values in a finite set (alphabet) $\mathcal{A}$; we recall from Chap. 1 the following:

- The language of u is the set $\mathcal{L}(u)$ of finite words that occur in u; we denote by $\mathcal{L}_n(u)$ the set of words of length n that occur in u. The complexity function of u is the function p_u which, to each integer n, associates the number Card $\mathcal{L}_n(u)$ of distinct words of length n that occur in u.
- The complexity p_u is an increasing function. If u is eventually periodic, then p_u is bounded. If there is an n such that $p_u(n+1) = p_u(n)$, then u is eventually periodic.

We deduce that, if u is not eventually periodic, we must have $p_u(n) \geq n+1$, because p_u has strictly increasing integral values, and $p_u(1)$ must be at least 2, otherwise there would be only one letter and u would be constant. In this chapter, we will be interested in non-periodic sequences with the smallest complexity:

Definition 6.1.1. *A sequence u is called* Sturmian *if it has complexity $p_u(n) = n+1$. We will denote by Σ' the set of Sturmian sequences.*

We will define, in Sec. 6.2 (Definition 6.2.6), the related set Σ of biinfinite Sturmian sequences; as we will see, this set is slightly more complicated to define, but it is technically easier to use in many proofs.

These sequences are also called *sequences of minimal complexity*, or *sequences with minimal block growth.*

Historical remark. The original definition of Hedlund and Morse (see [303]) was slightly different: first, they considered biinfinite sequences such that all **1**'s are separated by strings of **0**'s. For such a sequence, they defined a n-chain as a word occuring in the sequence, starting and ending in **1**, and containing exactly $n+1$ **1**'s (and hence exactly n strings of **0**'s). They defined Sturmian sequences as sequences such that the length of 2 n-chains differ at most by 1. This definition comes naturally for cutting sequences (see Exercise 6.1.16). They proved that this definition, which includes some periodic sequences, is equivalent to "balanced", to be defined below. The present simpler definition was only given 30 years later, by Coven and Hedlund (see [123], [122]).

The words that occur in a Sturmian sequence cannot disappear:

Proposition 6.1.2. *A Sturmian sequence is* recurrent, *that is, every word that occurs in the sequence occurs an infinite number of times.*

Proof. Suppose that a word U, of length n, occurs in a Sturmian sequence u a finite number of times, and does not occur after rank N. Let v be the sequence defined by $v_k = u_{k+N}$. It is clear that the language of v is contained in that of u, and does not contain U. Hence we must have $p_v(n) \leq n$. This implies that v is eventually periodic, and hence so is u; this is a contradiction. ∎

Since we have $p_u(1) = 2$, Sturmian sequences are sequences over two letters; in this chapter, we fix the alphabet $\mathcal{A} = \{\mathbf{0}, \mathbf{1}\}$. The definition of Sturmian sequences implies a very easy consequence, which we will use several times:

Lemma 6.1.3. *If u is Sturmian, then exactly one of the words* **00**, **11** *does not occur in u.*

Proof. We have $p_u(2) = 3$, so there are exactly three words of length 2 occurring in u. By the previous proposition, **0** and **1** each occur an infinite number of times in u, which implies that **01** and **10** both occur in u. But **00** and **11** are the two other words of length 2, and exactly one of them must occur. ∎

Definition 6.1.4. *We say that a Sturmian sequence is of* type 0 *if* **1** *is* isolated, *that is, if* **11** *does not occur in the sequence. This is equivalent to saying that* **0** *occurs more frequently than* **1**. *We say that the sequence is of type* 1 *if* **00** *does not occur. We denote by Σ'_0 (respectively Σ'_1) the set of Sturmian sequences of type 0 (respectively of type 1).*

Every element of Σ' is either of type 0 or of type 1; it is not immediately clear that Sturmian sequences exist at all, Chap. 5 gives the simplest example:

Example 6.1.5. The Fibonacci sequence $u = \mathbf{010010100}\ldots$, fixed point of the Fibonacci substitution

$$\sigma : \mathbf{0} \mapsto \mathbf{01}$$
$$\mathbf{1} \mapsto \mathbf{0}$$

is a Sturmian sequence. For a proof, see Exercise 5.4.11 and the preceding corollary, in Chap. 5.

There is another useful characterization of Sturmian words, for which we shall need the following notation and definition:

Notation 6.1.1 *If U is a finite word over the alphabet $\mathcal{A}$, we denote by $|U|$ the length of U, and $|U|_{\mathbf{a}}$ the number of occurrences of the letter* **a** *in U.*

Definition 6.1.6. *A sequence u over the alphabet $\{\mathbf{0}, \mathbf{1}\}$ is* balanced *if, for any pair of words U, V of the same length occurring in u, we have $||U|_{\mathbf{1}} - |V|_{\mathbf{1}}| \leq 1$.*

This means that, for words of $\mathcal{L}_n(u)$, the number of occurrences of **1** can take at most two consecutive values; we note that this property is very strong: this is the smallest possible number of values for non-periodic sequences. As we show below, in that case, this property is equivalent to Sturmian. We will often use the following technical lemma:

Lemma 6.1.7. *If the sequence u is not balanced, there is a (possibly empty) word W such that $\mathbf{0}W\mathbf{0}$ and $\mathbf{1}W\mathbf{1}$ occur in u.*

Proof. If u is not balanced, we can find two words A and B of length n such that $|A|_{\mathbf{1}} - |B|_{\mathbf{1}} > 1$. We first prove that we can suppose $|A|_{\mathbf{1}} - |B|_{\mathbf{1}} = 2$. Call A_k (respectively B_k) the suffix of length k of A (respectively B), and $d_k = |A_k|_{\mathbf{1}} - |B_k|_{\mathbf{1}}$. We have $d_n > 1$, $d_0 = 0$, and a short case study proves that $|d_{k+1} - d_k|$ is 0 or 1. But then, an intermediate value argument for integer valued functions shows that there is k such that $d_k = 2$. Suppose now that A and B are words of minimal length with this property. Write $A = \mathbf{a}_0\mathbf{a}_1 \ldots \mathbf{a}_{n-1}$ and $B = \mathbf{b}_0\mathbf{b}_1 \ldots \mathbf{b}_{n-1}$. We must have $\mathbf{a}_0 = \mathbf{a}_{n-1} = \mathbf{1}$ and $\mathbf{b}_0 = \mathbf{b}_{n-1} = \mathbf{0}$, otherwise we could find a shorter pair by removing some prefix (the first letter if the two initial letters are the same, a longer prefix if $\mathbf{a}_0 = \mathbf{0}$ and $\mathbf{b}_0 = \mathbf{1}$). It is then easy to prove by induction that for all the remaining letters, we have $\mathbf{a}_k = \mathbf{b}_k$, which proves the lemma. ∎

Theorem 6.1.8. *A sequence u is Sturmian if and only if it is a non-eventually periodic balanced sequence over two letters.*

Proof. We first prove the "if" part, by contradiction. Suppose that u is not Sturmian, we will show it is not balanced. Let n_0 be the smallest integer such that $p_u(n_0+1) \geq n_0+3$. We have $n_0 \geq 1$, since $p_u(1) = 2$. Because $p_u(n_0) = n_0 + 1$, there are at least two words U and V of length n_0 that can be extended on the right in two ways. Since n_0 is the smallest integer with this property, U and V differ only in the first letter; hence there is a word W such that $U = \mathbf{0}W$ and $V = \mathbf{1}W$. We have proved that $\mathbf{0}W\mathbf{0}$ and $\mathbf{1}W\mathbf{1}$ occur in u, hence this sequence is not balanced.

We now prove the converse: suppose that u belongs to Σ'; this implies, by definition, that it is a non-eventually periodic sequence on two letters. We do a proof by contradiction, and suppose that it is not balanced. By Lemma 6.1.7, there is a word W such that $\mathbf{1}W\mathbf{1}$ and $\mathbf{0}W\mathbf{0}$ occur in u. We consider such a word of minimal length $n+1$, and we write $W = w_0w_1 \ldots w_n$.

Remark that, under this assumption, if we have a pair of words U, V of the same length such that $||U|_\mathbf{1} - |V|_\mathbf{1}| \geq 2$, then their length is at least $n+3$, that is, $\mathbf{0}W\mathbf{0}$ and $\mathbf{1}W\mathbf{1}$ is a pair of non-balanced words of minimal length; for we can apply the same argument as in the proof of Lemma 6.1.7 to show that such a pair contains a pair $\mathbf{0}W'\mathbf{0}$ and $\mathbf{1}W'\mathbf{1}$.

The word W cannot be empty, otherwise $\mathbf{00}$ and $\mathbf{11}$ occur in u, which is impossible by Lemma 6.1.3. For the same reason, we must have $w_0 = w_n$; more generally, we must have $w_k = w_{n-k}$, that is, W is a *palindrome*, otherwise, letting $w_k = \mathbf{0}$, $w_{n-k} = \mathbf{1}$, we see that $\mathbf{0}w_0 \ldots w_{k-1}\mathbf{0}$ and $\mathbf{1}w_{n-k+1} \ldots w_n\mathbf{1}$ is a non-balanced pair of smaller length, which is impossible. Recall that a palindrome is a word that stays the same when it is read backwards; for example, *in girum imus nocte et consumimur igni* is a classical palindrome in Latin, *a man, a plan, a canal: Panama!* a classical palindrome in English.

But now, we know that there are $n+2$ words of length $n+1$; W can be extended in two ways on the right and on the left, and all the others can be extended in only one way. Exactly one of $\mathbf{0}W$ and $\mathbf{1}W$ (suppose it is $\mathbf{0}W$) can be extended in two ways on the right (if a word can be extended in two ways on the right, all its suffixes can also). So, the words $\mathbf{0}W\mathbf{0}$, $\mathbf{0}W\mathbf{1}$ and $\mathbf{1}W\mathbf{1}$ occur in u, but not $\mathbf{1}W\mathbf{0}$. Let i be the rank of an occurrence of $\mathbf{1}W\mathbf{1}$ in u. We prove:

Lemma 6.1.9. *The word* $\mathbf{0}W$ *cannot occur in* $u_iu_{i+1} \cdots u_{i+2n+3}$.

Proof of Lemma 6.1.9. The length of $u_iu_{i+1} \ldots u_{i+2n+3}$ is $2n+4$, the length of $\mathbf{1}W\mathbf{1}$ is $n+3$, and the length of $\mathbf{0}W$ is $n+2$. Hence this lemma exactly means that the first letter of an occurrence of $\mathbf{0}W$ cannot occur in an occurrence of $\mathbf{1}W\mathbf{1}$ in u. Suppose the beginning of $\mathbf{0}W$ overlaps $\mathbf{1}W\mathbf{1}$: $\mathbf{0}$ can obviously not be the first or last letter in $\mathbf{1}W\mathbf{1}$; if it is w_k, the overlap means that $\mathbf{0}w_0 \ldots w_{n-k} = w_kw_{k+1} \cdots w_n\mathbf{1}$; but this implies that $w_k = \mathbf{0}$ and $w_{n-k} = \mathbf{1}$, and we get a contradiction, since W is a palindrome. ∎

End of the proof of Theorem 6.1.8. It is immediate that there are exactly $n+3$ words of length $n+2$ occurring in $u_i u_{i+1} \ldots u_{i+2n+3}$; but there are $n+3$ words of length $n+2$, and $\mathbf{0}W$ does not occur, as we just proved, so at least one word occurs twice. But all these words can be extended in a unique way on the right (only $\mathbf{0}W$ can be extended in two ways on the right), hence u is eventually periodic: this is a contradiction. ■

From this characterization of Sturmian sequences as balanced sequences, we get the following proposition.

Proposition 6.1.10. *The* frequency *of* $\mathbf{1}$ *in a Sturmian sequence* u*, defined as the limit of* $\frac{|u_0 u_1 \ldots u_{n-1}|_{\mathbf{1}}}{n}$ *when* n *tends to infinity, is well defined, and is irrational.*

Proof. Let a_n be the minimum number of $\mathbf{1}$ that occur in a word of length n occurring in u. Since $|u_0 u_1 \ldots u_{n-1}|_{\mathbf{1}}$ is either a_n or $a_n + 1$, it is enough to prove that the limit of a_n/n exists and is irrational.

A word of length $kq + r$ can be split in k words of length q and one word of length r, and we get the inequality $ka_q \leq a_{kq+r} \leq k(a_q + 1) + r$. If we consider $n > q^2$, we can write $n = kq + r$, with $k \geq q$ and $0 \leq r < q$; since $r < k$, we check that $\frac{r}{n} < \frac{k}{n} \leq \frac{1}{q}$, hence the second part of the inequality gives $\frac{a_n}{n} \leq \frac{a_q}{q} + \frac{2}{q}$. One checks also immediately that the quantity $ra_q - n$ is negative, since $n \geq a_n \geq ka_q > ra_q$; hence the first part of the inequality above implies that $\frac{n}{q}(a_q - 1) = ka_q + \frac{1}{q}(ra_q - n) \leq a_n$, and, after dividing by n, we get:

$$\frac{a_q}{q} - \frac{1}{q} \leq \frac{a_n}{n} \leq \frac{a_q}{q} + \frac{2}{q}.$$

It follows that the sequence $(a_n/n)_{n\in\mathbb{N}}$ is a Cauchy sequence, and hence converges to some limit α.

Suppose that this limit is rational, equal to $\frac{p}{q}$. From the inequality $ka_n \leq a_{kn} < a_{kn} + 1 \leq k(a_n + 1)$, we deduce that, if n divides n', then $\frac{a_n}{n} \leq \frac{a_{n'}}{n'} < \frac{a_{n'}+1}{n'} \leq \frac{a_n+1}{n}$. In particular, the sequence $\left(\frac{a_{2^n q}}{2^n q}\right)_{n\in\mathbb{N}}$ is increasing, and the sequence $\left(\frac{a_{2^n q}+1}{2^n q}\right)_{n\in\mathbb{N}}$ is decreasing; we get the inequality:

$$\frac{a_q}{q} \leq \frac{a_{2^n q}}{2^n q} < \frac{a_{2^n q} + 1}{2^n q} \leq \frac{a_q + 1}{q}.$$

But this sequence must converge to $\frac{p}{q}$; this is only possible if $a_q = p$ and $a_{2^n q} = 2^n p$ for all n, or $a_q + 1 = p$, and $a_{2^n q} + 1 = 2^n p$ for all n.

We prove that it is impossible to have $a_{2^n q} = 2^n a_q$ for all n: since the sequence is not periodic, there is at least one word U of length q in the language of u such $|U|_1 = a_q + 1$. Since u is recurrent, this word occurs an infinite number of times, so that it must occur in two positions congruent mod q. Hence, we can find a word of length $2^n q$ that can be split in words of length q, with at least two occurrences of U; the number of **1** in this word is at least $2^n a_q + 2$, so that $a_{2^n q} > 2^n a_q$. A similar proof shows that $a_{2^n q} + 1$ cannot be equal to $2^n(a_q + 1)$ for all n, and we get a contradiction. ∎

From this theorem, it follows that the language of u is not so easy to compute:

Corollary 6.1.11. *If u is a Sturmian sequence, the language $\mathcal{L}(u)$ is not* regular *(or equivalently, it cannot be recognized by a finite automaton).*

Proof. Suppose that $\mathcal{L}(u)$ is regular; we can apply the *pumping lemma* (see also Chap. 3); this lemma says the following: for any regular language, there is an integer N such that any word U of length larger than N can be split as AWB, where W is a nonempty word with the following property: for any n, AW^nB belongs to $\mathcal{L}(u)$. Hence W^n also belongs to $\mathcal{L}(u)$; the frequency of **1** in W^n is equal to $|W|_1/|W|$, which is rational, and it tends to the frequency of **1** in the sequence u. By the preceding proposition, we get a contradiction. ∎

It is possible to define the balance property for sets of finite words; the following easy result, proved in [303] (see also Chap. 3), will be used in later exercises:

Exercise 6.1.12. We will say that a set E of finite words is balanced if, for any pair of words U, V in E, and for any words U', V' of same length that occur in U, V, we have $||U'|_1 - |V'|_1| \leq 1$. Prove that a balanced set of words of length n contains at most $n + 1$ distinct words. *(Hint: induction.)*

Exercise 6.1.13. As a consequence, prove that if u and v are two Sturmian sequences with the same frequency, they have the same language. *(Hint: show that the minimum number of* **1** *in a word of length n is the same for sequences u and v. Deduce that $\mathcal{L}_n(u) \cup \mathcal{L}_n(v)$ is a balanced set.)*

Exercise 6.1.14. Prove the converse: the frequency of **1** in a Sturmian sequence depends only on the language, not on the sequence itself, and explain how one can compute the frequency, knowing only the language. Deduce that every Sturmian sequence is minimal, that is, every word that occurs in the sequence occurs with bounded gaps (see Chaps. 1 and 5). *(Hint: suppose that U occurs in u with arbitrarily large gaps. Prove that there exists in the orbit closure of u a sequence v where U does not occur. Deduce that v is eventually periodic, and get a contradiction.)*

We will give another proof of the minimality in Proposition 6.3.16.

6.1.2 Rotation sequences

We now generalize the example of the Fibonacci sequence. We first need some definitions.

Notation 6.1.2 *Let x be a real number. The* integral part *of x is the integer $[x] = \sup\{n \in \mathbb{Z} | n \leq x\}$. The* fractional part *of x is the real number $\{x\} = x - [x]$. The* ceiling *of x is the integer $\lceil x \rceil = \inf\{n \in \mathbb{Z} | n \geq x\}$.*

Definition 6.1.15. *A* rotation sequence *is a sequence u such that there is an irrational number $\alpha \in [0,1]$ and a real number β such that:*

$$(\forall n \in \mathbb{N})\,(u_n = [(n+1)\alpha + \beta] - [n\alpha + \beta]) \tag{6.1}$$

$$\text{or } (\forall n \in \mathbb{N})\,(u_n = \lceil (n+1)\alpha + \beta \rceil - \lceil n\alpha + \beta \rceil) \tag{6.2}$$

The number α will be called the angle *associated with the rotation sequence, and β the* initial point.

Rotation sequences occur in a number of classical situations. The simplest one is the natural coding associated with a rotation. Consider the rotation:

$$R : \mathbb{T}^1 \to \mathbb{T}^1 \qquad x \mapsto x + \alpha \quad \bmod 1$$

where $\mathbb{T}^1$ is the circle, identified to $\mathbb{R}/\mathbb{Z}$, or to the interval $[0,1[$, by the function $e^{2i\pi x}$. We consider the two intervals I_0, I_1 on $\mathbb{T}^1$ delimited by 0 and $1-\alpha$. We denote by ν the *coding function* defined by $\nu(x) = \mathbf{0}$ if $x \in I_0$, $\nu(x) = \mathbf{1}$ otherwise; then the rotation sequence defined by α and β is just the sequence $\nu(R^n(\beta))$, that is, the coding of the positive orbit of β under the rotation R, see Fig. 6.2 below.

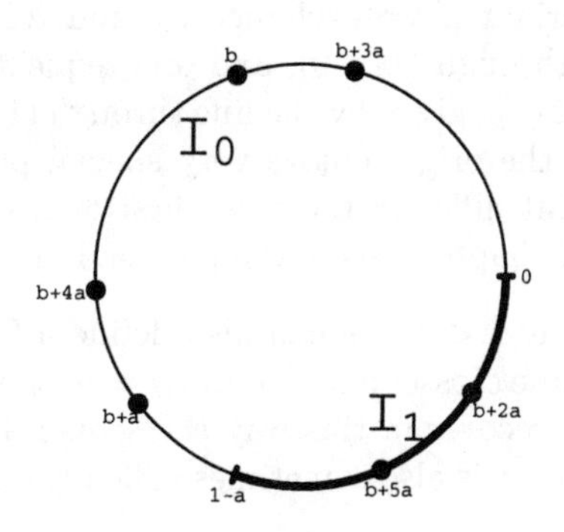

Fig. 6.2. The orbit of β for the rotation of angle α.

Remark. Of course, this definition is not precise enough: to properly define ν, we need to state what happens at the endpoints of the two intervals. We will always take the intervals closed on one side and open on the other. If we take $I_0 = [0, 1-\alpha[$, $I_1 = [1-\alpha, 1[$, this corresponds to choosing the integral part in the definition of the rotation sequence. If we take the other choice $I_0 =]0, 1-\alpha]$ and $I_1 =]1-\alpha, 1]$, it corresponds to the choice of the ceiling function. This choice is irrelevant, except if there is a nonnegative integer n such that $\beta + n\alpha$ is an integer, that is, if β is congruent to $-n\alpha$ mod. 1. In that case, α and β define two possible rotation sequences, depending on this choice. We will see later that the rotation sequence defined by the angle α and the initial point α, and the two rotation sequences defined by the angle α and the initial point $1 - \alpha$, play an important role.

In most proofs, we will only consider rotation sequences defined using $[x]$; it is usually immediate to check each time that the proof works also for $\lceil x \rceil$.

Exercise 6.1.16. Prove the following alternative characterization: consider the line $y = \alpha x + \beta$ on the plane, with α irrational. This line cuts all the horizontal lines $y = n$, and all the vertical lines $x = n$. This defines a so-called "*cutting sequence*": write **0** each time the line cuts a vertical line, and **1** each time it cuts a horizontal line. Prove that the cutting sequence is a rotation sequence associated with the angle $\frac{\alpha}{1+\alpha}$ and the initial point $\frac{\beta}{1+\alpha}$ (we need to be more precise in the case where the line goes through an integral point, but this happens at most once; in this case, we write **01** or **10** as we like).

We recover here the way to obtain a Sturmian sequence that was given at the very beginning of the chapter, in Fig. 6.1. This figure explains many things: it is clear that it is essentially invariant under a central symmetry, proving that the associated language is palindromic. It shows that it is completely natural to consider biinfinite sequences, since the line extends in both directions. It shows that, for a given direction, there are exactly two sequences that are invariant if we exchange u_n and u_{-n}, given by the lines through $(0, 1/2)$ and through $(1/2, 0)$, and one sequence that is invariant if we exchange u_n and u_{-1-n}, given by the line through $(1/2, 1/2)$. It also shows that the line through the origin enjoys very special properties, since it defines two sequences that differ in their two first terms. All these properties will occur later in this chapter, (see Example 6.4.3, and also in Chap. 9).

Remark. In the case $\alpha < 1$, we can also define a Sturmian sequence by defining $u_n = \mathbf{1}$ if the line crosses a horizontal $y = p$ for x between n and $n+1$, $u_n = \mathbf{0}$ otherwise. We recover in this way the sequence of the definition. In this setting, the number α is also sometimes called the *slope* of the sequence, and β the *intercept*.

An other classical example is the *square billiards*: we shoot a ball in a square billiards, with initial irrational slope a. The ball bounces on the sides of the billiards according to the laws of elastic shock, and we write **0** for a vertical side, and **1** for a horizontal side. This can be reduced in a straightforward way to a cutting sequence.

Since we will prove later that all Sturmian sequences are of this type, it is important to keep these examples in mind, as they can give intuition on the properties of Sturmian sequences.

Proposition 6.1.17. *Every rotation sequence is a Sturmian sequence.*

Proof. Because they are simple and instructive, we will give two proofs.

We first prove that a rotation sequence is balanced. From the definition of the sequence, we can compute the number of **1** in $u_k \dots u_k + n$:

$$|u_k \dots u_{k+n}|_{\mathbf{1}} = [(n+k+1)\alpha + \beta] - [k\alpha + \beta].$$

And it is easy to check that, for a fixed n, this number can take only two values, $[n\alpha]$ or $[n\alpha] + 1$, therefore the sequence is balanced.

It is a little more subtle to prove that the sequence is Sturmian. Consider the sequence as defined by the coding of the orbit of β for the rotation R with respect to the partition $\mathcal{P}$ in two intervals. The letter u_n tells us in which element of the partition $\mathcal{P}$ is $R^n\beta$, or equivalently in which element of $R^{-n}\mathcal{P}$ is β. The first n letters will give the position of β with respect to the partition intersection of $\mathcal{P}, \dots R^{1-n}\mathcal{P}$. In the same way, words of length n in the sequence will give the position of iterates of β with respect to this partition. But an easy computation shows that this partition is defined on the circle by the $n+1$ points $0, -\alpha, \dots, -n\alpha$, so that it consists of $n+1$ intervals (since α is irrational, the $n+1$ points are distinct). Hence, there can be no more than $n+1$ different words of length n. This proof is less straightforward than the first, but it can be generalized to more complicated cases (cf. [37, 45, 46]). ■

We can build in this way a large number of Sturmian sequences. In fact, we get in this way **all** Sturmian sequences: the converse of the proposition is true, and it is one of the main points of Secs. 6.3 and 6.4 to prove this result and to determine explicitly the angle and initial point of a given Sturmian sequence.

Exercise 6.1.18. It is possible to prove that Sturmian sequences are rotation sequences in an elementary way, as it is done in the original paper of Hedlund and Morse [303], by using the results of Exercise 6.1.12 and 6.1.13, along the following lines.

1. The sequence v of angle α and initial point 0 is Sturmian, with frequency α.
2. If u is an arbitrary Sturmian sequence of frequency α, there is a sequence n_k such that $S^{n_k}v$ tends to u. *(Hint: use Exercise 6.1.13.)*
3. There is a real number β such u is the rotation sequence (or one of the two rotation sequences) with angle α and initial point β.

Our approach will give us a more complete information on α and β.

Exercise 6.1.19. Prove that the language associated with a Sturmian sequence is *palindromic*, that is, the reverse word of any word in $\mathcal{L}_u$ also belongs to $\mathcal{L}_u$; this property is very easy for a rotation sequence (for example, one can use the obvious symmetry on the cutting sequence starting from the origin), but there is a purely combinatorial proof using Exercise 6.1.12. *(Hint:*

if we consider a balanced set of words, and add to it one of these words read backwards, it stays balanced; but a balanced set of words of length n contains at most $n + 1$ words.)

6.1.3 Dynamical systems associated with Sturmian sequences

We have defined in Chap. 1 the dynamical system associated with a sequence. Let us recall the main points: we consider the set of all sequences on $\{\mathbf{0}, \mathbf{1}\}$, with the product topology (this is a compact set, as product of finite sets), and the shift S on this set: the image $v = Su$ of a sequence u is the sequence v defined by $v_i = u_{i+1}$. In other words, v is u, with first letter u_0 deleted. With a sequence u, we associate the set X_u which is the closure of $\{S^n u | n \in \mathbb{N}\}$, the orbit of u for the shift. The set X_u is by construction shift-invariant, and (X_u, S) is the dynamical system associated with the shift. It is easy to show that the language of any sequence in X_u is included in the language of u, but in general it could be strictly smaller. The dynamical system associated with a Sturmian sequence has special properties:

Theorem 6.1.20. *The system associated with a Sturmian sequence is one-to-one, except one point that has two preimages.*

Proof. We first prove that the system is onto. Let v be a point in X_u; if v is of the form $S^n u$, with $n > 0$, then it is the image of $S^{n-1}u$. If it is not in the orbit of u, then there is a sequence n_k, with $\lim_{k\to\infty} n_k = \infty$, such that $S^{n_k}u$ tends to v; from $S^{n_k - 1}u$, we can, by compacity, extract a sequence that has a limit w, and this sequence w satisfies $Sw = v$. The only difficult point is to prove that u has a preimage. But we have seen that the sequence u is recurrent, that is, every word that occurs in u occurs an infinite number of times. Hence, the initial word of length k of u occurs again at some position n_k of u, with $n_k > 0$, and the sequence $S^{n_k}u$ tends to u. For the same reason as before, u also has a preimage in X_u.

We now prove that there is exactly one point with two preimages. The proof we just gave shows that any word in $\mathcal{L}_n(u)$ can be extended on the left, since it appears past the initial position. Since there are $n + 1$ words of length n, exactly one of them can be extended in two different ways; we call this word L_n. It is clear that any prefix of L_{n+1} can be extended on the left in two ways, so L_n is the prefix of length n of L_{n+1}. If a sequence u has two preimages, all its prefixes are the L_n; but there is exactly one such sequence in X_u. ■

Definition 6.1.21. *We call the L_n's the* left special words *of $\mathcal{L}_n(u)$, and the sequence l which has the L_n's as prefixes is called the* left special sequence *of X_u.*

Exercise 6.1.22. Define the *right special word* R_n, and prove that R_n is the reverse word of L_n. Prove that, if a_n is the minimum number of $\mathbf{1}$ in a word of length n occurring in the Sturmian sequence u, then $|\mathbf{0}L_n|_{\mathbf{1}} = a_{n+1}$.

We can find explicitly all the preimages of l, and this will be useful later:

Proposition 6.1.23. *The sequence l has two preimages $\mathbf{0}l$ and $\mathbf{1}l$, and two preimages of order 2, $\mathbf{10}l$ and $\mathbf{01}l$.*

In other words, the preimages of l differ only on two letters.

Proof. The preimages of l are of course $\mathbf{0}l$ and $\mathbf{1}l$. Suppose now that $\mathbf{1}$ is isolated in l (that is, $\mathbf{11}$ does not occur). Then the only preimage of $\mathbf{1}l$ has to be $\mathbf{01}l$. Because $\mathbf{1}$ is isolated, we can suppose that l starts with a string of n $\mathbf{0}$'s followed by a $\mathbf{1}$. If the preimage of $\mathbf{0}l$ is $\mathbf{00}l$, then the language of u contains a string of $n+2$ $\mathbf{0}$'s, in $\mathbf{00}l$, and a string of n $\mathbf{0}$'s surrounded by $\mathbf{1}$ in $\mathbf{1}l$, which is impossible since u is balanced. ∎

Exercise 6.1.24. For each n, m, prove by induction that the words $R_n\mathbf{01}L_m$ and $R_n\mathbf{10}L_m$ belong to the language of u. *(Hint: prove that if a_n is the minimum number of $\mathbf{1}$ in a word of length n in u, we have $a_{i+j} = a_i + a_j$ or $a_{i+j} = a_i + a_j + 1$; then use Exercise 6.1.13.)*

Prove that, for each $n \geq 3$, the special sequence has two preimages of order n, $R_{n-2}\mathbf{10}l$ and $R_{n-2}\mathbf{01}l$.

When we consider cutting sequences, the meaning of the proposition becomes clear: the special sequence corresponds to the line through $(0,0)$, which can be coded in two ways at this point; in all the other points, the coding is unique, and when we establish that all Sturmian sequences are cutting sequences, we have an immediate proof of the proposition and the exercise.

Remark. The special words have many other combinatorial properties. Some of them, the bispecial words, are at the same time left special and right special, hence they are palindromes; one can prove that any word of $\mathcal{L}(u)$ is contained in a bispecial word, and give rules to generate them (the so-called Rauzy rules): this is another way to obtain the results of Sec. 6.3.

Exercise 6.1.25. Find the special word of the Fibonacci substitution. Consider now the substitution σ defined by $\sigma(\mathbf{0}) = \mathbf{010}$ and $\sigma(\mathbf{1}) = \mathbf{10}$. Find its special word, in terms of the fixed words of the substitution. How is σ related to the Fibonacci substitution? *(Hint: show that, for any sequence u, the image by σ of $\mathbf{01}u$ and $\mathbf{10}u$ differ only on the first two letters; show that σ has 2 fixed points that differ only in the first 2 letters.)*

Exercise 6.1.26. Prove that the set Σ' is not closed in the space of all sequences, and that its closure is the set of balanced sequences. Prove that the complement of Σ' in its closure is countable. *(Hint: the set of balanced sequences is closed, and the set of periodic balanced sequences is countable.)*

6.2 Biinfinite Sturmian sequences

In this section, we consider biinfinite sequences; this will allow some simplifications in the following sections. We will in a first part define balanced biinfinite sequences, and Sturmian biinfinite sequences. In the second part, we will show that, for a definition in terms of complexity, we need to take some precautions in the case of biinfinite sequences. In a third part, we will consider dynamical systems generated by biinfinite sequences, and their relations with the systems considered in the previous section.

6.2.1 Balanced biinfinite sequences

We consider now sequences $u = (u_n)_{n\in\mathbb{Z}} \in \{\mathbf{0},\mathbf{1}\}^{\mathbb{Z}}$. For such a sequence, we need to define a special position to differentiate shifted sequences: the *initial letter* of a biinfinite sequence u is u_0.

We can define, in the same way as before, the words occurring in the sequence, and the definition of balanced words still makes sense:

Definition 6.2.1. *A biinfinite sequence u over the alphabet $\{\mathbf{0},\mathbf{1}\}$ is* balanced *if and only if, for any pair of words U, V of the same length occurring in u, we have $||U|_{\mathbf{1}} - |V|_{\mathbf{1}}| \leq 1$.*

The definition of periodicity is the same for infinite or biinfinite sequences; however, there is a difference for eventual periodicity. We say that a sequence u is *positively eventually periodic* if there exist an integer $p > 0$ and an integer $N \in \mathbb{Z}$ such that, for all $i > N$, $u_{i+p} = u_i$. The sequence is *negatively eventually periodic* if, for all $i < N$, $u_{i-p} = u_i$.

Lemma 6.2.2. *A balanced sequence is positively eventually periodic if and only if it is negatively eventually periodic.*

Proof. As we saw in the preceding section, if we consider a sequence U_n of words occurring in u, such that $|U_n| = n$, then the frequency $\frac{|U_n|_{\mathbf{1}}}{n}$ of $\mathbf{1}$ in U_n converges to some limit α. If we take another sequence of words V_n occurring in u, since the sequence u is balanced, the limit is the same; in particular, in that case, $\lim_{n\to\infty} \frac{|u_0 u_1 \ldots u_{n-1}|}{n} = \lim_{n\to\infty} \frac{|u_{-n+1} \ldots u_{-1} u_0|}{n} = \alpha$.

As we proved in the preceding section, α is rational if and only if the sequence u is positively eventually periodic (in that case, if W is a word corresponding to the period, the sequence $\frac{|u_0 u_1 \ldots u_{n-1}|}{n}$ converges to $\frac{|W|_{\mathbf{1}}}{|W|}$). But the symmetric proof, using the sequence $(u_{-n})_{n\in\mathbb{N}}$, shows that α is rational if and only if the sequence u is negatively eventually periodic. This proves the result. ∎

As a consequence, it suffices to say that a balanced sequence is eventually periodic, in which case, it must be simultaneously positively and negatively eventually periodic.

Exercise 6.2.3. Prove that, if u is balanced and eventually periodic, the period, and the words that occur infinitely often, are the same in the positive and the negative direction.

Definition 6.2.4. *A biinfinite sequence over the alphabet* $\{\mathbf{0}, \mathbf{1}\}$ *is* Sturmian *if and only if it is balanced and not eventually periodic.*

Remark. In the original paper [303], Sturmian biinfinite sequences are defined as balanced sequences; however, we do not need two different words for the same concept, and eventually periodic balanced sequences turn out to give some problems in the framework of the next section; this is the reason of the present definition.

6.2.2 Complexity of biinfinite sequences

We can as before define the complexity $p_u(n)$ of a biinfinite sequence u as the number of distinct words of length n that occur in u. The function p_u is still increasing; however, the basic property is no more true: there are eventually periodic biinfinite sequences of unbounded complexity. The simplest such sequence is $\mathbf{0}^\infty \mathbf{1} \mathbf{0}^\infty$, that is, the biinfinite sequence u defined by $u_0 = \mathbf{1}$ and $u_n = \mathbf{0}$ if $n \neq 0$.

It is nevertheless possible to define Sturmian biinfinite sequences in terms of complexity.

Proposition 6.2.5. *A biinfinite sequence u is Sturmian if and only if it is a sequence of complexity $n+1$ that is not eventually periodic.*

Remark. The statement can seem ambiguous; when we say that the biinfinite sequence u is not eventually periodic, this can have two different meanings for a general sequence: we can require, either that the sequence is not eventually periodic in one direction, or that it is not eventually periodic in both directions. In this case however, there is no ambiguity, since we proved that, for a balanced sequence, or for a sequence of complexity $n+1$, positive eventual periodicity is equivalent to negative eventual periodicity.

Proof. Suppose that u is not positively eventually periodic and of complexity $n+1$; then the restricted sequence $(u_n)_{n\in\mathbb{N}}$ is an infinite sequence of complexity at most $n+1$ (some words could have disappeared), that is, not eventually periodic. Hence it is of complexity exactly $n+1$, so it is Sturmian, and balanced, by the results in Sec. 6.1. But then, since all the words that occur in the biinfinite sequence also occur in the positive infinite sequence, the original sequence u is balanced, and not eventually periodic, so it is Sturmian. The symmetric proof works if u is not negatively eventually periodic.

Conversely, suppose that the biinfinite sequence u is Sturmian. Then, the positive infinite sequence is balanced and not eventually periodic, so that it is of complexity $n+1$. The same is true for all sequences $(u_{N+n})_{n\in\mathbb{N}}$, for all

$N \in \mathbb{Z}$; hence all these sequences have the same language, and the union of all these languages is the language of the biinfinite sequence u. Hence this biinfinite sequence has complexity $n+1$. ∎

Definition 6.2.6. *We denote by Σ the set of biinfinite Sturmian sequences.*

Exercise 6.2.7. Prove that there are infinitely many eventually periodic biinfinite sequences of complexity $n+1$, and that this set is countable.

Exercise 6.2.8. Prove that there are sequences of complexity $n+1$ that are not balanced. *(Hint: $\mathbf{0}^\infty\mathbf{1}^\infty$.)*

The methods of the next section will allow us to describe completely the set of biinfinite sequences of complexity $n+1$; we can then show that, if they are not Sturmian, then they are ultimately periodic in both directions, and that the frequencies of $\mathbf{1}$ in the future and the past are either equal (and the sequence is balanced), or Farey neighbors, that is, rational numbers $\frac{p}{q}, \frac{p'}{q'}$ such that $|pq' - p'q| = 1$ (and the sequence is not balanced); we have given above examples of both types.

6.2.3 Dynamical systems generated by Sturmian biinfinite sequences

We consider the shift S on $\{\mathbf{0},\mathbf{1}\}^{\mathbb{Z}}$, defined by $Su = v$, where $v_n = u_{n+1}$ for all $n \in \mathbb{Z}$. The difference with the preceding section is that this shift is invertible, since $S^{-1}u = w$, where $w_n = u_{n-1}$.

As in the preceding section, we can associate with a Sturmian sequence u its orbit closure X_u, that is, $X_u = \overline{\{S^n u | n \in \mathbb{Z}\}}$. The set X_u is invariant by S, and the shift on X_u is one-to-one, because it is on the total space $\{\mathbf{0},\mathbf{1}\}^{\mathbb{Z}}$. But X_u is also S^{-1}-invariant, and this implies that the shift on X_u is onto.

The set $\{\mathbf{0},\mathbf{1}\}^{\mathbb{Z}}$ has a natural product topology, given by the metric defined $d(u,v) = 1/2^k$, where $k = \inf\{n \in \mathbb{N} | u_n \neq v_n \text{ or } u_{-n} \neq v_{-n}\}$. Hence we can consider (X_u, S) as a topological dynamical system for the induced topology.

Definition 6.2.9. *A* Sturmian system *is a dynamical system (X_u, S), where X_u is the orbit closure of a Sturmian biinfinite sequence.*

Exercise 6.2.10. Prove that a Sturmian dynamical system is recurrent and minimal.

There is a natural projection Π from $\{\mathbf{0},\mathbf{1}\}^{\mathbb{Z}}$ to $\{\mathbf{0},\mathbf{1}\}^{\mathbb{N}}$. This projection sends a biinfinite Sturmian system X_u to an infinite Sturmian system $(X_u)'$. While the projection from the total space is highly non-injective, its restriction to X_u is almost one-to-one, as we show in the next exercise.

Exercise 6.2.11. Prove that the projection $\Pi : X_u \to (X_u)'$ is at most 2-to-one; prove that it is one-to-one except on a countable set, and describe this set.

We can define a different pseudo-metric on X_u in the following way:

Definition 6.2.12. *We set* $\delta(u,v) = \limsup_{n\to\infty} \frac{\#\{i\in\mathbb{Z} \mid -n\leq i\leq n, u_i\neq v_i\}}{2n+1}$

This pseudo-metric turns out to have nice properties.

Exercise 6.2.13. 1. Prove that the superior limit in the preceding definition is in fact a limit.
2. Prove that, if $\delta(u,v) = 0$, then either $u = v$, or u and v differ in exactly two positions.
3. Prove that, in the last case, u and v project to the orbit of the special point.
4. Prove that δ is shift invariant.
5. Prove that the quotient space of X_u by the relation $\delta(u,v) = 0$ is isometric to a circle, and the shift is semi-conjugate to a rotation on that circle.

In the language of ergodic theory, this shift is a 2-to-1 extension of a circle rotation, which is 1-to-1 except on a countable set. In particular, it is measurably isomorphic to a rotation.

Remark. We know that the properties in Exercise 6.2.13 must be true, because every Sturmian sequence is a rotation sequence, and it is easy in that case to compute δ; however, we do not know a direct combinatorial proof of the last question (that is, the combinatorial proof asked for in this exercise!); such a proof could be interesting, since it could extend to other symbolic systems, such as the systems generated by some substitution.

We see that pairs of biinfinite sequences that differ only in two points play a particular role; they are preimages of the orbit of the special sequence used in the last section. It is not difficult to show that, for each $n \in \mathbb{Z}$, there are exactly two sequences that differ exactly in n and $n+1$.

Definition 6.2.14. *The* fixed sequences *of a Sturmian system X_u are the two sequences that differ exactly at indices 0 and 1.*

These two sequences play an important role in the next section; they project to the preimages of order two of the special sequence.

In the next section, we will define a renormalization procedure for Sturmian sequences, and show that fixed sequences renormalize to fixed sequences; we will then be able to produce these sequences as fixed points for an infinite product of substitutions, hence the name.

6.3 Coding sequences for Sturmian sequences and Sturmian systems

A Sturmian sequence contains very little information: among the 2^n possible words of length n, it uses only $n+1$ words (in particular, this proves that the topological entropy of this sequence is zero). For example, if the sequence is of type 0, every **1** is isolated, so the next letter must be a **0**, and gives no new information. Hence, it must be possible to recode this sequence in a more compact way. The basic idea is to group, in a sequence of type 0, any **1** with an adjacent **0**, to obtain a new sequence. It turns out that a small miracle occurs: this new sequence is again Sturmian, and the procedure can be iterated; with the given Sturmian sequence, we can associate an infinite coding sequence, and the initial word of length n of the Sturmian sequence is, for a typical Sturmian sequence, defined by an initial word of the coding sequence of length of order $\log n$.

The procedure is simple, but there are difficulties: first of all, the coding procedure is not canonical: for a sequence of type 0, we can group every **1** with the preceding or the next **0**, that is, we can use two different substitutions, and similarly for sequences of type 1. We will see below that there is also a problem for the coding of the initial letter, and that we can recode using a suffix or a prefix; this gives a total of 16 possible coding procedures. The choice of the procedure is arbitrary, and several of them can be convenient for different purposes, as we shall see.

A second problem is that, if we consider one-sided infinite sequences, the coding is not well defined for all sequences. We will see in exercises that the sequences in the positive orbit of the special sequence admit two possible codings. However, this is not really a serious problem: for a given Sturmian system, it arises only on a countable set; but, this makes exact proofs and statement quite cumbersome. This difficulty disappears for biinfinite sequences; for that reason, from now on, we will only consider biinfinite sequences, except in exercises.

A third problem is that, even for biinfinite sequences, there are pairs of sequences (the fixed sequences of the system and their negative orbit) that admit the same coding. The underlying reason is that these pairs of sequences correspond to the same point for the associated circle rotation.

An important question is then to identify all possible coding sequences for Sturmian sequences. It turns out that the admissible coding sequences form a sofic system, defined by a simple automaton, and even a shift of finite type for a well-chosen coding procedure. (We recall that a sofic system is a set of sequences that are given as labels of infinite paths on a finite graph with labelled edges, while a subshift of finite type is a set of sequences defined by a finite number of forbidden words. Is is easy to check that a subshift of finite type is sofic, while the converse is not true: for example, the set of sequences that contain only runs of **0**'s of even length is a sofic system, given by a very simple graph, and it is not of finite type. More details shall be found in Chap. 7.)

A last remark: it is possible to code a given sequence, or the language associated with this sequence; coding the language is the same thing as associating a coding sequence with the dynamical system generated by the given sequence. We will first, since it is easier, recode Sturmian systems; a system is completely defined by its frequency α, and in Sec. 6.4, we will see that the coding we get is closely related to the continued fraction expansion of α. The coding sequence for a rotation sequence defines the angle α and the initial point β, and we will give arithmetical interpretations of these codings in Sec. 6.4.

6.3.1 Recoding a Sturmian sequence

From now on, and for the rest of this chapter, unless specifically stated, we consider only biinfinite Sturmian sequences: some basic propositions in the sequel are false for one-sided sequences, and the recoding process is more delicate to define for such sequences (although this is possible, as indicated in exercises).

As we remarked at the beginning, in a Sturmian sequence, one of the letters is always isolated. We recall that the sequence is of type 0 if **1** is isolated, of type 1 otherwise. It is clear that any Sturmian sequence of type 0 can be written using only the words **0** and **10**. We will make this formal:

Definition 6.3.1. *We denote by σ_0 the substitution defined by $\sigma_0(\mathbf{0}) = \mathbf{0}$, $\sigma_0(\mathbf{1}) = \mathbf{10}$, and by σ_1 the substitution defined by $\sigma_1(\mathbf{0}) = \mathbf{01}$, $\sigma_1(\mathbf{1}) = \mathbf{1}$. We will also later use the substitutions τ_0 and τ_1 defined in a symmetric way by $\tau_0(\mathbf{0}) = \mathbf{0}$, $\tau_0(\mathbf{1}) = \mathbf{01}$, and $\tau_1(\mathbf{0}) = \mathbf{10}$, $\tau_1(\mathbf{1}) = \mathbf{1}$.*

These substitutions extend in a natural way to finite words (as morphism of the free monoid), to infinite sequences, and to biinfinite sequences:

Definition 6.3.2. *Let σ be a substitution, and let v be a biinfinite sequence; the image $u = \sigma(v)$ of v by σ is the only biinfinite sequence such that, for all positive integers n, $\sigma(v_0 v_1 \dots v_n)$ is a prefix of $u_0 u_1 \dots$, and $\sigma(v_{-n} \cdots v_{-2} \cdots v_{-1})$ is a suffix of $\dots u_{-2} u_{-1}$.*

Lemma 6.3.3. *If u is Sturmian of type 0 (respectively type 1), then there exists a unique v such that either $u = \sigma_0(v)$, or $u = S\sigma_0(v)$ (respectively $u = \sigma_1(v)$ or $u = S\sigma_1(v)$).*

Exercise 6.3.4. Give a formal proof of this lemma; remark that an image $u = \sigma_0(v)$ can never satisfy $u_{-1}u_0 = \mathbf{10}$, hence the shift $S\sigma_0(v)$ is necessary in that case; remark also that the lemma is false for infinite sequences: a sequence u of type 0 whose first letter is **0** can be written both as $\sigma_0(v)$, with $v_0 = \mathbf{0}$, and as $S\sigma_0(v)$, with $v_0 = \mathbf{1}$. This is the main advantage of working with biinfinite sequences.

The remarkable fact is that v is Sturmian:

Lemma 6.3.5. *A biinfinite sequence v is Sturmian, if and only if $\sigma_0(v)$ is Sturmian.*

Proof. It is easy to prove that v is positively eventually periodic if and only if $\sigma_0(v)$ is positively eventually periodic. Hence we can restrict to the case where both v and $\sigma_0(v)$ are not eventually periodic.

Suppose that v is Sturmian. If $\sigma_0(v)$ is not Sturmian, by Lemma 6.1.7, there is a (possibly empty) word W such that $\mathbf{1}W\mathbf{1}$ and $\mathbf{0}W\mathbf{0}$ occur in $\sigma_0(v)$. But $\sigma_0(v)$, by construction, contains only isolated $\mathbf{1}$'s. Hence we can write $W = \mathbf{0}V\mathbf{0}$, and $V\mathbf{0} = \sigma_0(V')$. Then, taking inverse images, v must contain $\mathbf{1}V'\mathbf{1}$ and $\mathbf{0}V'\mathbf{0}$, which is impossible.

Suppose that v is not Sturmian; again, there is a (possibly empty) word W such that $\mathbf{0}W\mathbf{0}$ and $\mathbf{1}W\mathbf{1}$ occur in v. Since the sequence v is biinfinite, it contains the word $\mathbf{a0}W\mathbf{0}$, where $\mathbf{a}$ is an arbitrary letter. Then $\sigma_0(v)$ contains the words $\sigma_0(\mathbf{1})\sigma_0(W)\sigma_0(\mathbf{1}) = \mathbf{10}\sigma_0(W)\mathbf{10}$ and $\sigma_0(\mathbf{a})\sigma_0(\mathbf{0})\sigma_0(W)\sigma_0(\mathbf{0}) = \sigma_0(\mathbf{a})\mathbf{0}\sigma_0(W)\mathbf{0}$. Because, in any case, $\sigma_0(\mathbf{a})$ ends with $\mathbf{0}$, we see that $\sigma_0(v)$ contains the words $\mathbf{10}\sigma_0(W)\mathbf{1}$ and $\mathbf{00}\sigma_0(W)\mathbf{0}$: therefore, it is not balanced. ■

Using Lemma 6.3.5, we can define a "recoding map" on Sturmian sequences:

Definition 6.3.6. *We denote by $\Phi : \Sigma \to \Sigma$ the map defined by $\Phi(u) = v$, where v is the unique sequence such that $u = \sigma_i(v)$, if it exists, or else the unique sequence such that $u = S\sigma_i(v)$, with $i = 0$ or 1.*

Note that a similar definition for one-sided sequence would be more difficult, since we loose in that case the unicity of the recoded sequence; see Exercise 6.3.11 at the end of this section.

Definition 6.3.7. *We denote by Σ_0 (respectively Σ_1) the set of Sturmian sequences of type* 0 *(respectively* 1*).*

We denote by $\Sigma_0^{\mathbf{0}}$ (respectively $\Sigma_0^{\mathbf{1}}$, $\Sigma_1^{\mathbf{0}}$, $\Sigma_1^{\mathbf{1}}$) the set of Sturmian sequences of type 0 such that the initial letter is **0** *(respectively of sequences of type 0 with initial letter* **1***, of type 1 with initial letter* **0***, of type 1 with initial letter* **1***).*

This partition is nicely related to the map Φ:

Exercise 6.3.8. Prove the following properties:

1. $u = \sigma_0(\Phi(u))$ if and only if $u \in \Sigma_0$ and u and $\Phi(u)$ have the same initial letter.
2. $u = S\sigma_0(\Phi(u))$ if and only if $u \in \Sigma_0$ and u and $\Phi(u)$ have a different initial letter, in which case $u_0 = \mathbf{0}$ and $\Phi(u)_0 = \mathbf{1}$.

3. $\Phi(\Sigma_0^{\mathbf{0}}) = \Phi(\Sigma_1^{\mathbf{1}}) = \Sigma$, and Φ restricted to $\Sigma_0^{\mathbf{0}}$ (respectively $\Sigma_1^{\mathbf{1}}$) is one-to-one.
4. $\Phi(\Sigma_0^{\mathbf{1}}) = \Sigma_0^{\mathbf{1}} \cup \Sigma_1^{\mathbf{1}}$, and Φ restricted to $\Sigma_0^{\mathbf{1}}$ is one-to-one.

This exercise proves, among other things, that the partition by the four sets $\Sigma_\varepsilon^{\mathbf{a}}$ is a Markov partition for the map Φ (see Chap. 7 for a precise definition of Markov partitions).

Associated with these partitions, we can define maps that will be useful in the next subsections:

Definition 6.3.9. *We denote by τ (for "type") the map $\tau : \Sigma \to \{0,1\}$ defined by $\tau(u) = 0$ if u is of type 0, $\tau(u) = 1$ if u is of type 1.*

We denote by η the map $\eta : \Sigma \to \{\mathbf{0}, \mathbf{1}\}$, $u \mapsto u_0$ that sends u to its initial letter.

We denote by γ the map defined by $\gamma(u) = (\tau(u), \eta(u))$. This map tells us in which of the sets $\Sigma_i^{\mathbf{j}}$ is the sequence u.

It is easy to prove that the fixed sequences, as defined in the previous section, behave nicely under the map Φ.

Exercise 6.3.10. Prove that, if u and u' are a pair of fixed sequences that differ exactly at indices 0 and 1, the same is true for the recoded sequences $\Phi(u)$ and $\Phi(u')$.

It is also possible to define a recoding map for one-sided infinite sequences, but there are some technical problems, as shown in the following exercise.

Exercise 6.3.11. 1. Prove that, for any one-sided sequence u of type 0, there exists a unique sequence v such that $u = \sigma_0(v)$, and that Sv is Sturmian.
2. Prove that v is not always Sturmian.
3. Prove that, if u, of type 0, is not a special sequence, there exists a unique Sturmian sequence v such either $u = \sigma_0(v)$ or $u = S\sigma_0(v)$.
4. Prove that if u is a special sequence of type 0, there exists a special sequences v such that $u = \sigma_0(\mathbf{0}v) = S\sigma_0(\mathbf{1}v)$.

It is now possible, by iterating the map Φ and coding with respect to a partition (that is, by using maps τ or γ), to associate a coding sequence with any biinfinite Sturmian sequence. We will do so in the next subsection for the fixed sequences, giving a very simple rule for their generation, and for Sturmian systems; we consider general Sturmian sequences in the following subsection.

6.3.2 Coding sequences for Sturmian systems

We can recode Sturmian systems, by using the following lemma:

Lemma 6.3.12. *Let u and u' two Sturmian sequences in the same Sturmian system X; if u and u' are recoded respectively to v and v', v and v' belong to the same Sturmian system. In particular, v and v' have the same type.*

Proof. We can suppose that u and u' are of type 0. It is enough to prove that, if V is the initial word of v of length n, it occurs in v'. But V must occur infinitely often in v, otherwise $S^k v$ should be of complexity at most $P(n) = n$ for some k, and should be eventually periodic. Hence we can suppose that V is included in some word W, occurring in v, that starts with $\mathbf{1}$; the word $\sigma_0(W)$ occurs in u, so it occurs in u', and, because it begins with $\mathbf{1}$ and ends with $\mathbf{0}$, it can be recoded in only one way by σ_0, as W. This proves that W, and V, occur in v'. ∎

This lemma shows that, for any Sturmian sequence, u, the coding sequence $(\tau(\Phi^n(u)))_{n\in\mathbb{N}}$ only depends on the Sturmian system X_u.

Definition 6.3.13. *The* (additive) coding sequence *of a Sturmian system X, is the sequence $\tau(\Phi^n(u)))_{n\in\mathbb{N}}$, for any sequence $u \in X$.*

This definition raises two natural questions: what are the *admissible sequences*, that is, the sequences that can occur as coding sequence for a Sturmian system? Is a Sturmian system completely defined by its additive coding sequence? We have a complete and constructive answer for the latter question:

Proposition 6.3.14. *A Sturmian system is completely defined by its additive coding sequence.*

Proof. We will be a little more precise, and prove that at least one (in fact both) of the fixed points of the system is completely defined by the additive coding sequence.

Let $(i_n)_{n\in\mathbb{N}}$ be the coding sequence, with $i_n = 0$ or 1. Using Exercise 6.3.10, we see that the two fixed points u, u' (with $u_0 = \mathbf{0}$, $u'_0 = \mathbf{1}$) of the system can be written $u = \sigma_{i_0}\sigma_{i_1}\ldots\sigma_{i_n}(v)$ and $u' = \sigma_{i_0}\sigma_{i_1}\ldots\sigma_{i_n}(v')$, where v (respectively v') is the fixed point of the recoded system that has same initial letter as u (respectively u'). It is then enough to prove that an arbitrarily long initial word of one of the 2 sequences u or u' is determined, independently of v and v'.

Indeed, define the sequences of finite words $U_{n+1} = \sigma_{i_0}\sigma_{i_1}\ldots\sigma_{i_n}(\mathbf{0})$ and $U'_{n+1} = \sigma_{i_0}\sigma_{i_1}\ldots\sigma_{i_n}(\mathbf{1})$. It is immediate that the words U_n (respectively U'_n) are prefix of u (respectively u'), and it is easy to prove that the sequence $|U_n| + |U'_n|$ is strictly increasing; hence at least one of these sequences of words has a length that tends to infinite, so that it completely defines the corresponding fixed word. The other one is obtained from the first by exchanging the initial $\mathbf{0}$ and $\mathbf{1}$. But then the language, hence the system, is completely determined. ∎

Exercise 6.3.15 (Rauzy rules). There is a very simple way to compute these initial words: let i_n be a sequence of 0's and 1's which is not eventually constant, and let U_n and U'_n be the sequence of words defined by:

- $U_0 = \mathbf{0}$, $U'_0 = \mathbf{1}$;
- $U_{n+1} = U_n$, $U'_{n+1} = U'_n U_n$ if $i_n = 0$,
- $U_{n+1} = U_n U'_n$, $U'_{n+1} = U'_n$ if $i_n = 1$.

1. Prove that U_n (respectively U'_n) are initial words of the fixed point that has $\mathbf{0}$ (respectively $\mathbf{1}$) as initial letter in the Sturmian system with coding sequence (i_n).
2. Prove that, except for U_0 or U'_0, all the words U_n and U'_n are suffixes of the same infinite negative sequence, and that this sequence is a Sturmian sequence with two right extensions.
3. Prove that if we remove from U_n or U'_n a prefix of length 2, the remaining part is a palindrome; the words that we obtain in this way are the bispecial factors.

A first application of this coding is the following:

Proposition 6.3.16. *The system generated by a Sturmian sequence is minimal.*

Proof. Recall that a system is minimal if it does not contain a nonempty closed invariant subset, or equivalently, if the orbit of any element is dense.

If the system generated by u is not minimal, it contains a sequence v whose orbit is not dense. Then, there is a word U that occurs in u, but not in v. This implies that $p_v(n) < n + 1$ for some n, so that v is eventually periodic. Because the language of v is contained in that of u, there is a finite nonempty word W such that W^n occurs in u for all n. We will prove that this is impossible.

Since the sequence u is balanced, it cannot contain sequences of $\mathbf{0}$ or $\mathbf{1}$ of arbitrary length. Hence, W cannot be constant. Suppose that u is of type $\mathbf{0}$; after a cyclic permutation, we can suppose that the word W begins with $\mathbf{1}$ and ends with $\mathbf{0}$, and W can be recoded in a unique way in a word W' which is strictly shorter. Consider the sequence u' obtained by recoding u: its language contains all the powers of W'. We can iterate the procedure, and at each step we reduce the length of the word whose powers all belong to the language. After a finite number of steps, we get a word of length 1, so that the corresponding Sturmian sequence contains arbitrarily long sequences of $\mathbf{0}$ or $\mathbf{1}$: we just saw that this is impossible (we recover here the results of Exercise 6.1.14). ■

Exercise 6.3.17. Prove directly, using Exercise 6.1.12, that a Sturmian dynamical system is minimal, and deduce that every point in this system is a Sturmian sequence, with the same frequency.

It is now easy to find the admissibility condition for additive coding sequences, thus solving the question above:

Proposition 6.3.18. *A sequence with value in* $\{0, 1\}$ *is an admissible coding sequence for a Sturmian system if and only if* 0 *and* 1 *occur an infinite number of times, or equivalently if and only if it is not eventually constant.*

Proof. Consider a Sturmian sequence of type 0; let k be the minimum number of **0**'s between two **1**'s occurring in the sequence. Then the sequence can be recoded exactly k times using σ_0, to a sequence of type 1. By the same argument, any string of 0 or 1 in the additive coding sequence must be of finite length, and the sequence cannot be eventually constant.

Conversely, given a not eventually constant sequence, consider the set of words $U_n = \sigma_{i_0}(\sigma_{i_1}(\cdots(\sigma_{i_n}(\mathbf{0}))\cdots))$ of Exercise 6.3.15; it is easy to prove that these words are prefix of each other, and define a Sturmian sequence. The associated system has the given sequence $(i_n)_{n\in\mathbb{N}}$ as coding sequence. ■

We can rewrite in another way the additive coding of X. If we group strings of 0 and 1, we can write the sequence as $0^{a_0}1^{a_1}\ldots 0^{a_{2n}}1^{a_{2n+1}}\ldots$.

Definition 6.3.19. *This sequence (a_n) of integers, all strictly positive except maybe a_0, is called the* multiplicative coding sequence *for the system.*

We will see in the next part that this sequence is closely related to the frequency α: it is in fact the continued fraction expansion of $\frac{1-\alpha}{\alpha}$. The first sequence is related to the additive algorithm for the continued fraction expansion of the same number, this is the reason for the name.

The following is an immediate consequence of the previous proposition:

Proposition 6.3.20. *A sequence (a_n) of integers is an admissible multiplicative coding sequences if and only if $a_0 \geq 0$ and $a_n > 0$ for $n \geq 1$.*

The first integer can be 0, because of our convention that a_0 is the number of leading 0's, and the sequence could be of type 1.

Exercise 6.3.21. It is also possible to consider balanced eventually periodic sequences. Prove that the associated dynamical system is countable, with one finite orbit and one or two countable orbits asymptotic to the finite orbit in both directions, and that, with this system, one can associate an additive coding sequence that is eventually constant; prove that the frequency of such a system is rational. *(Hint: you can start by considering the sequences defined in a self-evident notation by: $\mathbf{0}^\infty$, $\mathbf{0}^\infty\mathbf{1}\mathbf{0}^\infty$, and $(\mathbf{01})^\infty$, $(\mathbf{01})^\infty(\mathbf{10})^\infty$, $(\mathbf{10})^\infty(\mathbf{01})^\infty$.)*

6.3.3 Coding sequences for Sturmian sequences

We saw in the preceding subsection that any Sturmian system can be defined in a unique way by its additive coding sequence, related to a sequence of substitutions. We would like to do the same thing with a Sturmian sequence, that is, to associate with it a coding sequence that defines it in a unique way.

Definition 6.3.22. *Let u be a Sturmian sequence. The* additive coding sequence *of u is the sequence $(\gamma(\Phi^n(u)))_{n\in\mathbb{N}}$.*

The two usual questions arise: does the additive coding sequence completely define the Sturmian sequence? What are the admissible coding sequences?

Proposition 6.3.23. *The additive coding sequence completely defines the Sturmian sequence, except if it is in the negative orbit of the fixed points; pairs of preimages of same rank of the two fixed points of a system have the same coding.*

The admissible sequences are the non-eventually constant sequences of vertices for infinite paths in the graph of Fig. 6.3, where we have represented all the possible transitions, labeling each edge by the corresponding transformation (we have given each vertex the name of the set $\Sigma_\varepsilon^{\mathbf{a}}$ instead of $(\varepsilon, \mathbf{a})$).

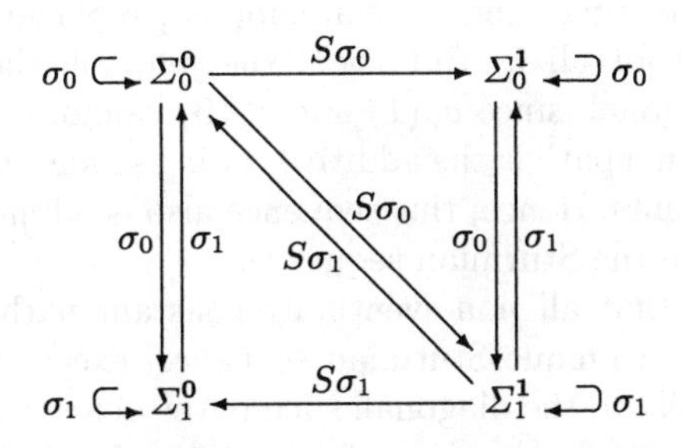

Fig. 6.3. The transition graph for additive coding.

Proof. Let us first examine the admissibility condition.

It is clear that if a sequence u is equal to $S\sigma_0\Phi(u)$, then $\gamma(u) = (0, \mathbf{0})$ (it is of type 0, because we use substitution σ_0, and it begins with $S\sigma_0(\mathbf{1}) = \mathbf{0}$), and we have $\gamma(\Phi(u)) = (0, \mathbf{1})$ or $(1, \mathbf{1})$. A similar study for the other cases proves that the possible arrows must be as indicated on Fig. 6.3. Hence, all admissible coding sequences must correspond to paths in the graph.

As we saw above, there must be an infinite number of changes of type in the sequence $(\Phi^n(u))$; this is equivalent to saying that the path must not be eventually constant, since any path staying on the upper or lower level is eventually constant.

It remains to prove that any non-eventually constant path in the graph is admissible.

Consider now an infinite path in the graph, and the corresponding sequence $(\varepsilon_n, \mathbf{a}_n)$. From this sequence, we deduce a sequence c_n, with $c_n = \sigma_{\varepsilon_n}$ or $S\sigma_{\varepsilon_n}$ (if there exists a sequence u with coding sequence $(\varepsilon_n, \mathbf{a}_n)$, we must have $\Phi^n(u) = c_n\Phi^{n+1}(u)$).

The sequence of words $c_0c_1 \dots c_n(\mathbf{a}_{n+1})$ defines an increasingly large part $u_{i_n} \dots u_0 \dots u_{j_n}$ of u. If both i_n and j_n are unbounded, the Sturmian sequence u is completely defined by $(\varepsilon_n, \mathbf{a}_n)$; hence it exists, and the sequence is admissible.

If i_n is bounded, then after a finite time there are no more shifts, and c_n is eventually of the form σ_0 or σ_1. Without loss of generality, we can suppose that all the c_n are of this form, or equivalently, that all the $\Phi^n(u)$ have same initial letter, and $(\mathbf{a}_n)$ is constant. But then, it is clear that u must be one of the two fixed points (the choice of the fixed point is given by the letter). Remark that, formally, only the positive part of the sequence is defined by the coding; however, the fixed point has a unique left extension, since it is in the negative orbit of the special sequence, and so it cannot be also in its positive orbit. Hence in that case also, the sequence is admissible.

There remains to check the case of a bounded sequence (j_n). It is then constant after a finite time, and we can suppose, replacing u by a suitable $\Phi^n(u)$, that $j_n = 0$ for all n. But then, the path on the graph must be restricted to the diagonal, since $\sigma_0(\mathbf{1})$ and $\sigma_1(\mathbf{0})$ cannot occur. It is easy to check that this diagonal path is the additive coding sequence of the preimages of the two special points. Hence, this sequence also is admissible, but it does not completely define the Sturmian sequence.

We have proved that all non-eventually constant paths are admissible, and that they define a unique Sturmian sequence, except in the case when they restrict eventually to the diagonal; such paths define a pair of preimages of same order of the two fixed points of some Sturmian system. ■

Remark. The basic idea of the theorem is that, in a Sturmian system, there is a one-to-one correspondence between Sturmian sequences and coding sequences, except on a countable set (negative orbit of the special sequence), but the precise details can seem puzzling. We will make the picture more clear in the next section: with a Sturmian system, we will associate a rotation on the circle. With any point of the circle, we associate a unique Sturmian sequence and a unique coding sequence, except for the orbit of 0; each point of the positive orbit of 0 corresponds to a pair of Sturmian sequences, and to a pair of coding sequences that are eventually, one of the form $(\varepsilon_n, \mathbf{0})$ and the other one of the form $(\varepsilon_n, \mathbf{1})$; each point of the negative orbit of 0 corresponds to a pair of Sturmian sequences and a unique coding sequence.

Exercise 6.3.24. We had to exclude from consideration eventually constant coding sequences; it would be nicer to allow also these sequences, so that the set of admissible sequences should be defined by an automaton. Explain how this could be done, by allowing all balanced sequences, including eventually periodic sequences.

Remark. Let $(\varepsilon_n, \mathbf{a}_n)$ be an additive coding sequence for a Sturmian sequence u, and consider the natural projection of this sequence to (ε_n); the latter sequence is an additive coding sequence for a Sturmian system X, and it is immediate to check that u belongs to X.

It is also possible to characterize an orbit of the Sturmian system in this way:

Lemma 6.3.25. *Let u be a Sturmian sequence, and $v = S^k u$. Then there exists $k' \leq k$ such that $\Phi(v) = S^{k'}\Phi(u)$ is a recoded sequence for u. Moreover, if u is of type 0 (respectively 1), we have $k' = k$ if and only if $u_0 \dots u_{k-1} = \mathbf{0} \dots \mathbf{0}$ (respectively $u_0 \dots u_{k-1} = \mathbf{1} \dots \mathbf{1}$).*

Proof. Suppose that u is of type 0. Then there is a unique k' such that $\sigma_0(S^{k'}\Phi(u)) = v$ or $S\sigma_0(S^{k'}\Phi(u)) = v$, so that $S^{k'}\Phi(u)$ is $\Phi(v)$. A simple case study shows that $k' < k$, except if all the first letters of u are $\mathbf{0}$. ∎

Proposition 6.3.26. *Let u be a Sturmian sequence, and $v = S^k u$. Then $\Phi^n(u)$ coincides eventually with $\Phi^n(v)$, except if u is in the negative orbit of a fixed point, and v in the positive orbit of this fixed point.*

Proof. By the preceding lemma, we have $\Phi^n(v) = S^{k_n}\Phi^n(u)$, where k_n is a decreasing sequence of integers. If it tends to 0, the proof is finished. Otherwise, it is constant, equal to k after a rank N. This means that, for $n > N$, u^n begins with k $\mathbf{0}$'s if it is of type 0, and with k $\mathbf{1}$'s if it is of type 1. It suffices to consider the case of a sequence of type 0 that recodes to a sequence of type 1 (this must occur infinitely often in the family) to check that we must have $k = 1$. We must then have $\Phi^n(v) = \sigma_i \Phi^{n+1}(v)$ for $n \geq N$, hence $\Phi^N(v)$ is a fixed point of its system. We conclude that v must be in the positive orbit of a fixed point, and u in the negative orbit of the same fixed point. ∎

6.3.4 Multiplicative coding sequences for Sturmian sequences

As in Sec. 6.3.2, we can rearrange the additive coding sequence in blocks of symbols of same type 0 or 1. The idea is to "accelerate" the coding map Φ.

Definition 6.3.27. *We define the map $\Psi : \Sigma \to \Sigma$ by $\Psi(u) = \Phi^{n_u}(u)$, where n_u is the smallest strictly positive integer such that $\Phi^n(u)$ is not of the same type as u.*

It is clear from the graph in Fig. 6.3 that, if u is of type 0, we have $u = \sigma_0^a \Psi(u)$ if u and $\Psi(u)$ have same initial letter, and $u = \sigma_0^n S \sigma_0 \sigma_0^p$ if they have different initial letter; in the latter case, the initial letter of u is $\mathbf{0}$, and the initial letter of $\Psi(u)$ is $\mathbf{1}$.

We can simplify this expression, using the following exercise:

Exercise 6.3.28. Prove that, if v is a sequence with initial letter $\mathbf{1}$, we have the equality $\sigma_0^n(Su) = S^{n+1}\sigma_0^n(u)$.

Hence, if u is of type 0, we can always write $u = S^k \sigma_0^a \Psi(u)$, with $0 \leq k \leq a$, and a similar property for Sturmian sequences of type 1; this is a more convenient way to denote the blocks.

Definition 6.3.29. *The first multiplicative coding sequence of a Sturmian sequence u is the unique sequence (a_n, k_n) such that:*

- *if u is of type 0, $\Psi^n(u) = S^{k_n} \sigma_{\varepsilon_n}^{a_n} \Psi^{n+1}(u)$, with $\varepsilon_n = n \mod 2$;*
- *if u is of type 1, $(a_0, k_0) = (0, 0)$ and $\Psi^n(u) = S^{k_{n+1}} \sigma_{\varepsilon_n}^{a_{n+1}} \Psi^{n+1}(u)$, with $\varepsilon_n = n + 1 \mod 2$.*

This expression gives a different role to sequences of type 0 and 1; the reason is that the sequence (a_n, k_n) does not indicate the type of the initial substitution. If one insists on keeping the symmetry, it is possible to add to the coding sequence the type of the Sturmian sequence, and we can then suppose $a_0 > 0$.

The properties of this multiplicative coding sequences are summarized in the following theorem:

Theorem 6.3.30. *The admissible multiplicative coding sequences are the double sequences of integer (a_n, k_n), with $a_0 \geq 0$, $a_n > 0$ if $n > 0$, and $0 \leq k_n \leq a_n$, that are labels of paths in the infinite graph of Fig. 6.4.*

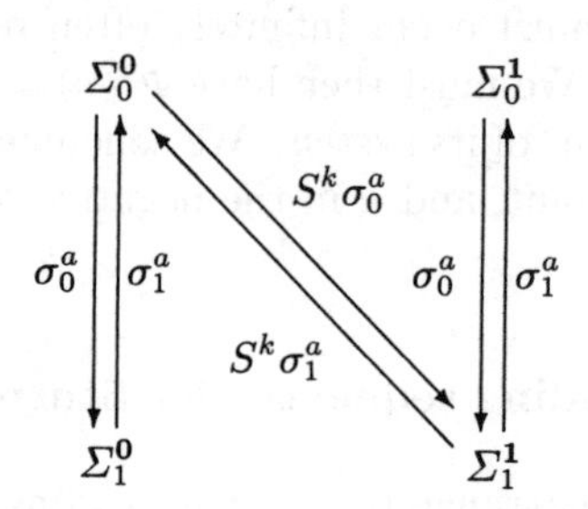

Fig. 6.4. The transition graph for multiplicative coding.

Each Sturmian sequence admits a unique multiplicative coding sequence.
Each coding sequence defines a unique Sturmian sequence, except the sequences that satisfy $k_n = 0$ for all n, which define the two fixed points of a system, and the sequences that satisfy eventually $k_n = a_n$, which define two preimages of the same order of the two fixed points.

Proof. The admissibility condition is immediate by looking at the graph for additive coding sequence: it suffices to consider all finite paths that stay on one level, except for the last state. The unicity condition is then just a reformulation of the additive case. ■

Exercise 6.3.31. Reformulate the definition of the coding sequence by defining a Markov partition for the map Ψ (it may be convenient to change slightly the definition to make it more symmetric with respect to type).

We can reformulate the admissibility condition:

Proposition 6.3.32. *A sequence (a_n, k_n) is an admissible multiplicative coding sequence if it satisfies $a_0 \geq 0$, $a_n > 0$ if $n > 0$, $0 \leq k_n \leq a_n$, and the additional condition: in the sequence (k_n), any maximal block of 0, except maybe the initial block, is of even length.*

Proof. After coding by $S^k \sigma_0^a$, with $k > 0$, we must be in state $(1,1)$. If we recode by σ_1^b, we get to the state $(0,1)$, and it is clear on the graph that the next coding symbol can only be of the type σ_0^c; the proof is then clear by induction. ■

The set of admissible coding sequences forms what we call a sofic system (on an infinite alphabet), since it is defined by the set of admissible paths on a graph with a finite number of vertices; it is however not a finite type system (that is, a symbolic system where we can check whether a sequence is admissible by looking at subwords of bounded lengths), since we might need to consider arbitrarily long words to check whether a given sequence is admissible. We will make a small modification to improve this.

As we remarked at the beginning of this section, when we recode a sequence u of type 0, it is not always possible to write $u = \sigma_0(v)$; we have then the choice to write $u = S\sigma_0(v)$, or $u = S^{-1}\sigma_0(w)$, where $v_0 = \mathbf{1}$ and $w = Sv$. If we recode n times, and if we call σ the product of the n substitutions involved, the first viewpoint leads to $u = S^k\sigma(v)$, and the second to $u = S^{-b}\sigma(w)$, where (if $k > 0$) $w = Sv$. If $v_0 = \mathbf{a}$, we check that $k + b = |\sigma(\mathbf{a})|$.

We can build another coding, the second multiplicative coding sequence, using this idea; we just state the theorem, and explain the relation to the first coding sequence:

Theorem 6.3.33. *For any Sturmian sequence u, we can define a sequence (a_n, b_n), with $b_n \leq a_n$, and a family of recoded Sturmian sequences $w^{(n)}$, such that $w^{(2n)} = S^{-b_{2n}}\sigma_0(w^{(2n+1)})$ and $w^{(2n+1)} = S^{-b_{2n+1}}\sigma_1(w^{(2n+2)})$. The coding sequence (a_n, b_n) is uniquely defined. The sequence (a_n, b_n) uniquely defines the sequence u, except if b_n is eventually 0, in which case it defines two preimages of the same order of the two fixed points.*

The set of admissible sequences is completely characterized by the following properties:

- *all the a_n are strictly positive integers, except maybe the first one;*
- *the integers b_n satisfy $b_n \leq a_n$;*
- *if $b_{n+1} = a_{n+1}$, then $b_n = 0$.*

These admissibility conditions are summarized in the graph of Fig. 6.5.

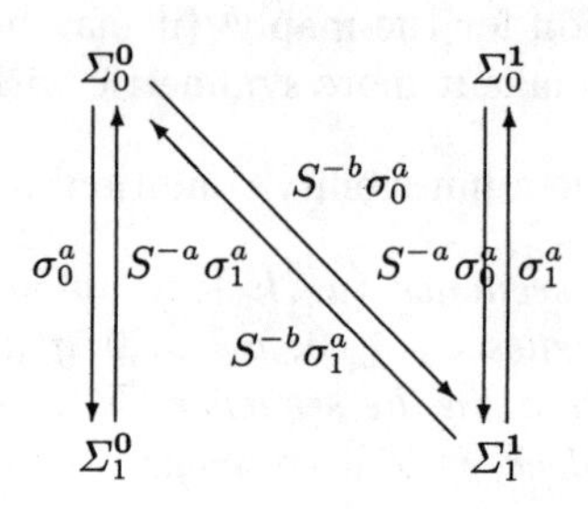

Fig. 6.5. The transition graph for the second multiplicative coding.

We leave the proof as an exercise, since it is completely similar to the preceding one, the basic idea being to study precisely the first step of the coding process.

Definition 6.3.34. *The sequence (a_n, b_n) is called the second multiplicative coding of u.*

The interest of this second coding is that the admissibility condition is now of finite type; we see that, as for the first multiplicative coding, there is a one-to-one relation between Sturmian sequences and coding sequences, except on the orbit of the fixed sequences. It is in fact easy to compute one of the coding sequences, knowing the other, by the next proposition:

Proposition 6.3.35. *Let u be a Sturmian sequence not in the positive orbit of the fixed sequences, and let (a_n, k_n) and (a_n, b_n) be the first and second multiplicative coding sequence of u. Let $N = \inf\{n \in \mathbb{N} | k_n \neq 0\}$.*

- *For $n < N$, we have $b_n = k_n = 0$.*
- *$b_N = 1 + a_N - k_N \neq 0$.*
- *For $n > N$ and $k_n \neq 0$, we have $b_n = a_n - k_n$.*
- *For $n > N$ and $k_n = 0$, if $k_{n-1} = b_{n-1} = 0$, then $b_n = a_n$; otherwise, $b_n = 0$.*

Proof. The idea is to consider the family of recoded sequences $v^{(n)}$ and $w^{(n)}$ associated with the two coding processes; it is easy to prove that $w^{(n)} = v^{(n)}$ for $n < N$, and $w^{(n)} = Sv^{(n)}$ for $n \geq N$. The condition then follows, by studying the first letter of v_n. ■

Exercise 6.3.36. Define a map and a partition associated with this coding; prove that the map is infinite-to-one.

Exercise 6.3.37. Explain how one could define a similar coding for infinite (one-sided) Sturmian sequences. *(Hint: it does not make sense in this setting* *to write* $u = S^{-b}\sigma(w)$*; however, it is possible to write, in a unique way under suitable conditions,* $u = X\sigma(w)$*, where* $X = \mathbf{0}^b$ *or* $X = \mathbf{1}^b$*, depending on the type of* u*. One obtains in this way what is sometime called a prefix coding of* u*.)*

6.3.5 The dual additive coding for Sturmian sequences

In this section, we will define another coding that will be useful in Secs. 6.4 and 6.6. We leave the proofs to the reader.

As we stated at the beginning of this part, the choice of the two substitutions σ_0, σ_1 is somewhat arbitrary. We could use the substitutions τ_0, such that $\tau_0(\mathbf{0}) = \mathbf{0}$, and $\tau_0(\mathbf{1}) = \mathbf{01}$, and τ_1, such that $\tau_1(\mathbf{0}) = \mathbf{10}$, and $\tau_1(\mathbf{1}) = \mathbf{1}$ (this last one amounts to grouping any isolated $\mathbf{0}$ in a sequence of type 1 with the preceding $\mathbf{1}$, instead of the following $\mathbf{1}$). We can then recode any sequence of type 1 as $\tau_1(v)$ or $S\tau_1(v)$.

It is then possible to associate with each Sturmian sequence a *dual coding sequence*, using now $(\sigma_0, S\sigma_0, \tau_1, S\tau_1)$, as we did before.

It is now possible to associate with the pair (σ_0, τ_1) a dual multiplicative coding: first group the substitutions in blocks $\tau_1^{a_n}$ and $\sigma_0^{a_n}$; then recode the given sequence as $u = S^{k_n}\sigma_0^{a_n}(v)$ or $u = S^{k_n}\tau_1^{a_n}(v)$, depending on the case. Now, define a sequence b_n in the following way (this will be justified by the geometry in Sec. 6.6):

First consider the case when u is of type 0, so that $u = S^k\sigma_0^a(v)$:

- if $v_0 = \mathbf{0}$ (hence $u_0 = \mathbf{0}$ and $k = 0$) we define $b = a$;
- if $v_0 = 1$ then $b = \sup(0, k-1)$.

Then consider the case when u is of type 1, so that $u = S^k\tau_1^a(v)$:

- if $v_0 = \mathbf{1}$ (hence $u_0 = \mathbf{1}$ and $k = 0$) we define $b = a$;
- if $v_0 = \mathbf{0}$ then $b = \sup(0, a-1-k)$.

Exercise 6.3.38. Prove that, in this way, one can associate with any Sturmian sequence a double sequence of integers (a_n, b_n), where (a_n) is the coding sequence for the Sturmian system, and b_n satisfies $0 \leq b_n \leq a_n$, and $b_n = a_n$ implies $b_{n+1} = 0$.

Remark. For this coding, the admissibility condition is a condition of finite type, which is exactly similar to the condition we obtained in the last section, except that the direction is reversed; this is why we call this the dual coding. The next section gives an arithmetic and dynamic natural interpretation to this strange coding.

Exercise 6.3.39 (Review Exercise). The reader may feel confused by this variety of recoding sequences; we can organize them in a more systematic way.

We saw before that any coding map will use, for sequences of type 0, one of the two substitutions σ_0, τ_0, and the shift S or its inverse, and similarly for sequences of type 1. This gives 16 elementary coding maps.

We will denote $\Phi_{\varepsilon_0,\delta_0,\varepsilon_1,\delta_1}$, with all subscripts equal $+1$ or -1, the coding map given, on sequences of type 0, by the use of σ_0 if $\varepsilon_0 = 1$ and τ_0 if $\varepsilon_0 = -1$, and $S\sigma_0$ or $S\tau_0$ if $\delta_0 = 1$, or $S^{-1}\sigma_0$ or $S^{-1}\tau_0$ if $\delta_0 = -1$, and similarly for sequences of type 1, so that, for example, the map Φ defined above is equal to $\Phi_{1,1,1,1}$.

It is quite obvious that these coding are not all completely different, and we can make this more precise.

Let E be the flip that exchanges $\mathbf{0}$ and $\mathbf{1}$ (it is the trivial substitution $\mathbf{0} \mapsto \mathbf{1}, \mathbf{1} \mapsto \mathbf{0}$).

1. Prove that E is a one-to-one map on the set of all Sturmian sequences, and that it conjugates $\Phi_{\varepsilon_0,\delta_0,\varepsilon_1,\delta_1}$ to $\Phi_{\varepsilon_1,\delta_1,\varepsilon_0,\delta_0}$.
   Let R be the *retrogression* map, that is, the map that takes u the the *retrograde* sequence defined by $v_n = u_{-1-n}$ (this map reads u backwards, and exchange the positive and the strictly negative part of the sequence; for many reasons, it turns out to be useful to have no fixed letter, as one would get by the natural definition $v_n = u_{-n}$).
2. Prove that R is a one-to-one map on the set of Sturmian sequences, and that it preserves all Sturmian systems.
3. Prove that R conjugates $\Phi_{\varepsilon_0,\delta_0,\varepsilon_1,\delta_1}$ and $\Phi_{-\varepsilon_0,-\delta_0,-\varepsilon_1,-\delta_1}$.
4. Prove that we obtain in this way 6 classes of coding maps globally invariant by flip and exchange, 2 of them with 4 members, and 4 of them with 2 members.
5. Prove that the coding sequence of a Sturmian system does not depend on the choice of the coding map.

Remark. See [206], where an expansion of Sturmian sequences of the same type as above is used to get an explicit formula which computes the supremum of all real numbers $p > 0$ for which there exist arbitrarily long prefixes which are pth-powers.

6.3.6 Dynamical systems on coding sequences

The set of additive or multiplicative coding sequences is invariant by the shift; in this way, we can define a dynamical system, by shifting coding sequences. It is important to note that this new dynamical system is completely different from the Sturmian dynamical systems we considered in previous sections.

Let us denote by Γ the set of additive coding sequences; there is an almost one-to-one map from the set Σ of biinfinite Sturmian sequences to γ. This map conjugates the map Φ on Σ to the shift on Γ. It is easy to check that Φ is at least 2-to-1, and one can prove that it is of strictly positive topological entropy, and ergodic for a suitable measure (this is clear on Γ, since it is an irreducible shift of finite type).

One can also consider the shift on the set of all Sturmian sequences; we saw above that it is a union of disjoint Sturmian systems. To each Sturmian system, we can associate the coding sequences with a given sequence of substitutions; it is then possible to write explicitly the map corresponding on coding sequences to the Sturmian shift. This is what is called an adic system, and it is somewhat similar to an odometer; in particular, it satisfies a kind of commutation relation with the shift on coding sequences.

Exercise 6.3.40. Let (a_n, k_n) be the first multiplicative coding sequence of a Sturmian sequence u; write down explicitly the multiplicative coding sequence for Su. Do the same for the second multiplicative coding sequence, and for the dual multiplicative coding.

6.4 Sturmian sequences. Arithmetic properties and continued fractions

The main purpose of this section is to make effective the coding sequence we defined above.

More precisely, if we are given a Sturmian sequence, we have explained how to obtain, using the map Φ, its coding sequence. We now associate with a given multiplicative coding sequence (a_n, b_n), a rotation sequence whose angle and initial point are defined by explicit arithmetic formulas in term of a_n and b_n, using continued fraction expansion and Ostrowski expansion.

This proves in a constructive way that every Sturmian sequence is a rotation sequence.

In the first subsection, we study the induction of a rotation on a subinterval, and show that it is directly related to the recoding of Sturmian sequences. We show that there are several possibilities for induction, in one-to-one relation with the different recoding processes we already explained. We give explicit arithmetic formulas for a particular induction, obtained by iteration of the induction on the image of the largest continuity interval.

In Secs. 6.4.2 and 6.4.3, we give without proofs the basic facts on continued fractions and Ostrowski expansion; we apply these formulas in Sec. 6.4.4 to recover the angle and initial point of a rotation sequence from its coding sequence. We pay particular attention there to the possibility of multiple coding, and its geometric and arithmetic meaning.

We survey in Sec. 6.4.5 some other possibilities of induction algorithm, and in particular Rauzy induction and its link with the dual multiplicative algorithm.

6.4.1 Induction of rotations and recoding of rotation sequences

To make the following computations easier, we make a slight change of notation; we consider a rotation as an exchange of two intervals, and we renormalize these intervals so that the larger interval has length one. We can suppose without loss of generality that the largest interval is on the left (that is, the sequence is of type 0), and we denote by θ the length of the smallest one.

To be more formal, we are now studying the map R:

$$R: \quad [-1,\theta[\to[-1,\theta[$$
$$x \mapsto x+\theta \text{ if } x<0$$
$$x \mapsto x-1 \text{ if } x \geq 0.$$

For the corresponding standard rotation, defined on $[0,1[$, the angle α is given by $\alpha = \frac{\theta}{1+\theta}$.

The map R has two continuity intervals, which we will denote by $I_0 = [-1,0[$ and $I_1 = [0,\theta[$; we also give names to their images $J_0 = R(I_0) = [\theta-1,\theta[$ and $J_1 = R(I_1) = [-1,\theta-1[$.

Definition 6.4.1. *The* induced map *of R on J_0, denoted by $R_{|J_0}$, is the map defined on J_0 by $R_{|J_0}(x) = R^{n_x}(x)$, where n_x is the first return time of x to J_0, $n_x = \inf\{n>0 | R^n(x) \in J_0\}$.*

This map is directly related to the previous section; to make this relation explicit, we first define itineraries with respect to a partition.

Definition 6.4.2. *Let $T: X \to X$ be a map, and $(X_i)_{i\in\mathcal{A}}$ be a partition of X.*

For any point $x \in X$, the itinerary *of x under T with respect to the partition $(X_i)_{i\in\mathcal{A}}$ is the sequence u, with values in the set $\mathcal{A}$ of indices of the partition, defined by $u_n = i$ if $T^n(x) \in X_i$.*

Remark. It is immediate to show that the itinerary of any point under R with respect to the partition (I_0, I_1) (or J_0, J_1) is a *rotation sequence*, in the sense of Sec. 6.1.2.

Exercise 6.4.3. Explain how the angle and initial point are changed when we apply to the sequence the flip E and the retrogression R defined in the previous section.

Use this to prove that, in any Sturmian system generated by a rotation sequence, there are exactly two sequences that are invariant under retrogression (these are the biinfinite equivalent of the palindromes of even length; there is also one sequence that is invariant by the map that takes u to v defined by $v_n = u_{-n}$, it is the biinfinite equivalent of palindromes of odd length).

Give another characterization of these sequences: they are the coding of an orbit of the square billiards that hits the middle of a side; the other invariant sequence, generated by odd palindromes, is the coding of an orbit through the center of the square, and the fixed sequences are the two possible coding of an orbit through a corner; it is almost a palindrome, up to the two initial letters.

Lemma 6.4.4. *The map $R_{|J_0}$ is again a rotation (or an exchange of 2 intervals).*

If x is a point of J_0, let u be the itinerary of x under the map R with respect to the partition I_0, I_1, and v be the itinerary of x under the induced map $R_{|J_0}$ with respect to the partition $(I_0 \cap J_0, I_1 \cap J_0)$.

We have $u = \sigma_0(v)$.

If x is in J_1, then both $R(x)$ and $R^{-1}(x)$ are in J_0; in that case, let u be the itinerary of x with respect to the partition (I_0, I_1), and v (respectively w) be the itinerary of $R^{-1}(x)$ (respectively $R(x)$) under the induced map $R_{|J_0}$ with respect to the partition $(I_0 \cap J_0, I_1 \cap J_0)$.

We have $u = S\sigma_0(v) = S^{-1}\sigma_0(w)$.

Exercise 6.4.5. Prove the lemma.

Remark. We see that the induction is the exact analogue of the recoding process defined in Sec. 6.3.1. Sequences of type 0 correspond to rotations where the interval on the left (I_0) is the longest (this is the case we considered above), and sequences of type 1 to rotations for which the interval on the right is the largest. When we iterate the induction, we alternate between sequences of type 0 and sequences of type 1.

We have the choice for the induction interval; we described above the induction on J_0, but we could also induce on I_0, using substitution τ_0.

We see also that we have the choice of images or preimages for points out of the induction interval; this corresponds to recoding using either the shift, or its inverse.

Exercise 6.4.6. Give a geometric model for the coding maps $\Phi_{\varepsilon_0,\delta_0,\varepsilon_1,\delta_1}$ defined in the previous section, in terms of induction; characterize in each case the unique point that is in the intersection of all the induction domains. In the case under study, it is, in the convention of this section, the point 0, and this is the reason for choosing this convention. In other cases, like $\Phi_{1,1,-1,1}$, it is the left extremity of the interval, and it would be more convenient to keep the initial convention of defining the rotation on $[0, 1[$.

It is now natural to iterate this induction operation as many times as possible.

Theorem 6.4.7. *let R be, as above, the rotation of parameter θ on $[-1, \theta[$. Let $a = \left[\frac{1}{\theta}\right]$. Let R' be the map defined by*

$$
\begin{aligned}
R' : \quad & [-\{\tfrac{1}{\theta}\}, 1[\to [-\{\tfrac{1}{\theta}\}, 1[\\
& x \mapsto x + 1 \text{ if } x < 0 \\
& x \mapsto x - \{\tfrac{1}{\theta}\} \text{ if } x \geq 0.
\end{aligned}
$$

For any $y \in [-\{\frac{1}{\theta}\}, 1[$, and any integer b such that $0 \leq b \leq a$, with $b = 0$ if $y > 1 - \{\frac{1}{\theta}\}$, define $x = \theta(y - b)$.

If u is the itinerary of x under R with respect to the partition (I_0, I_1), and if v is the itinerary of y under R' with respect to the natural partition $([-\{\frac{1}{\theta}\}, 0[, [0,1[)$, we have:

$$u = S^{-b}\sigma_0^a(v).$$

Proof. The proof is an immediate consequence of the previous lemma, once the proposition is rewritten in the good order: x should be given first, and y computed from x.

First remark that we can induce on the image of the left continuity interval a times; each time, the left interval decreases by θ, until it becomes of length $1 - a\theta < \theta$; we can then no more use the same induction (the return time would be larger than 2). If x is in the induction interval, and if u, v are its itineraries with respect to R and its induced map, iteration of the previous proposition shows that $u = \sigma_0^a(v)$. Otherwise, let b be the return time of x to the induction interval, and $z = x + b\theta$ its first image; it is easy to prove that $z < (a+1)\theta - 1$ (See Fig. 6.6), and that, for the itinerary u, v of x, y, we have $u = S^{-b}\sigma_0^a(v)$.

It remains to remark that, to renormalize, we must divide by θ, and we recover the formulas of the theorem. ■

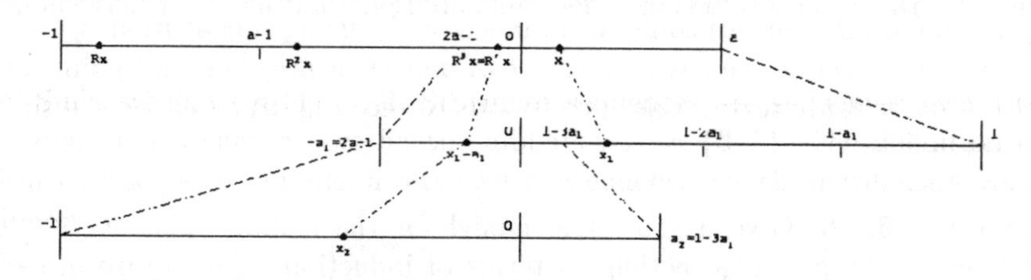

Fig. 6.6. Two steps of induction.

Exercise 6.4.8. Explain what happens in the variants, when we use preimages instead of images for those points that are out of the induction interval, or when we iterate induction on I_0.

Exercise 6.4.9. Explain how we can iterate the construction of the previous theorem. *(Hint: we can either change the type of induction, inducing now as many times as possible on J_1, or conjugate the induced map by $x \mapsto -x$, so that we recover the initial situation. This translates to symbolic dynamics by exchanging* **0** *and* **1**.*)*

6.4.2 Continued fractions

We are interested here only in the combinatorics of the usual continued fraction, not in its approximation and diophantine property. We will give a very brief exposition, without proofs, to recall the arithmetic expressions we need. For more details, see for instance [197, 245].

First of all, we recall that, for any sequence $(a_n)_{n>0}$ of strictly positive integers, the sequence $(x_n)_{n>0}$ defined by

$$x_1 = \frac{1}{a_1} \qquad x_2 = \frac{1}{a_1 + \frac{1}{a_2}} \qquad x_n = \frac{1}{a_1 + \frac{1}{a_2 + \frac{1}{\dots + \frac{1}{a_n}}}}$$

converges to a real number x.

Reciprocally, to any irrational number $x \in [0, 1]$, there exists a unique sequence $(a_n)_{n>0}$ of strictly positive integers such that x is the limit of the sequence defined above. We denote this by:

$$x = \frac{1}{a_1 + \frac{1}{a_2 + \frac{1}{\dots + \frac{1}{a_n + \dots}}}}.$$

It is also customary to note: $x = [0; a_1, \dots, a_n, \dots]$. We can generalize this expansion to all irrational numbers by replacing the initial 0 by $a_0 = [x]$, hence the presence of the initial 0; as we will only consider numbers in the interval $[0, 1]$, we always have $a_0 = 0$.

Definition 6.4.10. *Let $x = [0; a_1, \dots, a_n, \dots]$. The integers a_n are called the* partial quotients *of x.*

The rational numbers $\frac{p_n}{q_n} = \frac{1}{a_1 + \frac{1}{a_2 + \frac{1}{\dots + \frac{1}{a_n}}}}$ *are called the* convergents *of* x.

Exercise 6.4.11. Prove that the convergents satisfy the following relation: $p_n = a_n p_{n-1} + p_{n-2}$, $q_n = a_n q_{n-1} + q_{n-2}$ for $n > 2$.

Exercise 6.4.12. Prove that any rational number can be also written as a continued fraction, but now the sequence is finite, and it is not uniquely defined: to each rational number correspond exactly two continued fraction expansions, of the form $[0; a_1, \dots, a_n, 1]$ and $[0; a_1, \dots, a_n + 1]$.

We will use the following facts:

Proposition 6.4.13. *Let* $T :]0,1[\to [0,1[, x \mapsto \{\frac{1}{x}\}$, *and* $F :]0,1[\to \mathbb{N}, x \to [\frac{1}{x}]$.

For any irrational number $x \in]0,1[$, *the partial quotients of* x *are given by* $a_n = F(T^{n-1}(x))$.

Define $\beta_n = (-1)^n T(x) T^2(x) \dots T^n(x)$. *We have:*

$$\forall n \in \mathbb{N}, \ q_n\alpha - p_n = \beta_n.$$

Remark. There is a nice way to explain the continued fraction algorithm, which is directly related to the previous subsection. Consider two intervals, of respective length α and β; we suppose that $\alpha < \beta$, and that the ratio is $x = \frac{\alpha}{\beta}$. Now subtract α from β as many times (that is, $a_1 = [\frac{\beta}{\alpha}]$ times) as you can. The remaining interval is now smaller than α; subtract it from α, obtaining a new integer a_2, and iterate. It is not difficult to check that we obtain in this way the continued fraction expansion of x. If we start with a normalized interval $\beta = 1$, we can check that the successive lengths are the absolute values of the numbers $q_n\alpha - p_n = \beta_n$ defined above.

In the rational case, the algorithms terminates after a finite number of steps; we can suppose that both intervals are of integer length, and in that case this is exactly Euclid's algorithm.

This algorithm, however, gives information only on the lengths of the intervals; the induction process gives also information on orbits of particular points, and we will see that we can account for that by a small change in this algorithm: instead of deleting some intervals, we stack them over other intervals, thus building what is called “Rokhlin towers” (see Chap. 5).

6.4.3 The Ostrowski numeration system

We first define the Ostrowski numeration system on the integers. For more details, see [195, 217, 236, 237, 309, 318, 413, 436] and see also the survey [74].

Let $(a_n)_{n>0}$ be a sequence of strictly positive integers, and let q_n be the sequence recursively defined by $q_0 = 1$, $q_1 = a_1$, $q_n = a_n q_{n-1} + q_{n-2}$ for $n \geq 2$ (the q_n are the denominators of the convergents of the number $\theta = [0; a_1, \dots, a_n, \dots])$. We have the following proposition:

Proposition 6.4.14 (The Ostrowski numeration system on the integers). *Any integer* $k \geq 0$ *can be written as* $k = \sum_{n=1}^{N} b_n q_{n-1}$, *where the coefficients* b_n *are integers such that* $0 \leq b_n \leq a_n$, *and* $b_n = a_n$ *implies* $b_{n-1} = 0$.

Proof. It is clear from definition that the sequence q_n is strictly increasing. We apply the greedy algorithm: let N be the largest integer such that $q_{N-1} \leq k$. Let $b_N = [\frac{k}{q_{N-1}}]$. It is clear that $b_N \leq a_N$, otherwise we should have $q_N < (a_N + 1)q_N \leq k$. The integer $k - a_N q_{N-1}$ is strictly smaller than q_{N-1}, and if $b_N = a_N$, it must be strictly smaller than q_{N-2}, otherwise we should have $q_N = a_N q_{N-1} + q_{N-2} \leq k$. The result is then clear by induction. ■

Exercise 6.4.15. Prove that the expansion is unique, and that all possible expansions are completely characterized by the nonstationary Markov condition $0 \leq b_n \leq a_n$, and $b_n = a_n$ implies $b_{n-1} = 0$.

This numeration system translates to a numeration system on the real numbers in the following way. We fix as above a sequence $(a_n)_{n>0}$ of partial quotients corresponding to a real irrational number θ. We define $\theta_0 = \theta$, $\theta_n = T^n(\theta)$, where T is the Gauss map $T(x) = \{\frac{1}{x}\}$, and $\beta_n = (-1)^{n+1}\theta_0\theta_1 \ldots \theta_n$.

Theorem 6.4.16 (The Ostrowski numeration system on real numbers). *Any real number $x \in [-1, \theta]$ can be written as $x = \sum_{n=1}^{\infty} b_n \beta_n$, where the coefficients b_n are integers such that $0 \leq b_n \leq a_n$, and $b_n = a_n$ implies $b_{n-1} = 0$.*

Exercise 6.4.17. Prove the theorem; examine the unicity of the expansion, and the characterization of admissible expansions. *(Hint: it can be useful to prove first the recurrence relation $\beta_n = a_n\beta_{n-1} + \beta_{n-2}$.)*

Remark. Note that there are infinitely many Ostrowski systems, one for each sequence of partial quotients.

It is also possible to define a dual Ostrowski numeration system:

Theorem 6.4.18 (The dual Ostrowski numeration system on real numbers). *Any real number $x \in [0, 1+\theta]$ can be written $x = \sum_{n=1}^{\infty} b_n |\beta_n|$, where the coefficients b_n are integers such that $0 \leq b_n \leq a_n$, and $b_n = a_n$ implies $b_{n+1} = 0$.*

Proof. The proof here is simpler than in the preceding theorem; the sequence $|\beta_n|$ is a strictly decreasing sequence of positive real numbers. Hence, we can apply a kind of greedy algorithm: we consider the smallest n such that $|\beta_n| < x$, and we subtract $|\beta_n|$ from x as many times as possible. The numbers $|\beta_n|$ satisfy a recurrence relation that immediately implies the Markov condition. ■

This dual numeration system, as the preceding one, is associated with a numeration system on the integers, but this time on negative and positive integers:

Proposition 6.4.19. *Any integer $k \in \mathbb{Z}$ can be written in a unique way $k = \sum_{n=1}^{N} b_n(-1)^n q_{n-1}$, where the coefficients b_n are integers such that $0 \leq b_n \leq a_n$, and $b_n = a_n$ implies $b_{n+1} = 0$.*

Exercise 6.4.20. Prove this proposition.

Remark. It is important for the sequel, particularly for Sec. 6.6, to remark that, in the dual numeration system, the Markov condition, compared to that of the original numeration system, has changed direction: when $b_n = a_n$, we obtain a condition on the next coefficient, and not the previous coefficient.

6.4.4 Sturmian sequences and arithmetic

Consider now, in the notation defined at the beginning of this section (rotation on the interval $[-1, \theta[$), a rotation sequence u of type 0, of unknown angle θ and initial point x; that is, the sequence is given as the itinerary of the point x under the rotation with respect to the partition (I_0, I_1). We want to recover the angle and initial point from the symbolic sequence u.

We associate with the sequence u its second multiplicative coding (a_n, b_n) as defined in Sec. 6.3.4. We can then recover θ and x as follows:

Theorem 6.4.21. *Let u be a rotation sequence which is the itinerary of a point x under the rotation of angle θ on $[-1, \theta[$. Let (a_n, b_n) be the second multiplicative coding of u. We have:*

$$\theta = [0; a_1, a_2, \ldots, a_n, \ldots],$$

that is, (a_n) is the continued fraction expansion of θ.

Moreover, if we define as above $\theta_0 = \theta$, $\theta_n = T^n(\theta)$, where T is the Gauss map $T(x) = \{\frac{1}{x}\}$, and $\beta_n = (-1)^{n+1}\theta_0\theta_1 \ldots \theta_n$, we have:

$$x = \sum_{n=1}^{\infty} b_n \beta_n,$$

that is, the sequence (b_n) is the Ostrowski expansion of x with respect to the sequence (a_n).

Proof. We know by hypothesis that u is a rotation sequence. The second multiplicative coding process gives a family $u^{(n)}$ of sequences, such that $u^{(n)} = S^{-b_{n+1}} \sigma_0^{a_{n+1}}(u^{(n+1)})$ or $u^{(n)} = S^{-b_{n+1}} \sigma_1^{a_{n+1}}(u^{(n+1)})$, depending on the parity of n. With each of these sequences are associated an angle θ_n and an initial point x_n; but Theorem 6.4.7 gives us the relation between these quantities, namely $\theta_n = \frac{1}{a_{n+1}+\theta_{n+1}}$ and $x_n = \theta_n(x_{n+1} - b_{n+1})$; the proof of the theorem follows. ∎

We obtain now a constructive proof that any Sturmian sequence is a rotation sequence (for more details, see [41], and also [315, 436]):

Theorem 6.4.22. *Every Sturmian sequence is a rotation sequence, whose angle and initial point are explicitly determined by the second multiplicative coding.*

Proof. Let u be a Sturmian sequence, and let (a_n, b_n) be its second multiplicative coding.

Consider the numbers θ whose continued fraction is (a_n), and x whose Ostrowski expansion with respect to (a_n) is (b_n). These define a rotation sequence, whose second multiplicative coding is, by construction, the same as that of u. Hence this rotation sequence is equal to u. ∎

Remark. Strictly speaking, we should have discussed the orbit of the fixed point, since for these sequences, the additive or multiplicative coding does not completely define the sequence; however, the two sequences obtained differ only in two positions, and if one is a rotation sequence, so does the other, the only difference being whether one uses right-closed or left closed intervals.

6.4.5 Arithmetic of the dual coding

We defined in the end of the previous section a dual multiplicative coding; this can be interpreted in term of the dual Ostrowski expansion.

Theorem 6.4.23. *Let u be a Sturmian sequence of type 0, and let (a_n, b_n) be its dual arithmetic coding. Let $\theta = [0; a_1, \ldots, a_n, \ldots]$, and let y be the positive real number whose dual Ostrowski expansion, with respect to the sequence (a_n), is (b_n). The sequence u can be obtained as the coding, under the rotation of angle θ on $[0, 1+\theta[$, of the point x defined by $x = y$ if $u_0 = 0$, and $x = 1+y$ if $u_0 = 1$.*

Exercise 6.4.24. Prove this theorem; we must use here a different kind of induction, called Rauzy induction, where we always induce on the longest of I_0 and J_1. A geometric proof of this fact will be given in Sec. 6.6

Exercise 6.4.25. Work out the arithmetic formulas associated with the various coding maps we defined in the previous section, and their multiplicative counterpart. It might be useful to determine first the point whose expansion is 0: this is the intersection point of the successive induction domains that was the subject of Exercise 6.4.6.

6.5 Sturmian substitutions. Dynamical interpretations

6.5.1 Some classical theorems on periodic continued fractions

The numbers whose continued fraction expansion is periodic are specially interesting, and we have two classical theorems:

Theorem 6.5.1 (Lagrange's theorem). *The continued fraction expansion of x is eventually periodic if and only if x is a quadratic number.*

The "only if" part is easy, and analogues of it are easily proved for all known generalized continued fraction algorithms; one way to prove it is that, if the expansion is periodic, one can write $x = R(x)$, where R is a rational fraction with integer coefficients; one can also write a matrix form of the continued fraction expansion, and look at eigenvectors of a matrix in $SL(2, \mathbb{Z})$; the "if" part is more difficult, and does not generalize easily to multidimensional algorithms, although some cases are known.

Theorem 6.5.2 (Galois' theorem). *The continued fraction of x is purely periodic if and only if x is a quadratic integer in $[0,1]$ whose conjugate is smaller than -1.*

6.5.2 Sturmian systems with periodic coding

We are interested in Sturmian systems and sequences whose coding is periodic; these are linked with quadratic numbers, by the two previous theorems.

Let X be a Sturmian system with periodic coding; we can suppose that its multiplicative coding is given by a sequence $(a_n)_{n>0}$ of period d, that is, $a_{n+d} = a_n$. We can suppose that d is even (otherwise, just consider the double of the period). In that case, it is easy to check that the map Ψ^d, defined in Sec. 6.3 preserves Ω. (If d were odd, the map Ψ^d would send X on the flipped system $E(X)$.)

Another way to explain that is to remark that, for any element of X, we can consider the second multiplicative coding $(a_n, b_n)_{n\in\mathbb{N}}$; the sequence (a_n) depends only on the system, and it is periodic of period d. The sequence (b_n) satisfies a Markov condition, and it is immediate to check that the set of coding sequences for elements of X is not invariant by S, but it is invariant by S^d; grouping the coding sequence in blocks of length d, we obtain a shift of finite type.

We can be more explicit: let us denote by σ the composed substitution $\sigma = \sigma_0^{a_1}\sigma_1^{a_1}\sigma_0^{a_3}\dots\sigma_1^{a_d}$. Then the fixed points of the system X are the fixed points of σ; the system X is the substitutive dynamical system associated with σ. Any element $u \in X$ can be written in a unique way $u = S^k\sigma(v)$, with $k < |\sigma(v_0)|$; we have of course $v = \Psi^d(u)$, and the map Ψ^d can be seen as some kind of inverse of σ on X.

Exercise 6.5.3. Prove that the sequence $(k_n)_{n\in\mathbb{N}}$ obtained by iteration completely defines the initial Sturmian sequence u, except if k_n is ultimately zero, in which case we obtain points in the positive orbits of the two fixed points.

Exercise 6.5.4. Instead of giving just numbers k_n, one could give the corresponding prefixes of $\sigma(v_0)$.

1. Explain what kind of prefixes can arise.
2. Characterize admissible sequences of prefixes.
3. Explain how one could code with suffixes.

6.5.3 One-dimensional dynamical systems associated with Sturmian substitutions

We just explained how, on a Sturmian system with periodic coding, one can define two dynamics: the shift, and the map Ψ^d, inverse of the substitution σ.

We have seen in previous sections, and proved in Sec. 6.4, that there is a nice geometric model for the shift on a Sturmian system: a rotation on an interval, coded by the two continuity intervals. It is natural to look for a geometric model for the map Ψ^d in this framework. In fact, Sec. 6.4.1 hints at the possible model: we showed in this section that a variant of the coding map can be given as a map on an interval, of the form $x \mapsto \frac{x}{\theta} + b$, where b is an integer. The problem here is that the domain and the image of this map are different intervals, the domain corresponding to the initial rotation, and the image corresponding to the induced rotation.

If however the system has periodic coding of even period d, after inducing d times we recover the initial rotation, and the composed map is now from an interval to itself. We see that this composed map is locally an homothety of constant factor $\lambda = \theta_0\theta_1 \ldots \theta_{d-1}$, where $\theta_0 = \theta$, the angle of the system, and $\theta_{n+1} = \left\{\frac{1}{\theta_n}\right\}$.

Exercise 6.5.5. Prove that λ is the smallest eigenvalue of the matrix associated with the substitution σ, and that it is a quadratic integer.

The map Ψ^d, reduced to the Sturmian system, is in fact associated with a generalized λ-shift, and we leave to the reader the proof of the following theorem, which is the essential result of this subsection:

Theorem 6.5.6. *Let X be a Sturmian system of type 0, with periodic coding of even period d. Let $R : [-1, \theta[\to [-1, \theta[$ be the corresponding rotation of angle θ, as defined in Sec. 6.4.1. Let $\sigma = \sigma_0^{a_1}\sigma_1^{a_1}\sigma_0^{a_3}\ldots\sigma_1^{a_d}$ be the corresponding substitution, and λ its smallest eigenvalue.*

There is a finite partition of the domain of R in intervals $K_1, \ldots, K_n$ a finite set of quadratic numbers $d_1, \ldots, d_n$ in $\mathbb{Q}[\lambda]$, and a map $F : [-1, \theta[\to [-1, \theta[$, $x \mapsto \frac{x}{\lambda} + d_i$ if $x \in K_i$ such that the rotation sequence u (respectively v) associated with x (respectively $F(x)$) under R satisfy $v = \Psi^d(u)$. Furthermore, the image of any of the intervals K_i under F is either I_0 or I_1.

Exercise 6.5.7. Characterize explicitly the set of digits d_i. *(Hint: it is interesting to use here the prefixes of images of the letters under the substitution σ.)*

Exercise 6.5.8. Prove in this setting that the Ostrowski system reduces to an expansion in powers of λ, with digits in a finite set of quadratic integers; characterize the set of admissible expansions (it satisfies a Markov condition). This can be another way to answer the preceding exercise.

Exercise 6.5.9. Explain what happens if we change the type of the induction, and how the corresponding map F is changed.

The most simple example is of course the golden number. This does not fall exactly under the framework of this section, since the minimal period is 1, so it is not even. It is however interesting to work out this case.

Exercise 6.5.10. We consider the Fibonacci substitution defined by $\sigma(\mathbf{0}) = \mathbf{01}$, $\sigma(\mathbf{1}) = \mathbf{0}$.

1. Show that the corresponding dynamical system is the rotation by ϕ, the golden number.
2. Show that it can be coded by the set of sequences $(\varepsilon_n) \in \{0,1\}$ such that $\varepsilon_i \varepsilon_{i+1} = 0$ (that is, the word 11 is not allowed).
3. Show that the induction map admits as geometric model the map $[0,1[\to [0,1[\ x \mapsto \{\phi x\}$.
4. Show that this map admits a unique invariant measure absolutely continuous with respect to Lebesgue measure, and compute this invariant measure. *(Hint: the density is constant on intervals.)*
5. Prove that all real numbers in [0,1[can be written in exactly one way as $x = \sum_{n=1}^{\infty} \varepsilon_n \phi^{-n}$, with $(\varepsilon_n) \in \{0,1\}$ such that $\varepsilon_i \varepsilon_{i+1} = 0$; explain the relation with preceding questions.
6. Prove that all real numbers in $[-1, \phi[$ can be written in exactly one way as $x = \sum_{n=0}^{\infty} \varepsilon_n (-1)^n \phi^{-n}$, with $(\varepsilon_n) \in \{0,1\}$ such that $\varepsilon_i \varepsilon_{i+1} = 0$; what is the associated dynamical system?

6.5.4 Two-dimensional dynamical systems related to Sturmian substitutions

The coding sequences are infinite sequences, with admissibility conditions of finite type. The shift on these sequences is not a one-to-one map, and the geometric model we just gave is far from being injective.

It is natural to try to extend the coding sequences to biinfinite sequences, and to look for a one-to-one system that projects on the given one. This is a very simple example of what is called *natural extension* , see [361]; there is a general abstract way to define the natural extension as an inductive limit, but we would like to obtain more concrete models.

For a system of finite type, it is natural to consider biinfinite sequences subject to the same finite type condition. This gives us a first model, as a two-sided shift of finite type.

We will see in Sec. 6.6 how we can associate with this biinfinite coding sequence a pair of Sturmian sequences (u^+, u^-) with same initial letter, by considering the negative part of the coding as the dual coding for the sequence u^-. The shifted coding sequences is then associated with $(\Psi^d(u^+), S^k \tau(u^-))$ for a suitable k and τ.

Another way, more geometric, to understand two-sided coding sequences is the following: since the coding sequence describes what occurs under induction, extending this sequence on the left means that we are trying to "exduce" the given system, that is, to find a larger system of which it is induced on a suitable set. There is in general no clear way to do this; however, it is possible in the present case to find a nice representation of this natural extension as a toral automorphism.

We just give here the idea of the construction, since the details are quite involved. More details can be found in Sec. 6.6, where we work out the non-periodic case, and where the combinatorics is simpler, since we use the additive map Φ.

With a Sturmian system X, one can associate a rotation R on two intervals I_0, I_1; if X has periodic coding, we can find two intervals I'_0, I'_1 such that the induced map of R on these smaller intervals is conjugate by an homothety to the initial map. The map Ψ^d corresponds to stacking subintervals over I'_0, I'_1; if we know at which level in the stack is the corresponding point, we can recover the initial point, and "exduce" the transformation. However, if we have a finite stack, we can exduce only a finite number of times.

The idea is then to consider two rectangles, of respective basis I_0, I_1, to cut these rectangles in slices and to stack these slices over I'_0, I'_1. If we take for heights the coordinates of an appropriate eigenvector, the final pair of rectangles will be the image of the initial one by a measure-preserving dilation.

We show below the figure corresponding to the simplest example, period 2, continued fraction expansion $(1,1)$, matrix $\begin{pmatrix} 2 & 1 \\ 1 & 1 \end{pmatrix}$, substitution $\sigma(\mathbf{0}) = \mathbf{010}$, $\sigma(\mathbf{1}) = \mathbf{10}$. The figure represents three pairs of rectangle; at each step, we cut out a part of the largest rectangle, and stack it on the smaller. The initial figure is made of two squares, of respective side 1 and the golden number. It is then easy to compute that it is the image of the third by a dilation. If we compose with this dilation, we obtain a map from a pair of rectangles to itself, and one can prove that, after suitable identification on the boundary, this map becomes a toral automorphism corresponding to the matrix of the substitution σ.

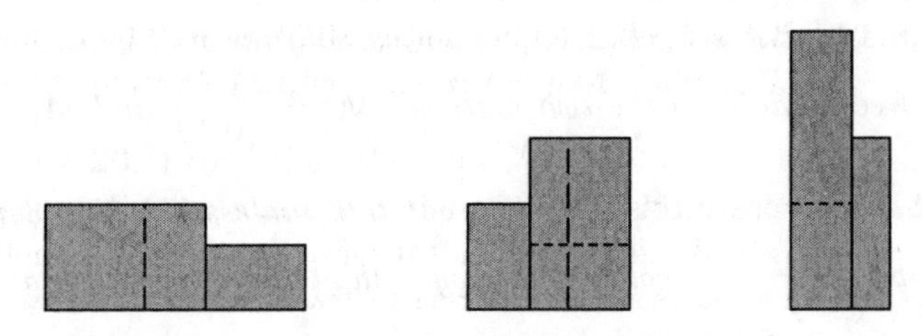

Fig. 6.7. The Fibonacci automorphism.

Exercise 6.5.11. (sequel of Exercise 6.5.10) We consider the set Ω of biinfinite sequence $(\epsilon_n)_{n\in\mathbb{Z}}$, taking values in $\{0,1\}$, and such that there are no two consecutive ones.

1. let $F:\Omega\to\mathbb{R}^2$ defined by $F(\varepsilon)=(\sum_0^\infty \varepsilon_n(-1)^n\phi^{-n},\sum_{-1}^{-\infty}\varepsilon_n\phi^n$. What is the image of F?
2. Prove that the image of F is a fundamental domain for a plane lattice. Deduce an almost one-to-one map $G:\Omega\to\mathbb{T}^2$.
3. Prove that the shift on Omega is conjugate by G to an hyperbolic toral automorphism. (This is just an arithmetic reformulation of the geometric fact shown by Fig. 6.7.)

6.5.5 Similarity of toral automorphisms: the Adler-Weiss theorem

We can recover in this way the Markov coding that was the basis of Adler-Weiss theorem on similarity of toral automorphism with the same entropy. For more details, see [6, 7]. It is based on a few simple lemmas:

Lemma 6.5.12. *Any hyperbolic matrix in $SL(2,\mathbb{Z})$ is conjugate, in $SL(2,\mathbb{Z})$, to a matrix with nonnegative coefficients.*

Proof. Since the matrix is hyperbolic, it has a contracting and an expanding direction, which both have irrational slope.

But it is easy to prove that, for any two independent vectors U,V with irrational slope which form an oriented basis, it is possible to find a basis of $\mathbb{Z}^2$ such that U is in the positive cone, and V is in the second quadrant.

Let us do this for a basis formed of an expanding and a contracting vector; it is clear that, after the change of basis, the image of the positive cone by the map is included in the positive cone, hence the matrix in the new basis is nonnegative. ∎

But nonnegative matrix in $SL(2,\mathbb{Z})$ have quite special properties.

Definition 6.5.13. *We denote by $SL(2,\mathbb{N})$ the set of nonnegative matrices in $SL(2,\mathbb{Z})$, that is, the set of matrices with nonnegative integral coefficients and determinant 1.*

Lemma 6.5.14. *The set $SL(2,\mathbb{N})$, endowed with the multiplication on matrices, is a free monoid on the two matrices* $\mathbf{M}_0=\begin{pmatrix}1&1\\0&1\end{pmatrix}$ *and* $\mathbf{M}_1=\begin{pmatrix}1&0\\1&1\end{pmatrix}$.

Exercise 6.5.15. Prove this. *(Hint: prove that, unless it is the identity, for any element $\begin{pmatrix}a&b\\c&d\end{pmatrix}$, one column is bigger than the other, i.e., $a\ge b$ and $c\ge d$, or $a\le b$ and $c\le d$; deduce that this matrix can be decomposed in a unique way as a product of the two given matrices.)*

We can now associate with that matrix a substitution: with the matrix $\mathbf{M}$ decomposed as

$$\mathbf{M} = \mathbf{M}_0^{a_1}\mathbf{M}_1^{a_2}\dots\mathbf{M}_1^{a_d},$$

we associate the substitution σ defined as $\sigma_0^{a_1}\sigma_1^{a_2}\dots\sigma_1^{a_d}$.

Exercise 6.5.16. Prove that we can now define a fundamental domain for the torus, made of a pair of rectangles with sides parallels to the eigendirections; the substitution dynamical system associated with σ corresponds to the rotation, first return map on the basis of the rectangles of the flow along the unstable direction; the natural extension of the shift on coding sequences is the initial toral automorphism.

Show how one can define a dual substitution, associated with the flow along the stable direction.

In this way, one can associate in a canonical way a Markov coding for any hyperbolic automorphism of the two-dimensional torus. Adler and Weiss then proceeded to prove, using this coding, that any two automorphisms with same entropy are measurably isomorphic (see for instance [7, 6]).

Note that this is a much deeper part: the following exercise shows that there are automorphisms with same entropy that are not conjugate in $SL(2,\mathbb{Z})$, and one can prove that these automorphisms cannot be topologically conjugate (one must know that, for an automorphism of the two-dimensional torus, the entropy is the logarithm of the dominant eigenvalue, and remark that this only depends on the trace).

Exercise 6.5.17. Give an example of 2 elements of $SL(2,\mathbb{N})$ with same trace that are not conjugate in $SL(2,\mathbb{Z})$. *(Hint: if they are conjugate, they are conjugate mod 2; consider the matrices* $\begin{pmatrix} 1 & 2 \\ 2 & 5 \end{pmatrix}$ *and* $\begin{pmatrix} 2 & 7 \\ 1 & 4 \end{pmatrix}$*.)*

6.6 Natural extension for the recoding of Sturmian sequences

In the preceding section, we worked out a natural extension for the map Ψ^d on a Sturmian system with periodic coding of period d.

It is tempting to work out a similar result for the map Φ on the set Σ of all Sturmian sequences.

From the formal viewpoint, there is no difficulty: in Sec. 6.3.3, we showed that the sequences $(\gamma(\Phi^n(u)))_{n\in\mathbb{N}}$ give a symbolic model of Φ as a one-sided shift of finite type on a finite alphabet; to obtain the natural extension, it suffices to consider the two-sided shift satisfying the same condition.

This abstract version of the natural extension is not very satisfying; we will give below, first a geometric model, then a symbolic model as a map on the set of pairs of Sturmian sequences with same initial. For more details, see for instance [388, 38, 42], and also [48].

6.6.1 Geometric model for the natural extension: the pairs of rectangles

We use here the idea of Sec. 6.4. We showed there that, with any Sturmian sequence, one can associate first a rotation, on a normalized interval $[-1, \alpha]$ or $[-\alpha, 1]$ depending on the type of the sequence, and a point in this interval, such that the given Sturmian sequence is the orbit of the given point under the rotation of angle α or $-\alpha$.

We showed also that the map Φ amounts to an inducing and stacking operation; if one can remember the height of stacking, and not only the resulting induced map, it is possible to recover from the sequence $\Phi(u)$ the initial sequence u. However, if one considers only stacks of integer heights, one can only recover a finite number of preimages.

It works better if one considers real heights for the stacks.

Let us now consider pairs of rectangles, such that:

- one rectangle has width 1, and the other irrational width $\alpha < 1$;
- the total area is 1;
- both rectangles have a common lower vertex in $(0, 0)$, and empty intersection;
- there is a distinguished point (x, y).

We can now define a map on the set of all these pairs of marked rectangles, by cutting a slice of width α of the larger rectangle, stacking it on the narrower rectangle, and renormalizing by a measure-preserving dilation so that the size of the new larger rectangle becomes 1 (see Fig. 6.8).

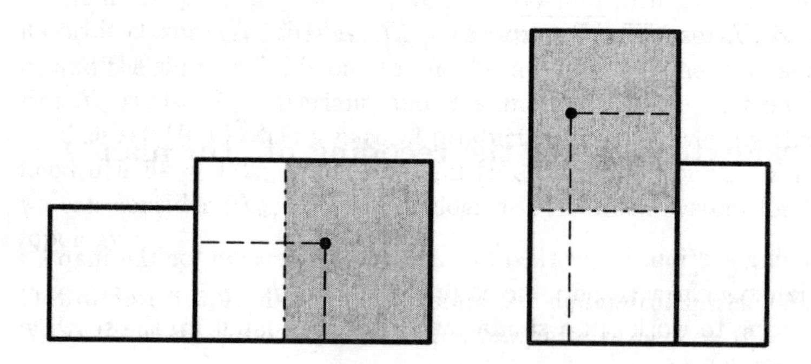

Fig. 6.8. The stacking map.

It is clear, for geometric reasons, that this map is one-to-one (one can now "unstack", using the rectangle of larger height), and it obviously projects to Φ, just by forgetting the heights.

Definition 6.6.1. *We denote by* Δ_a *the set of* 6*-tuples* $(a,b,c,d,x,y) \in \mathbb{R}^6$ *such that:* a,b,c,d *are all nonnegative,* $\sup(a,b) = 1$, $ad+bc = 1$, $-a \leq x \leq b$, *and* $0 \leq y < d$ *if* $x < 0$, $O \leq y < c$ *if* $x \geq 0$.

We denote by $\Delta_{a,0}$ *the set of elements of* Δ_a *such that* $a = 1 > b$, *and* $\Delta_{a,1}$ *its complement (elements of* Δ_a *such that* $a < b = 1$*).*

Of course, these equations just define the set of marked rectangles, with four variables for the shape of the rectangles, and two variables for the marked point; this set is really of dimension 4, because of the two relations $\sup(a,b) = 1$ and $ad + bc = 1$ (the total area is 1); remark that, to recover traditional notations, d is the height of the rectangle of width a, not of the rectangle of width b. Hence, in each of the two sets $\Delta_{a,0}$, $\Delta_{a,1}$, we can reduce to four coordinates, namely, the width and height of the narrower rectangle, and the coordinates of the marked points. We leave it as an exercise to prove that, in these coordinates, the analytic form of the map is given as follows:

Definition 6.6.2. *We denote by* $\tilde{\Phi}$ *the map defined on* $\Delta_{a,1}$ *by:*

- *if* $a < 1/2$;

$$(a,d,x,y) \mapsto \left(\frac{a}{1-a}, (d+1-ad)(1-a), \frac{x}{1-a}, y(1-a)\right) \quad \text{if } x < 1-a;$$

$$(a,d,x,y) \mapsto \left(\frac{a}{1-a}, (d+1-ad)(1-a), \frac{x-1}{1-a}, (y+d)(1-a)\right) \text{ if } x \geq 1-a;$$

- *if* $a > 1/2$;

$$(a,d,x,y) \mapsto \left(\frac{1-a}{a}, (1-ad)a, \frac{x}{a}, ya\right) \quad \text{if } x < 1-a;$$

$$(a,d,x,y) \mapsto \left(\frac{1-a}{a}, (1-ad)a, \frac{x-1}{a}, (y+d)a\right) \quad \text{if } x \geq 1-a;$$

- *the case* $a = 1/2$ *is degenerate.*

The map $\tilde{\Phi}$ *is defined by similar formulas on* $\Delta_{a,0}$.

Remark that it is very easy to prove that the map preserves the Lebesgue measure (for the coordinates (a,d,x,y) on $\Delta_{a,1}$), since the Jacobian is 1; the total mass of this measure is infinite.

Exercise 6.6.3. Strictly speaking, this definition is correct only on a set of measure one; explain what should be done, in term of strict and large inequality and of irrational numbers, so that the map be defined everywhere.

We can of course iterate this map; we can define pair of rectangles of type 0 (if the widest rectangle is on the left), and of type 1 (if it is on the right), and we can define a "multiplicative map" $\tilde{\Psi}$ by iterating until we change type. It is easy to compute this iterated map:

Definition 6.6.4. *We denote by Δ_m the set of marked pairs of rectangles such that the widest rectangle is also the highest one (that is, in the above notations, $a > b$ is equivalent to $d > c$); we denote by $\Delta_{m,0}$ the subset such $a = 1$, $\Delta_{m,1}$ the subset such that $b = 1$.*

Definition 6.6.5. *The map $\tilde{\Psi}$ is defined on $\Delta_{m,1}$ by:*

$$(a,d,x,y) \mapsto \left(\left\{\frac{1}{a}\right\}, a - da^2, \frac{x}{a}, ya\right) \quad \text{if } x < a\left\{\frac{1}{a}\right\};$$

$$(a,d,x,y) \mapsto \left(\left\{\frac{1}{a}\right\}, a - da^2, -\left\{\frac{1-x}{a}\right\}, ya + da + (a - da^2)\left\lfloor\frac{1-x}{a}\right\rfloor\right)$$

$$\text{if } x \geq l\left\{\frac{1}{l}\right\},$$

and similarly on $\Delta_{m,0}$.

Exercise 6.6.6. Prove that this map preserves Lebesgue measure, and that this invariant measure in finite, of measure $2\ln 2$. *(Hint: by definition of $\Delta_{m,1}$, one has $d < c$, hence $d(a+1) < 1$.)*

Exercise 6.6.7. Prove that this map factors over the classical continued fraction map, and also over a two-dimensional continued fraction, and over a natural extension of the usual continued fraction.

Exercise 6.6.8. Prove that the maps $\tilde{\Phi}$ and $\tilde{\Psi}$ are one-to-one (except on a set of measure zero, as explained in Exercise 6.6.3 above), and compute the reciprocal maps.

Exercise 6.6.9. Explain how one can obtain, using map $\tilde{\Psi}$, a two-sided sequence (a_n, k_n) of nonnegative integers, and how one can recover the initial coordinates from this two-sided sequence. *(Hint: difficult; this sequence is related to, but not equal to, the Ostrowski expansion considered in Sec. 6.4.)*

6.6.2 Symbolic model for the natural extension: pairs of Sturmian sequences

We can at this point recover the initial Sturmian sequence. The idea is to identify, by a rotation, the upper and lower sides of the pair of rectangles, and also the free left and right sides. The quotient space is easily seen to be a torus (a more formal way to see it is proposed in the next exercise).

Exercise 6.6.10. Prove that any pair of rectangles that share a lower vertex at $(0,0)$ and whose interior do not intersect tile the plane periodically, and express the group Γ of the tiling as a function of the height and width of the rectangles.

Deduce that there is a natural identification on the boundary such that, modulo this identification, the pair of rectangles becomes isomorphic to the torus $\mathbb{R}^2/\Gamma$.

This torus comes with two natural flows: a vertical flow $((x, y) \mapsto (x, y+t))$ and a horizontal flow $((x, y) \mapsto (x+t, y)$; any vertical trajectory cuts the two rectangles, and we obtain in this way a sequence u_n, itinerary of this trajectory with respect to the two rectangles; it is an easy exercise to prove that this sequence is in fact a rotation sequence.

One can do the same with the horizontal flow, and obtain in this way another Sturmian sequence v; remark that we have, by definition, $u_0 = v_0$.

We can then give another symbolic model for the natural extension of the map Φ, as a map on the set of pairs of Sturmian sequences (u, v) with same initial, that takes (u, v) to (u', v'), where (u', v') are the Sturmian sequences that correspond to the marked point (x, y) in the new "stacked" domain; one has clearly $u' = \Phi(u)$.

Exercise 6.6.11. Explain how one can compute v' from u and v. *(Hint: use Fig.6.8.)*

It is obviously possible to iterate this map, and we obtain in this way a natural extension of the map Ψ.

Exercise 6.6.12. Explain how one can give Markov symbolic dynamics for the maps $\tilde{\Phi}$ and $\tilde{\Psi}$, and their symbolic counterparts.

Show that u (respectively v) determines the future (respectively the past) of this symbolic dynamics.

Make explicit the dual algorithm that determines v as a function of the past symbolic dynamics.

Remark. This is the origin of the dual coding exposed in Sec. 6.3.5. It seems quite difficult to find this dual coding in a purely combinatorial way, without any geometric insight.

6.6.3 The geodesic flow on the modular surface

It is clear that, if we do not know exactly the sequence u, but only the corresponding Sturmian system, we cannot determine the abscissa of the marked point x, but only the width of the rectangles, and similarly for the sequence v.

This amounts to forgetting (x, y) in the above formulas for maps Φ and Ψ; we thus obtain a map on a subset of the plane that is a natural extension of the continued fraction map.

There is a nice way to represent this: we consider the space of lattices of covolume one of the plane; this set is naturally isomorphic to $SL(2, \mathbb{Z}) \backslash SL(2, \mathbb{R})$.

Exercise 6.6.13. Prove that $SL(2, \mathbb{R})$ is isomorphic to the set of oriented basis of $\mathbb{R}^2$ such that the unit square has area 1.

Prove that $SL(2, \mathbb{Z})$ is the group of automorphisms of $\mathbb{Z}^2$.

Deduce that $SL(2, \mathbb{Z}) \backslash SL(2, \mathbb{R})$ is the space of lattices of the plane of covolume 1.

There is a natural action on this set of lattices, by diagonal matrices $g_t = \begin{pmatrix} e^{\frac{t}{2}} & 0 \\ 0 & e^{-\frac{t}{2}} \end{pmatrix}$.

It is possible to prove that each lattice admits a fundamental domain shaped as a pair of rectangles; those lattices for which one of the rectangles has area 1 form a section of the flow, and the map $\tilde{\Phi}$ is in fact, if we forget the marked point, a first return map to this section.

Exercise 6.6.14. Work out the section corresponding to the map Ψ, find the return time to this section and prove that the space $SL(2,\mathbb{Z})\backslash SL(2,\mathbb{R})$ has finite volume.

Exercise 6.6.15. The flow g_t is usually called the *geodesic flow* on the modular surface, for a reason explained in this exercise.

One consider the set $\mathbb{H} = \{x + iy \in \mathbb{C} | y > 0\}$ (Poincaré half plane), endowed with the metric $\frac{dx^2+dy^2}{y^2}$.

1. Prove that the group $SL(2,\mathbb{R})$ acts on $\mathbb{H}$ by:
$$\begin{pmatrix} a & b \\ c & d \end{pmatrix} .z = \frac{az+b}{cz+d}.$$
2. Prove that it is an action by isometries (that is, $z \mapsto M.z$ is an isometry).
3. Prove that $SL(2,\mathbb{R})$ acts transitively on $\mathbb{H}$ and on $T^1\mathbb{H}$, deduce that $T^1\mathbb{H}$ is isomorphic to $PSL(2,\mathbb{R}) = SL(2,\mathbb{R})/\{Id, -Id\}$ (remark that the action of $-Id$ is trivial).
   We will fix the isomorphism by associating the identity matrix and the vertical unit vector at i.
4. Prove that the curve ie^t is a geodesic.
5. Prove that the right action of the group $\{g_t | t \in \mathbb{R}\}$ defined above on $SL(2,\mathbb{R})$ is the geodesic flow on the Poincaré half plane.
6. Prove that the quotient of $\mathbb{H}$ by the action of $SL(2,\mathbb{Z})$ is a surface with one cusp and two singular points (conic points of angle $\frac{\pi}{2}$ and $\frac{Pi}{3}$). This surface is called the modular surface.
   Hence the action of the one-parameter group g_t turns out to be exactly the geodesic flow on the unit tangent bundle of the modular surface. Another model of the map Φ could be given in this context.

For more details, see also [38, 62, 387, 388].

6.6.4 The scenery flow

It is possible to find a similar model for the maps Φ and Ψ. The idea is to consider, not only lattices in the plane, but also cosets of these lattices.

This can be formalized by taking, not linear groups, but affine groups; we define the group $SA(2,\mathbb{R})$ of measure preserving affine maps of the plane, of the form $(x, y) \mapsto (x, y).\mathbf{M} + \mathbf{v}$, where $\mathbf{M}$ is a matrix of determinant 1, and $\mathbf{v}$ is an element of $\mathbb{R}^2$. It is easy to check that the subset $SA(2, \mathbb{Z})$ of elements with integral coefficients is a subgroup, and we can consider the left quotient $SA(2, \mathbb{Z}) \backslash SA(2, \mathbb{R})$.

Exercise 6.6.16. Compute the product of the elements $(\mathbf{M}, \mathbf{v})$ and $(\mathbf{M}', \mathbf{v}')$ of $SL(2, \mathbb{R})$ (we use the same convention as in the preceding paragraph).

Show that $SA(2, \mathbb{R})$ is isomorphic to a subgroup of $SL(3, \mathbb{R})$, by

$$(\mathbf{M}, \mathbf{v}) \mapsto \begin{pmatrix} \mathbf{M} & \begin{matrix} 0 \\ 0 \end{matrix} \\ \mathbf{v} & 1 \end{pmatrix}.$$

The group $SA(2, \mathbb{R})$ acts on the right on the quotient $SA(2, \mathbb{Z}) \backslash SA(2, \mathbb{R})$. In particular, one can study the action of the flow g_t, or *scenery flow*.

The maps $\tilde{\Phi}$ and $\tilde{\Psi}$ turn out to be the first return map of the scenery flow to a suitable section. For more details, see [39, 42].

We will just remark here that this flow is Anosov, and that it admits a number of different interpretations (Teichmüller flow on the twice punctured torus, renormalizations on Sturmian quasi-crystals, see [39]).

6.7 Miscellaneous remarks

6.7.1 Problems of orientation

A recurrent problem in the theory of continued fractions or Sturmian systems is whether one should use any invertible maps, or only orientation preserving invertible maps.

For example, the flip substitution E: $\mathbf{0} \mapsto \mathbf{1}, \mathbf{1} \mapsto \mathbf{0}$ reverses orientation, as does the Fibonacci substitution $\mathbf{0} \mapsto \mathbf{01}, \mathbf{1} \mapsto \mathbf{0}$.

In this chapter, we have chosen to work only in the orientation preserving case, that is, with substitutions whose abelianization belongs to $SL(2, \mathbb{Z})$. This is the reason why, for example, we consider only continued fractions with even periods.

It is perfectly possible to work with orientation reversing maps, and this can simplify some results. For example, it is known that two numbers are equivalent modulo maps $z \mapsto \frac{az+b}{cz+d}$, with $ad - bc = \pm 1$, if and only if their continued fraction expansion coincide eventually, up to a power of the shift, and that they are equivalent modulo $SL(2, \mathbb{Z})$ if and only if their continued fraction expansion coincide eventually, up to an even power of the shift.

In the same way, one could use only the maps σ_0, τ_0 and E, and consider only Sturmian sequences of type 0; each time one obtains sequences of type 1, we just need to apply E to recover a sequence of type 0.

6.7.2 The circle as completion of $\mathbb{N}$ for an adic distance

In Sec. 6.4, we defined the Ostrowski expansion of integers and of real numbers; the relation between these two expansions might have seemed unclear to the reader, and we can give here a nice explanation.

Suppose that a sequence $(a_n)_{n\in\mathbb{N}}$ of strictly positive integers is given, and consider the associated Ostrowski expansion of the integers.

We define an adic distance on the set of integers in the following way: let $n = \sum \varepsilon_i q_i$ and $m = \sum \delta_i q_i$ be two integers, we define $d(n,m) = 0$ if $n = m$, and otherwise $d(n,m) = 2^{-k}$, where k is the smallest integer such that $\varepsilon_k \neq \delta_k$.

Exercise 6.7.1. Prove that, for this distance, we have $lim_{i\to\infty} q_i = 0$; prove that $\mathbb{N}$ is not a complete space for this distance.

Exercise 6.7.2. Prove that the map $n \mapsto n+1$ is continuous for the topology induced by this distance.

The space $\mathbb{N}$ is not complete, but it can be completed:

Exercise 6.7.3. Prove that the completion of $\mathbb{N}$ for the Ostrowski distance is a circle, and that the addition of 1 on $\mathbb{N}$ extends to a rotation on this circle.

In fact, this Ostrowski topology on $\mathbb{N}$ is none else than the topology induced by the embedding of $\mathbb{N}$ on the circle as the orbit of 0; the fact that the sequence q_n tends to 0 is related to the fact that, if the q_n are the denominators of the convergents of α, $|q_n\alpha - p_n|$ tends to 0, that is, $R_\alpha^{q_n}(x)$ tends to x for any point x on the circle.

6.7.3 The 16 possible additive codings and their relations

As we explained in Sec. 6.3.5, there are exactly 16 possible additive algorithms. We determined in Exercise 6.3.39 some relations between these algorithms, due to the diverse symmetries given by the flip and the retrogression.

There is another relation: it seems that each algorithm has a dual algorithm, that can be used to build a natural extension. However, the correct formalism to compute this is at the moment unclear.

The best way seems to be to obtain a suitable fundamental domain, adapted to the algorithm under study. We show in Fig. 6.9 several possibilities; each one determine an algorithm and a dual algorithm.

When one tries to compute the dual algorithm, a strange difficulty appears: for some algorithms, the convention of Definition 6.3.2, that is, if $u = \sigma(v)$, then u_0 is the first letter of $\sigma(v_0$ is not convenient, and it is better to define u_0 as the last letter of $\sigma(v_0$; there is no particular reason to prefer one of these conventions.

The problems are obviously more difficult for the many possible multiplicative algorithms; the dual multiplicative algorithm related to Ostrowski expansion and its dual is the only one for which, guided by arithmetics, we have succeeded to find explicitly the dual algorithm.

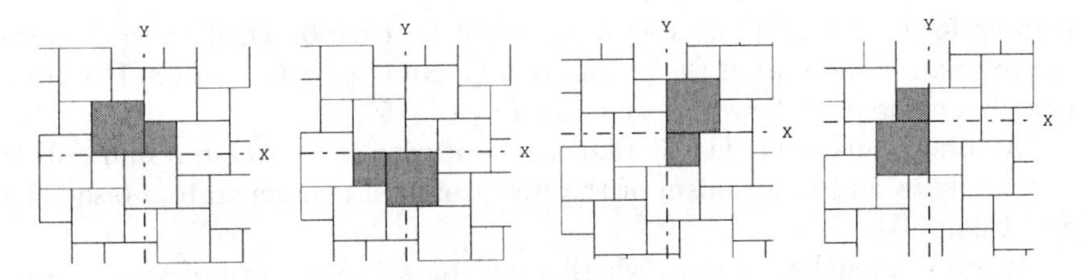

Fig. 6.9. Some examples of fundamental domains.

6.7.4 Sturmian substitutions and automorphisms of free groups

In Sec. 6.5, we could have raised a more general question: what are the substitutions whose fixed points are Sturmian? These substitutions are completely known. For more details, see [68, 67, 69] and Chap. 9. They have very remarkable properties.

Definition 6.7.4. *A substitution σ is* Sturmian *if the image by σ of any Sturmian sequence is a Sturmian sequence.*

Exercise 6.7.5. Prove that the fixed point of a primitive substitution is Sturmian if and only if this substitution is Sturmian.

Exercise 6.7.6. Prove that the composition of two Sturmian substitutions is a Sturmian substitution.

Hence, the Sturmian substitutions form a monoid, the Sturm monoid, which is very well understood. One knows a presentation by generators and relations (see for instance [69]; a proof of this result is given in Chap. 9).

In particular, σ_0, τ_0 and the flip E generate the Sturm monoid.

The submonoid generated by τ_0 and E is called the *monoid of standard morphisms*; it preserves all infinite special words. The submonoid generated by σ_0 and E preserves all fixed words; it contains, as a submonoid of order 2, the monoid generated by σ_0 and $\sigma_1 = E\sigma_0 E$; these are the particular morphisms we considered in the preceding sections. They have the advantage to have determinant 1, and to preserve the fixed words.

Note that there are easy tests to check whether a substitution is Sturmian:

Proposition 6.7.7 (see [68]). *A primitive substitution σ is Sturmian if and only if the word $\sigma(\mathbf{10010010100101})$ is a balanced word.*

Proposition 6.7.8 (see [448]). *A substitution σ over a two-letter alphabet is Sturmian if and only if the words $\sigma(\mathbf{01})$ and $\sigma(\mathbf{10})$, which have the same length, differ only in 2 consecutive indices, where **01** is replaced by **10**.*

This last property has a nice geometric interpretation: with a word on two letters, one can associate a path in the plane, starting from 0, and going one step up for each **1** and one step to the right for each **0**. Then the morphism is Sturmian if and only if the paths corresponding to $\sigma(\mathbf{01})$ and $\sigma(\mathbf{10})$ differ exactly on the boundary of a unit square (see [157]).

Another remarkable fact is that a substitution is Sturmian if and only if it extends to an isomorphism of the free group on two generators (see [448] and Chap. 9).

It is not immediately clear whether a word is a fixed word of some substitution; for example, we proved in Exercise 6.1.25 that, if we denote by u the Fibonacci word, fixed point of the Fibonacci substitution, its two preimages of order two, $\mathbf{01}u$ and $\mathbf{10}u$, are also fixed points of a substitution.

It is easy to prove that there can be only a countable number of fixed points of primitive substitutions, since there is only a countable number of substitutions, and a primitive substitution has a finite number of fixed points. One can also prove that, in the case of Sturmian substitutions, the fixed point of a substitution belongs to a system corresponding to a periodic continued fraction expansion (i.e., a system with a particular quadratic slope). The following question is thus natural: what are the Sturmian sequences that are fixed points of a substitution? For an answer, see [125, 464].

6.7.5 The problem of the Gauss measure

When we work on the set Σ of all biinfinite Sturmian sequences, we would like to speak of sets of measure 0 (for example, to say that almost all sequences are completely defined by their coding sequence, or that the union of orbits of the fixed points is of measure 0).

The problem is that there is no obvious natural measure on Σ.

One could use the map Φ, and look for an invariant measure for this map, using the fact that it can be coded as a shift of finite type.

However, the most interesting measure should be the pullback of the Gauss measure, and this can probably be obtained as a Gibbs measure for the shift of finite type associated with Φ.

7. Spectral theory and geometric representation of substitutions

From geometry to symbolic dynamics. As explained in Chap. 5, symbolic dynamical systems were first introduced to better understand the dynamics of geometric maps. Indeed, by coding the orbits of a dynamical system with respect to a cleverly chosen finite partition indexed by the alphabet $\mathcal{A}$, one can replace the initial dynamical system, which may be difficult to understand, by a simpler dynamical system, that is, the shift map on a subset of $\mathcal{A}^{\mathbb{N}}$.

This old idea was used intensively, up to these days, particularly to study dynamical systems for which past and future are disjoint, such as toral automorphisms or pseudo-Anosov diffeomorphisms of surfaces. These systems with no memory, whose entropy is strictly positive, are coded by subshifts of finite type, defined by a finite number of forbidden words. Some very important literature has been devoted to their many properties (see [264]). The partitions which provide a good description for a topological dynamical system, leading to a subshift of finite type, are called Markov partitions (a precise definition will be given in Sec. 7.1).

Self-Similar dynamics: where substitutions naturally appear. For a dynamical system, it is a usual problem to try to understand the local structure of its orbits. A classical method to study this problem is to consider the first return map (Poincaré map) over an appropriate neighborhood of a given point. For some systems such as toral quadratic rotations or some interval exchanges with parameters living in a quadratic extension, the system defined by the first return map on some subset is topologically conjugated to the original system. One can say that the original dynamical system has a *self-similar structure.* A basic idea is that, in general, as soon as self-similarity appears, a substitution is hidden behind the original dynamical system.

A significant example is the addition φ of the golden ratio α on the one-dimensional torus $\mathbb{T}$: the first return map of this map on the interval $[1-\alpha, 1[$ is the addition of α over $\mathbb{T}$, up to a reversal of the orientation and a renormalization (see Chap. 6).

Indeed, let us choose the natural partition of $\mathbb{T}$ by intervals of continuity $\mathbb{T} = I_2 \cup I_1$, with $I_2 = [0, 1-\alpha[$ and $I_1 = [1-\alpha, 1[$. The first return map

[1] This chapter has been written by A. Siegel

over I_1, denoted by φ_1, is conjugate to the initial map. The conjugacy maps the initial partition onto a partition $J_1 \cup J_2$ of I_1. An easy computation gives $J_1 = [1-\alpha, 2-2\alpha[$ and $J_2 = [2-2\alpha, 1[$. It is immediate to check that J_1 is included in I_1, its image is I_2 and its second image is included in I_1, whereas J_2 is included in I_1 and is immediately mapped into the return subset I_1. The Fibonacci substitution $1 \mapsto 12$ and $2 \mapsto 1$ naturally appears here.

More precisely, for any point $x \in J_1$, if we denote by w the symbolic sequence given by the orbit of x with respect to the partition $I_1 \cup I_2$, and by v the symbolic sequence for the induced map with respect to the partition $J_1 \cup J_2$, it is clear that w is the image of v by the Fibonacci substitution. In particular, the symbolic sequence of the unique fixed point of the conjugacy is the fixed point of the Fibonacci substitution. It is then easy to prove that all the symbolic sequences associated with points in $\mathbb{T}$ belong to the symbolic dynamical system generated by the substitution (see also Chap. 6).

This situation is quite general: in case of self-similarity, under suitable hypotheses, the trajectories (with respect to the partition) of points in the return subset, before they come back into the subset, define a substitution. In that case, the codings of the trajectories of points of the full system belong to the symbolic system associated with this substitution (see Chaps. 1 and 5).

Is this a good representation? We have defined a coding map from the geometric system onto the substitutive system. The question is: how far is this map from being a bijection?

For the example of the toral addition of the golden ratio, we can define an inverse map, from the symbolic system onto the torus. It is proved that this map is continuous, 2-to-1, and 1-to-1 except on a countable set (see Chap. 6); this is the best possible result, given the fact that one of the sets is connected and the other one a Cantor set.

For other examples, the question can be much more difficult.

It is natural then to focus on the reverse question: given a substitution, which self-similar actions are coded by this substitution?

For the Morse substitution, it was proved that the symbolic dynamical system associated with this substitution is a two-point extension of the dyadic odometer, that is, the group $\mathbb{Z}_2$ of 2-adic integers (see Chap. 5).

The three-letter equivalent of the Fibonacci substitution is the Tribonacci substitution $1 \mapsto 12$, $2 \mapsto 13$, $3 \mapsto 1$. G. Rauzy, with methods taken from number theory, proved in 1981 that the symbolic dynamical system associated with this substitution is measure-theoretically isomorphic, by a continuous map, to a domain exchange on a self-similar compact subset of $\mathbb{R}^2$ called the *Rauzy fractal* [349]. Tiling properties of the Rauzy fractal bring an isomorphism between the substitutive system and a translation on the two-dimensional torus.

Considering these examples, we see the connection between the search for a geometric interpretation of symbolic dynamical systems and understanding whether substitutive dynamical systems are isomorphic to already known dynamical systems or if they are new. Since substitutive dynamical systems are deterministic, i.e., of zero entropy, they are very different from a subshift of finite type. Hence, the following question is natural: which substitutive dynamical systems are isomorphic to a rotation on a compact group? More generally, what is their maximal equicontinuous factor ?

Some history. A precise answer was obtained for substitutions of constant length during the seventies [228, 229, 280, 134]. We know that the maximal equicontinuous factor of a substitutive system of constant length l is a translation on the direct product of the adic group $\mathbb{Z}_l$ and a finite group. There exists a measure-theoretic isomorphism between such a substitutive system and its maximal equicontinuous factor, if and only if the substitution satisfies a combinatorial condition called the *coincidence condition.*

It is natural but more difficult to study substitutions of nonconstant length. G. Rauzy first tackled this question with a complete study of the system associated with the Tribonacci substitution [349]. Then, B. Host made a significant advance by proving that all eigenfunctions of primitive substitutive dynamical systems are continuous [214]. Thus, the two main dynamical classifications (up to measure-theoretic isomorphism and topological conjugacy) are equivalent for primitive substitutive systems. In the continuation of this, it was proved that the spectrum of a substitutive system can be divided into two parts. The first part has an arithmetic origin, and depends only on the incidence matrix of the substitution. The second part has a combinatorial origin, and is related to the return words associated with the fixed point of the substitution [173].

Many papers deal with conditions for a substitutive dynamical system to have a purely discrete spectrum [214, 215, 265, 266, 267, 438, 291, 410, 411, 412, 205]. Some are necessary conditions, others are sufficient conditions. In particular, P. Michel, B. Host and independently A. N. Livshits, whose work was restated and generalized by M. Hollander, defined a combinatorial condition on two-letter substitutions called *coincidence condition*, which generalizes the condition of F. M. Dekking for constant length substitutions, for two-letter primitive substitutions. This condition is equivalent with the purely discrete spectrum property.

Except for the Tribonacci substitution [349] and some examples developed by B. Solomyak [410, 411], these works do not give an explicit realization of the maximal equicontinuous factor. For every unimodular substitution of Pisot type on d letters, P. Arnoux and S. Ito build explicitly a self-similar compact subset of $\mathbb{R}^{d-1}$ called generalized *Rauzy fractal* [43]. They generalize the coincidence condition to all nonconstant length substitutions, and prove that this condition is sufficient for the system first to be semi-topologically conjugate to a domain exchange on the Rauzy fractal of the substitution, second to admit as a topological factor a minimal translation on the $(d-1)$-dimensional torus. The techniques they use do not provide results for a measure-theoretic isomorphism between the substitutive system and its toral topological translation factor, as is the case for a two-letter alphabet.

An alternative construction of the Rauzy fractal associated with a substitution of Pisot type can be obtained by generalizing the techniques used by G. Rauzy [349, 352], as developed by V. Canterini and A. Siegel [101, 102]. This construction is based on the use of formal power series, and provides new proofs for the results of P. Arnoux and S. Ito in [43]. Moreover, this point of view allows one to give a combinatorial necessary and sufficient condition for a substitutive unimodular system of Pisot type to be measure-theoretically isomorphic to its toral topological translation factor [397]. This has consequences for the construction of explicit Markov partitions for toral automorphisms, the main eigenvalue of which is a Pisot number [396]. An interesting feature of those Markov partitions is that their topological and geometrical properties can be studied, such as connectedness [100], fractal boundary [288, 290, 289, 44, 207] or simple connectedness [223].

Description of the chapter. The aim of this chapter is to provide a more precise description of the above results. We have no claim to be exhaustive in our exposition since each of the notions we refer to requires a lot of related material to be exposed in detail.

The aim of Sec. 7.1 is to provide a brief introduction to subshifts of finite type, related to Markov partitions and self-similarity. The aim of Sec. 7.2 is to illustrate the deep relationship between substitutive dynamical systems and shifts of finite type, via the tool of adic transformations.

The entire Sec. 7.3 is devoted to the spectral theory of substitutive dynamical systems. Section 7.3.2 presents an overview of the general spectral theory of substitutive systems of nonconstant length. In Sec. 7.3.3 the main attention is devoted to the spectral properties of systems associated with the extensively studied class of substitutions of Pisot type.

In Sec. 7.4 we expose in details the construction and the properties of the Rauzy fractal defined by G. Rauzy to get a geometric realization of the system associated with the Tribonacci substitution. As a fifth part, in Sec. 7.5 we explore in details the construction and properties of the Rauzy fractal for any substitution of Pisot type. A special attention is devoted to their tiling properties. Finally, the aim of Sec. 7.6 is to explain how a geometrical representation of substitutions of Pisot type provides explicit Markov partitions for some Pisot toral automorphisms.

7.1 Shifts of finite type: introduction

Our investigation of the properties of substitutive dynamical systems will need the use of the most well known class of symbolic dynamical systems, that is, shifts of finite type. These shifts are introduced in Sec. 7.1.1. Geometrically, shifts of finite type are related to Markov partitions. Section 7.1.2 is devoted to them. These Markov partitions can be seen as a generalization of self-similar tilings, defined in Sec. 7.1.3.

7.1.1 Shifts of finite type

While they have been around implicitly for a long time, the systematic study of shifts of finite type began with Parry [322], Smale [409] and R. Williams [459]. A complete presentation can be found in [264].

Definition. Let $\mathcal{A}$ be a finite set. Endowed with the shift map, a subset X of $\mathcal{A}^{\mathbb{Z}}$ is a *bilateral shift of finite type* if and only if there exists a finite set $\mathcal{F} \subset \mathcal{A}^\star$ of finite words over $\mathcal{A}$ such that X is the set of biinfinite words in $\mathcal{A}^{\mathbb{Z}}$ which have no factor in $\mathcal{F}$. The subshift X is also denoted $X_{\mathcal{F}}$.

Equipped with the topology on $\mathcal{A}^{\mathbb{Z}}$, a shift of finite type is a symbolic dynamical system, since it is compact and invariant under the shift. One defines similarly the notion of *unilateral shift of finite type.*

Example 7.1.1. The bilateral fullshift $\{1, \dots, d\}^{\mathbb{Z}}$ over a d-letter alphabet is a shift of finite type with no forbidden word.

Example 7.1.2. The *golden mean shift of finite type* is the set of all binary sequences with no consecutive 1's. The associated set of forbidden words is $\mathcal{F} = \{11\}$.

Example 7.1.3. The *even shift*, that is, the set of all binary sequences (that is sequences over $\{0, 1\}$) so that between any two 1's there are an even number of 0's, is not a shift of finite type: it cannot be described by a finite number of constraints.

Higher presentation. Let (X, S) be a symbolic dynamical system over the alphabet $\mathcal{A}$. A basic construction defines a new system by looking at the blocks of consecutive letters in the language of X, and by considering them as letters from a new alphabet.

Indeed, let n be an integer. The new alphabet is $\mathcal{A}^{[n]} = \mathcal{L}_n(X)$, that is, the set of blocks of length n that appear in at least one element of X. Define the mapping $\beta_n : X \to (\mathcal{A}^{[n]})^{\mathbb{Z}}$ by $\beta_n(w)_i = w_i \dots w_{i+n-1}$ for all $w \in X$. Let $X^{[n]} = \beta_n(X)$ and let $S^{[n]}$ denote the shift map over $(\mathcal{A}^{[n]})^{\mathbb{Z}}$.

Then $(X^{[n]}, S^{[n]})$ is a symbolic dynamical system, called the *n-th higher presentation of* (X, S). If (X, S) is a shift of finite type, $(X^{[n]}, S^{[n]})$ is also a shift of finite type.

Example 7.1.4. The 2-th higher presentation of the golden mean shift of finite type (Example 7.1.2) is the shift of finite type over the alphabet $\mathcal{A}^{[2]} = \{a = 10,\ b = 01,\ c = 00\}$, associated with the following set of forbidden words: $\mathcal{F}^{[2]} = \{aa, bb, bc, ca\}$.

Graphs. A *finite graph* $\mathcal{G}$ consists of a finite set of *vertices* or *states* $\mathcal{V}$ together with a finite set of *edges* $\mathcal{E}$. If e is an edge, $i(e) \in \mathcal{V}$ is the *initial vertex* and $t(e) \in \mathcal{V}$ is the *terminal vertex*. Two vertices can be connected by more than one edge.

The *adjacency matrix* of a finite graph $\mathcal{G}$ has size the cardinality of $\mathcal{V}$. If $(i,t) \in \mathcal{V}^2$, its coefficient of index (i,t) is equal to the number of edges in $\mathcal{E}$ with initial state i and terminal state t. A finite graph is thus uniquely determined by its adjacency matrix (up to graph isomorphism).

Shift associated with a graph. Let $\mathbf{A}$ be a matrix with coefficients in $\{0,1\}$; the *bilateral shift associated with the finite graph* $\mathcal{G}$ of adjacency matrix $\mathbf{A}$ is defined as the following set of biinfinite paths in the graph $\mathcal{G}$, endowed with the shift map:

$$X_{\mathcal{G}} = X_{\mathbf{A}} = \{((e_n)_{n\in\mathbb{Z}} \in \mathcal{E}^{\mathbb{Z}};\ \forall n \in \mathbb{Z},\, t(e_n) = i(e_{n+1})\}.$$

One can easily check that this set is closed and hence compact, and invariant under the shift map in $\mathcal{V}^{\mathbb{Z}}$.

The *unilateral shift* is defined as the set of right paths in $\mathcal{G}$, that is, as the set of paths indexed by $\mathbb{N}$.

Example 7.1.5. The shift associated with the matrix $\begin{pmatrix} 1 & 1 \\ 1 & 0 \end{pmatrix}$ is the subset of $\{a,b,c\}^{\mathbb{N}}$ that contains all the labels of walks in the graph shown in Fig. 7.1.

Example 7.1.6. The bilateral full-shift $\{1,\ldots,d\}^{\mathbb{Z}}$ over a d-letter alphabet is the shift associated with the matrix $\mathbf{A} = [d]$ of size 1 associated with the graph shown in Fig. 7.2.

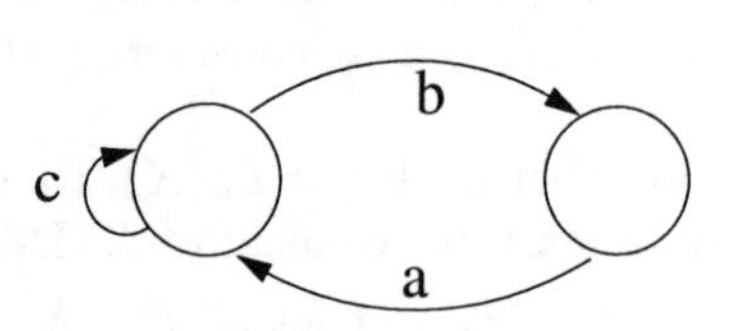

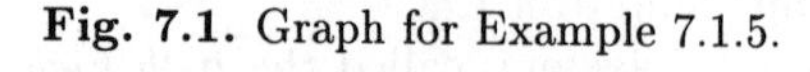
Fig. 7.1. Graph for Example 7.1.5.

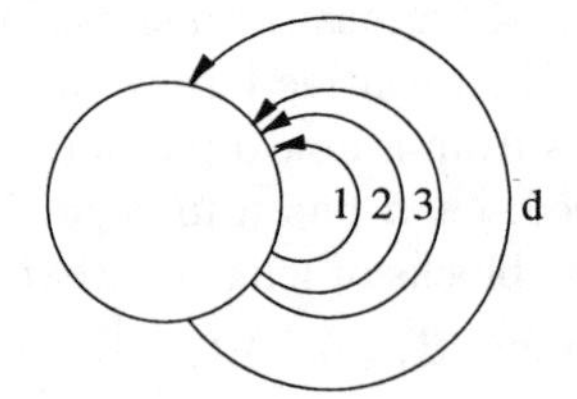

Fig. 7.2. Graph for the fullshift over d letters.

Relationship between the two categories of subshifts. One should remark that any shift associated with a graph is a shift of finite type: the set of forbidden words simply consists of pairs of edges ef such that $t(e) \neq i(f)$.

On the contrary, not every shift of finite type is the shift associated with a graph: for instance, the golden mean shift (Example 7.1.2) cannot be described by a graph.

However, any shift of finite type can be recoded, using a higher block presentation, to be the shift associated with a graph:

Theorem 7.1.7 (see [264]). *Let $(X_{\mathcal{F}}, S)$ be a shift of finite type. Let M denotes the maximum of the lengths in $\mathcal{F}$. Then there exists a graph $\mathcal{G}$ such that the shift associated with $\mathcal{G}$ is the M-th higher presentation of (X, S).*

Example 7.1.8. Up to a recoding of the labels ($a = 10$, $b = 01$, $c = 00$), the 2-th higher presentation of the golden mean shift described in Example 7.1.4, is the shift associated with the matrix $\begin{pmatrix} 1\,1 \\ 1\,0 \end{pmatrix}$, given in Example 7.1.5.

This results means that shifts of finite type are essentially the same as shifts associated with graphs (up to a recoding). From now on, we identify shifts of finite type and shifts associated with a graph.

Labelled graphs. An *automaton* $(\mathcal{G}, t)$ (also called *labelled graph* by the set $\mathcal{A}$) consists of a finite graph $\mathcal{G} = (\mathcal{V}, \mathcal{E})$ together with a labeling map of the edges $t : \mathcal{E} \to \mathcal{A}$.

Sofic systems. Endowed with the shift map, a subset of $\mathcal{A}^{\mathbb{Z}}$ or $\mathcal{A}^{\mathbb{N}}$ is called a (unilateral or bilateral) *sofic system* if it is defined as the labeling of infinite (unilateral or bilateral) paths of an automaton. Endowed with the topology over $\mathcal{A}^{\mathbb{Z}}$ or $\mathcal{A}^{\mathbb{N}}$, such a system is still a symbolic dynamical system.

Example 7.1.9. The set of all binary sequences so that between any two 1's there is an even number of 0's is a sofic system, associated with the automaton shown in Fig. 7.3. We recall that it is not a shift of finite type.

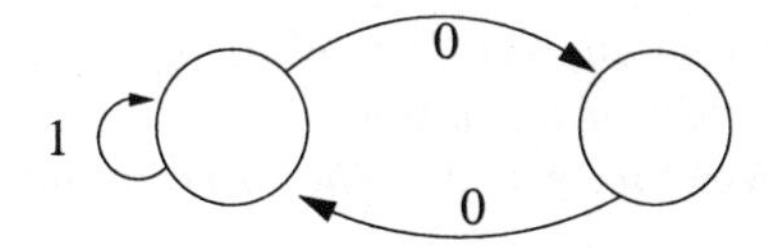

Fig. 7.3. The graph of Example 7.1.9.

Connection between shifts of finite type and sofic systems. A shift of finite type is nothing else that a sofic system associated with an automaton whose all edges have different labels. If so, the automaton is said to be *deterministic.*

These two categories of symbolic dynamical systems are fundamentally connected:

Theorem 7.1.10 (see [264]). *Any sofic system is a topological factor of a shift of finite type, and conversely.*

7.1.2 Markov partitions

As we already explained, one of the main sources of interest in symbolic dynamics is its use in representing other dynamical systems. Indeed, to describe the orbits of an invertible map f on a space $\mathcal{E}$, one may divide the space X into a finite number d of pieces $\mathcal{E}_i$, and then track the orbit of a point $z \in \mathcal{E}$ by keeping a record of which of these pieces $f^n(z)$ lands in. This defines a symbolic sequence $x \in \{1 \dots d\}^{\mathbb{Z}}$ defined by $x_n = i$, if $f^n(z) \in \mathcal{E}_i$. In short, the pieces are said to be a *Markov partition of* $\mathcal{E}$ if, first the set of all symbolic sequences associated with the points of $\mathcal{E}$ is a shift of finite type, and, secondly, the mapping from $\mathcal{E}$ onto the subshift which maps any point on its coding is precise enough. We refer the reader to [264, 241, 242, 394] for more details on the concepts introduced in this section.

Precise definitions. A *topological partition* of a metric space $\mathcal{E}$ is a finite collection $\Lambda = \{\mathcal{E}_1, \dots, \mathcal{E}_d\}$ of disjoint open sets whose closure covers $\mathcal{E}$, that is, $\mathcal{E} = \cup \overline{\mathcal{E}_i}$.

Let $(\mathcal{E}, f)$ be a dynamical system. If f is invertible, the ***bilateral symbolic dynamical system*** associated with a topological partition Λ is the set $X_{\mathcal{E}}$ endowed with the shift map:

$$X_{\mathcal{E}} = \{(x_n)_{n \in \mathbb{Z}} \in \{1, \dots, d\}^{\mathbb{Z}};\ \exists z \in \mathcal{E},\ \forall n \in \mathbb{Z},\ f^n(z) \in \mathcal{E}_{x_n}\}.$$

If f is non-invertible, the ***one-sided symbolic dynamical system*** associated with a topological partition Λ consists of sequences indexed by $\mathbb{N}$, corresponding to the nonnegative orbits.

Definition 7.1.11. *Let $(\mathcal{E}, f)$ be an invertible (respectively non-invertible) dynamical system. A topological partition $\Lambda = \{\mathcal{E}_1, \dots, \mathcal{E}_d\}$ of $\mathcal{E}$ is said to be a* Markov partition *of $\mathcal{E}$ if*

- *the bilateral (respectively one-sided) symbolic dynamical system $(X_{\mathcal{E}}, S)$ associated with Λ is a shift of finite type;*
- *for every symbolic sequence $x \in X_{\mathcal{E}}$, the intersection*

$$\bigcap_{-\infty}^{+\infty} \overline{f^{-k}(\mathcal{E}_{x_k})} \left(\text{respectively } \bigcap_{k=0}^{+\infty} \overline{f^{-k}(\mathcal{E}_{x_k})} \right)$$

consists of exactly one point.

Remarks.

- By definition of the dynamical system $X_{\mathcal{E}}$, the intersection above is never empty. The fact that it consists of exactly one point means that the coding mapping from $\mathcal{E}$ onto $X_{\mathcal{E}}$ is one-to-one.
- There are number of variants on the definition of Markov partition in the literature, many involving geometry of the pieces $\mathcal{E}_i$. This one is simple to state but is somewhat weaker than other variants.

Proposition 7.1.12 (see [264]). *Let $\Lambda = \{\mathcal{E}_1, \ldots, \mathcal{E}_d\}$ be a Markov partition of the invertible (respectively non-invertible) dynamical system $(\mathcal{E}, f)$. Let $X_{\mathcal{E}}$ be the bilateral (respectively one-sided) symbolic dynamical system associated with Λ.*

Then the inverse map $\varphi : X_{\mathcal{E}} \to \mathcal{E}$ of the coding mapping is well-defined as follows:

$$\forall x \in X_{\mathcal{E}},\ \{\varphi(x)\} = \bigcap_{-\infty}^{+\infty} \overline{f^{-k}(\mathcal{E}_{x_k})} \ \left(\textit{respectively} \ \bigcap_{k=0}^{+\infty} \overline{f^{-k}(\mathcal{E}_{x_k})} \right).$$

The map φ is continuous, onto, and realizes a commutative diagram between the shift map on $X_{\mathcal{E}}$ and f on $\mathcal{E}$, that is, $\varphi \circ S = f \circ \varphi$.

Example. The most simple example of a Markov partition is the doubling map on the circle: the torus $\mathbb{T} = \mathbb{R}/\mathbb{Z}$ is provided with the non-invertible action $f(z) = 2z \bmod 1$. Subdivide $\mathbb{T}$ into two equal subintervals $I_1 = [0, 1/2)$ and $I_2 = [1/2, 1)$. The alphabet here shall be $\mathcal{A} = \{1, 2\}$. For every $z \in \mathbb{T}$ we define $x_n \in \{1, 2\}$ by $f^n(z) \in I_{x_n}$. The sequence $x = (x_n)_{n \geq 0}$ is an element of the full-shift $\{1, 2\}^{\mathbb{N}}$. Notice that x is nothing else that the sequence of digits of z in its binary expansion. The action of f corresponds to the shift map: the action of multiplying by 2 an element of $\mathbb{T}$, and taking the rest modulo 1 simply shifts the binary digits to the left and delete the first digit. Hence, $\mathbb{T} = I_1 \cup I_2$ realizes a Markov partition for f.

More generally, Markov partitions can be seen as generalizations of the binary system (see [243]).

General interest of Markov partitions. Markov partitions allow a good combinatorial understanding of symbolic dynamical systems. They are partitions of a space $\mathcal{E}$ which describe by means of a shift of finite type, the dynamics of a map acting on $\mathcal{E}$, this description preserving properties such as density of periodic points, transitivity or mixing. The existence of Markov partitions allows one to deduce from the properties of the shift of finite type a "quasi-immediate" description of certain dynamical properties of the system.

For instance, Adler and Weiss show in [7] that topological entropy is a complete invariant for the measure-theoretic isomorphism of continuous ergodic automorphisms of the torus $\mathbb{T}^2$. Their proof is based on the construction of an explicit Markov partition of the torus $\mathbb{T}^2$ with respect to the automorphism.

The intersection property. More precisely, Adler and Weiss define a partition of $\mathbb{T}^2$ which is natural relatively to a linear automorphism f . The key to prove that this partition is a Markov partition is the following *intersection property*:

- the image under f of the intersection of a cylinder $\mathcal{E}_i$ with an affine subspace directed with the stable direction of f is included in a intersection of the same type,

- the image under f^{-1} of the intersection of a cylinder $\mathcal{E}_i$ with an affine subspace directed with the unstable direction of f is included in a intersection of the same type.

In other variants on the definition of Markov partition, this property appears to be a hypothesis in the definition.

Existence of Markov partitions. R. Bowen in [87] shows that for every hyperbolic diffeomorphism of a manifold $\mathcal{E}$, there exists a Markov partition of $\mathcal{E}$. His proof is not constructive.

One should remark that, except for the one and two dimensional-cases, Markov partitions for toral automorphisms are made of pieces that cannot be simple parallelograms. Indeed, R. Bowen proved in [89] that for hyperbolic automorphisms of $\mathbb{T}^3$, the boundary of the elements of the Markov partition are typically fractals: their Haussdorf dimension is not an integer.

Explicit Markov partitions. Markov partitions remained abstract objects for a long time: except for linear automorphisms of the two-dimensional torus and some automorphisms of $\mathbb{T}^3$ [36, 61, 288, 290, 289], explicit constructions of Markov partitions for toral hyperbolic automorphisms were obtained only recently. For precise details about results on this subject, see [243].

Making use of β-developments defined by A. Bertrand-Mathis (see [179] for a general reference), B. Praggastis [335] builds tilings which induce Markov partitions for toral hyperbolic automorphisms whose dominant eigenvalue is a unimodular Pisot number. These constructions are particularly simple and explicit for the automorphisms associated with the companion matrix of a polynomial of the type $x^n - d_1x^{n-1} - \cdots - d_n$, with $d_1 \geq d_2 \geq \cdots \geq d_n = 1$. Indeed, the β-development associated with the Pisot roots of such polynomials admits very special properties [180] that are intensively used by B. Praggastis.

More generally, on the one hand, S. Leborgne [261, 260], on the other hand, R. Kenyon and A. Vershik [244], study the case of hyperbolic automorphisms of $\mathbb{T}^n$, using algebraic properties of the eigenvalues of the automorphism to code its action. Just as with B. Praggastis, those codings are based on a numeration system with a non-integral basis.

Thus, with an equivalent of β-developments, S. Leborgne represents hyperbolic automorphisms by sofic systems. Independently, choosing particular digits in expansions with a non-integral basis, R. Kenyon et A. Vershik construct a representation by a shift of finite type. This representation is not totally explicit, since the digits are chosen at random.

Relation with substitutions. We will see in Sec. 7.6 how a geometrical representation of substitutive systems of Pisot type provides a Markov partition for the incidence matrix of the substitution, the Markov partition being described by the combinatorial representation in terms of subshifts exposed in Sec. 7.2.2.

7.1.3 Self-similarity

Markov partitions usually generate tilings with a specific property: self-similarity. Let us set up precise definitions about this notion.

Tilings. A collection Λ of compact subsets of $\mathbb{R}^n$ is a *tiling* of a set $\mathcal{E}$ which is the closure of its interior if and only if:

- $\mathcal{E} = \cup_{C \in \Lambda} C$,
- every element of Λ is the closure of its interior,
- every compact subset of $\mathcal{E}$ intersects a finite number of elements of Λ,
- elements of Λ have disjoint interior.

Periodic tilings. The tiling is said to be a *periodic tiling* of $\mathbb{R}^n$ modulo $\mathcal{L}$, where $\mathcal{L}$ is a lattice, if and only if

- $\mathbb{R}^n = \cup_{C \in \Lambda, \mathbf{z} \in \mathcal{L}} (C + \mathbf{z})$,
- for all $\mathbf{z} \in \mathcal{L}$, $(C + \mathbf{z}) \cap \mathrm{Int}\, C' \neq \emptyset$ implies $\mathbf{z} = \mathbf{0}$ and $C = C'$.

Self-similar sets. A tiling is said to be *self-similar* if it is invariant under the action of a given expanding diagonal map in $\mathbb{R}^n$, the image of each tile under the mapping being an exact union of original tiles. More precise and general definitions can be found in [241, 242, 430].

A self-similar tiling can be seen as a Markov partition for the expanding map associated with the tiling, and can be seen as a generalization of the decimal system, through the theory of the representation of numbers in non-standard bases (for more details, see [243]).

A compact subset $\mathcal{E}$ of $\mathbb{R}^n$ is said to be *self-similar* if there exists a finite partition of $\mathcal{E}$ whose translates under a given lattice generate a self-similar and periodic tiling of $\mathbb{R}^n$.

7.2 Substitutive dynamical systems and shifts of finite type

The spectral theory of substitutive systems of constant length was developed during the seventies (see Sec. 7.3). One of the main problems raised by the generalization of this theory to substitutions of nonconstant length was to find a general way to decide positions in which any factor of the fixed point appears, that is, to "desubstitute" the substitutive system. The important notion of recognizability deals with this problem. As it will be explained in Sec. 7.2.1, it took quite a long time to be correctly defined and to get general results about it.

The problem of recognizability being solved, an important part of the results on spectral theory of nonconstant length substitutive systems was obtained by using a deep relationship between substitutive dynamical systems and an other important class of symbolic dynamical systems, that is, shifts of finite type. In Sec. 7.2.2 we focus on this relationship.

In Sec. 7.2.3, we expose how the notion of recognizability allows one to make explicit the results of Sec. 7.2.2, thanks to the useful tool of prefix-suffix expansion.

7.2.1 Recognizability

Let σ be a primitive substitution over the alphabet $\mathcal{A}$, and let u be a biinfinite fixed point of σ. We thus have $u = \sigma(u)$. Let E_1 be the following set of lengths

$$E_1 = \{0\} \cup \{|\sigma(u_0 \dots u_{p-1})|, |\sigma(u_{-p} \dots u_{-1})|;\ p > 0\}.$$

Since $u = \sigma(u)$, for every factor $W = u_i \dots u_{i+|W|-1}$ of u, there exists a rank j, a length l, a suffix S of $\sigma(u_j)$ and a prefix P of $\sigma(u_{j+l+1})$ such that

$$W = S\sigma(u_{j+1}) \dots \sigma(u_{j+l})P,$$

and such that $E_1 \cap \{i, \dots, i + |W| - 1\} = (i - k) + E_1 \cap \{k, \dots, k + |W| - 1\}$, with $k = |\sigma(u_0 \dots u_j)| - |S|$, which means that it is equivalent to cut u with respect to σ at the ranks i and k.

We say that $[S, \sigma(u_{j+1}), \dots, \sigma(u_{j+l}), P]$ is the *1-cutting at the rank i* of W, and that W comes from the word $u_j \dots u_{j+l+1}$, which will be called the *ancestor word* of W.

By minimality, the factor W appears infinitely many times in the sequence u. The question of the unicity of the 1-cutting of W is thus natural. Let us briefly survey the existing literature on this question.

In [281], J. C. Martin calls *rank one determined* substitutions for which every long enough factor of a fixed point admits a 1-cutting and an ancestor word independent from the rank of apparition of the factor. The author claims that any substitution on a two-letter alphabet which is not shift-periodic is rank one determined. His proof is not convincing.

Later, B. Host [214] and M. Queffélec [339] introduce the notion of *recognizable substitution*, that is, substitutions for which the 1-cutting of any long enough factor is independent of the rank of occurrence of the factor, except maybe for a suffix of the word, the length of this suffix being bounded. More precisely, a substitution σ is said to be *unilaterally recognizable* if there exists $L > 0$ such that $u_i \dots u_{i+L-1} = u_j \dots u_{j+L-1}$, with $i \in E_1$ implies $j \in E_1$.

B. Host proves in [214] that this property is equivalent to the fact that the image $\sigma(X_\sigma)$ under σ of the substitutive dynamical system X_σ associated with σ is an open set. G. Rauzy announced that he had a short proof of the fact that a constant length substitution which is one-to-one on the set of letters and which is not shift-periodic is unilaterally recognizable [339]. Nobody could check this proof. However, a new short proof in this case was recently obtained in [35]. Concerning substitutions of nonconstant length, the question remained unsettled.

In [305], B. Mossé studies the question of unilateral recognizability. She proves that a substitution is not necessarily unilaterally recognizable: a sufficient condition for a substitution not to be unilaterally recognizable is that for every couple of distinct letters (a, b), $\sigma(a)$ is a strict suffix of $\sigma(b)$, or conversely.

Example 7.2.1. The substitution $1 \mapsto 1112$ and $2 \mapsto 12$ is not unilaterally recognizable.

To get a general result of recognizability, B. Mossé introduces a new notion of recognizable substitution, for which the 1-cutting of any long enough factor is independent from the rank of apparition of the factor, except maybe for a suffix and a prefix of the factor, the lengths of those prefixes and suffixes being bounded. More precisely, a substitution σ is said to be ***bilaterally recognizable*** if there exists $L > 0$ such that $u_{i-L} \ldots u_{i+L} = u_{j-L} \ldots u_{j+L}$, with $i \in E_1$ implies $j \in E_1$.

The following theorem shows that this notion of recognizability is the right one:

Theorem 7.2.2 (Mossé [305, 306]). *Let σ be a primitive substitution with a non-periodic fixed point u. Then the substitution σ is bilaterally recognizable, and the ancestor word of every factor of u is unique except on the end of the factor: there exists $L > 0$ such that if $u_{i-L} \ldots u_{j+L} = u_{i'-L} \ldots u_{j'+L}$, then $u_i \ldots u_j$ and $u_{i'} \ldots u_{j'}$ have the same 1-cutting and the same ancestor word at ranks i and i'.*

Thus, if a substitution is primitive and not shift-periodic, then one can always *desubstitute* any factor of the fixed point, except possibly on the ends of the factor.

Remark. There exist nontrivial substitutions which are shift-periodic, for instance $1 \mapsto 121$, $2 \mapsto 21212$. It is not so easy to recognize if a substitution is not shift-periodic. However, this problem is solved thanks to an algorithm (see [198, 320]).

If we just suppose from now on that σ is a primitive substitution which is not shift-periodic, σ has not necessarily a fixed point but has at least a periodic point. Consequently, a power of σ is bilaterally recognizable. Since the dynamical systems associated with a substitution or with any of its iterations are identical, we deduce from Theorem 7.2.2 that any word of the bilateral symbolic dynamical system X_σ associated with the substitution σ can be desubstituted in a unique biinfinite word:

Corollary 7.2.3. *Let σ be a primitive substitution which is not shift-periodic. Let X_σ be the substitutive dynamical system associated with σ. Then we have*

$$\forall w \in X_\sigma,\ \exists v \in X_\sigma,\ v \text{ unique};\ w = S^k \sigma(v), \text{ and } 0 \le k < |\sigma(v_0)| \quad (7.1)$$

This corollary means that any word w in X_σ can by cut or desubstituted in a unique way in the following form:

$$w = \ldots \mid \underbrace{\ldots}_{\sigma(v_{-1})} \mid \underbrace{w_{-k}\ldots w_{-1}.w_0\ldots w_l}_{\sigma(v_0)} \mid \underbrace{\ldots}_{\sigma(v_1)} \mid \underbrace{\ldots}_{\sigma(v_2)} \mid \cdots \tag{7.2}$$

where the word $v = \ldots v_{-n}\ldots v_{-1}v_0v_1\ldots v_n\ldots$ belongs to X_σ.

Note that this property is satisfied only by two-sided substitutive dynamical systems, and is not satisfied by a one-sided substitutive dynamical system. A counter-example is again given by the substitution defined by $1 \mapsto 1112$ and $2 \mapsto 12$.

This notion of recognizability is an important hypothesis used by most of the authors to get results about the spectrum of substitutive systems of nonconstant length. Before B. Mossé's work, most authors had to suppose that the substitutions they considered were recognizable.

7.2.2 Markov compactum and adic transformation

The object of this section is to focus on a deep relationship between shifts of finite type and substitutive dynamical systems, which allows one to transpose some of the properties of shifts of finite type to substitutive dynamical systems. The results of this section will be explicitly illustrated in Sec. 7.2.3, by using the notion of recognizability exposed in Sec. 7.2.1.

Markov compactum. At the beginning of the eighties, A. M. Vershik defined a new type of dynamical systems, called *adic systems*, their support being called Markov compacta [437]. In short, a *Markov compactum* is the set of paths in an infinite labeled graph. One can define arbitrary a partial order between the labels of the edges of the graph, and deduce a partial order between the elements of the Markov compactum. The choice of the successor with respect to this partial order is a continuous map called *adic transformation.*

A motivation for the introduction of this new family of systems is the question of the approximation of ergodic systems. Indeed, every automorphism of a Lebesgue space with an ergodic invariant measure is measure-theoretically isomorphic to an adic transformation. One can refer to [437] for more details about these systems.

Stationary Markov compactum. As a particular case of Markov compactum one can define the notion of *stationary Markov compactum.* A Markov compactum is said to be stationary if its graph is a tree and each level of the tree has the same structure. This implies that a stationary Markov compactum can be defined as a shift of finite type. More precisely, let $\mathbf{M}$ be a $d \times d$ matrix, with entries 0 or 1. The stationary Markov compactum associated with $\mathbf{M}$ is the set of the infinite paths in the infinite graph with levels indexed by $0, 1, 2 \ldots$, and having d vertices on each level. The edges connect vertices of the i-th level with vertices of the $i+1$-th level according to the adjacency matrix $\mathbf{M}$. Thus, the Markov compactum is the following set:

$$X_{\mathbf{M}} = \{(x_i)_{i\geq 0} \in \{1 \ldots d\}^{\mathbb{N}};\ \ \forall i \geq 0,\ X_{x_i, x_{i+1}} = 1\}.$$

There exists a canonical map which acts on this compactum: the one-sided shift map defined by $T((x_i)_{i\geq 0}) = (x_i)_{i\geq 1}$. This map is a dynamical system with positive entropy, the properties of which are quite well known.

Partial ordering on the Markov compactum. The edges issued from a vertex i are totally ordered by the number of the exit vertex of the edge. From this partial order on vertices can be deduced the following partial order on the paths $X_{\mathbf{M}}$:

$$(x_i)_{i\geq 0} \prec (y_i)_{i\geq 0} \text{ if there exists } i_0 \text{ such that } \begin{cases} x_i = y_i, & \forall i > i_0, \\ x_n < y_n. \end{cases}$$

In other words, a path $x = (x_i)_{i\geq 0}$ is said to precede a path $y = (y_i)_{i\geq 0}$ if they differ in finitely many terms and in the last such term the edge of x precedes the edges of y.

The Markov compactum is said to be *proper* if $X_{\mathbf{M}}$ contains exactly one maximal point and one minimal point with respect to the partial ordering.

Adic transformation. One may check that any non-maximal path $x \in X_{\mathbf{M}}$ admits an immediate successor $\tau(x)$, that is, the minimum of y in $X_{\mathbf{M}}$ preceded by x. This defines the *adic transformation* τ on $X_{\mathbf{M}}$, except on the set of the maximal elements which is finite when $\mathbf{M}$ is primitive.

Theorem 7.2.4 ([437, 438]). *The adic transformation associated with a primitive matrix is minimal and uniquely ergodic. The unique invariant measure is equivalent to the probability measure on $X_{\mathbf{M}}$ which is of maximal entropy for the shift map.*

Note that the probability measure which is of maximal entropy for the shift map is well-known: by primitivity, the matrix $\mathbf{M}$ admits a dominant eigenvalue α, and right and left eigenvectors $\mathbf{u} = (u_i)_i$, and $\mathbf{v} = (v_i)_i$ respectively, with positive coefficients.

One can normalize these vectors so that $< \mathbf{v}, \mathbf{u} >= 1$. One sets $p_i = v_i\, u_i$ and $p_{i,j} = a_{i,j} u_j / (\alpha u_i)$. The matrix $\mathbf{P}$ defined by $\mathbf{P} = (p_{i,j})$ is thus *stochastic*, and (p_i) is a right eigenvector associated with the eigenvalue 1.

The probability measure μ defined as follows over the cylinders of the shift of finite type is easily seen to be shift invariant, and is of *maximal entropy* (see for instance [444]):

$$\mu([a_0, a_1, \ldots, a_n]) = p_{a_0} p_{a_0, a_1} p_{a_1, a_2} \cdots p_{a_{n-1}, a_n}.$$

Remark. Though the unique invariant measure for the adic transformation on $X_{\mathbf{M}}$ is equivalent to the probability measure on $X_{\mathbf{M}}$ which is of maximal entropy for the shift map, these maps do not have the same entropy: the maximal entropy of the shift map is strictly positive, though the entropy of the adic transformation is equal to zero.

Stationary Markov compactum and adic transformations. A. N. Livshits proves in [266] that the stationary adic transformation associated with a primitive $d \times d$ matrix $\mathbf{M}$ is measure-theoretically isomorphic to the dynamical system associated with the substitution defined on $\{1, \ldots, d\}$ by $i \mapsto 1^{\mathbf{M}_{1,i}} 2^{\mathbf{M}_{2,i}} \ldots d^{\mathbf{M}_{d,i}}$, as soon as the substitution σ is recognizable. More generally, a proof of the fact that every primitive substitutive dynamical system is isomorphic to an adic transformation is sketched in [438], the definition of adic transformation being adapted to a slightly more general context than stationary Markov compactum.

These results were completed by A. Forrest [174], who proved the following result in terms of Bratteli diagrams, which is a slightly different version of Markov compactum: he proves that the measure-theoretic isomorphism of Theorem 7.2.4 turns to be a topological conjugacy.

Theorem 7.2.5 (Forrest [174, 155]). *Each adic transformation on a stationary proper Markov compactum is topologically conjugate either to a primitive substitutive dynamical system or to an odometer. Conversely, each primitive substitutive dynamical system or odometer is topologically conjugate to an adic transformation on a stationary proper Markov compactum.*

The proof of A. Forrest is not constructive. It was completed and restated in the above terms by F. Durand, B. Host and C. F. Skau in [155]. In these papers, the authors give a new proof using substitutions, return words, and more precisely results about recognizability stated in Sec. 7.2.1 (for a definition of return words, see Chap. 3 and Sec. 7.3.2). Their approach provides an algorithm which computes for any substitutive dynamical system an adic transformation on a Markov compactum which is conjugate to a substitutive system. One can also refer to [216] for more details.

7.2.3 Desubstitution and prefix-suffix expansion

The computation of an adic transformation conjugated to a given substitutive dynamical system in [155] is not very natural and not very easy. Using the properties of recognizability, G. Rauzy initiated in [349] and [352] the idea of representing a substitutive dynamical system directly thanks to (7.1) and its interpretation given in (7.2). Let us illustrate this approach in more detail.

Desubstitution. It will be recalled that a consequence of the works of B. Mossé is that every word in X_σ can be expanded in the following form:

$$w = \dots | \underbrace{\dots}_{\sigma(y_{-1})} | \underbrace{w_{-k} \dots w_{-1}.w_0 \dots w_l}_{\sigma(y_0)} | \underbrace{\dots}_{\sigma(y_1)} | \underbrace{\dots}_{\sigma(y_2)} | \dots$$

with $\dots y_{-n} \dots y_{-1} y_0 y_1 \dots y_n \dots$ in X_σ.

Let $p = w_{-k} \dots w_{-1}$ denote the prefix of $\sigma(y_0)$ of length k and $s = w_1 \dots w_l$ the suffix of length l. The word w is completely determined by the word y and the decomposition of $\sigma(y_0)$ as $\sigma(y_0) = pw_0s$. Let $\mathcal{P}$ be the finite set of all such decompositions:

$$\mathcal{P} = \{(p, a, s) \in \mathcal{A}^\star \times \mathcal{A} \times \mathcal{A}^\star;\ \exists b \in \mathcal{A} \text{ and } \sigma(b) = pas\}.$$

Example 7.2.6. For the Fibonacci substitution $1 \mapsto 12$ and $2 \mapsto 1$, one gets:

$$\mathcal{P} = \{(\varepsilon, 1, 2), (1, 2, \varepsilon), (\varepsilon, 1, \varepsilon)\}.$$

For the substitution $1 \mapsto 1112$ and $2 \mapsto 12$, one gets:

$$\mathcal{P} = \{(\varepsilon, 1, 112), (1, 1, 12), (11, 1, 2), (111, 2, \varepsilon), (\varepsilon, 1, 2), (1, 2, \varepsilon)\}.$$

Thus, Theorem 7.2.2 implies that we can define, on the one hand, a continuous *desubstitution map* on X_σ (which sends w to y), on the other hand, a partition corresponding to the decomposition of $\sigma(y_0)$.

Theorem 7.2.7. *For any primitive substitution σ the following* desubstitution map θ *and the* prefix-suffix coding map γ *are well defined and continuous on X_σ:*

$$\theta : \begin{cases} X_\sigma \to X_\sigma \\ w \ \mapsto y \end{cases} \quad \text{such that } w = S^k\sigma(y) \text{ and } 0 \le k < |\sigma(y_0)|,$$

$$\gamma : \begin{cases} X_\sigma \to \mathcal{P} \\ w \ \mapsto (p, w_0, s) \end{cases} \quad \text{such that } \sigma(y_0) = pw_0s \text{ and } |p| = k.$$

Prefix-suffix expansion. In [101] and independently in [210], the itineraries of the points of X_σ under the desubstitution according to the partition defined by γ are studied. More precisely, the *prefix-suffix expansion* is the map Γ defined on X_σ by:

$$\forall w \in X_\sigma, \quad \Gamma(w) = (\gamma(\theta^i w))_{i \ge 0} = (p_i, a_i, s_i)_{i \ge 0} \in \mathcal{P}^{\mathbb{N}}.$$

Exercise 7.2.8. Let $w \in X_\sigma$ and $\Gamma(w) = (p_i, a_i, s_i)_{i \ge 0}$ be its prefix-suffix development. Suppose that nor all the prefixes p_i neither all the suffixes s_i are empty from a certain rank. Prove that w and $\Gamma(w)$ satisfy the following relationship:

$$w = \dots \sigma^n(p_n) \dots \sigma(p_1) p_0 \dot{a}_0 s_0 \sigma(s_1) \dots \sigma^n(s_n) \dots$$

Important questions are

- the identification of the image of Γ,
- the injectivity of Γ,
- the identification of an action on $\Gamma(X_\sigma)$ such that Γ is a measure-theoretic isomorphism between that action and the shift map S on X_σ.

The image and the injectivity of Γ: prefix-suffix shift of finite type. The definition of Γ leads us to define the shift of finite type described by the automaton, called *prefix-suffix automaton* of the substitution σ, which has $\mathcal{A}$ as a set of vertices and $\mathcal{P}$ as a set of label edges: there is an edge labeled by $e = (p, a, s)$ from a to b if and only if $pas = \sigma(b)$.

The *prefix-suffix shift of finite type* is the set $\mathcal{D}$ of labels of infinite walks in this automaton. By definition, it is the support of a shift of finite type.

Example 7.2.9. The prefix-suffix automaton of the Fibonacci substitution $1 \mapsto 12$, $2 \mapsto 1$ is shown in Fig. 7.4.

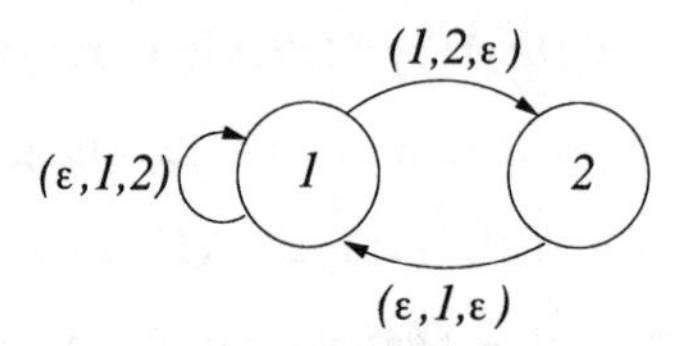

Fig. 7.4. Prefix-suffix automaton for the Fibonacci substitution.

The prefix-suffix of finite type for the Fibonacci substitution is the subset of $\{(\varepsilon, 1, 2), (1, 2, \varepsilon), (\varepsilon, 1, \varepsilon)\}^{\mathbb{N}}$ which consists of the labels of paths of the automaton.

It is proved in [101] and [210] that Γ is onto $\mathcal{D}$ and almost everywhere one-to-one on X_σ.

Prefix-suffix partial order. We explained before that a stationary Markov compactum is nothing else than the support of a shift of finite type, provided with a partial order deduced from a partial order on the labels of the edges of the subshift.

The preceding construction associates with a substitution the support of a shift of finite type, that is, the prefix-suffix shift of finite type $\mathcal{D}$. Thus, to define a Markov compactum on this set, we only need a partial order on the set $\mathcal{P}$ of labels of $\mathcal{D}$. However, the substitution σ provides a very natural partial order: two labels (p, a, s) and (q, b, r) are comparable for this order if they issue from the same letter, i.e., $pas = qbr = \sigma(c)$ for a letter c. The order of comparison comes from the length of the prefix p. More precisely, let us define

$$(p, a, s) \prec (q, b, r) \text{ if there exists } c \in \mathcal{A} \text{ such that } \begin{cases} pas = qbr = \sigma(c), \\ |p| < |q|. \end{cases} \tag{7.4}$$

As explained is Sec. 7.2.2, this order on the set of labels $\mathcal{P}$ induces an order on the support $\mathcal{D}$ of a shift of finite type, obtained as a lexicographical order from the right: a path $e = (e_i)_{i \geq 0} \in \mathcal{D}$ is said to precede a path

$f = (f_i)_{i \geq 0} \in \mathcal{D}$ if they differ in finitely many terms and in the last such term i_0, e_{i_0} precedes f_{i_0} for the partial order defined at (7.4):

$$(e_i)_{i \geq 0} \prec (f_i)_{i \geq 0} \text{ if there exists a rank } i_0 \text{ such that } \begin{cases} \forall i > i_0,\ e_i = f_i \\ e_{i_0} \prec f_{i_0} \end{cases}. \tag{7.5}$$

Remark. Denote $e_{i_0} = (p_{i_0}, a_{i_0}, s_{i_0})$ and $f_{i_0} = (q_{i_0}, b_{i_0}, r_{i_0})$. As $e_{i_0+1} = f_{i_0+1}$, we have $p_{i_0} a_{i_0} s_{i_0} = q_{i_0} b_{i_0} r_{i_0}$, and e_{i_0} is comparable with f_{i_0} for the prefix-suffix partial order. The condition $e_{i_0} \prec f_{i_0}$ means that the length of the prefix p_{i_0} is less than the length of the prefix q_{i_0}.

Prefix-suffix adic transformation. Following the construction described in Sec. 7.2.2, the prefix-suffix partial order on the support $\mathcal{D}$ of a shift of finite type provides an adic transformation, called *prefix-suffix adic transformation* and denoted by τ. This map is an immediate successor transformation on the set of non-maximal points of $\mathcal{D}$.

Exercise 7.2.10. 1. Prove that a point $e = (p_i, a_i, s_i)_{i \geq 0} \in \mathcal{D}$ is maximal for the prefix-suffix partial order if and only if all the suffixes s_i are empty.

2. Let $e = (p_i, a_i, s_i)_{i \geq 0} \in \mathcal{D}$ be a non-maximal point in $\mathcal{D}$. Let i_0 be the smallest integer such that $s_{i_0} \neq \varepsilon$. Prove that $\tau(e)$ is the unique path $f = (q_i, b_i, t_i)_{i \geq 0} \in \mathcal{D}$ defined by:

$$\begin{aligned} &\forall i > i_0, (q_i, b_i, t_i) = (p_i, a_i, s_i), \\ &i = i_0 \quad q_i b_i t_i = p_i a_i s_i \text{ and } |q_i| = |p_i| + 1, \\ &i < i_0 \quad q_i = \varepsilon \text{ and } b_i t_i = \sigma(b_{i+1}). \end{aligned}$$

Exercise 7.2.11. Consider the substitution $1 \mapsto 21$, $2 \mapsto 13$, $3 \mapsto 1$. Let u be the periodic point defined by

$$u = \lim_{n \to \infty} \sigma^{2n}(1.1) = \ldots 1321212111321.1321212111 \ldots.$$

1. Prove that the prefix-suffix expansion of u is:

$$\Gamma(u) = m = (\varepsilon, 2, 1)(\varepsilon, 1, 3)(\varepsilon, 2, 1)(\varepsilon, 1, 3) \ldots.$$

2. Prove that

$$\begin{aligned} \tau(m) &= (2, 1, \varepsilon)(\varepsilon, 1, 3)(\varepsilon, 2, 1)(\varepsilon, 1, 3) \ldots \\ \tau^2(m) &= (\varepsilon, 1, \varepsilon)(1, 3, \varepsilon) \quad - \quad - \quad \ldots \\ \tau^3(m) &= (\varepsilon, 1, 3)(\varepsilon, 2, 1)(2, 1, \varepsilon) \quad - \quad \ldots. \end{aligned}$$

3. Prove that

$$\tau^7(m) = (\varepsilon, 1, 3)(\varepsilon, 2, 1)(\varepsilon, 1, \varepsilon)(1, 3, \varepsilon)[(\varepsilon, 2, 1)(\varepsilon, 1, 3)]^\infty.$$

The prefix-suffix adic transformation on the set $\mathcal{D}$ provides an example of Markov compactum as defined in Sec. 7.2.2. The following theorem means that this explicit Markov compactum realizes explicitly the main result in [438] which states that any primitive substitutive dynamical system is measure-theoretically isomorphic to an adic transformation on a Markov compactum.

The conjugacy map is not a topological conjugacy map but a semi-conjugacy, contrary to the one obtained in [155]. But this conjugacy map can be deduced in a simple, natural and explicit way from the substitution.

Theorem 7.2.12 ([101, 210]). *Let σ be a primitive substitution which is not shift-periodic and (X_σ, S) the dynamical system generated by σ. The prefix-suffix expansion is a continuous mapping onto the shift of finite type $\mathcal{D}$. This map is one-to-one except on the orbit of periodic points of σ.*

This prefix-suffix expansion is a semi-topological conjugacy between the shift map S on X_σ and an adic transformation on $\mathcal{D}$, considered as a Markov compactum when the set $\mathcal{P}$ of edges is provided with the natural prefix-suffix partial ordering coming from the substitution.

It can be noted that the prefix-suffix automaton which has allowed us to define the Markov compactum $\mathcal{D}$ is the most complete form of a large class of automata appearing in literature.

It was introduced in [101] in order to extend and to make more precise the prefix automaton (sometimes also called "automate à la provençale"), and its dual suffix automaton, defined by G. Rauzy in the seminal papers [349, 352] with the final goal to represent substitutive dynamical systems of Pisot type by domain exchanges in the Euclidean space. The difference between the prefix automaton and the prefix-suffix automaton is that the subshift generated by the second one is of finite type, while the one generated by the first automaton is only sofic. This has important consequences for the injectivity of geometric representations of substitutive dynamical systems defined thanks to the subshifts (see [349, 352, 102] and Sec. 7.4).

Moreover, the prefix-suffix automaton can be found in a coded form among the number theory works which followed [349]: first in a work by J.-M. Dumont and A. Thomas [150] about numeration scales associated with certain substitutions; then among their development by V. Sirvent [403] who proves some properties of the Rauzy fractal (see also Sec. 7.4). We can also mention the definition by T. Kamae [230] of colored tilings associated with a weighted substitution, the layout of the tiles following rules being close to the transition rules of the prefix automaton, but in a non-explicit way.

To finish this brief survey, the prefix-suffix automaton also appears in the context of language theory: the prefix-suffix automaton projects, on the one hand, onto the automaton defined by A. Maes [275] in his works about decidability of arithmetical theory, on the other hand, onto the automaton defined by P. Narbel [310] to study the boundary of a language using the description with trees of the set of paths in his automaton.

7.3 Spectral theory of substitutive dynamical systems

Substitutive dynamical systems were introduced by W. H. Gottschalk [190], as examples of symbolic dynamical systems, the study of which was initialized in [302, 303]. The attention first focused on systems associated with a substitution of *constant length.* Let us recall that a substitution is said to be of *constant length* if the image of any letter of the alphabet under the substitution contains the same number of letters. We give in Sec. 7.3.1 some results about these substitutions.

At the end of the seventies, the spectral theory of constant length substitutive dynamical systems was quite well known, and people started to focus on spectral theory of nonconstant length substitutive dynamical systems. We will give in the next section the main results about this theory. Thanks to recognizability and Markov compactum, they got the results on the spectral theory of substitutive systems presented in Sec. 7.3.2. A special attention was devoted to the class of substitutions of Pisot type. This will be the object of Sec. 7.3.3.

7.3.1 Constant length substitutions

First topological and ergodic properties. They were obtained at the beginning of the seventies: T. Kamae proves in [228] that the condition of primitivity is sufficient for the minimality of the system. Then he gives a sufficient condition for the system to have a purely discrete spectrum [229]. B. G. Klein shows that primitive substitutive systems of constant length are uniquely ergodic and that their topological entropy equals zero [248].

Partial results about the maximal equicontinuous factor. In [124], E. M. Coven and M. S. Keane focus on a class of constant length substitutions over a two-letter alphabet which generate a minimal system, related to Toeplitz substitutions. They prove that the maximal equicontinuous factor of these systems is the n-adic group $\mathbb{Z}_n$, where n is the length of the substitution. If n is not prime, $\mathbb{Z}_n$ is defined to be the product of the p-adic groups $\mathbb{Z}_p$, for all primes p that divide n .

J. C. Martin generalizes this to substitutions of constant length n over a finite alphabet [280]: if the substitution is one-to-one on the set of letters, the maximal equicontinuous factor of the substitutive dynamical system is the addition of $(1,1)$ on $\mathbb{Z}_n \times \mathbb{Z}/m\mathbb{Z}$, where m in an integer depending on the substitution. J. C. Martin then gives some conditions for the factor to be isomorphic to the system: some are necessary conditions, the others are sufficient.

Determining the maximal equicontinuous factor. Finally, F. M. Dekking generalizes these results for the constant length case, by introducing new methods:

Theorem 7.3.1 (Dekking [134]). *Let σ be a non-periodic substitution of constant length n. Let u be a periodic point for σ. We call* height *of the substitution the greatest integer m which is coprime with n and divides all the strictly positive ranks of occurrence of the letter u_0 in u. The height is less that the cardinality of the alphabet.*

The maximal equicontinuous factor of the substitutive dynamical system associated with σ is the addition of $(1,1)$ on the abelian group $\mathbb{Z}_n \times \mathbb{Z}/m\mathbb{Z}$.

F. M. Dekking gives an algorithmic method for computing the height of a substitution.

Exercise 7.3.2. Show that the Morse substitution has height 1. Deduce the maximal equicontinuous factor of the system associated with this substitution.

Exercise 7.3.3.
1. Give the height of the substitution $1 \mapsto 121$, $2 \mapsto 312$, $3 \mapsto 213$.
2. Prove that the spectrum (*defined in Chap. 1*) of the system associated with this substitution is

$$\left\{ e^{2in\pi/3^m + 2ik\pi/2};\ n, k \in \mathbb{Z}, m \in \mathbb{N}, \right\}.$$

The question of isomorphism for substitutions of height 1. F. M. Dekking also introduces the combinatorial condition which later will be called *coincidence condition*: a constant length substitution σ satisfies this condition if there exist two integers k, n such that the images of any letter of the alphabet under σ^k has the same n-th letter. Generalizing [229], F. M. Dekking proves that a substitution of height 1 satisfies this condition if and only if the associated substitutive dynamical system is measure-theoretically isomorphic to its maximal equicontinuous factor, that is, if the system has a purely discrete spectrum:

Theorem 7.3.4 (Dekking [134]). *Let σ be a substitution of constant length and of height 1. The substitutive dynamical system associated with σ has a purely discrete spectrum if and only if the substitution σ satisfies the condition of coincidence.*

Exercise 7.3.5. Prove that the substitutive dynamical system associated with the Morse substitution is not isomorphic to a translation on $\mathbb{Z}_2$.

Note that this result was already shown in Chap. 5.

Pure base of a substitution. If the height h of a primitive substitution σ of constant length n is different from 1, the *pure base* of σ is the substitution η defined as follows:

- The alphabet $\mathcal{A}_1$ of η is the collection of all blocks of length h appearing at position kh in a periodic point u for σ.

- For every block $W \in \mathcal{A}_1$ of length h, $\eta(W) = W_1W_2\ldots W_n$, with $W_i \in \mathcal{A}_1$ and $W_1W_2\ldots W_n = \sigma(W)$.

One should notice that η is primitive, of constant length n, and that its height is 1. Up to this tool, the result of Dekking can be generalized to all substitutions of constant length.

Theorem 7.3.6 (Dekking [134]). *Let σ be a substitution of constant length. The substitutive dynamical system associated with σ has a purely discrete spectrum if and only if its pure base satisfies the condition of coincidence.*

Exercise 7.3.7. Prove that the pure base of $1 \mapsto 121$, $2 \mapsto 312$, $3 \mapsto 213$ is given by the substitution $a \mapsto aab$, $b \mapsto aba$. Has this substitution a purely discrete spectrum?

7.3.2 Nonconstant length substitutions

The first partial results about the spectrum of substitutions of nonconstant length were obtained by studying a class of nonconstant length substitutive systems which are measure-theoretically isomorphic to constant length substitutive systems [124, 134]. Later, F. M. Dekking et M. S. Keane proved that strongly mixing substitutive dynamical systems do not exist, but weak-mixing is possible [138]. P. Michel proved in [291] that the dynamical system associated with a primitive substitution of nonconstant length is uniquely ergodic, see also 1. M. Queffélec studies in detail the spectral type of substitutive systems in [339, 341]. Most of the results cited above appear in this last reference.

These general topological and ergodic properties having been studied, authors focused on the explicit description of the spectrum of a substitutive dynamical system. They also took interest in obtaining conditions for a substitutive system to have a purely discrete spectrum, partially continuous spectrum, or continuous spectrum. The description of the known results is the aim of this section.

Rational eigenvalues and the characteristic polynomial of the incidence matrix. J. C. Martin obtains a partial result about the spectrum of nonconstant length substitutions on a two-letter alphabet in [281]. His work suffers from a too weak definition of the property of recognizability (see Sec. 7.2.1). If we take into account the results about recognizability proved later, his results become:

Theorem 7.3.8 (Martin [281]). *Let σ be a not shift-periodic and primitive substitution on a two-letter alphabet. Let $X^2 - tX + d$ be the characteristic polynomial of the incidence matrix of σ and let $r \in \mathbb{Q}$. Let i_0 be the greatest positive integer such that:*

- *the prime factors of* i_0 *divide both* $|\sigma(1)|$ *and* $|\sigma(2)|$,
- i_0 *is prime with* r,
- i_0 *divides* $|\sigma(1)| - |\sigma(2)|$.

Then $\exp(2\pi i r)$ *is an eigenvalue of the dynamical system associated with* σ *if and only if there exist integers* k, m, n *such that* $r = k/i_0 + m/(pgcd(d,t))^n$.

A consequence of this theorem is the characterization of p-adic measure-theoretic factors of substitutive systems over a two-letter alphabet:

Exercise 7.3.9. 1. Let σ be a not shift-periodic and primitive substitution over a two-letter alphabet. Let p be a prime number. Show that the addition of 1 on $\mathbb{Z}_p$ is a measure-theoretic factor of the substitutive dynamical system associated with σ if and only if p divides the determinant and the trace of the characteristic polynomial of the incidence matrix of σ.
2. Find the p-adic factors of the systems defined by the substitutions $1 \mapsto 1112$, $2 \mapsto 12$, and $1 \mapsto 11222$, $2 \mapsto 1222$.

In particular, there exist non-unimodular substitutive systems which do not admit any p-adic factor. Such a behavior is totally different from the case of constant length substitutions. This result has been generalized by F. Durand:

Theorem 7.3.10 (Durand [154]). *Let* σ *be a primitive substitution, the incidence matrix* $\mathbf{M}_\sigma$ *of which has an irreducible characteristic polynomial* $\chi_{\mathbf{M}_\sigma}$. *Let* p *be a prime number. The addition of 1 over* $\mathbb{Z}_p$ *is a measure-theoretic factor of the substitutive dynamical system associated with* σ *if and only if* p *divides all the coefficients of* $\chi_{\mathbf{M}_\sigma}$ *except the dominant one, that is, if and only if* $\mathbf{M}_\sigma$ *is nilpotent modulo* p.

A new proof of this result is given in [395] by studying the ramifications of the determinant of the incidence matrix of σ in the Galois extension of its dominant eigenvalue.

Coboundaries: a full description of the spectrum. In [214], B. Host introduces a good notion of recognizability and thanks to this notion proves the following important result.

Theorem 7.3.11 (Host [214]). *Let* σ *be a not shift-periodic and primitive substitution. Each eigenfunction of the substitutive dynamical system associated with* σ *is continuous.*

To be more precise, since we deal with functions in $\mathcal{L}^2$, Theorem 7.3.11 means that any class of eigenfunctions contains a continuous eigenfunction.

Exercise 7.3.12. Let σ be a not shift-periodic and primitive substitution. Show that a minimal toral rotation or the addition of 1 in the p-adic integer set $\mathbb{Z}_p$ is a topological factor of the substitutive dynamical system associated with σ if and only if it is a measure-theoretic factor of this system.

B. Host deduces from Theorem 7.3.11 a complete description of the spectrum of the nonconstant length substitutions, which are recognizable, and one-to-one on the letters. This description is a generalization of [134] and [281]. Here, $\mathbb{U} \subset \mathbb{C}$ denotes the unit circle.

Definition 7.3.13. *A* coboundary *of a substitution σ is defined as a map $h : \mathcal{A} \to \mathbb{U}$ such that there exists a map $f : \mathcal{A} \to \mathbb{U}$ with $f(b) = f(a)h(a)$, for every word ab of length 2 which belongs to the language of the substitution.*

In the most simplest cases the only coboundary is the trivial one, that is, the constant function equal to 1. However, there exist some substitutions with nontrivial coboundaries:

Exercise 7.3.14. 1. Prove that the substitution $1 \mapsto 12$, $2 \mapsto 13$, $3 \mapsto 1$ has no nontrivial coboundary. *(Hint: note that the language of this substitution contains* 12, 21 *and* 11.*)*
2. Prove that any primitive and not shift-periodic substitution over two letters has no nontrivial coboundary.
3. Find a nontrivial coboundary for the substitution $1 \mapsto 1231$, $2 \mapsto 232$, $3 \mapsto 3123$.

In Sec. 7.5.3 we give a method to construct substitutions with nontrivial coboundaries.

Theorem 7.3.15 (Host [214]). *Let σ be a not shift-periodic and primitive substitution over the alphabet $\mathcal{A}$. A complex number $\lambda \subset \mathbb{U}$ is an eigenvalue of (X_σ, S) if and only if there exists $p > 0$ such that for every $a \in \mathcal{A}$, the limit $h(a) = \lim_{n\to\infty} \lambda^{|\sigma^{pn}(a)|}$ is well defined, and h is a coboundary of σ.*

Exercise 7.3.16. Let σ be the substitution defined by $1 \mapsto 12121$ and $2 \mapsto 122$.

1. Prove that $|\sigma^{n+2}(a)| = 5|\sigma^{n+1}(a)| + 4|\sigma^n(a)|$, for $a = 1, 2$.
2. Deduce that $2|\sigma^n(2)| - |\sigma^n(1)| = 1$, for every nonnegative n.
3. Conclude that the system associated with σ is weakly mixing (that is, 1 is its only eigenvalue; see Chap. 1).

Since the constant function equal to one 1 is always a coboundary, a sufficient condition for λ to be an eigenvalue is the following:

Corollary 7.3.17. *Let σ be a not shift-periodic and primitive substitution. If there exists p such that $\lambda \in \mathbb{C}$ satisfies $\lim \lambda^{|\sigma^{pn}(a)|} = 1$ for every letter a of the alphabet, then λ is an eigenvalue of the substitutive dynamical system associated with σ. We say that λ is an* eigenvalue associated with the trivial coboundary.

Let $\mathbf{1}$ be the column vector whose all coordinates are equal to 1.

Exercise 7.3.18. Let $\lambda = \exp(2i\pi\beta)$ of modulus 1. Prove that $\lim \lambda^{|\sigma^n(a)|} = 1$ for every letter a, if and only if $\beta {}^t\mathbf{M}_\sigma{}^n \mathbf{1}$ tends to zero modulo $\mathbb{Z}^d$.

B. Host provides a method to identify the set of eigenvalues associated with the trivial coboundary.

Proposition 7.3.19 (Host [214]). *Let σ be a primitive d-letter substitution. Let X_σ be the substitutive system associated with σ and $\mathbf{M}_\sigma$ be its incidence matrix.*

The complex $\exp(2i\pi\beta)$ *is an eigenvalue for X_σ associated with the trivial coboundary if and only if $\beta\mathbf{1} = \mathbf{x}_1 + \mathbf{x}_2$, where there exist two integers p and q such that ${}^t\mathbf{M}_\sigma{}^{pn}\mathbf{x}_1$ tends to zero in $\mathbb{R}^d$ and $\mathbf{x}_2 \in {}^t\mathbf{M}_\sigma^{-q}\mathbb{Z}^d$.*

An important example of application can be found among substitutions of Pisot type, which will be treated in Sec. 7.3.3.

In [173], S. Ferenczi, C. Mauduit et A. Nogueira give a new simplified version of the proofs of Theorems 7.3.11 and 7.3.15. They deduce from these new proofs a more explicit characterization of the spectrum of primitive substitutive dynamical systems, in terms of polynomials (see Theorem 7.3.27).

Combinatorial conditions: a mix of coincidences and return words. The parallel between substitutive systems and adic systems presented in the preceding section, and the ideas developed by B. Host in [214], led many authors to give some necessary and sufficient conditions for a substitutive dynamical system to have a purely discrete spectrum, a partially continuous spectrum or to be weakly mixing (see the survey in [438]).

For instance, A. N. Livshits stated a relationship between the fact that a substitutive system has a purely discrete spectrum and the existence of a code of the system made of modified return words on one side, coincidences on letters on the other side. To precise this, let us introduce the following notions:

Definition 7.3.20. *Let σ be a primitive and not shift-periodic substitution defined over the alphabet $\mathcal{A}$.*

*For any letter a, the a-*blocks *are the elements of the set Δ_a:*

$$\Delta_a = \left\{ (a_1 \dots a_{r-1}, a_2 \dots a_r) \text{ with } \left\{ \begin{array}{l} a_1 \dots a_r \in \mathcal{L}(X_\sigma) \\ i = 1, r, \forall n > 0, |\sigma^n(a_i)| = |\sigma^n(a)| \\ i \neq 1, r, \exists n > 0, |\sigma^n(a_i)| \neq |\sigma^n(a)| \end{array} \right. \right\}$$

Note that if the substitution is of constant length, then $\Delta_a = \{(b, b),\ bb \in \mathcal{L}(X_\sigma)\}$. Moreover, the set Δ_a contains the *return words* of the substitution, as introduced by F. Durand in [152, 151].

Definition 7.3.21. *Let σ be a primitive and not shift-periodic substitution. A* return word *is a word $W = a_1 \dots a_{k-1} \in \mathcal{L}(X_\sigma)$ such that there exists a letter a_k with*

- $a_1 \ldots a_{k-1} a_k \in \mathcal{L}(X_\sigma)$,
- *for every large enough n,* $\begin{cases} \sigma^n(a_k) = \sigma^n(a_1), \\ \forall j \neq 1, k,\ \sigma^n(a_j) \neq \sigma^n(a_1). \end{cases}$

The associated return time is defined to be $r_n(W) = |\sigma^n(a_1 \ldots a_{k-1})|$.

A. N. Livshits gives the following characterization of substitutive systems with a purely discrete spectrum. This characterization was announced in [265] and [438], and proved in [267]. As defined in Chap. 1, $\mathbf{l} : \mathcal{A} \to \mathbb{N}^d$ denotes the abelianization map.

Theorem 7.3.22 (Livshits [265]). *Let σ be a primitive and not shift-periodic substitution defined over the alphabet* $\mathcal{A}$.

Suppose that there exists a letter a and a finite set Δ of pairs (W_{i_k}, W_{j_k}) *with* $\mathbf{l}(W_{i_k}) = \mathbf{l}(W_{j_k})$, *which satisfy*

$$\forall (V_1, V_2) \in \Delta \cup \Delta_a,\ \exists m > 0,$$

$$\text{such that } \begin{cases} \sigma^m(V_1) = W_{l_1} \ldots W_{l_s} \\ \sigma^m(V_2) = W_{j_1} \ldots W_{j_s} \end{cases} \text{with } \begin{cases} \forall r,\ (W_{l_r}, W_{j_r}) \in \Delta \cup \Delta_a \\ \exists r,\ W_{l_r} = W_{j_r} \end{cases}.$$

Then the substitutive dynamical system associated with σ has a purely discrete spectrum.

As before, examples of application of this theorem can be found among substitutions of Pisot type, which will be treated in detail in the next section (see in particular Theorem 7.3.35).

The converse of Theorem 7.3.22 can be expressed under the following form.

Theorem 7.3.23 (Livshits [265]). *Let σ be a primitive and not shift-periodic substitution. Let* $c \neq d$ *and a be letters such that* $ac, ad \in \mathcal{L}(X_\sigma)$.

Suppose that there exist a finite set of words $\{W_i\}$ *and an increasing sequence of integers* m_i *satisfying for all i:*

$$\begin{cases} \sigma^{m_i}(c) = W_{l_1} \ldots W_{l_s} A \\ \sigma^{m_i}(d) = W_{k_1} \ldots W_{k_s} B \end{cases} \text{with } \begin{cases} A = \varepsilon \text{ or } B = \varepsilon, \\ \forall r,\ \mathbf{l}(W_{l_t}) = \mathbf{l}(W_{k_t}), \\ \forall r,\ W_{l_t} \neq W_{k_t}. \end{cases}$$

Then the substitutive dynamical system associated with σ has a partially continuous spectrum.

For instance the system associated with the substitution $1 \mapsto 23$, $2 \mapsto 12$ and $3 \mapsto 23$ admits a continuous spectral component. However, the substitution is of constant length, so that the system admits $\mathbb{Z}_2$ as a factor (Theorem 7.3.1) and is not weakly mixing.

Another consequence of this theorem is that the Morse substitution system has a partially continuous spectrum, which has been studied: the maximal spectral type is known to be a Riesz product (see the bibliography of

[339]). Finally, the substitutive system associated with the Morse substitution is measure-theoretically isomorphic to a two-point extension of its maximal equicontinuous factor, that is, the addition of 1 on $\mathbb{Z}_2$ (see also Chap. 5).

Finally, A. N. Livshits gives a sufficient condition for a substitutive dynamical system to be weakly mixing, that is, to have no nontrivial eigenvalue.

Theorem 7.3.24 (Livshits [266]). *Let σ be a not shift-periodic and primitive substitution.*

If every eigenvalue of the incidence matrix of σ has a modulus greater than or equal to 1, then the substitutive dynamical system associated with σ has no irrational eigenvalue.

If there exists an element of the language of σ, denoted $W = a_1 \dots a_k = a_1 V$ with $a_1 = a_k$, such that for all integers n, the lengths $|\sigma^n(V)|$ are coprime to each other for $n > N$, then the substitutive dynamical system associated with σ has no rational eigenvalue (except 1).

Note that in the previous theorem, W is a return word as soon as a_1 does not appear is V before the last rank. As an application, the substitution $1 \mapsto 112$ and $2 \mapsto 12222$ introduced in [138] is weakly mixing [265].

Polynomial conditions. B. Solomyak makes the above theorem somewhat more precise by giving a characterization of irrational eigenvalues of substitutive systems in a more explicit way than Host's one.

Theorem 7.3.25 (B. Solomyak [410]). *Let σ be a primitive substitution such that its incidence matrix $\mathbf{M}_\sigma$ admits an irreducible characteristic polynomial. Let α_k, $k = 1 \dots m$, be the eigenvalues of $\mathbf{M}_\sigma$ such that $|\alpha_k| \geq 1$.*

The substitutive system associated with σ admits at least one irrational eigenvalue if and only if there exists a polynomial $P \in \mathbb{Z}[X]$ such that $P(\alpha_j) = P(\alpha_k)$ for every $j, k \leq m$.

If so, $\exp(2\pi i P(\alpha_1))$ and $\exp(2\pi i P(\alpha_1)^n)$ are eigenvalues of the substitutive system.

Remark.

- Note that the numbers $\exp(2i\pi n P(\alpha_1))$ are eigenvalues as soon as the number $\exp(2\pi i P(\alpha_1))$ is an eigenvalue.
- Theorem 7.3.25 does not state that any irrational eigenvalue is of the form $\exp(2\pi i P(\alpha_1))$.

For instance, the spectrum of the system associated with the substitution $1 \mapsto 1244$, $2 \mapsto 23$, $3 \mapsto 4$, $4 \mapsto 1$ contains the set $\exp(2\pi i \mathbb{Z}\sqrt{2})$.

Theorem 7.3.25 produces other sufficient conditions for a substitutive system to have nontrivial rational spectrum, in particular in the Pisot case (see Sec. 7.3.3).

Exercise 7.3.26 ([410]).

- Let s denotes the smallest prime divisor of the cardinality d of the alphabet of a primitive substitution σ. Suppose that the number of eigenvalues of $\mathbf{M}_\sigma$ greater than 1 is strictly greater that d/s. Show that the substitutive dynamical system associated with σ has no irrational eigenvalue.
- For any prime p, any substitutive dynamical system with irrational eigenvalues over a p-letter alphabet is of Pisot type.

In [173], S. Ferenczi, C. Mauduit and A. Nogueira got interested in the relationship between coboundaries and return words.

Proposition 7.3.27 (Ferenczi, Mauduit, Nogueira [173]). *Let σ be a not shift-periodic and primitive substitution and (X_σ, S) be the substitutive dynamical system associated with σ.*

A complex number λ with modulus 1 is an eigenvalue of the substitutive dynamical system (X_σ, S) if and only if $\lim_{n\to\infty} \lambda^{r_n(W)} = 1$ for every return word W.

This proposition is illustrated for the Chacon sequence in Chap. 5.

Using techniques from the study of substitutive normal sets [282], the authors give a constructive version of the proofs of B. Host [214]. They get a characterization of eigenvalues which generalizes (no more hypothesis of irreducibility) and gives a partial converse to Theorem 7.3.25. This characterization is too long to be cited here. A consequence of it is the following:

Theorem 7.3.28 (Ferenczi, Mauduit, Nogueira [173]). *Let σ be a primitive and not shift-periodic substitution and α_k, $k = 1 \ldots r$, be the eigenvalues of $\mathbf{M}_\sigma$ such that $|\alpha_k| \geq 1$. Let β be irrational.*

- *If $\lambda = \exp(2\pi i\beta)$ is an eigenvalue of X_σ associated with the trivial coboundary – that is, $\lim \lambda^{|\sigma^n(a)|} = 1$ for every letter a – then there exists a polynomial $P \in \mathbb{Z}[X]$ such that $P(\alpha_i) = \beta$ for every i.*
- *If $P \in \mathbb{Z}[X]$ is a polynomial with $P(\alpha_i) = \beta$ for every i, then there exists an integer k such that $\exp(2\pi i k P(\alpha_i))$ is an eigenvalue of the substitutive system.*

Thus, an integer polynomial which is constant over the set of eigenvalues of $\mathbf{M}_\sigma$ of modulus greater than one, does not provide directly an eigenvalue for the system: the constant may have to be multiplied by an integer to become the argument of an eigenvalue.

As an application, the spectrum of the system associated with the substitution considered above $1 \mapsto 1244$, $2 \mapsto 23$, $3 \mapsto 4$, $4 \mapsto 1$, is exactly the set $\exp(2\pi i\mathbb{Z}\sqrt{2})$: Theorem 7.3.25 implied that the spectrum contains this set. The converse is stated by Theorem 7.3.28.

Conclusion. The spectrum of a primitive dynamical system can be divided into two parts. The first part is the set of complex numbers associated with the trivial coboundary; according to Theorem 7.3.28, these eigenvalues depend only on the incidence matrix of the substitution, and, more precisely, for the irrational eigenvalues, on the matrix eigenvalues which are greater than 1. In this sense, this class of eigenvalues is quite natural.

In a more exotic and surprising way, the substitutive system can have some eigenvalues which are associated with a nontrivial coboundary, and not associated with the trivial coboundary. Such eigenvalues are not described by Theorem 7.3.28. These eigenvalues depend heavily on the return times and more generally on the combinatorics of the substitution, which in other words gives them a non-commutative aspect.

There exist some unpublished partial results about a substitution with rational noncommutative eigenvalues. We do not know any example of substitution with irrational noncommutative eigenvalue.

7.3.3 Substitutions of Pisot type

Some of the general results of Sec. 7.3.2 have special consequences for the spectral theory of Pisot type substitutive systems.

Eigenvalues associated with the trivial coboundary. As claimed by B. Host in [215], Theorem 7.3.11 and Theorem 7.3.15 allow one to determine the spectrum associated with the trivial coboundary, in the unimodular case.

Proposition 7.3.29 ([215]). *Let σ be a unimodular substitution of Pisot type. The group of eigenvalues of X_σ associated with the trivial coboundary is generated by the frequencies of letters in any word of the system, that is, by the coordinates of a right normalized eigenvector associated with the dominant eigenvalue of the incidence matrix of the substitution.*

Hints to deduce this result from Theorem 7.3.15 and Lemma 7.3.19 are the following:

Exercise 7.3.30. Let σ be a unimodular substitution of Pisot type over d letters. Let $\mathbf{M}_\sigma$ be the incidence matrix of σ and $\mathbf{1}$ be the column vector whose all coordinates are equal to 1. Let $\mathbf{u}_1$ be an expanding eigenvector of $\mathbf{M}_\sigma$, normalized so that the sum of its coordinates is equal to one.

1. Let $\mathbf{x} \in \mathbb{R}^d$. Show that ${}^t\mathbf{M}_\sigma{}^n\mathbf{x}$ tends to zero in $\mathbb{R}^d$ if and only if $\mathbf{x}$ is orthogonal to $\mathbf{u}_1$.
2. Show that ${}^t\mathbf{M}_\sigma{}^{-1}\mathbb{Z}^d = \mathbb{Z}^d$.
3. Deduce that $\exp(2i\pi\beta)$ is an eigenvalue of X_σ associated with the trivial coboundary if and only β is an integer combination of the coordinates of $\mathbf{u}_1$.
4. Note that the coordinates of $\mathbf{u}_1$ are the frequencies of letters in any point in the system and deduce Proposition 7.3.29.

Some basic number theory allows one to precise Proposition 7.3.29. We recall that the characteristic polynomial of a Pisot type substitution is always irreducible over $\mathbb{Q}$ (see Chap. 1).

Exercise 7.3.31. Let σ be a substitution of Pisot type and α be the dominant eigenvalue of its incidence matrix.

1. Prove that the coordinates of a dominant right eigenvector of $\mathbf{M}_\sigma$, normalized so that the sum of its coordinates is equal to one, belong to $\mathbb{Z}[\alpha]$ *(Hint: these coordinates are described by a system of equations with rational coefficients.)*
2. Prove that these coordinates are rationally independent *(Hint: the eigenvalues of the incidence matrix are simple and algebraic conjugates; the eigenvectors are linearly independent.)*
3. If σ is unimodular, show that the group of eigenvalues associated with the trivial coboundary is $\exp(2\pi i\mathbb{Z}[\alpha])$.

The result of Exercise 7.3.31 can be obtained independently and generalized (out of the unimodular context) as a consequence of Theorem 7.3.25:

Proposition 7.3.32. *Let α denote the dominant eigenvalue of the incidence matrix of a substitution of Pisot type. The spectrum of the substitutive dynamical system associated with σ contains the set* $\exp(2\pi i\mathbb{Z}[\alpha])$.

In particular, substitutive dynamical systems of Pisot type are never weakly mixing.

Proposition 7.3.32 means that Pisot type substitutions are quite close to constant length substitutions with respect to the fact that both are never weakly mixing: constant length substitutive dynamical systems always have nontrivial rational eigenvalues whereas Pisot type substitutive dynamical systems always have irrational eigenvalues. But if a substitution is neither of Pisot type, nor of constant length, then completely different behaviors can occur, such as the absence of spectrum for which we gave sufficient conditions in Theorem 7.3.24. An example of such a substitution is $1 \mapsto 112$, $2 \mapsto 12222$.

Since $\mathbb{Z}[\alpha]$ is of rank $d-1$, we deduce from Proposition 7.3.32 and Exercise 7.3.12 that:

Corollary 7.3.33. *Any Pisot type substitutive dynamical system over d letters admits as a topological factor a minimal translation on the torus $\mathbb{T}^{d-1}$.*

Rational spectrum. In [412], M. Solomyak focuses on the rational part of the spectrum of unimodular Pisot type substitutive dynamical systems:

Theorem 7.3.34 (M. Solomyak [412]). *The rational spectrum of a unimodular substitutive dynamical system of Pisot type is finite.*

Note that this is not true in general for non-unimodular substitutions, as stated in Theorem 7.3.10.

A specific class of examples: β-numeration. A special attention was devoted by B. Solomyak in [411, 410] to substitutions of Pisot type whose incidence matrix has a specific form, such as to be a companion matrix.

For instance, the dynamical system associated with the unimodular substitution of Pisot type $1 \mapsto 123\ldots d$, $2 \mapsto 1$, $3 \mapsto 2,\ldots$, $d \mapsto d-1$ has a purely discrete spectrum, and is measure-theoretically isomorphic to the rotation, over $(d-1)$-dimensional torus, by a normalized right dominant eigenvector for the incidence matrix.

From a similar point of view, substitutions defined by $1 \mapsto 1^{a_1}2$, $2 \mapsto 1^{a_2}3$, $\ldots$, $d-1 \mapsto 1^{a_{d-1}}d$ and $d \mapsto 1^{a_d}$ have a very specific matrix. The numeration systems associated with the dominant eigenvalue of such a matrix has been studied in the context of β-numeration. From the properties of β-expansion, B. Solomyak deduces in [411] that substitutions of Pisot type which are of the above form have a purely discrete spectrum.

Two-letter substitutions: the coincidence condition. M. Hollander and B. Solomyak restate in [205] the proof of the results of A. N. Livshits and apply them to the case of Pisot type substitutions over a two-letter alphabet.

He proves that the systems associated with such substitutions have a purely discrete spectrum if the substitution satisfies a combinatorial condition which generalizes the coincidence condition introduced by F. M. Dekking for substitutions of constant length (see Sec. 7.3.1). Indeed, a substitution over the alphabet $\{1, 2\}$ is said to satisfy the *coincidence condition* if there exist two integers k, n such that $\sigma^n(1)$ and $\sigma^n(2)$ have the same k-th letter, and the prefixes of length $k-1$ of $\sigma^n(1)$ and $\sigma^n(2)$ have the same image under the abelianization map.

This definition appeared in literature before [205] (see [339]). However, the following results, although conjectured, were not proved before.

Theorem 7.3.35 (Hollander, Solomyak [205]). *Let σ be a substitution of Pisot type over a two-letter alphabet which satisfies the coincidence condition. Then the substitutive dynamical system associated with σ has a purely discrete spectrum.*

Remark. The condition of coincidence and the condition of no nontrivial coboundary are distinct: we already stated that two-letter substitutions have no nontrivial coboundary, whereas there exist two-letters substitutions with no coincidence, for instance the Morse sequence. ∎

Up to recently, it was not known whether there existed a Pisot type two-letter substitution which did not satisfy the coincidence condition. A preprint from M. Barge and B. Diamond answers that question, by proving the following:

Theorem 7.3.36 (Barge, Diamond [54]). *Any substitution of Pisot type over a two-letter alphabet satisfies the coincidence condition.*

The proof is a very nice mix of combinatorics and geometry.

7.4 The Rauzy fractal

The spectral theory of substitutive dynamical systems implies a better understanding of these systems. In particular, under some conditions such as unimodularity or Pisot type, these symbolic dynamical systems are not weakly mixing, and their spectrum is the spectrum of a toral rotation. An important remaining question is whether these systems have a purely discrete spectrum. To answer that question, G. Rauzy developed the idea of looking for a geometric representation of substitutive dynamical systems, as a rotation on a suitable space. This representation being explicit, the problem of purely discrete spectrum may be explored through the question of the representation's injectivity.

Following that point of view, the remainder of this chapter is devoted to the reverse of the question that initiated the interest for substitutions: indeed, as symbolic dynamical systems were first introduced to understand better the dynamics of a given geometric map, one can ask for the reverse, that is, which geometric actions are coded by a given substitution?

7.4.1 Where symbolic dynamics codes geometry...

Geometric representation. M. Queffélec defines in [339] (pp. 140) the notion of geometric representation:

Definition 7.4.1. *A* geometric representation *of a symbolic dynamical system (X, S) is a continuous map φ from X* onto *a geometric dynamical system $(\mathcal{X}, T)$, on which there exists a partition indexed by the alphabet $\mathcal{A}$, such that every word in X_u is the itinerary of a point of $(\mathcal{X}, T)$ with respect to the partition.*

It is thus natural to ask which systems admit such a geometric representation, and, if so, how precise the representation is, that is, what is the degree of injectivity of the representation. The representation is considered to be precise if the map φ is a semi-topological conjugacy.

Nontrivial examples of representations. There have been many partial answers given to Queffélec's question: in [49], G. Rauzy and P. Arnoux introduce the so-called *Arnoux-Rauzy sequences*, introduced in Chap. 1; they show that each such sequence codes the orbit of a point under an exchange of six intervals on the circle.

In [85], M. Boshernitzan and I. Kornfeld give explicitly a conjugacy map between the dynamics of the restriction to the limit of a piecewise translation and the dynamics of a substitutive system.

In [165], S. Ferenczi obtains a geometric realization of the Chacon sequence as exduction of a triadic rotation (see also Chap. 5).

Recently, J. Cassaigne, S. Ferenczi and L. Q. Zamboni have exhibited an example of a non-substitutive Arnoux-Rauzy sequence whose properties are very far from those of substitutive Arnoux-Rauzy sequences: in particular, their example cannot be geometrically realized as a natural coding of a toral translation [109] (we will come back to this topic in Chap. 12).

In [406], V. F. Sirvent gives a geometrical realization of the Tribonacci substitution as a dynamical system defined on a geodesic lamination on the hyperbolic disc. This result is generalized in the same paper, for all Arnoux-Rauzy sequences.

Representations by a translation on a compact abelian group. The spectral theory of substitutive dynamical systems provides some information about the geometric realizations by a translation on a compact group that one may get: if a substitutive dynamical system is measure-theoretically isomorphic to a translation on a compact group, this translation must have in its spectrum all the eigenvalues of the dynamical system. As a consequence, the compact group must be a n-dimensional torus if the spectrum of the system is irrational and of rank n. The compact group must admit as a factor, either finite groups, if the spectrum contains rational numbers, or p-adic groups, if there are enough rational eigenvalues (as is the case for constant length substitutions). In a general case, the compact group may be a solenoid.

According to the spectral theory of Pisot type substitutive dynamical systems presented in Sec. 7.3.3, and under this hypothesis of no non-trivial coboundary, unimodular Pisot type substitutive systems are good candidates to have a purely discrete spectrum, that is, to be measure-theoretically isomorphic to a toral translation, this translation being obtained with the frequencies of occurrences of letters in any word of the system (see Proposition 7.3.29). A method to test this hypothesis is to try to explicit a conjugacy between the shift map on the substitutive system and the toral translation formerly described. Then, the problem will be to study the injectivity of the conjugacy.

The idea of representing geometrically substitutive systems of Pisot type was developed first by G. Rauzy [349], within the framework of discrepancy for some real sequences. The aim of this section is to expose Rauzy's work about the Tribonacci substitution and the development given by many authors to this original and fundamental example.

7.4.2 The Tribonacci substitution

In 1981, G. Rauzy generalized in [349] the dynamical properties of the Fibonacci substitution to a three-letter alphabet substitution, called Tribonacci substitution or Rauzy substitution, and defined by

$$\sigma(1) = 12 \qquad \sigma(2) = 13 \qquad \sigma(3) = 1.$$

The incidence matrix of this substitution is $\mathbf{M}_\sigma = \begin{pmatrix} 1 & 1 & 1 \\ 1 & 0 & 0 \\ 0 & 1 & 0 \end{pmatrix}$.

Its characteristic polynomial is $X^3 - X^2 - X - 1$, and its set of roots consists of a real number $\beta > 1$ and two complex conjugates α and $\overline{\alpha}$. Since it satisfies the equation $\alpha^3 = \alpha^2 + \alpha + 1$, the complex number α is called a *Tribonacci number*, in reference to the Fibonacci equation $X^2 = X + 1$.

In particular, the incidence matrix $\mathbf{M}$ admits as eigenspaces in $\mathbb{R}^3$ an expanding one-dimensional direction and a contracting plane.

7.4.3 Geometric construction of the Rauzy fractal

Let u denote a biinfinite word which is a periodic point for σ, for instance $u = \sigma^{3\infty}(1) \cdot \sigma^{\infty}(1)$. This infinite word u is embedded as a broken line in $\mathbb{R}^3$ by replacing each letter in the periodic point by the corresponding vector in the canonical basis $(\mathbf{e}_1, \mathbf{e}_2, \mathbf{e}_3)$ in $\mathbb{R}^3$.

An interesting property of this broken line is that it remains at a bounded distance of the expanding direction of $\mathbf{M}_\sigma$, turning around this line. When one projects the vertices of the broken line on the contracting plane , parallel to the expanding directing, then one obtains a bounded set in a two-dimensional vector space. The closure of this set of points is a compact set denoted by $\mathcal{R}$ and called the *Rauzy fractal* (see Fig. 7.5).

To be more precise, denote by π the linear projection in $\mathbb{R}^3$, parallel to the expanding directing, on the contracting plane, identified with the complex plane $\mathbb{C}$. If $u = (u_i)_{i \in \mathbb{Z}}$ is the periodic point of the substitution, then the Rauzy fractal is

$$\mathcal{R} = \overline{\left\{ \pi \left(\sum_{i=0}^{n} \mathbf{e}_{u_i} \right) ; \ n \in \mathbb{Z} \right\}}.$$

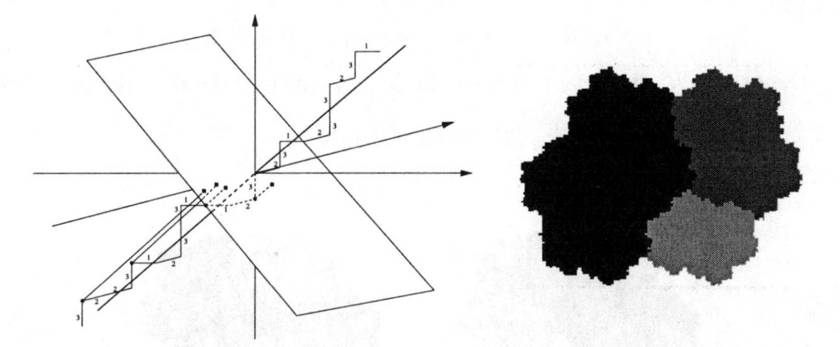

Fig. 7.5. The projection method to get the Rauzy fractal for the Tribonnacci substitution.

Three subsets of the Rauzy fractal can be distinguished. Indeed, for each letter $j = 1, 2, 3$, the *cylinder* $\mathcal{R}_j$ is defined to be the closure of the set of origins of any segment on the broken line which is parallel to the canonical vector $\mathbf{e}_j$:

$$\mathcal{R}_j = \overline{\left\{ \pi \left(\sum_{i=0}^{n} \mathbf{e}_{u_i} \right);\ n \in \mathbb{Z},\ u_n = j \right\}}.$$

The union of these three cylinders covers the compact $\mathcal{R}$, and G. Rauzy proved in [349] that their intersection has measure zero:

Proposition 7.4.2 (Rauzy [349]). *The cylinders constitute a measurable partition of the Rauzy fractal:*

$$\mathcal{R} = \mathcal{R}_1 \cup \mathcal{R}_2 \cup \mathcal{R}_3, \quad \textit{the union being disjoint in measure.}$$

7.4.4 Dynamics over the Rauzy fractal

One can note that it is possible to move on the broken line, from a vertex to the following one, thanks to a translation by one of the three vectors of the canonical basis. In the contracting plane, this means that each cylinder $\mathcal{R}_i$ can be translated by a given vector, i.e., $\pi(\mathbf{e}_i)$, without going out of the Rauzy fractal:

Exercise 7.4.3. Prove that for every letter $i = 1, 2, 3$, the following inclusion is true:

$$\mathcal{R}_i - \pi(\mathbf{e}_i) \subset \mathcal{R}.$$

Dynamics. Thus, the following map φ, called a *domain exchange* (see Fig. 7.6) is defined for any point of the Rauzy fractal which belongs to a unique cylinder. According to Proposition 7.4.2, the cylinders intersect on a set of measure zero, so that this map is defined almost everywhere on the Rauzy fractal:

$$\forall x \in \mathcal{R}, \quad \varphi(x) = x - \pi(\mathbf{e}_i), \quad \text{if } x \in \mathcal{R}_i.$$

Denote by $\mathcal{R}_i'$ the image of the cylinder $\mathcal{R}_i$ through the domain exchange:

$$\mathcal{R}_i' = \mathcal{R}_i - \pi(\mathbf{e}_i).$$

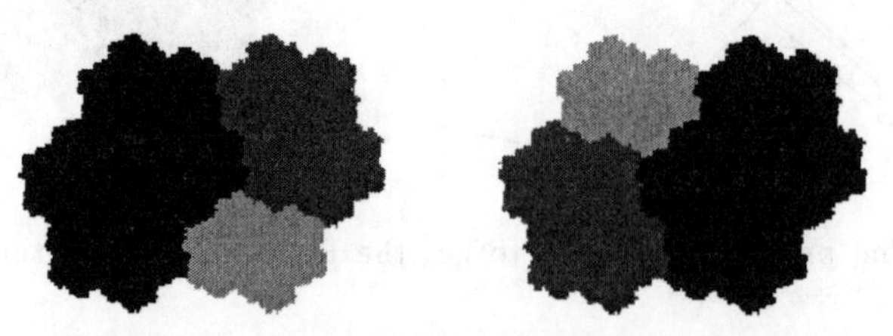

Fig. 7.6. Domain exchange over the Rauzy fractal.

Symbolic dynamics. It is natural to code, up to the partition defined by the 3 cylinders, the action of the domain exchange φ over the Rauzy fractal $\mathcal{R}$. G. Rauzy proved in [349] that the coding map, from $\mathcal{R}$ into the three-letter alphabet full shift $\{1,2,3\}^{\mathbb{Z}}$ is almost everywhere one-to-one. Moreover, this coding map is onto the substitutive system associated with the Tribonacci substitution. Thus we have the following result:

Theorem 7.4.4 (Rauzy, [349]). *The domain exchange φ defined on the Rauzy fractal $\mathcal{R}$ is semi-topologically conjugate to the shift map on the symbolic dynamical system associated with the Tribonacci substitution.*

7.4.5 Geometric interpretation of the substitution: self-similarity

Number theory. Note that G. Rauzy obtains these results by proving that the Rauzy fractal can be defined numerically as the complex set of power series in α whose digits consist in all the sequences of 0 and 1 with no three consecutive ones:

$$\mathcal{R} = \left\{ \sum_{i\geq 0} \varepsilon_i \, \alpha^i ; \;\; \varepsilon_i \in \{0,1\}; \, \varepsilon_i\varepsilon_{i+1}\varepsilon_{i+2} = 0 \right\} \subset \mathbb{C}. \tag{7.6}$$

The condition $\varepsilon_0 = 0$ produces the cylinder $\mathcal{R}_1$, whereas the cylinder $\mathcal{R}_2$ is defined by the condition $\varepsilon_0\varepsilon_1 = 10$, and $\mathcal{R}_3$ by $\varepsilon_0\varepsilon_1 = 11$.

The Tribonacci substitutive dynamical system X_σ has a special property: since the letter 1 appears only at the beginning of $\sigma(1)$, $\sigma(2)$ and $\sigma(3)$, we have

$$[1] = \sigma\left(X_\sigma\right), \quad [2] = S\left(\sigma^2(X_\sigma)\right), \quad [3] = S\left(\sigma[2]\right).$$

The abelianization of the shift map S on the broken line is the translation to the following vertex, which projects on the contracting plane as the domain exchange φ. In the same point of view, the abelianization of the substitution σ is the matrix $\mathbf{M}$, which projects on the contracting plane as a complex multiplication by the Tribonacci number α. So that we get:

$$\mathcal{R}_1 = \alpha\mathcal{R}, \quad \mathcal{R}_2 = \alpha^2\mathcal{R} + 1 \quad \mathcal{R}_3 = \alpha^3\mathcal{R} + \alpha + 1. \tag{7.7}$$

This means that the Rauzy fractal $\mathcal{R}$ is partitioned by contractions of itself. We say that $\mathcal{R}$ has a *self-similar structure.*

Exercise 7.4.5. Deduce (7.7) from (7.6).

7.4.6 Tilings : toral translation as a factor

The domain exchange φ is defined only almost everywhere. This problem is solved when one quotients the contracting plane by the lattice $\mathcal{L} = \mathbb{Z}\,\pi(\mathbf{e}_1 -$

$\mathbf{e}_3) + \mathbb{Z}\,\pi(\mathbf{e}_2 - \mathbf{e}_3)$. Indeed, this quotient maps the contracting plane onto a two-dimensional torus; the 3 vectors $\pi(\mathbf{e}_1)$, $\pi(\mathbf{e}_2)$ and $\pi(\mathbf{e}_3)$ map onto the same vector on the torus. Thus, the projection of the domain exchange φ on the quotient is a toral translation.

G. Rauzy proved in [349] that the restriction of the quotient map to the Rauzy fractal is onto and almost everywhere one-to-one. Finally, we get that the domain exchange on the Rauzy fractal, which is known to be semi-topologically conjugate to the Tribonacci substitutive dynamical system, is also measure-theoretically isomorphic to a minimal translation on the two-dimensional torus.

More geometrically, the fact that the quotient map is one-to-one means that any translate of $\mathcal{R}$ by a vector of the lattice $\mathcal{L}$ intersects $\mathcal{R}$ on a set of measure zero. The fact that the restriction of the quotient map to the Rauzy fractal is onto means that the union of these translated Rauzy fractal recovers the plane. In other words, the Rauzy fractal generates a periodic tiling of the plane (see Fig. 7.7).

Fig. 7.7. The periodic and autosimilar tiling generated by the Rauzy fractal.

The self-similarity of $\mathcal{R}$ implies that the induced tiling is quasi-periodic. This is analogous to the well understood connection between Sturmian sequences, tilings of the line and quasi-crystals (see [39]): through methods similar to those used by G. Rauzy in [349], E. Bombieri and J. Taylor use substitutions to exhibit a connection between quasi-crystals and number theory in [82] (see also [52, 243, 385, 430] and [439] for a study of the Rauzy fractal as a quasicrystal).

Finally, by mixing dynamics, self-similarity and number theory, we get the three following equivalent results:

Theorem 7.4.6 (Rauzy, [349]).

- Geometry. *The Rauzy fractal generates a self-similar periodic tiling of* $\mathbb{C}$.
- Dynamics. *The symbolic dynamical system generated by the Tribonacci substitution is measure-theoretically isomorphic to a toral translation.*

- Spectral theory. *The Tribonacci substitutive dynamical system has a purely discrete spectrum.*

Additional geometric properties. Using methods introduced by F. M. Dekking in [136], S. Ito and M. Kimura obtain in [221] an alternative construction of the Rauzy fractal and use it to compute the Hausdorff dimension of the boundary of the Rauzy fractal, which is strictly larger than 1 (see Chap. 8). V. Sirvent obtains equivalent results with a different method in [400]. By using numeration systems, A. Messaoudi obtains in [288, 290, 289] additional properties of the fractal $\mathcal{R}$ including an explicit boundary parametrization.

7.5 Geometric realization of substitutions of Pisot type

The case of the Tribonacci substitution being quite well understood, the question of its generalization to a larger class of substitutions is thus natural. Which substitutions generate self-similar and periodic tilings? Which ones code the action of a toral translation? From a spectral point of view, a general formulation for these questions is: which substitutions generate a dynamical system with a purely discrete spectrum?

7.5.1 Generalized Rauzy fractals

For the Tribonacci substitution, the two main properties that allow us the construction of the Rauzy fractal are the following:

- The incidence matrix of the substitution has a one-dimensional expanding direction.
- The broken line corresponding to any periodic point of the substitution remains at a bounded distance from the expanding direction.

Generalized Rauzy fractals. These properties are satisfied as soon as the modulus of all but one of the eigenvalues of the substitution incidence matrix is strictly less than 1. This means that the dominant eigenvalue of the matrix is a *Pisot number* and the substitution is *of Pisot type.*

Definition 7.5.1. *Let σ be a unimodular Pisot type substitution over d letters. Let $\alpha_1 > 1, \ldots, \alpha_r$ be the real eigenvalues of the incidence matrix of σ, $\alpha_{r+1}, \overline{\alpha_{r+1}}, \ldots, \alpha_{r+s}, \overline{\alpha_{r+s}}$ be its complex eigenvalues. Let $\mathbf{v}_1, \ldots \mathbf{v}_{r+s} \in \mathbb{C}^d$ be right eigenvectors of the incidence matrix associated with each of the eigenvalues α_i, $1 \leq i \leq r+s$.*

For any finite word W, we define the following vector in $\mathbb{R}^{r-1} \times \mathbb{C}^s$:

$$\delta(W) = \sum_{k=1}^{|W|} \left((\mathbf{v}_2)_{W_i}, \ldots, (\mathbf{v}_{r+s})_{W_i}\right) \in \mathbb{R}^{r-1} \times \mathbb{C}^s .$$

Let $u = (u_i)_{i\in\mathbb{Z}}$ *be a periodic point for* σ. *Then the* generalized Rauzy fractal *or* Rauzy fractal *associated with the substitution* σ *is the following compact set:*

$$\mathcal{R} = \overline{\{\delta\,(u_0 \ldots u_i)\,;\ i \in \mathbb{N}\}} \subset \mathbb{R}^{r-1} \times \mathbb{C}^s.$$

Examples of generalized Rauzy fractals are shown in Fig. 7.8.

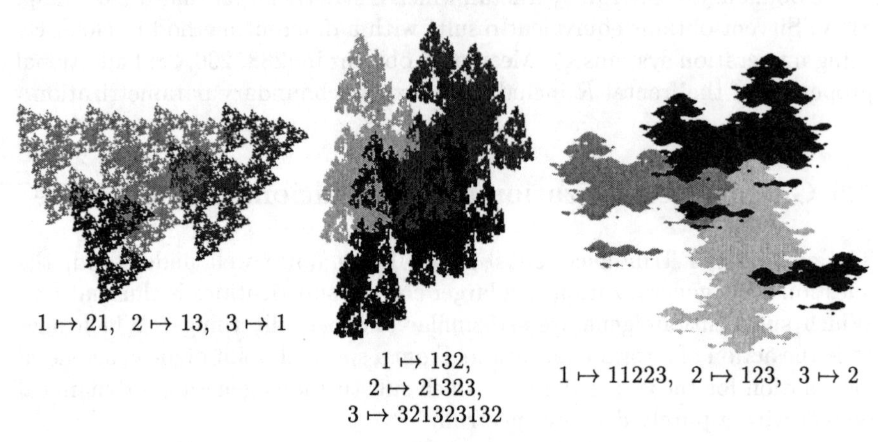

Fig. 7.8. Examples of generalized Rauzy fractals.

Since $r + 2s = d$, $\mathbb{R}^{r-1} \times \mathbb{C}^s$ is a $\mathbb{R}$-vector space of dimension $d - 1$, so that the Rauzy fractal associated with a unimodular substitution of Pisot type can be considered as a part of $\mathbb{R}^{d-1}$. Let us illustrate this point for the Tribonacci substitution, by proving that the Rauzy fractal defined in Sec. 7.4.3 and the one defined above are identical:

Exercise 7.5.2. Let σ be the Tribonacci substitution and α be the Tribonacci number, with $|\alpha| < 1$.

1. Prove that the vector $\mathbf{u}_\alpha = (a_1, a_2, a_3) = (1/\alpha, 1/\alpha^2, 1/\alpha^3)$ is a normalized right eigenvector of $\mathbf{M}_\sigma$ associated with α.
   Then δ is a mapping from $\{1,2,3\}^*$ on $\mathbb{C}$: $\delta(W_1 \ldots W_n) = \sum_{1\leq i\leq n} a_{W_i}$.
2. Determine a left eigenvector $\mathbf{v}_\alpha$ of $\mathbf{M}_\sigma$ associated with α and which satisfies ${}^t\mathbf{v}_\alpha \mathbf{u}_\alpha = 1$.
3. Let $\mathbf{w}_1$, $\mathbf{w}_2$ be its real and imaginary parts: $\mathbf{v}_\alpha = \mathbf{w}_1 + i\mathbf{w}_2$. Show that the family $\{\mathbf{w}_1, \mathbf{w}_2\}$ is a real basis of the contracting plane of $\mathcal{H}$ of $\mathbf{M}_\sigma$.
4. Let $q_{\mathcal{H}} : \mathbb{C} \to \mathcal{H}$ be following $\mathbb{R}$-linear embedding:

$$\forall z = x + iy \in \mathbb{C},\ q_{\mathcal{H}}(z) = 2x\mathbf{w}_1 - 2y\mathbf{w}_2.$$

Prove that up to this identification, the map δ coincides with the linear projection $\pi : \mathbb{R}^3 \to \mathcal{H}$ parallel to the expanding direction of $\mathbf{M}_\sigma$:

$$\forall\, W \in \{1,2,3\}^*,\ q_{\mathcal{H}}(\delta(W)) = \pi(\mathbf{l}(W)).$$

5. Let $\mathcal{R}$ be the generalized Rauzy fractal of σ given in Definition 7.5.1 and $\mathcal{R}_1$ be the Rauzy fractal as defined in Sec. 7.4.3. Prove that $q_{\mathcal{H}}(\mathcal{R}) = \mathcal{R}_1$.

The result proved in this exercise is general:

Lemma 7.5.3. *Let σ be a substitution of Pisot type over a d-letter alphabet. Let π be the linear projection parallel to the expanding direction of $\mathbf{M}_\sigma$, defined from $\mathbb{R}^3$ onto the contracting hyperplane $\mathbf{M}_\sigma$. There exists a linear embedding $q_{\mathcal{H}} : \mathbb{R}^{r-1} \times \mathbb{D}^s \to \mathcal{H}$ such that: $\forall W \in \mathcal{A}$, $q_{\mathcal{H}}(\delta(W)) = \pi(\mathbf{l}(W))$.*

This implies that Definition 7.5.1 generalizes the Rauzy fractal defined by G. Rauzy for the Tribonacci substitution:

Corollary 7.5.4. *Let $\mathcal{R}$ be the generalized Rauzy fractal of σ. Let $\mathcal{R}_1 = \{\pi\left(\sum_{i=0}^{n} \mathbf{e}_{u_i}\right);\ n \in \mathbb{Z}\}$, where u is a periodic point for σ. Then $\mathcal{R}$ and $\mathcal{R}_1$ are linearly isomorphic: $q_{\mathcal{H}}(\mathcal{R}) = \mathcal{R}_1$.*

The reason that led us to define generalized Rauzy fractals by a numerical method instead of a geometric method is that the map δ allows a better understanding of the action of the shift map and of the substitution σ on the Rauzy fractal. Indeed, the following relationships are easy to get:

$$\forall\, W_1, W_2 \in \mathcal{A}^*, \quad \delta(W_1 W_2) = \delta(W_1) + \delta(W_2);$$

$$\forall\, W \in \mathcal{A}^*,\ \delta(\sigma(W)) = \begin{pmatrix} \alpha_2 & & (0) \\ & \ddots & \\ (0) & & \alpha_{r+s} \end{pmatrix} \delta(W). \tag{7.8}$$

For instance, such equalities are the main argument to get the formulas of Sec. 7.4.5 that describe the self-similar structure of the Rauzy fractal for the Tribonacci substitution.

Cylinders. As for the Tribonacci substitution, let us define the *cylinders*:

$$\forall i = 1 \ldots d, \quad \mathcal{R}_i = \overline{\{\delta\left(u_0 \ldots u_j\right);\ j \in \mathbb{N};\ u_j = i\}}.$$

Exercise 7.5.5. Prove that the Rauzy fractal is the union of cylinders:

$$\mathcal{R} = \bigcup_{i=1}^{d} \mathcal{R}_i.$$

Remark. Note that we do not know at the moment whether the union is in general disjoint in measure. We will see in the next sections that this is one of the main problems raised by the study of generalized Rauzy fractals.

7.5.2 Topological properties of generalized Rauzy fractals

Many authors investigated topological properties of Rauzy fractals: V. F. Sirvent and Y. Wang prove in [409] that generalized Rauzy fractals have nonempty interior and are the closure of their interior.

Conditions for simple connectedness are given in [224] and in [187]. Conditions for connectedness are given by V. Canterini in [100, 101], and restated in a more algorithmic way in [398]. An estimation for the Haussdorf dimension of the boundary of Rauzy fractals is computed in [163].

C. Holton and L. Q. Zamboni define in [208, 211, 210] real and complex representations associated with every substitution, the incidence matrix of which has a nonzero eigenvalue of modulus less than one. Their work gives explicit bounds for the Haussdorf dimension of the projections on axes of the Rauzy fractal associated with a substitution. More precisely, for any $k = 2, \ldots, r+s$, they prove that the projection of the Rauzy fractal on the axis associated with the eigenvalue α_k, that is, the k-coordinate of the Rauzy fractal in $\mathbb{R}^{r-1} \times \mathbb{D}^s$ (this axis can be $\mathbb{R}$ or $\mathbb{C}$), admits a Haussdorf dimension which is less than $-\log(\alpha_1)/\log(\alpha_k)$.

Thus, for any $k = 2, \ldots, r+s$, the projection is a Cantor set. If $k = s+1, \ldots, s+d$ and $|\alpha_k|^2\alpha_1 < 1$, then the projection has nonempty interior.

Let us also mention that Rauzy fractals appear in [96], as continuous factors of odometers, derived from sequence of cutting time of an interval transformation restricted to a wild attractor.

Finally, some precise results exist about the n-dimensional generalization of the Tribonacci substitution: the n-bonacci substitution is the n-letter alphabet substitution defined by

$$1 \mapsto 12, \quad 2 \mapsto 13, \quad \ldots \quad n-1 \mapsto 1n, \quad n \mapsto 1.$$

The spectral study of this substitution was made by B. Solomyak (see Sec. 7.3.3). In [400, 402], V. F. Sirvent studies dynamical and geometrical properties of the Rauzy fractals associated with these substitutions. Other properties, mainly arithmetical, of the substitutions are given in [405, 403, 408]. In [406] are studied the dynamical and geometrical properties which are common to all the k-bonacci substitutions.

The construction of the Rauzy fractal being generalized, the question is to obtain conditions for a substitutive system to be semi-topologically conjugate to a domain exchange on the Rauzy fractal associated with the substitution, that is, to generalize the results of [350].

The problem of domain exchange. As for the Tribonacci substitution, one can move along the broken line associated with a periodic point of a substitution thanks to a translation by one of the canonical vectors. This can be expressed in the following way:

Exercise 7.5.6. Let $\mathcal{R}$ be the Rauzy fractal associated with a unimodular substitution of Pisot type, as defined in Definition 7.5.1. Let $\mathcal{R}_i$ be the cylinders, the union of which is equal to $\mathcal{R}$.

Prove that for every $i = 1 \dots d$, the following inclusion is satisfied:

$$\mathcal{R}_i - \delta(i) \subset \mathcal{R}. \tag{7.9}$$

The inclusion (7.9) means that one may define dynamics over the Rauzy fractal, in terms of a domain exchange, as soon as the cylinders $\mathcal{R}_i$ are disjoint up to a set of zero measure. Unfortunately, whereas we know that the union of the d cylinders $\mathcal{R}_i$ is equal to $\mathcal{R}$ (Exercise 7.5.5), nothing allows us to say that these cylinders partition $\mathcal{R}$ up to a set of zero measure.

A necessary condition for this is the substitution to be unimodular: if the substitution is not unimodular, the substitutive system may have a p-adic component and the cylinders should intersect on a nonzero measure set.

Unfortunately again, the restriction to the class of unimodular substitution is still not sufficient to define dynamics on the Rauzy fractal of a substitution: for substitutions over a two-letter alphabet, B. Host proved explicitly that the coincidence condition (see Sec. 7.3.3) is necessary and sufficient for the cylinders to be disjoint in measure. This is developed in Sec. 7.5.4.

For substitutions over a finite alphabet, P. Arnoux and S. Ito give an alternative construction of the Rauzy fractal thanks to induction and successive approximations (see also Chap. 8). This allowed them to generalize the coincidence condition and to prove that this condition is sufficient for the cylinders to intersect on a set of zero measure. Their main results are presented in Sec. 7.5.4.

The problem of tilings. The main problem of the geometric realization is to know whether a substitutive dynamical system is isomorphic to a compact group translation. Again, for two-letter alphabet substitutions, B. Host answered that question in terms of coincidences (see Sec. 7.5.4).

Another approach to this problem was made in [102] by formalizing and generalizing the ideas of G. Rauzy in [352]: thanks to recognizability and formal power series, it is possible to define an explicit conjugacy between a substitutive system and the domain exchange over the Rauzy fractal. This conjugacy is the reverse from the map which maps the Rauzy fractal onto the symbolic space $\{1, \dots, d\}$, by coding the action of the domain exchange over the Rauzy fractal. This brings a characterization of substitutive systems with a purely discrete spectrum. It will be presented in Sec. 7.5.5.

7.5.3 The coincidence condition

As explained in Sec. 7.3, the spectrum of a substitutive dynamical system is related to the notion of coboundary introduced by B. Host. In two specific cases, that is, constant length substitutions and two-letter substitutions, the spectrum is purely discrete if and only if the substitution satisfies the combinatorial condition of coincidence (Theorem 7.3.4 and 7.3.35).

Let us point out the general definition of coincidences as introduced in [43], and its relationship with coboundaries. We will see later that the general condition of coincidence plays an important role in the problem of the domain exchange.

As before, $\mathbf{l} : \{1, \ldots, d\} \to \mathbb{N}^d$ is the abelianization map over $\{1, \ldots, d\}$.

Definition 7.5.7 (Arnoux, Ito [43]). *A substitution σ satisfies the* coincidence condition *on prefixes (respectively suffixes) if for every couple of letters (j, k), there exists a constant n such that $\sigma^n(j)$ and $\sigma^n(k)$ can be decomposed in the following way:*

$$\sigma^n(b_1) = p_1 a s_1 \ \text{ and } \ \sigma^n(b_2) = p_2 a s_2, \ \text{ with } \ \mathbf{l}(p_1) = \mathbf{l}(p_2)$$
$$(\text{respectively } \ \mathbf{l}(s_1) = \mathbf{l}(s_2))\,.$$

Example 7.5.8. The Tribonacci substitution satisfies the coincidence condition, whereas the Morse sequence does not satisfy this condition.

The coincidence condition is connected with the notion of coboundary introduced by B. Host and detailed in Sec. 7.3.2:

Lemma 7.5.9 (Host). *Let σ be a substitution with a nontrivial coboundary $g : \mathcal{A} \to \mathbb{U}$. Let f be the function of modulus 1 which satisfies $f(b) = g(a)f(a)$ as soon as the word ab belongs to the language of the fixed point of the substitution.*

If there exist two letters a and b and a rank k such that

- *$f(a) \neq f(b)$,*
- *$\sigma^k(a)$ begins with a and $\sigma^k(b)$ begins with b,*

then σ does not satisfy the coincidence condition on prefixes.

Proof. If σ satisfies the coincidence condition, there exist a multiple k_1 of k and a letter c such that

$$\sigma^{k_1}(a) = p_1 c s_1 \ \text{ and } \ \sigma^n(b) = p_2 c s_2, \ \text{ with } \ \mathbf{l}(p_1) = \mathbf{l}(p_2).$$

Let $p_1 = a u_1 \ldots u_{n-1}$ and $p_2 = b v_1 \ldots v_{n-1}$. Since g is a coboundary we have

$$\frac{f(c)}{f(b)} = g(v_{n-1}) \ldots g(v_1) = g(u_{n-1}) \ldots g(u_1) = \frac{f(c)}{f(a)}.$$

Thus $f(a) = f(b)$ which is impossible. ∎

By using the following method of B. Host to construct substitutions with nontrivial coboundaries, one gets a family of substitutions without coincidence.

Proposition 7.5.10 (Host). *Let σ be a primitive substitution over an alphabet $\mathcal{A}$. Let $\mathcal{L}_2(X_\sigma)$ be the set of words of length 2 in the language of the substitution.*

The following substitution τ on the alphabet $\mathcal{L}_2(X_\sigma)$ is primitive and has a nontrivial coboundary:

$$\forall\, ab \in \mathcal{L}_2(X_\sigma), \quad \textit{if} \begin{cases} \sigma(a) = a_1 \ldots a_k \\ \sigma(b) = b_1 \ldots b_l \end{cases},$$
$$\textit{then } \tau(ab) = (a_1a_2)(a_2a_3)\cdots(a_{k-1}a_k)(a_kb_1) \in \mathcal{L}_2(X_\sigma)^*.$$

Proof. For any letter $a \in \mathcal{A}$, choose a number $\psi(a) \neq 1$ of modulus 1. For any $(ab) \in \mathcal{L}_2(X_\sigma)$, let $f(ab) = \psi(a)$ and $g(ab) = \psi(b)/\psi(a)$.

Let (ab) and $(a'b')$ be two elements of $\mathcal{L}_2(X_\sigma)$ such that the word $(ab)(a'b')$ (of length 2 in $\mathcal{L}_2(X_\sigma)^*$) belongs to the language of τ. By construction, we have $b = a'$, so that

$$f(a'b') = \psi(a') = \psi(b) = g(ab)\psi(a) = g(ab)f(ab).$$

Thus, g is a nontrivial coboundary for τ. ■

Remark. This last method to get substitutions with nontrivial coboundaries shows that the notion of coboundary is the analogous of the notion of height in the constant length case (see Sec. 7.3.1). ■

Substitutions of Pisot type. This last method does not provide Pisot type substitutions with nontrivial coboundary: the characteristic polynomial of the substitution that is obtained is never irreducible. We still do not know whether there exists a Pisot type substitution which does not satisfy the coincidence condition.

A negative answer to that question was recently given by M. Barge and B. Diamond [54], for substitutions over a two-letter alphabet, as stated in Theorem 7.3.36. In the general context of substitutions of Pisot type over more than 2 letters, they prove that the coincidence condition is satisfied by at least two letters:

Theorem 7.5.11 (Barge, Diamond [54]). *Let σ be a substitution of Pisot type on the alphabet $\mathcal{A} = \{1 \ldots d\}$. There exist two distinct letters i, j for which there exist integers k, n such that $\sigma^n(i)$ and $\sigma^n(j)$ have the same k-th letter, and the prefixes of length $k-1$ of $\sigma^n(i)$ and $\sigma^n(j)$ have the same image under the abelianization map.*

This is the most complete result on coincidence that is know at the moment.

7.5.4 Condition for non-overlapping in generalized Rauzy fractals

Construction by successive approximations. P. Arnoux and S. Ito generalize in [43] the alternative construction given by S. Ito and M. Kimura [221] of the Rauzy fractal associated with the Tribonacci substitution. Indeed, they construct by successive approximations the Rauzy fractal for any unimodular Pisot type substitution. The construction is described precisely in Chap. 8.

Thanks to this construction and by generalizing some of the techniques of B. Host in [215], P. Arnoux and S. Ito prove that, under the condition of coincidence, the d cylinders of the generalized Rauzy fractal associated with a substitution over d letters are disjoint in measure. The first consequence of this is that the Rauzy fractal of such a substitution has a self-similar structure: a generalized Rauzy fractal satisfies formulas such as those stated in Sec. 7.4.5 for the Tribonacci substitution.

The second consequence is that, now, nothing prevents us to define a domain exchange on the Rauzy fractal, as it was the case for the Tribonacci substitution.

$$\forall x \in \mathcal{R}, \quad \varphi_\sigma(x) = x - \delta(i), \quad \text{if } x \in \mathcal{R}_i.$$

P. Arnoux and S. Ito prove that coding this action on the Rauzy fractal leads to the symbolic dynamical system generated by the substitution:

Theorem 7.5.12 (Arnoux, Ito [43], see Chap. 8). *Let σ be a unimodular Pisot type substitution over a d-letter alphabet which satisfies the condition of coincidence. Then the substitutive dynamical system associated with σ is measure-theoretically isomorphic to the exchange of d domains defined almost everywhere on the self-similar Rauzy fractal of σ.*

Remark. This result was previously proved by B. Host in an unpublished work [215] for substitutions over a two-letter alphabet.

Appropriate quotient: toral translation factor. The Euclidean space $\mathbb{R}^{d-1}$ projects onto a $d-1$-dimensional torus modulo the lattice $\mathcal{L} = \sum_{i=2}^{d} \mathbb{Z}(\delta(i) - \delta(d))$. Through the quotient map, all the vectors $\delta(i)$ map onto the same vector, so that the domain exchange φ_σ maps to a toral translation. In other words, the diagram shown in Fig. 7.5.4 commutes, where the symbol $\simeq$ on an arrow denotes a measure-theoretic isomorphism.

Thus the domain exchange can be factorized by an irrational translation. This implies a more geometrical version of Corollary 7.3.33:

Proposition 7.5.13 (Arnoux, Ito [43]). *Any unimodular Pisot type substitutive dynamical system which satisfies the coincidence condition admits as a continuous factor an irrational translation on the torus $\mathbb{T}^{d-1}$, the fibers being finite almost everywhere.*

Remark. The fibers are finite almost everywhere as the Rauzy fractal is bounded.

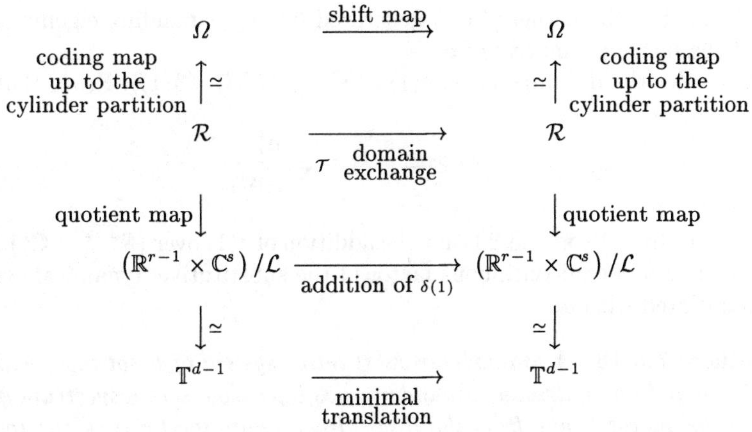

$$\begin{array}{ccc}
\Omega & \xrightarrow{\text{shift map}} & \Omega \\
\text{coding map up to the cylinder partition}\ \uparrow\simeq & & \simeq\uparrow\ \text{coding map up to the cylinder partition} \\
\mathcal{R} & \xrightarrow[\mathcal{T}\ \text{domain exchange}]{} & \mathcal{R} \\
\text{quotient map}\downarrow & & \downarrow\text{quotient map} \\
\left(\mathbb{R}^{r-1}\times\mathbb{C}^{s}\right)/\mathcal{L} & \xrightarrow[\text{addition of } \delta(1)]{} & \left(\mathbb{R}^{r-1}\times\mathbb{C}^{s}\right)/\mathcal{L} \\
\downarrow\simeq & & \simeq\downarrow \\
\mathbb{T}^{d-1} & \xrightarrow[\text{minimal translation}]{} & \mathbb{T}^{d-1}
\end{array}$$

Fig. 7.9. Relations between the different maps related to Rauzy fractals.

Cardinality of the fibers of the quotient map. According to a proof from B. Host, one can easily prove that the fibers are not only finite almost everywhere: they are constant almost everywhere.

Exercise 7.5.14. Let σ be a unimodular substitution of Pisot type. Let $\varphi : \Omega \to \mathbb{T}^{d-1}$ be a continuous toral representation of the substitutive dynamical system associated with σ, that is, there exists $\mathbf{t} \in \mathbb{T}^{d-1}$ such that $\varphi S(w) = \varphi(w) + \mathbf{t}$, for all $w \in \Omega$.

Let $\psi : \mathbb{T}^{d-1} \mapsto \mathbb{N} \cup \infty$ be the cardinality of the fibers:

$$\forall \mathbf{x} \in \mathbb{T}^{d-1},\ \psi(\mathbf{x}) = \operatorname{Card}\{w \in \Omega,\ \varphi(w) = \mathbf{x}\}.$$

1. Let $w_1, \ldots, w_n$ be n distinct elements in Ω and $\varepsilon \in \mathbb{R}$. Let $\mathcal{A}_{n,\varepsilon}$ be the set:
$$\mathcal{A}_{n,\varepsilon} = \{\mathbf{x} \in \mathbb{T}^{d-1};\ \forall i \leq n,\ d(\varphi(w_i), \mathbf{x}) < \varepsilon\}.$$
Prove that $\mathcal{A}_{n,\varepsilon}$ is an open set for the Borel topology.
2. Deduce that ψ is a measurable map for the Lebesgue measure over $\mathbb{T}^{d-1}$ and the discrete measure over $\mathbb{N}$.
3. Prove that ψ is constant almost everywhere.

A more explicit version of the tiling problem. Note that the toral translation obtained in the diagram below is precisely the maximal equicontinuous factor of the substitutive dynamical system, up to the hypothesis of no nontrivial coboundary:

Exercise 7.5.15. Let σ be a unimodular substitution of Pisot type with no nontrivial coboundary.

1. Extend δ as a linear map over $\mathbb{R}^d$.

2. Prove that the kernel of δ is generated by any expanding eigenvector $\mathbf{u}$ of the incidence matrix of σ.
3. Consider the decomposition $\delta(1) = \sum_{i\geq 2} \lambda_i(\delta(i) - \delta(1))$. Prove that

$$\forall i \geq 2, \quad \lambda_i = \frac{(\mathbf{u})_i}{\sum_j (\mathbf{u})_j}.$$

4. Deduce from Prop. 7.3.29 that the addition of $\delta(1)$ over $\left(\mathbb{R}^{r-1} \times \mathbb{C}^s\right)/\mathcal{L}$ is the maximal equicontinuous factor of the substitutive dynamical system associated with σ.

Corollary 7.5.16. *A unimodular substitutive system of Pisot type, with coincidence and no nontrivial coboundary, has a purely discrete spectrum if and only if the quotient map from the Rauzy fractal onto the torus is one-to-one. In particular, in this case, the Rauzy fractal generates a periodic tiling.*

An answer to the tiling problem? For two-letter substitutions, B. Host proves that the problem of domain exchange is equivalent to the tiling problem. Indeed, in an unfortunately unpublished work [215], B. Host explicitly realizes the inverse map of the coding map up to the partition by cylinder (that is, the arrow $\mathcal{R} \to X_\sigma$ on the lower commutative diagram). He states that there exists an explicit continuous mapping $f : X_\sigma \to \mathcal{R}$ such that

$$\exists\, a_0, a_1 \in \mathbb{R}; \quad \forall w \in X_\sigma,\ f(Sw) = f(w) + a_{w_0}. \tag{7.10}$$

An important fact is the following:

Lemma 7.5.17 (Host [215]). *Let σ be a substitution over a two-letter alphabet. Let $f : X_\sigma \to \mathcal{R}$ be a continuous function which satisfies (7.10). Then f is almost everywhere one-to-one if and only if its projection onto the torus modulo $(a_1 - a_0)$ is also almost everywhere one-to-one.*

As we know that f is almost everywhere one-to-one as soon as the substitution satisfies the coincidence condition (Theorem 7.5.12), we deduce that the dynamical system associated with a unimodular substitution of Pisot type with coincidence over a two-letter alphabet is explicitly semi-topologically conjugate to a toral translation. In other words:

Corollary 7.5.18 (Host [215]). *The generalized Rauzy fractal of a two-letter unimodular substitution of Pisot type with coincidence generates a periodic tiling of $\mathbb{R}$.*

Remark. The "toral" version of this corollary is included in Hollander's result in [205] (see Sec. 7.3.3). However, it was obtained much before Hollander's work. It is less general, since it depends on the unimodular condition, but it has the advantage of being constructive.

Unfortunately, in the general case, these techniques do not allow one to prove that the projection on the torus preserves the injectivity: Lemma 7.5.17 is not true when the alphabet contains strictly more than two letters.

P. Arnoux and S. Ito give in [43] a sufficient condition for the rotation factor to be isomorphic to the substitutive system: if there exists a set with nonempty interior which is included in all the domains (after renormalization) which appear in the construction by successive approximation of the Rauzy fractal of the substitution, then we get an isomorphism (see Chap. 8).

7.5.5 Rauzy fractals and formal power series: condition for tiling

Thus, the work of P. Arnoux and S. Ito generalizes [349] to a large class of substitutions, but two fundamental questions remain unsolved: first, the mapping from the substitutive dynamical system onto the Rauzy fractal which conjugates the shift map and the domain exchange is not explicit, and the same for the mapping onto the torus. Secondly, there is no result about the injectivity of the mapping onto the torus. Both questions were solved by G. Rauzy in [349] for the Tribonacci substitution. He gave in [352] ideas to generalize his work for any Pisot type substitution. This approach has been partially realized in [100] and [396], to obtain explicit representation mappings and conditions for injectivity of the toral mapping.

Prefix-suffix expansion. Indeed, in [101], the prefix-suffix expansion defined in Sec. 7.2.2 is used to produce a measure-theoretic isomorphism between the shift map on a substitutive system and an adding machine over the shift of finite type, considered as a Markov compactum in which the order is naturally defined by the substitution. This map is expressed under the form $\Gamma(w) = (p_i, a_i, s_i)_{i\geq 0}$ for all w in X_σ, so that we have:

$$w = \ldots \sigma^i(p_i)\sigma^{i-1}(p_{i-1})\ldots\sigma(p_1)p_0{\cdot}a_0s_0\sigma(s_1)\ldots\sigma^i(s_i)\ldots.$$

We already explained that this provides a new proof of A. V. Vershik's results, originally proved thanks to a stationary Markov compactum with a partial ordering on the labels of the edges. This ordering was arbitrary, so that the representation by the Markov compactum was not explicit. On the contrary, the definition of labels through prefixes of images of letters under the substitution induces a natural ordering between the edges and provides a well determined combinatorial representation.

Geometric realization of the substitutive system. If the substitution is of Pisot type over a d-letter alphabet, then this prefix-suffix expansion allows one to realize explicitly any element of X_σ as a point in $\mathbb{R}^{d-1}$ expressed in terms of power series. More precisely, a consequence of (7.5.1) is the following, where $\delta_2,\ldots,\delta_{r+s}$ denote the coordinates of the map δ:

$$\psi_\sigma(w) = \lim_{n\to+\infty} \delta(\sigma^n(p_n)\dots\sigma^0(p_0)) = \begin{pmatrix} \sum_{i\geq 0} \delta_2(p_i)\,{\alpha_2}^i \\ \vdots \\ \sum_{i\geq 0} \delta_{r+s}(p_i)\,{\alpha_{r+s}}^i \end{pmatrix} \in \mathcal{R}.$$

On each coordinate, this mapping coincides with the real or complex representations of C. Holton and L. Q. Zamboni in [207, 209]. Gathering all the one-dimensional realizations together is natural: indeed each of them is a different expression of the same formal numeration system (the one generated by the characteristic polynomial of the incidence matrix of the substitution).

It is proved in [102] that this representation map is the reverse map of the coding map, up to the cylinders partition, from the Rauzy fractal onto the substitutive dynamical system (see Theorem 7.5.12). In particular, this map projects onto the $(d-1)$-dimensional torus and satisfies a commutation relation with the translation described in the preceding section. This gives a new proof of Corollary 7.3.33 and 7.5.13 which makes the toral representation totally explicit.

This also gives a new proof of Theorem 7.5.13. The interest of this method lies in the fact the Rauzy fractal is obtained as the image of an explicit map, so that it makes it possible to derive various topological properties of the fractal, such as a criterion for the connectedness of the fractal [100], or a description of the boundary of the Rauzy fractal [397].

Application to the tiling question. The main interest of such a construction is that the explicit expression of the mapping from the substitutive dynamical system onto the Rauzy fractal on one side, and onto the torus one the other side, gives an efficient approach for a detailed study of the injectivity of the realizations. Indeed, in [397], the question of injectivity is not treated as usual from a measurable point of view [215, 43], but from a formal and algebraic point of view, based on two points. The first one is the combinatorial structure of the substitutive system, deduced from the prefix-suffix automaton. The second one is the algebraic structure of the system, produced by the incidence matrix of the substitution.

More precisely, one can note that the representation map ψ_σ onto the Rauzy fractal of a Pisot type substitution shall be expressed with a formal power series with coefficients into the splitting field of the characteristic polynomial of the incidence matrix of the substitution: the representation map is just obtained by gathering the set of finite values which can be taken for any Archimedean topology by the formal power series.

In that way, the study of the injectivity of the representation map is reduced to the study of sequences of digits such that the associated formal power series tends to zero for all the Archimedean metrics for which the power series has a limit. This implies that the sequences of digits are labels of paths in a finite automaton. The understanding of the structure of the automaton is fundamentally connected with the injectivity of the geometric (onto the Rauzy fractal) and toral realizations.

For instance, the injectivity in measure of the geometric representation of a unimodular Pisot type substitutive system can be characterized in terms of explicit combinatorial conditions on graphs. Contrary to the coincidence condition, it is a necessary and sufficient condition, but it is quite difficult to verify this condition quickly. Furthermore, this condition is satisfied by substitutions with coincidence, but we do not know whether equivalence holds.

The main interest of this approach is that, contrary to what happens with the coincidence condition, these methods also give results about the injectivity of the toral realization.

Theorem 7.5.19 (Siegel [397]). *The injectivity in measure of the toral representation of a unimodular Pisot type substitutive dynamical system deduced from the Rauzy fractal construction is equivalent to an explicit combinatorial condition which depends only on the definition of the substitution which generates the system.*

Thus one can test whether the spectrum of a given unimodular Pisot type substitutive dynamical system is purely discrete.

Corollary 7.5.20 (Siegel [397]). *There exists an explicit and effective sufficient combinatorial condition for a unimodular Pisot type substitutive dynamical system to be measure-theoretically isomorphic to a toral translation, that is, to have a purely discrete spectrum. This condition is necessary when the substitution has no non-trivial coboundary.*

For instance, the systems generated by the substitutions $1 \mapsto 12$, $2 \mapsto 31$, $3 \mapsto 1$, and $1 \mapsto 12$, $2 \mapsto 13$, $3 \mapsto 132$ have a purely discrete spectrum: both are measure-theoretically isomorphic to a translation on the two-dimensional torus, both conjugacy maps being explicit semi-topological conjugacies.

In more geometrical terms, the injectivity of the toral realization has the following interpretation in terms of tilings:

Corollary 7.5.21 (Siegel [397]). *There exists an explicit and effective combinatorial condition for the Rauzy fractal associated with a unimodular Pisot type substitution over d letters to generate a periodic tiling of* $\mathbb{R}^{d-1}$.

The condition is too long to be given here. The difficulty in checking whether this condition holds prevents one from obtaining general results about the injectivity of the toral realization. However, even if the condition is too long to be checked by hand, it can be checked in a finite time for any explicit example, thanks to the effective algorithm given in [397]. For instance, this algorithm allows one to state that the substitution $1 \mapsto 21$, $2 \mapsto 13$, $3 \mapsto 1$ generates a periodic tiling shown in Fig. 7.10. Notice that all the substitutions that have been tested give a positive answer to the tiling question.

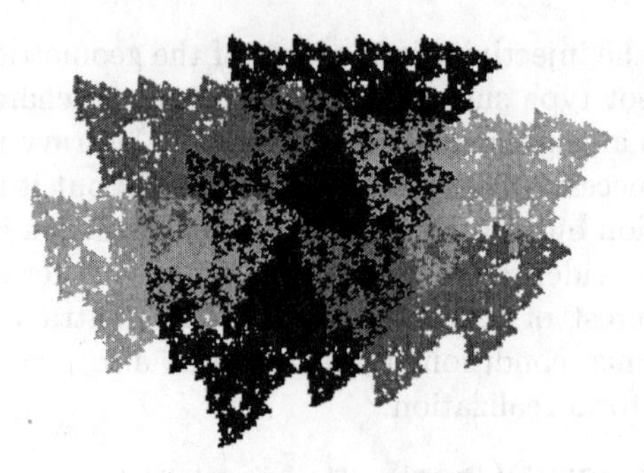

Fig. 7.10. Periodic tiling for the Rauzy fractal of $1 \mapsto 21$, $2 \mapsto 13$, $3 \mapsto 1$.

About unimodular substitutions of Pisot type, an important remaining question at the moment is to obtain a general sufficient condition, for instance in terms of coincidences, for a unimodular Pisot type substitutive system to have purely discrete spectrum, or equivalently, to be measure-theoretically isomorphic to a toral rotation.

7.6 Extensions and applications

7.6.1 Non-unimodular substitutions

The formal point of view of substitutive dynamical systems can be extended to non-unimodular substitutions. Indeed, in the unimodular case, the Rauzy fractal of a substitution of Pisot type over d letters is made of the set of finite values which can be taken for any Archimedean topology by the formal power series. But, in the non-unimodular case, there exists a new type of topology for which the formal power series converges: the p-adic topology. One can define a new Rauzy fractal, with p-adic components, which consists of the set of finite values which can be taken for any topology by the formal power series (Archimedean and p-adic). Then, one obtains a set in the product of $\mathbb{R}^{d-1}$ and finite extensions of p-adic fields $\mathbb{Q}_p$, where p are prime divisors of the determinant of the incidence matrix of the substitution [395].

The techniques developed in the unimodular case can be generalized:

- there exists a combinatorial condition on the substitution (in terms of graphs) which is sufficient for the substitutive system to code the action of a domain exchange in the new Rauzy fractal;
- there exists a lattice such that the quotient map modulo this lattice transposes the domain exchange into a translation over a compact group;
- there exists an automaton which decides whether the quotient map is almost everywhere one-to-one.

Thus, we get the following result:

Theorem 7.6.1 (Siegel [397]). *There exists an explicit and effective combinatorial condition for a Pisot type substitutive dynamical system to be measure-theoretically isomorphic to a toral translation, that is, to have a purely discrete spectrum.*

Example 7.6.2. The system generated by the substitution $1 \mapsto 1112$ et $2 \mapsto 12$, which is known to have a purely discrete spectrum (Theorem 7.3.35), is semi-topologically conjugate to a translation over the direct product of $\mathbf{Z}_2$ with two-adic solenoid $\mathcal{S}_2$(see [395, 397]).

The system generated by the substitution $1 \mapsto 2$, $2 \mapsto 3$, $3 \mapsto 11233$ has a purely discrete spectrum.

Many questions remain for non-unimodular Pisot substitutions: what is in general the structure of the maximal equicontinuous factor? Even in the two-dimensional case, where a condition for a purely discrete spectrum is known, the structure is difficult to understand. As a final question, we know that the substitution defined by $1 \mapsto 11222$ and $2 \mapsto 1222$ is isomorphic to a compact group translation. We know that this group contains a one-dimensional torus, and no p-adic factors. But what is precisely this group?

7.6.2 Substitutions and Markov partitions

The prefix-suffix expansion and the Rauzy fractal associated with a unimodular substitution of Pisot type over a d-letter alphabet, naturally lead to define a domain in $\mathbb{R}^d$.

This domain consists of finite unions of cylinders. The base of each cylinder is a sub-domain of the Rauzy fractal of the substitution embedded in the contracting hyperplane of the incidence matrix. Each cylinder is directed by the expanding direction of the matrix. Examples are given in Fig. 7.11 for the Fibonacci substitution and the Tribonacci substitution.

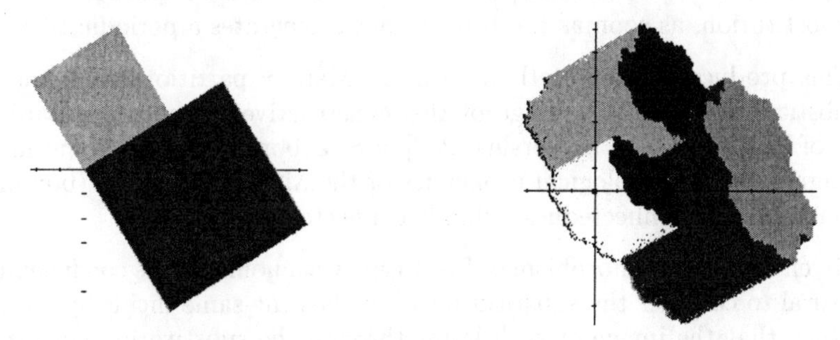

Fig. 7.11. Domains in $\mathbb{R}^d$ associated with the Fibonacci and Tribonacci substitutions.

If the Rauzy fractal of the substitution generates a tiling of $\mathbb{R}^{d-1}$ (see Corollary 7.5.21), then the domain is a fundamental domain identified with the d-dimensional torus $\mathbb{T}^d$ [396]. Illustrations are shown in Fig. 7.12 for the Fibonacci and Tribonacci substitutions.

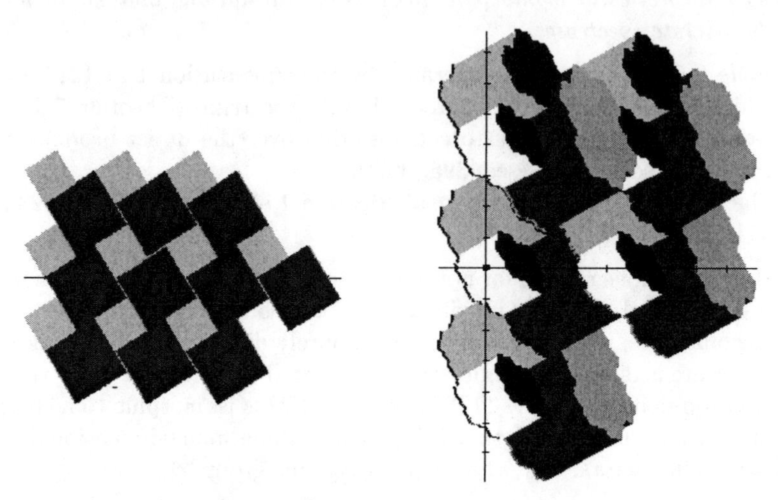

Fig. 7.12. Tilings of $\mathbb{R}^2$ and $\mathbb{R}^3$ associated with the Fibonacci and Tribonacci substitutions.

Moreover, the prefix-suffix automaton induces a natural partition of this fundamental domain such that the action of the toral automorphism associated with the incidence matrix of the substitution is coded, with respect to this partition, by the set of biinfinite walks of the prefix-suffix automaton.

Thus, the Rauzy fractal of a unimodular substitution of Pisot type together with the prefix-suffix automaton produce a completely explicit Markov partition for the toral automorphism associated with the incidence matrix of the substitution, as soon as the Rauzy fractal generates a periodic tiling.

This produces a new method to build Markov partition, with the use of substitutions. The advantage of this constructive method, compared to those of R. Kenyon and A. Vershik [244] or S. Leborgne [260], is to permit to characterize some topological properties of the Markov partition (boundary parametrization, connectedness, simple connectedness).

Given a toral automorphism of Pisot type with nonnegative coefficients, it is natural to consider the substitution which has the same incidence matrix, and such that the image of each letter through the substitution consists of letters in an increasing order. Such a substitution satisfies the coincidence condition, so that there exists a Rauzy fractal associated with the substitution. It remains to check that the Rauzy fractal generates a periodic tiling to finally obtain an explicit Markov partition of the toral automorphism considered at the beginning.

8. Diophantine approximations, substitutions, and fractals

In this chapter we shall show how to associate with a substitution σ a domain $\mathcal{R}_\sigma$ with fractal boundary, generalizing Rauzy's famous construction of the Rauzy fractal [349] (see also Chap. 7). The substitutions are of Pisot type and unimodular, and our method is constructive. In this way, we obtain a geometric representation of the substitution as a domain exchange and, with a stronger hypothesis, we get a rotation on a torus. In fact, there are two quite different types of dynamics which act on the set $\mathcal{R}_\sigma$: the first one (given by the shift) corresponds to the exchange of domains, while the dynamics of the substitution is given by a Markov endomorphism of the torus whose structure matrix is equal to the incidence matrix of the substitution. The first one has zero entropy, whereas the second one has positive entropy. The domain $\mathcal{X}_\sigma$ is interesting both from the viewpoint of fractal geometry, and of ergodic and number theory; see for instance [349, 352, 221, 223, 290, 289, 373] and Chap. 7.

We illustrate this study with some substitutions such as the "modified Jacobi-Perron substitutions". We also end each section by evoking the results in the one-dimensional case corresponding to those obtained in the two-dimensional case. We end this chapter by surveying some applications in Diophantine approximation. In particular, we give a geometric and symbolic interpretation of the natural extension of the modified Jacobi-Perron algorithm.

This chapter is based on the study of a few representative examples of generalized substitutions illustrating the general theory. We have chosed to suppress most of the proofs (which are often rather technical) and to replace them by figures to help the comprehension. Most of the proofs of the results mentioned in this chapter are to be found in [43, 44].

We recommend the reading of Chap. 7, and in particular Sec. 7.4, as a motivation for the results and the techniques developed in the present chapter. Indeed the present approach has been inspired by the alternative construction of the Rauzy fractal associated with the Tribonacci substitution given in [221], by constructing successive approximations of the Rauzy fractal.

[1] This chapter has been written by S. Ito

8.1 Substitutions and domains with fractal boundary

The aim of this section is to fix the framework in which we will be working in this chapter. We introduce in particular the domain of $\mathbb{R}^2$ with fractal boundary associated with a substitution over a three-letter alphabet. Everything extends in a natural way to substitutions over any finite alphabet, but to avoid cumbersome terminology and to help the visualization, we restrict ourselves to substitutions over 2 or 3 letters.

8.1.1 Notation and main examples

In this chapter we focus on substitutions over the three-letter alphabet $\mathcal{A} = \{1, 2, 3\}$. The canonical basis of $\mathbb{R}^3$ is denoted by $\{\mathbf{e_1}, \mathbf{e_2}, \mathbf{e_3}\}$.

Let us introduce the families of substitutions we will consider in this chapter.

Example 8.1.1 (Rauzy substitution [349]). The *Rauzy substitution* is defined by:

$$\sigma(1) = 12,\ \sigma(2) = 13,\ \sigma(3) = 1.$$

The incidence matrix of this substitution is

$$\mathbf{M}_\sigma = \begin{pmatrix} 1 & 1 & 1 \\ 1 & 0 & 0 \\ 0 & 1 & 0 \end{pmatrix}.$$

Example 8.1.2 (Modified Jacobi-Perron substitution [223]). For every $a \in \mathbb{N}^+$ and $\delta \in \{0, 1\}$, let us define the substitution $\sigma_{(a,\delta)}$ as follows:

$$\sigma_{(a,0)} : \begin{array}{l} 1 \longrightarrow \overbrace{11 \cdots 1}^{a \text{ times}} 2 \\ 2 \longrightarrow 3 \\ 3 \longrightarrow 1 \end{array}, \quad \sigma_{(a,1)} : \begin{array}{l} 1 \longrightarrow \overbrace{11 \cdots 1}^{a \text{ times}} 3 \\ 2 \longrightarrow 1 \\ 3 \longrightarrow 2 \end{array}.$$

The substitutions $\sigma_{(a,\delta)}$, for $a \in \mathbb{N}^+$ and $\delta \in \{0, 1\}$, are called *modified Jacobi-Perron substitutions* and are connected in a natural way with the modified Jacobi-Perron algorithm: for more details concerning this algorithm, see Sec. 8.6.

For every $a \in \mathbb{N}^+$, the incidence matrices of the modified Jacobi-Perron substitutions $\sigma_{(a,0)}$ and $\sigma_{(a,1)}$ are the following:

$$\mathbf{M}_{\sigma_{(a,0)}} = \begin{pmatrix} a & 0 & 1 \\ 1 & 0 & 0 \\ 0 & 1 & 0 \end{pmatrix} \text{ and } \mathbf{M}_{\sigma_{(a,1)}} = \begin{pmatrix} a & 1 & 0 \\ 0 & 0 & 1 \\ 1 & 0 & 0 \end{pmatrix}.$$

Throughout this chapter, substitutions will always be of *Pisot type and unimodular*. Let us recall that this means that the eigenvalues λ,λ' and λ'' of its incidence matrix $\mathbf{M}_\sigma$ satisfy $\lambda > 1 > \mid \lambda' \mid, \mid \lambda'' \mid > 0$ and $\det \mathbf{M}_\sigma = \pm 1$ (see Sec. 1.2.5). Notice that a substitution of Pisot type is *primitive*, that is, there exists N such that $\mathbf{M}_\sigma^N > 0$ (see Theorem 1.2.9).

Contracting plane. Let ${}^t(1, \alpha, \beta)$ and ${}^t(1, \gamma, \delta)$ be the right and left eigenvectors of $\mathbf{M}_\sigma$ for the dominant eigenvalue λ:

$$\mathbf{M}_\sigma \, {}^t(1, \alpha, \beta) = \lambda \, {}^t(1, \alpha, \beta) \quad \text{and} \quad {}^t\mathbf{M}_\sigma \, {}^t(1, \gamma, \delta) = \lambda \, {}^t(1, \gamma, \delta).$$

Let $\mathcal{P}_{\gamma,\delta}$ be the plane defined by

$$\mathcal{P}_{\gamma,\delta} := \left\{ \boldsymbol{x} \in \mathbb{R}^3 \mid \left\langle \boldsymbol{x}, \, {}^t(1, \gamma, \delta) \right\rangle = 0 \right\}. \tag{8.1}$$

Projection. The map $\pi : \mathbb{R}^3 \to \mathcal{P}_{\gamma,\delta}$ is defined to be the *projection* along the vector $(1, \alpha, \beta)$.

The lattice $\boldsymbol{L}_0$. We denote by $\boldsymbol{L}_0$ the lattice

$$\boldsymbol{L}_0 = \{ n\pi(\boldsymbol{e}_2 - \boldsymbol{e}_1) + m(\boldsymbol{e}_3 - \boldsymbol{e}_2) | m, n \in \mathbb{Z} \}.$$

Lemma 8.1.3. *The plane $\mathcal{P}_{\gamma,\delta}$ is the* contracting invariant plane *with respect to $\mathbf{M}_\sigma$. More precisely, we have $\mathbf{M}_\sigma \mathcal{P}_{\gamma,\delta} = \mathcal{P}_{\gamma,\delta}$, and letting $\boldsymbol{v}' = {}^t(1, \alpha', \beta')$ and $\boldsymbol{v}'' = {}^t(1, \alpha'', \beta'')$ be the right eigenvectors of $\mathbf{M}_\sigma$ associated respectively with eigenvalues λ' and λ'', we have:*

- *In the case where λ' and λ'' are real, then*

$$\boldsymbol{v}', \boldsymbol{v}'' \in \mathcal{P}_{\gamma,\delta} \text{ and } (\mathbf{M}_\sigma \boldsymbol{v}', \mathbf{M}_\sigma \boldsymbol{v}'') = (\boldsymbol{v}', \boldsymbol{v}'') \begin{pmatrix} \lambda' & 0 \\ 0 & \lambda'' \end{pmatrix}.$$

- *If λ' and λ'' are complex, then $\overline{\lambda'} = \lambda''$, and letting $\boldsymbol{u}' := \frac{1}{2}(\boldsymbol{v}' + \boldsymbol{v}'')$ and $\boldsymbol{u}'' := \frac{1}{2i}(\boldsymbol{v}' - \boldsymbol{v}'')$, we have*

$$\boldsymbol{u}', \boldsymbol{u}'' \in \mathcal{P}_{\gamma,\delta} \text{ and } \exists \theta \text{ with } (\mathbf{M}_\sigma \boldsymbol{u}', \mathbf{M}_\sigma \boldsymbol{u}'') = \frac{1}{\sqrt{\lambda}} (\boldsymbol{u}', \boldsymbol{u}'') \begin{pmatrix} \cos\theta & -\sin\theta \\ \sin\theta & \cos\theta \end{pmatrix}.$$

The proof is left as an exercise.

8.1.2 Domain with fractal boundary associated with a substitution

Let us associate with the fixed point u (and with the dynamical system (X_u, T) associated with u) a domain included in the contracting plane $\mathcal{P}_{\gamma,\delta}$ by projecting the broken lines in $\mathbb{R}^3$ corresponding to the images under the abelianization homomorphism (see Chap. 1) of the prefixes of the sequence u.

Let σ be a unimodular substitution of Pisot type such that $\sigma(1)$ begins with 1. We associate with the substitution σ a domain $\mathcal{R}_\sigma$ with a fractal boundary, which admits a natural decomposition into three sets $\mathcal{R}_i$, $i = 1, 2, 3$, as follows.

The set $\mathcal{R}_\sigma$ is the closure of $\bigcup_{N=1}^{\infty} Y_N$,

$$\text{with } Y_N = \left\{ -\pi(\textstyle\sum_{j=1}^{k} e_{u_j}) \,\middle|\, k = 1, \cdots, N \right\}. \quad (8.2)$$

For $i = 1, 2, 3$, the set $\mathcal{R}_i$ is the closure of $\bigcup_{N=1}^{\infty} Y_{N,i}$,

$$\text{with } Y_{N,i} = \left\{ -\pi(\textstyle\sum_{j=1}^{k} e_{u_j}) \,\middle|\, e_{u_k} = i,\, k = 1, \cdots, N \right\}.$$

The sign $-$ in the above formulas appears here for some technical reasons due to the very definition of the generalized substitutions we introduce later. This set $\mathcal{R}_\sigma$ will be called the *(generalized) Rauzy fractal* associated with the substitution σ.

For the Rauzy substitution (Example 8.1.1), and the modified Jacobi-Perron substitution (Example 8.1.2) with $a = 1$ and $\delta = 0$, the domains $\mathcal{R}_\sigma$ are shown on Fig. 8.1.

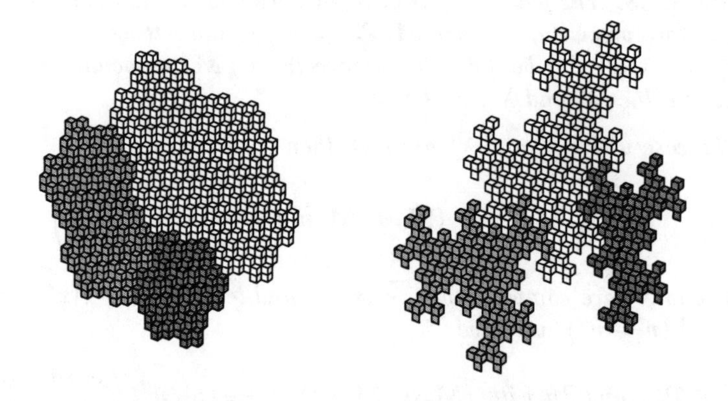

Fig. 8.1. The domain $\mathcal{R}_\sigma$ with fractal boundary for the Rauzy substitution and the modified Jacobi-Perron substitution $\sigma_{(1,0)}$.

Our aim now is to give an explicit device of construction for the set $\mathcal{R}_\sigma$ by introducing, in the next section, a notion of generalized substitution.

8.1.3 The two-dimensional case

While, in this chapter, we focus on substitutions over a three-letter alphabet, the results are satisfied by substitutions over a general alphabet $\{1, 2 \cdots, d\}$.

In particular, the simplest case of a two-letter alphabet is fundamental. At the end of each section of this chapter, we will survey the corresponding results in the two-dimensional case.

Let ${}^t(1,\alpha)$ and ${}^t(1,\gamma)$ be, respectively, the right and left eigenvectors of the matrix $\mathbf{M}_\sigma$ for the dominant eigenvalue λ:

$$\mathbf{M}_\sigma\ {}^t(1,\alpha) = \lambda\ {}^t(1,\alpha) \qquad {}^t\mathbf{M}_\sigma\ {}^t(1,\gamma) = \lambda\ {}^t(1,\gamma).$$

The *contracting invariant line* l_σ with respect to $\mathbf{M}_\sigma$ is defined as

$$l_\sigma = \left\{ \boldsymbol{x} \in \mathbb{R}^2 \ \middle|\ \left\langle \boldsymbol{x},\ {}^t(1,\gamma) \right\rangle = 0 \right\}.$$

The *projection* along ${}^t(1,\alpha)$ is also denoted $\pi : \mathbb{R}^2 \to l_\sigma$.

For any unimodular substitution σ of Pisot type such that $\sigma(1)$ begins with 1, one can obtain also some projective sets Y_N and $Y_{N,i}$ (for $i = 1,2$), and their limit $\mathcal{R}_\sigma$ and $\mathcal{R}_i$ (for $i = 1,2$): $\mathcal{R}_i$ is the closure of $\bigcup_{N=1}^{\infty} Y_{N,i}$ (for $i = 1,2$), and $\mathcal{R}_\sigma$ is the closure of $\bigcup_{N=1}^{\infty} Y_N$.

Example 8.1.4 (Continued fraction substitutions). For every $a \in \mathbb{N}^+$, let us define the substitutions σ_a over a two-letter alphabet by

$$\sigma_a(1) = \overbrace{11\cdots1}^{a\ times}2, \qquad \sigma_a(2) = 1.$$

The substitutions σ_a, with $a \in \mathbb{N}^+$, are called *continued fraction substitutions* (the definition of the continued fraction algorithm will be recalled in Sec. 8.6).

The matrix of σ_a is

$$\mathbf{M}_{\sigma_a} = \begin{pmatrix} a & 1 \\ 1 & 0 \end{pmatrix}.$$

In this example, the sets $\mathcal{R}_\sigma$ and $\mathcal{R}_i$, $i = 1,2$ associated with the continued fraction substitutions are intervals.

8.2 Generalized substitutions

8.2.1 Definitions

In this section, we construct a map Θ associated with the substitution σ: the map Θ is a generalized substitution, which acts over translated faces of the unit cube. It is obtained as a dual map of the substitution σ. For more details, see [43, 44]. The purpose of the introduction of such a device is, first, to get an explicit construction of an approximation of the contracting plane as a stepped surface by the iteration Θ^n, and secondly, to recover, in an explicit way after a suitable renormalization, the domain $\mathcal{R}_\sigma$ associated with the substitution σ.

We illustrate this notion through the examples of the Rauzy substitution and the modified Jacobi-Perron substitutions. We end this section by considering the two-dimensional case.

Basic squares. Let $\boldsymbol{x} \in \mathbb{Z}^3$ and $i \in \{1,2,3\}$. Translate the unit cube of $\mathbb{R}^3$ by the vector $\boldsymbol{x}$. Then the face of the translated cube which contains $\boldsymbol{x}$ and is orthogonal to the vector $\boldsymbol{e}_i$ is denoted $(\boldsymbol{x}, i^*)$. More precisely, let (j,k) be such that $(i,j,k) \in \{(1,2,3),(2,3,1),(3,1,2)\}$, then:

$$(\boldsymbol{x}, i^*) := \{\boldsymbol{x} + \lambda \boldsymbol{e}_j + \mu \boldsymbol{e}_k \mid 0 \le \lambda \le 1,\ \ 0 \le \mu \le 1\}.$$

Such a set is called a *basic square*. Illustrations of some basic squares are given in Fig. 8.2.

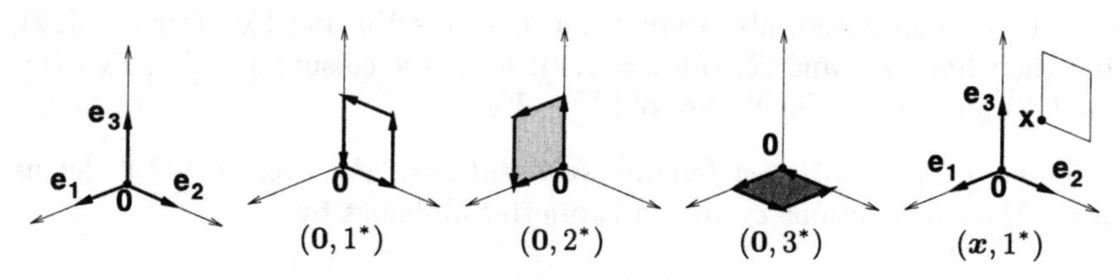

Fig. 8.2. The basic squares $(0, i^*)$, $i = 1,2,3$ and $(x, 1^*)$.

The $\mathbb{Z}$-module generated by the set of basic squares. Our next step will be now to define a linear map which acts on the set of basic squares. The idea is to substitute (according to a scheme given by the substitution σ) a basic square by a union of basic squares. When iterating such a device, overlap problems of basic squares may occur. We thus need to introduce a suitable formalism. Let us equip the set of basic squares with a structure of a $\mathbb{Z}$-module.

Let Λ denote the set of all the basic squares $(\boldsymbol{x}, i^*)$:

$$\Lambda = \{(\boldsymbol{x}, i^*) \mid \boldsymbol{x} \in \mathbb{Z}^3, i \in \{1,2,3\}\}.$$

Let $\mathcal{G}$ be the $\mathbb{Z}$-module generated by Λ:

$$\mathcal{G} = \{\textstyle\sum_{\lambda \in \Lambda} m_\lambda \lambda \mid m_\lambda \in \mathbb{Z},\ \#\{\lambda \mid m_\lambda \neq 0\} < +\infty\}.$$

The set $\mathcal{G}$ is the set of formal sums of weighted faces.

The lattice $\mathbb{Z}^3$ acts on $\mathcal{G}$ by translation:

$$\forall \boldsymbol{y}, \boldsymbol{x} \in \mathbb{Z}^3,\ \forall i \in \{1,2,3\}, \qquad \boldsymbol{y} + (\boldsymbol{x}, i^*) := (\boldsymbol{y} + \boldsymbol{x}, i^*).$$

Now, we can define a map Θ acting on the set $\mathcal{G}$.

Definition 8.2.1. *The* endomorphism Θ *of* $\mathcal{G}$ *associated with* σ *is defined as follows, where* Y *and* W *denotes prefixes and suffixes:*

$$\begin{cases} \forall i \in \mathcal{A}, & \Theta(\mathbf{0}, i^*) := \sum_{j=1}^{3} \sum_{W:\sigma(j)=Y\cdot i\cdot W} \left(\mathbf{M}_\sigma^{-1}(f(W)), j^*\right), \\ \forall \boldsymbol{x} \in \mathbb{Z}^3, \forall i \in \mathcal{A}, & \Theta(\boldsymbol{x}, i^*) := \mathbf{M}_\sigma^{-1}\boldsymbol{x} + \Theta(\mathbf{0}, i^*), \\ \forall \sum_{\lambda\in\Lambda} m_\lambda \lambda \in \mathcal{G}, & \Theta(\sum_{\lambda\in\Lambda} m_\lambda \lambda) := \sum_{\lambda\in\Lambda} m_\lambda \Theta(\lambda). \end{cases}$$

The formulation given above is quite cumbersome. For a better understanding of the map Θ, see Fig. 8.4 , 8.5, 8.6, and see also the next sections, where we apply the above formulas in some basic cases.

Note that $\mathbf{M}_\sigma^{-1}$ maps $\mathbb{Z}^3$ to $\mathbb{Z}^3$, since the substitution is unimodular.

Let us introduce the stepped surface associated with a plane on which the endomorphism Θ acts, in order to discuss the geometrical properties of Θ.

Stepped surface of a plane. Let $0 < \gamma, \delta < 1$. Let $\mathcal{P}_{\gamma,\delta}$ be the plane orthogonal to the vector ${}^t(1, \gamma, \delta)$:

$$\mathcal{P}_{\gamma,\delta} := \left\{\boldsymbol{x} \in \mathbb{R}^3 \mid \langle \boldsymbol{x}, {}^t(1,\gamma,\delta) \rangle = 0\right\}.$$

For each γ, δ, the *stepped surfaces* $\mathcal{S}^+_{\gamma,\delta}$ and $\mathcal{S}^-_{\gamma,\delta}$ associated with $\mathcal{P}_{\gamma,\delta}$ are defined as the surfaces consisting of basic squares which are the best below and above approximations of the plane $\mathcal{P}_{\gamma,\delta}$. More precisely, let $\mathbf{S}^+_{\gamma,\delta}$ and $\mathbf{S}^-_{\gamma,\delta}$ gather the faces of the translates with integer vertices of the unit cube which intersect $\mathcal{P}_{\gamma,\delta}$:

$$\mathbf{S}^+_{\gamma,\delta} := \left\{(\boldsymbol{x}, i^*) \mid \langle \boldsymbol{x}, {}^t(1,\gamma,\delta) \rangle > 0, \ \langle (\boldsymbol{x} - \boldsymbol{e}_i), {}^t(1,\gamma,\delta) \rangle \le 0\right\},$$

$$\mathbf{S}^-_{\gamma,\delta} := \left\{(\boldsymbol{x}, i^*) \mid \langle \boldsymbol{x}, {}^t(1,\gamma,\delta) \rangle \ge 0, \ \langle (\boldsymbol{x} - \boldsymbol{e}_i), {}^t(1,\gamma,\delta) \rangle < 0\right\}.$$

Then the stepped surfaces are unions of such basic squares:

$$\mathcal{S}^+_{\gamma,\delta} := \bigcup_{(\boldsymbol{x},i^*)\in\mathbf{S}^+_{\gamma,\delta}} (\boldsymbol{x}, i^*), \qquad \mathcal{S}^-_{\gamma,\delta} := \bigcup_{(\boldsymbol{x},i^*)\in\mathbf{S}^-_{\gamma,\delta}} (\boldsymbol{x}, i^*).$$

These two stepped surfaces are the best approximations of the plane $\mathcal{P}_{\gamma,\delta}$ by faces contained respectively in the closed upper and lower half-spaces $\{\boldsymbol{x} | \langle \boldsymbol{x}, {}^t(1,\gamma,\delta) \rangle \ge 0\}$ and $\{\boldsymbol{x} | \langle \boldsymbol{x}, {}^t(1,\gamma,\delta) \rangle \le 0\}$. The stepped surfaces are called *discretization* of the plane $\mathcal{P}_{\gamma,\delta}$, or *discrete planes.*

Let $\Gamma^+_{\gamma,\delta}$ and $\Gamma^-_{\gamma,\delta}$ be the subset of $\mathcal{G}$ consisting of all finite unions of elements of $\mathbf{S}^+_{\gamma,\delta}$ and $\mathbf{S}^-_{\gamma,\delta}$:

$$\Gamma^+_{\gamma,\delta} := \left\{\sum_{\lambda\in\mathbf{S}^+_{\gamma,\delta}} n_\lambda \lambda \;\middle|\; n_\lambda \in \{0,1\}, \ \#\{\lambda \mid n_\lambda = 1\} < +\infty\right\},$$

$$\Gamma^-_{\gamma,\delta} := \left\{\sum_{\lambda\in\mathbf{S}^-_{\gamma,\delta}} n_\lambda \lambda \;\middle|\; n_\lambda \in \{0,1\}, \ \#\{\lambda \mid n_\lambda = 1\} < +\infty\right\}.$$

Thus elements of $\Gamma^+_{\gamma,\delta}$ are finite parts of the stepped surface $\mathcal{S}^+_{\gamma,\delta}$. We call them *patches* of the stepped surface. The stepped surface has the following invariance property with respect to Θ:

Theorem 8.2.2 ([44]). *For each unimodular substitution σ of Pisot type, let Θ be the endomorphism of $\mathcal{G}$ associated with σ, and let $\Gamma^+_{\gamma,\delta}$ and $\Gamma^-_{\gamma,\delta}$ be the patches of the stepped surface associated with the contracting invariant plane $\mathcal{P}_{\gamma,\delta}$. Then $\Gamma^+_{\gamma,\delta}$ and $\Gamma^-_{\gamma,\delta}$ are invariant with respect to the endomorphism Θ, that is,*

$$\sum_{\lambda\in\mathbf{S}^+_{\gamma,\delta}} n_\lambda\lambda \in \Gamma^+_{\gamma,\delta} \quad \textit{implies} \quad \Theta\Big(\sum_{\lambda\in\mathbf{S}^+_{\gamma,\delta}} n_\lambda\lambda\Big) \in \Gamma^+_{\gamma,\delta},$$

$$\sum_{\lambda\in\mathbf{S}^-_{\gamma,\delta}} n_\lambda\lambda \in \Gamma^-_{\gamma,\delta} \quad \textit{implies} \quad \Theta\Big(\sum_{\lambda\in\mathbf{S}^-_{\gamma,\delta}} n_\lambda\lambda\Big) \in \Gamma^-_{\gamma,\delta}.$$

This theorem means that one can apply Θ to any finite part of the stepped surface and obtain a new finite part of the stepped surface. Let us define a partial relationship of inclusion between parts of the stepped surface.

Notation. For $\gamma = \sum_{\lambda\in\mathbf{S}^+_{\gamma,\delta}} n_\lambda\lambda$, $\delta = \sum_{\lambda\in\mathbf{S}^+_{\gamma,\delta}} m_\lambda\lambda \in \Gamma^+_{\gamma,\delta}$, $\gamma \succ \delta$ means that $n_\lambda \neq 0$ if $m_\lambda \neq 0$, i.e., the patch δ is a subpatch of γ (See Fig. 8.3).

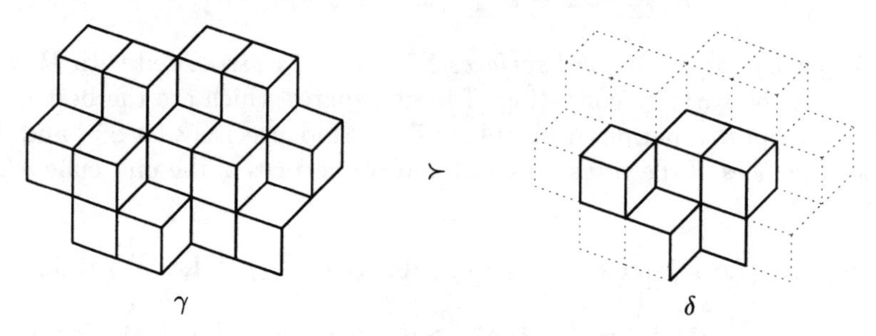

Fig. 8.3. Two patches γ and δ such that $\gamma \succ \delta$.

Let $\mathcal{U}'$ denote the above faces of the unit cube, that is, the set of faces which are in contact with the origin $(0,0,0)$, and $\mathcal{U}$ the below faces, that is, the set of faces which are in contact with the point $(1,1,1)$ (see for instance Fig. 8.5 below).

The next lemma means that the application of Θ to patches generates new patches which contain the initial half-unit cube. The proof is left as an exercise, and is a direct consequence of the definition of Θ.

Lemma 8.2.3. *Let*

$$\mathcal{U} := \sum_{i=1,2,3} (\mathbf{e}_i, i^*) \qquad \textit{and} \qquad \mathcal{U}' := \sum_{i=1,2,3} (\mathbf{0}, i^*).$$

Then we have

$$\Theta(\mathcal{U}) \succ \mathcal{U} \quad \textit{and} \quad \Theta(\mathcal{U}') \succ \mathcal{U}'.$$

8.2.2 First example: the Rauzy substitution

Let us recall that the Rauzy substitution is defined as $\sigma(1) = 12$, $\sigma(2) = 13$ and $\sigma(3) = 1$.

Let $\boldsymbol{f}_i$ be the column vectors of the matrix $\mathbf{M}_\sigma^{-1}$:

$$\mathbf{M}_\sigma^{-1} = \begin{pmatrix} 0 & 1 & 0 \\ 0 & 0 & 1 \\ 1 & -1 & -1 \end{pmatrix} = (\boldsymbol{f}_1, \boldsymbol{f}_2, \boldsymbol{f}_3).$$

Then the endomorphism Θ is defined by the following formulas:

$$\Theta : \begin{array}{l} (\mathbf{0}, 1^*) \longrightarrow (\mathbf{0}, 3^*) + (\boldsymbol{f}_2, 1^*) + (\boldsymbol{f}_3, 2^*) \\ (\mathbf{0}, 2^*) \longrightarrow (\mathbf{0}, 1^*) \\ (\mathbf{0}, 3^*) \longrightarrow (\mathbf{0}, 2^*) \end{array}.$$

The way Θ acts on the faces of the unit cube is shown in Fig. 8.4.

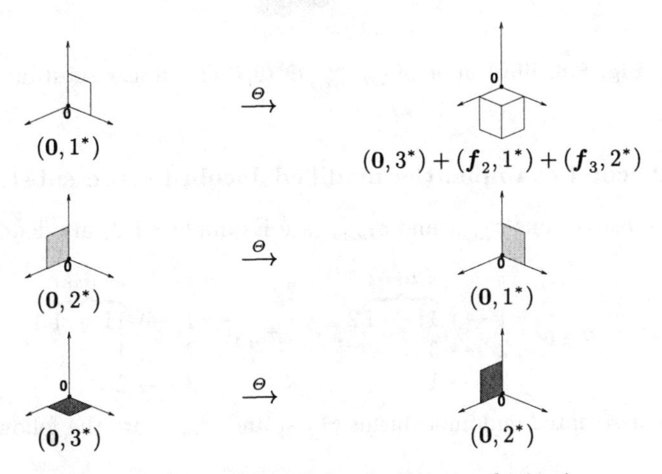

Fig. 8.4. Endomorphism Θ for the Rauzy substitution.

In Fig. 8.5, one sees the successive images of the half unit cube $\mathcal{U}' = (\mathbf{0}, 1^*) \cup (\mathbf{0}, 2^*) \cup (\mathbf{0}, 3^*)$ through Θ. The three different colors describe the successive images of each face $(\mathbf{0}, 1^*)$, $(\mathbf{0}, 2^*)$ and $(\mathbf{0}, 3^*)$. Note that even if the image of a unit face does not contain the same unit face, the image of $\mathcal{U}'$ does contain $\mathcal{U}'$.

As a consequence, the sequence of patches $\Theta^n(\mathcal{U}')$ is strictly increasing. We will see in the next sections that there exists a renormalization that makes the patches converge to a compact set in $\mathbb{R}^2$. The limit of this sequence may be able to be imagined in Fig. 8.6.

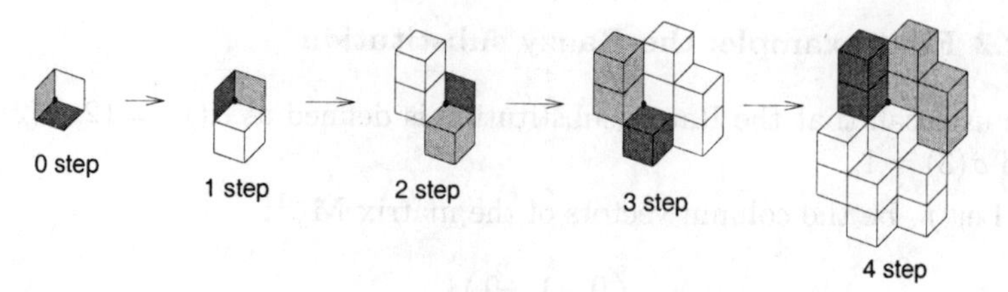

Fig. 8.5. Illustration of $\bigcup_{i=1,2,3} \Theta^n(0,i^*)$ for Rauzy substitution, $n = 0,1,2,3,4$.

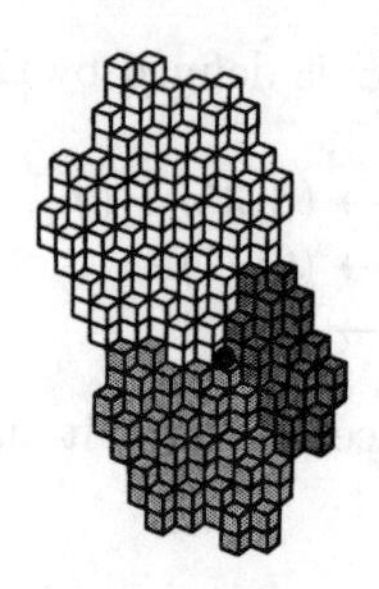

Fig. 8.6. Illustration of $\bigcup_{i=1,2,3} \Theta^8(0,i^*)$ for Rauzy substitution.

8.2.3 Second example: the modified Jacobi-Perron substitutions

These substitutions $\sigma_{(a,0)}$ and $\sigma_{(a,1)}$ (see Example 8.1.2) are defined by:

$$\sigma_{(a,0)} : \begin{array}{l} 1 \longrightarrow \overbrace{11\cdots1}^{a\ times}2 \\ 2 \longrightarrow 3 \\ 3 \longrightarrow 1 \end{array}, \qquad \sigma_{(a,1)} : \begin{array}{l} 1 \longrightarrow \overbrace{11\cdots1}^{a\ times}3 \\ 2 \longrightarrow 1 \\ 3 \longrightarrow 2 \end{array}.$$

The associated endomorphisms $\Theta_{(a,0)}$ and $\Theta_{(a,1)}$ are the following:

$$\Theta_{(a,0)} : \begin{array}{l} (\mathbf{0},1^*) \longrightarrow (\mathbf{0},3^*) + \sum_{1\le k\le a} ((e_1 - ke_3),1^*) \\ (\mathbf{0},2^*) \longrightarrow (\mathbf{0},1^*) \\ (\mathbf{0},3^*) \longrightarrow (\mathbf{0},2^*) \end{array},$$

$$\Theta_{(a,1)} : \begin{array}{l} (\mathbf{0},1^*) \longrightarrow (\mathbf{0},2^*) + \sum_{1\le k\le a} ((e_1 - ke_2),1^*) \\ (\mathbf{0},2^*) \longrightarrow (\mathbf{0},3^*) \\ (\mathbf{0},3^*) \longrightarrow (\mathbf{0},1^*) \end{array}.$$

The way $\Theta_{(a,\varepsilon)}$ (for $\varepsilon = 0,1$) acts on the faces of the unit cube is shown on Figs. 8.7 and 8.8. In Fig. 8.9, one sees the successive images of the half cube $\mathcal{U}' = (\mathbf{0},1^*) \cup (\mathbf{0},2^*) \cup (\mathbf{0},3^*)$ through Θ. Similarly as in the case of the Rauzy substitution, the sequence of patches $\Theta^n(\mathcal{U}')$ is strictly increasing, since $\Theta(\mathcal{U}')$ contains $\mathcal{U}'$. An idea of the limit of the sequence of renormalized patches is given in Fig. 8.10.

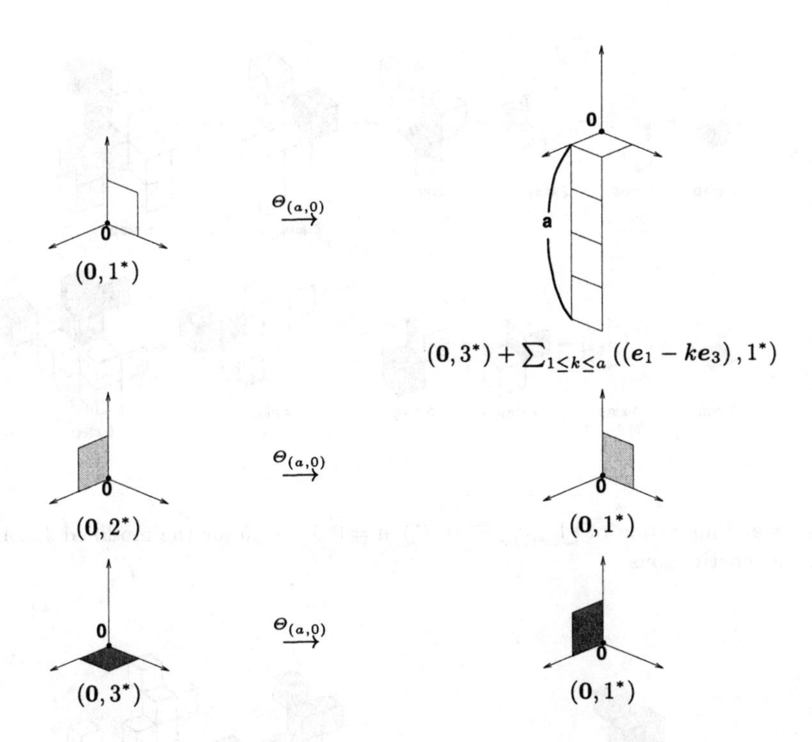

Fig. 8.7. The map Θ for the modified Jacobi-Perron substitutions $\sigma_{(a,0)}$.

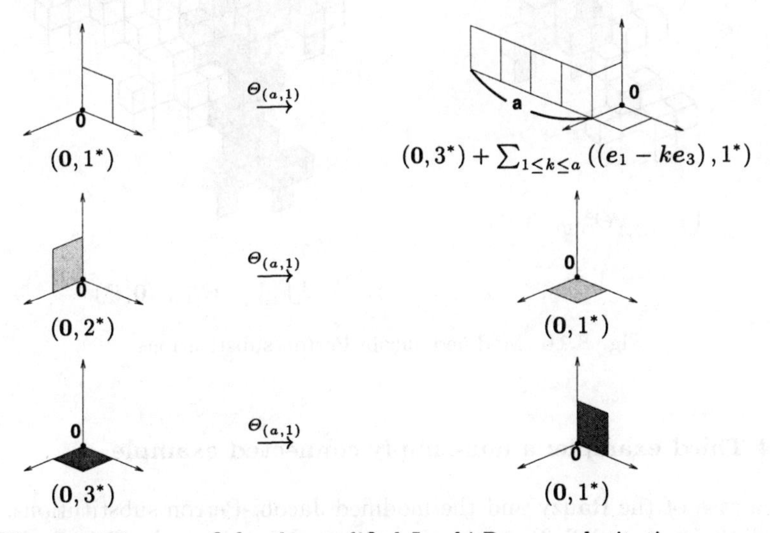

Fig. 8.8. The map Θ for the modified Jacobi-Perron substitutions $\sigma_{(a,1)}$.

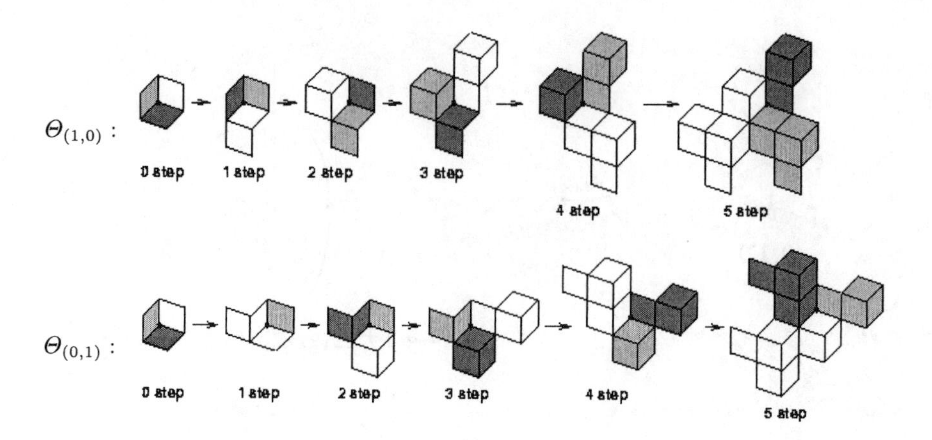

Fig. 8.9. Illustration of $\bigcup_{i=1,2,3} \Theta^n(o, i^*), n = 0, 1, \cdots, 5$ for the modified Jacobi-Perron substitutions.

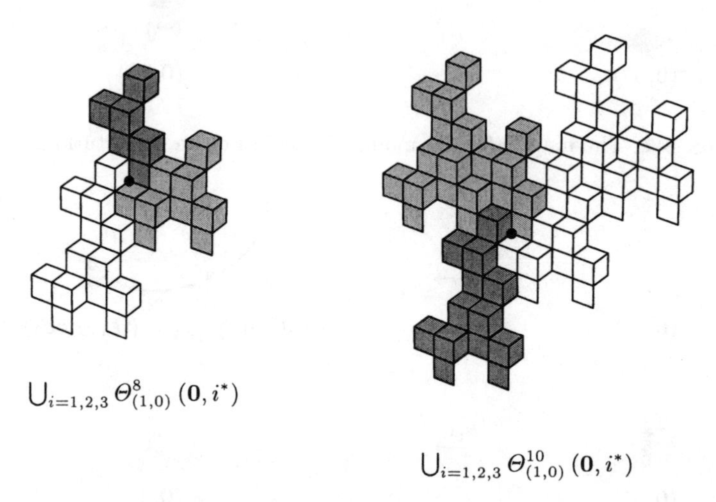

Fig. 8.10. Modified Jacobi-Perron substitutions.

8.2.4 Third example: a non-simply connected example

In the case of the Rauzy and the modified Jacobi-Perron substitutions, the sets $\Theta^n(\mathbf{0}, i^*)$, $(i = 1, 2, 3)$ are simply connected, but it can be remarked that, in general, the sets $\Theta^n(\mathbf{0}, i^*)$ (for $i = 1, 2, 3$) are not simply connected. The example below is characteristic. Geometrical properties of the sets $\Theta^n(\mathbf{0}, i^*)$, $i = 1, 2, 3$ are studied in [222, 223] and [185].

Example 8.2.4 (A non-simply connected example). Consider the substitution defined by:

$$1 \mapsto 12, \; 2 \mapsto 31, \; 3 \mapsto 1.$$

If $\mathbf{M}$ denotes the matrix of this substitution, let $\boldsymbol{f}_i$ be the columns of its inverse:

$$\mathbf{M}^{-1} = \begin{pmatrix} 0 & 1 & 0 \\ 0 & 0 & 1 \\ 1 & -1 & -1 \end{pmatrix} = (\boldsymbol{f}_1, \boldsymbol{f}_2, \boldsymbol{f}_3) \,.$$

The associated endomorphism Θ is the following:

$$\Theta : \begin{array}{l} (\mathbf{0}, 1^*) \longrightarrow (\mathbf{0}, 1^*) + (\mathbf{0}, 3^*) + (\boldsymbol{f}_3, 2^*) \\ (\mathbf{0}, 2^*) \longrightarrow (\boldsymbol{f}_1, 1^*) \\ (\mathbf{0}, 3^*) \longrightarrow (\mathbf{0}, 2^*). \end{array}$$

The way Θ acts on the faces of the unit cube is shown on Fig. 8.11. The successive images of the half unit cube $\mathcal{U}' = (\mathbf{0}, 1^*) \cup (\mathbf{0}, 2^*) \cup (\mathbf{0}, 3^*)$ under the action of Θ are given on Fig. 8.12. Here again, $\Theta(\mathcal{U}')$ contains $\mathcal{U}'$, hence the sequence of patches $\Theta^n(\mathcal{U}')$ converges, see Fig. 8.13. These figures show that for this substitution, the sets $\Theta^n(\mathbf{0}, i^*)$, for $i = 1, 2, 3$, are not simply connected.

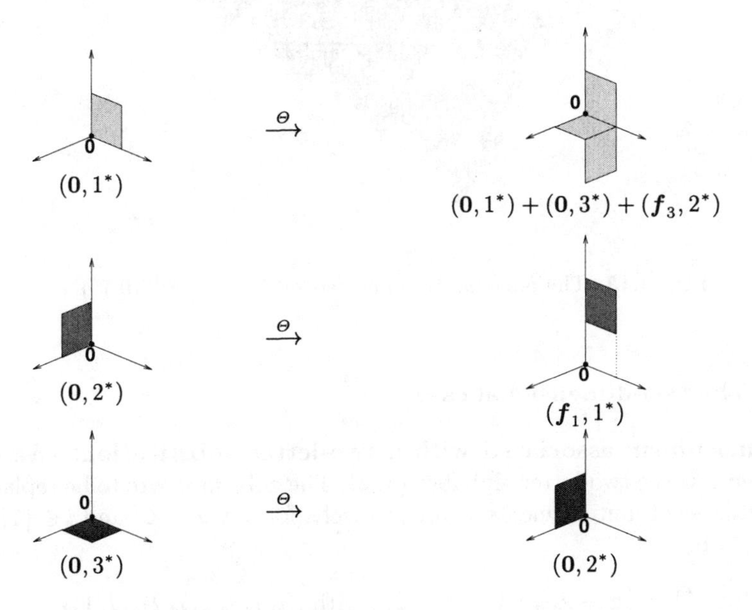

Fig. 8.11. The endomorphism Θ for a substitution which generates a non-simply connected stepped surface.

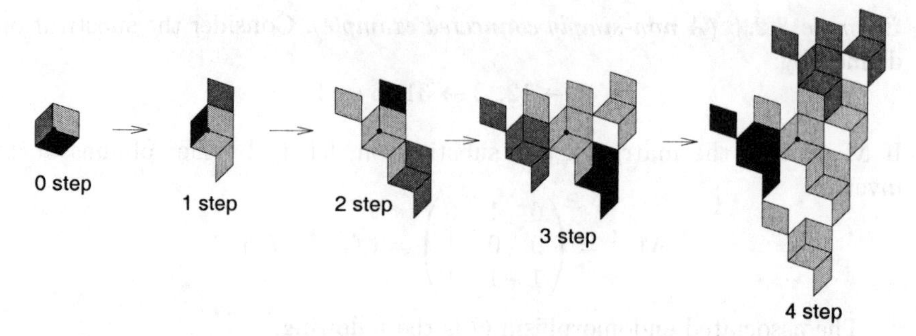

Fig. 8.12. Convergence of $\bigcup_{i=1,\ldots,n} \Theta^n(0,i^*)$ towards a non-simply connected stepped surface.

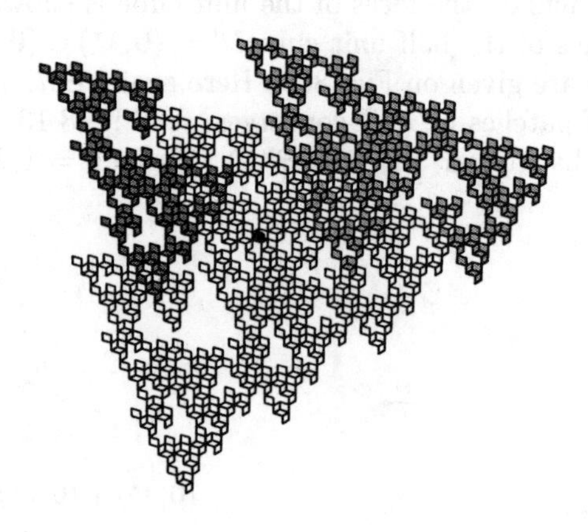

Fig. 8.13. The non-simply connected set $\bigcup_{i=1,2,3} \Theta^{10}(0,i^*)$.

8.2.5 The two-dimensional case

Endomorphism associated with a two-letter substitution. We will consider now the two-letter alphabet $\{1,2\}$. The cube faces are to be replaced by translates of unit segments. More precisely, for any $\boldsymbol{x} \in \mathbb{Z}^2$ and $i \in \{1,2\}$, let $(\boldsymbol{x}, i^*)$ be

$$(\boldsymbol{x}, i^*) = \{\boldsymbol{x} + \lambda \boldsymbol{e}_j \mid 0 \leq \lambda \leq 1\}, \text{ with } (i,j) \in \{(1,2),(2,1)\}.$$

By analogy with the three-dimensional case, Λ denotes the set which contains these segments and $\mathcal{G}$ the $\mathbb{Z}$-module generated by the translated segments:

$$\Lambda = \{(\boldsymbol{x}, i^*) | \ \boldsymbol{x} \in \mathbb{Z}^2, \ i \in \{1,2\}\};$$

$$\mathcal{G} = \left\{ \sum_{\lambda \in \Lambda} m_\lambda \lambda \mid m_\lambda \in \boldsymbol{Z},\ \#\{\lambda \mid m_\lambda \neq 0\} < +\infty \right\}.$$

The *endomorphism of $\mathcal{G}$ associated with σ*, denoted by Θ, is defined as in Definition 8.2.1.

Stepped curve of a line. For any $0 < \gamma < 1$, let l_γ be the line which admits ${}^t(1,\gamma)$ as a normal vector:

$$l_\gamma = \left\{ \boldsymbol{x} \mid \langle \boldsymbol{x},\ {}^t(1,\gamma) \rangle = 0 \right\}.$$

The *stepped curves* $\mathcal{S}^+$ and $\mathcal{S}^-$ associated with the line l_γ are the unions of segments which are the best approximations of the line l_γ.

$$\mathbf{S}^+ := \left\{ (\boldsymbol{x}, i^*) \mid \langle \boldsymbol{x},\ {}^t(1,\gamma) \rangle > 0,\ \langle (\boldsymbol{x} - \boldsymbol{e}_i),\ {}^t(1,\gamma) \rangle \leq 0 \right\},$$

$$\mathbf{S}^- := \left\{ (\boldsymbol{x}, i^*) \mid \langle \boldsymbol{x},\ {}^t(1,\gamma) \rangle \geq 0,\ \langle (\boldsymbol{x} - \boldsymbol{e}_i),\ {}^t(1,\gamma) \rangle < 0 \right\}.$$

We similarly obtain two stepped curves $\mathcal{S}^+$ and $\mathcal{S}^-$ by considering the union of the supports of the elements of $\mathbf{S}^+$ and $\mathbf{S}^-$, and the sets Γ^+ and Γ^- of the finite unions of elements of $\mathbf{S}^+$ and $\mathbf{S}^-$. Then Theorem 8.2.2 holds according to this framework.

Example 8.2.5 (Continued fraction substitutions). Let us recall that the continued fraction substitutions are defined as (see Example 8.1.4):

$$\sigma_a(1) = \overbrace{11\cdots1}^{a\ times}2 \qquad \sigma_a(2) = 1.$$

The endomorphism Θ_a associated with σ_a is the following:

$$\Theta_a : \begin{array}{l} (\mathbf{0}, 1^*) \longrightarrow (\mathbf{0}, 2^*) + \sum_{k=1}^{a} ((\boldsymbol{e}_1 - k\boldsymbol{e}_2), 1^*) \\ (\mathbf{0}, 2^*) \longrightarrow (\mathbf{0}, 1^*) \end{array}.$$

Figure 8.14 illustrates how Θ_a acts on the unit segments $(\mathbf{0}, 1^*)$ and $(\mathbf{0}, 2^*)$. In the special case $a = 1$, Fig. 8.15 illustrates the iteration of Θ_1 on the unit half square $(\mathbf{0}, 1^*) \cup (\mathbf{0}, 2^*)$.

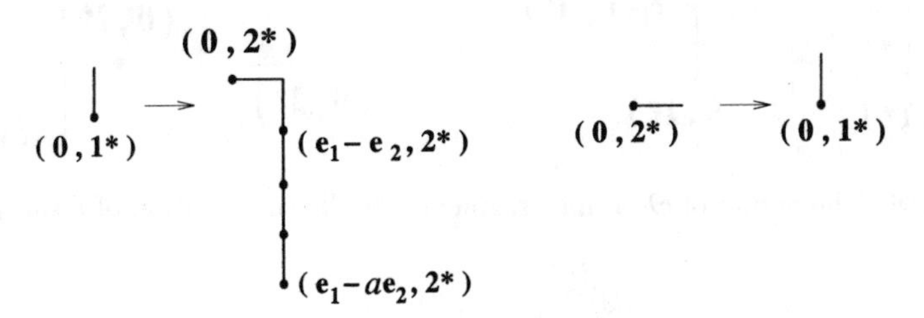

Fig. 8.14. The action of Θ on unit segments for the continued fraction substitutions.

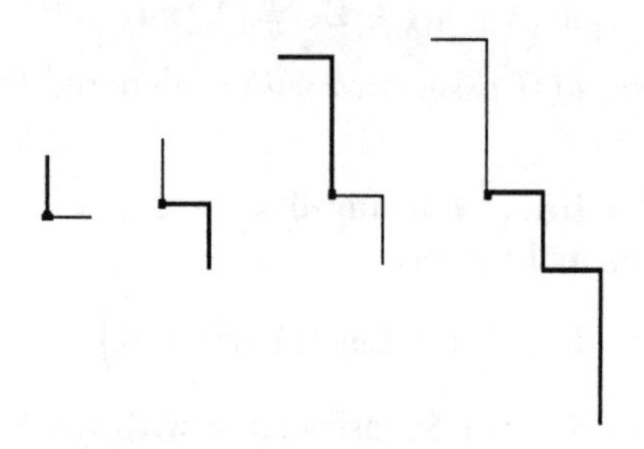

Fig. 8.15. $\bigcup_{i=1,2} {\Theta_1}^n(0, i^*)$, $n = 0, 1, 2, 3$.

Example 8.2.6. Consider the square of the Fibonacci substitution, that is:

$$1 \mapsto 121, \quad 2 \mapsto 12.$$

Let $\mathbf{M}$ denote the incidence matrix of this substitution, and let $\boldsymbol{f}_1, \boldsymbol{f}_2$ be the column vectors of its inverse:

$$\mathbf{M}^{-1} = \begin{pmatrix} 1 & -1 \\ -1 & 2 \end{pmatrix} = (\boldsymbol{f}_1, \boldsymbol{f}_2).$$

Then the endomorphism Θ associated with the substitution σ is the following:

$$\Theta : \begin{array}{l} (\mathbf{0}, 1^*) \longrightarrow (\mathbf{0}, 1^*) + (\boldsymbol{f}_1 + \boldsymbol{f}_2, 1^*) + (\boldsymbol{f}_2, 2^*) \\ (\mathbf{0}, 2^*) \longrightarrow (\boldsymbol{f}_1, 1^*) + (\mathbf{0}, 2^*) \end{array}.$$

Figure 8.16 illustrates how Θ acts on the unit segments $(\mathbf{0}, 1^*)$ and $(\mathbf{0}, 2^*)$ and Fig. 8.17 illustrates the iteration of Θ on the union $(\mathbf{0}, 1^*) \cup (\mathbf{0}, 2^*)$.

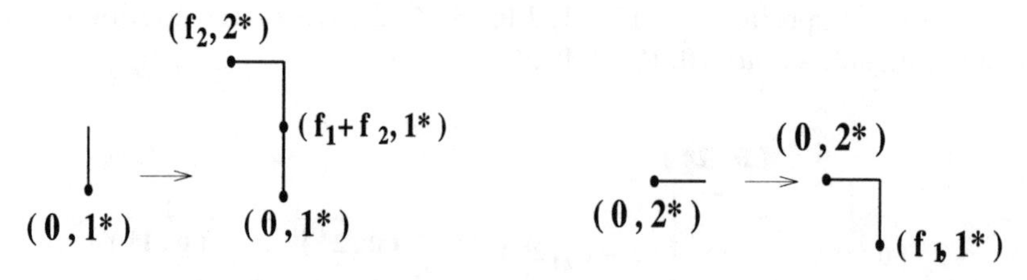

Fig. 8.16. The action of Θ on unit segments for the substitution of Example 8.2.6.

Simple connectedness. On Example 8.1.4 and Example 8.2.6, one can remark that the sets $\Theta^n(\mathbf{0}, i^*)$, $i = 1, 2$, are simply connected. This property does not hold any more in the general case as illustrated by the following example.

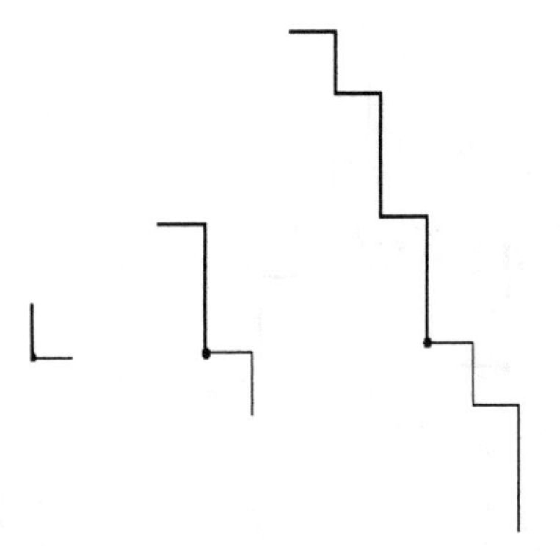

Fig. 8.17. The set $\bigcup_{i=1,2} \Theta^n(0, i^*), n = 0, 1, 2.$

Example 8.2.7 (A non-simply connected example). Consider the substitution defined by

$$1 \mapsto 112, \quad 2 \mapsto 21.$$

Note that this substitution has the same incidence matrix as the substitution in the previous example. We have just performed an exchange in the order of the letters, which induces non-simply connectedness.

Let $\boldsymbol{f}_1$, $\boldsymbol{f}_2$ still denote the column vectors of the inverse of the incidence matrix. The endomorphism Θ is given by

$$\Theta : \begin{array}{l} (\mathbf{0}, 1^*) \longrightarrow (\mathbf{0}, 2^*) + (\boldsymbol{f}_2, 1^*) + (\boldsymbol{f}_1 + \boldsymbol{f}_2, 1^*) \\ (\mathbf{0}, 2^*) \longrightarrow (\mathbf{0}, 1^*) + (\boldsymbol{f}_1, 2^*). \end{array}$$

The action Θ on unit segments is described in Fig. 8.18. The iteration of Θ on $(\mathbf{0}, 1^*) \cup (\mathbf{0}, 2^*)$ is illustrated in Fig. 8.19, where one can see that the sets $\Theta^n(\mathbf{0}, i^*)$, $i = 1, 2$, are not simply connected.

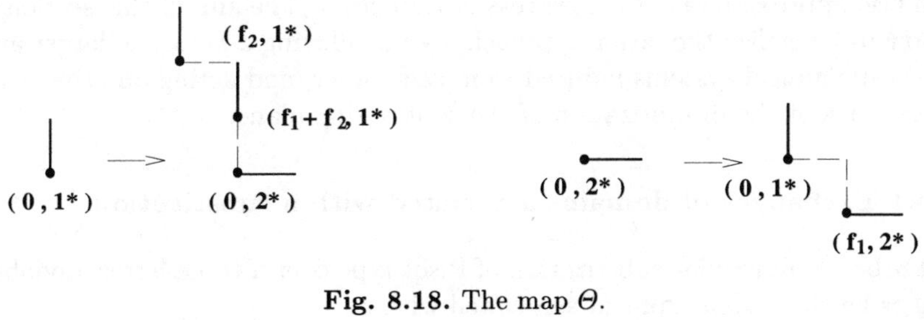

(f2, 1*)
(f1 + f2, 1*)
(0, 1*) —> (0, 2*)
(0, 2*) —> (0, 1*)
(f1, 2*)

Fig. 8.18. The map Θ.

In the case of two-letter substitutions, the simple connectedness property can be characterized (see also Chap. 9):

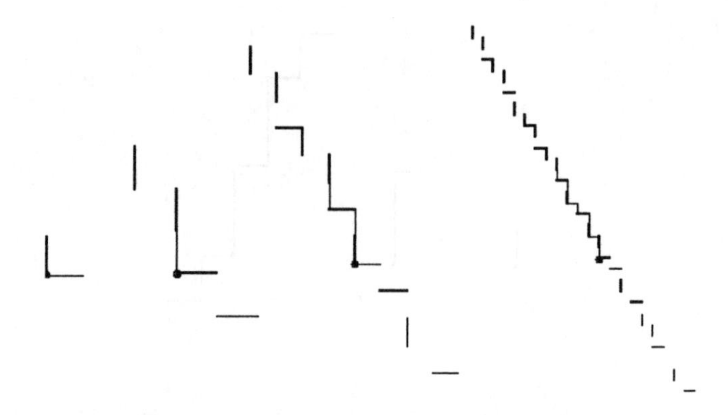

Fig. 8.19. The set $\bigcup_{i=1,2} \Theta^n(0, i^*), n = 0, 1, 2, 3$.

Theorem 8.2.8 (**[449, 157]**). *The sets $\Theta^n(0, i^*)$, $i = 1, 2$ are simply connected for every integer n if and only if the substitution σ is an* invertible substitution, *that is, if there exists an automorphism $\theta : \Gamma_2 \to \Gamma_2$ of the free group Γ_2 of rank 2 generated by $\{1, 2\}$, such that*

$$\sigma \circ \theta = \theta \circ \sigma = Id.$$

8.3 Dynamical systems associated with the stepped surface

There are two different methods of constructing the generalized Rauzy fractal $\mathcal{R}_\sigma$ associated with the substitution σ. First, one can use convergent series associated with the substitution, as is done in the seminal paper of Rauzy (this is also the approach developed in [101, 102], see Chap. 7). Secondly, one can also approximate it by a process of exduction. The aim of this section is to try to formalize the latter approach, by introducing a family of larger and larger dynamical systems induced from each other, and acting on arbitrarily large parts of the discretization of the contracting plane.

8.3.1 Exchanges of domains associated with a substitution

Let σ be a unimodular substitution of Pisot type over a three-letter alphabet and Θ be the endomorphism associated with σ.

The aim of this section is to construct a sequence of domains D_n which tile the contracting plane $\mathcal{P}_{\gamma,\delta}$ and a sequence of dynamical systems W_n on D_n. The domains D_n are obtained by iterating Θ; we thus obtain a sequence of larger and larger pieces of the contracting plane.

Some special domains in the contracting plane. For any letter $i = 1, 2, 3$, we define the sets $D_n^{(i)}$ and $D_n^{(i)'}$ as the projections on the contracting plane $\mathcal{P}_{\gamma,\delta}$, of the n-fold iterate of Θ of the two unit basic squares orthogonal to the vector $\boldsymbol{e}_i$, i.e., $(\boldsymbol{e}_i, i^*)$ and $(\mathbf{0}, i^*)$:

Definition 8.3.1. *Let*

$$D_n^{(i)} := \pi\left(\Theta^n(\boldsymbol{e}_i, i^*)\right), \text{ and } D_n^{(i)'} := \pi\left(\Theta^n(0, i^*)\right), \text{ for } i = 1, 2, 3.$$

We denote by D_n and D'_n the union of these domains. Thus, if $\mathcal{U}$ and $\mathcal{U}'$ are the above and below half unit cubes, then we have

$$D_n := \pi\left(\Theta^n(\mathcal{U})\right), \quad D'_n := \pi\left(\Theta^n(\mathcal{U}')\right).$$

Note that these unions are disjoint up to a set of zero measure and that the sets D_n and D'_n are subsets of the contracting plane $\mathcal{P}_{\gamma,\delta}$.

Inclusion relations between these domains. By definition, we have $\mathcal{U} \in \Gamma^+_{\alpha,\beta}$ and $\mathcal{U}' \in \Gamma^-_{\alpha,\beta}$. Thus, Theorem 8.2.2 and Lemma 8.2.3 imply that

$$\Theta^n(\mathcal{U}) \succ \Theta^{n-1}(\mathcal{U}) \quad \text{and} \quad \Theta^n(\mathcal{U}') \succ \Theta^{n-1}(\mathcal{U}').$$

Therefore, for every n, the domain D_n contains the domain D_{n-1}, and D'_n contains D'_{n-1}.

One can remark that $\pi(\mathcal{U})$ and $\pi(\mathcal{U}')$ are just the above and below projection of the unit cube on a plane which cross this cube, which implies that they are equal. We deduce from this (see [43, 44] for more details), that we have more generally:

$$D_n = D'_n.$$

This implies that we can partition (up to a set of zero measure) D_n either by the sets $D_n^{(i)}$, for $i = 1, 2, 3$, or by the sets ${D'}_n^{(i)}$, for $i = 1, 2, 3$. Furthermore, each part ${D'}_n^{(i)}$ of D_n is the image under a translation by a specified vector of the part $D_n^{(i)}$ of D_n. In dynamical terms, next theorem shows how to perform on D_n an exchange of domains, the domains being the sets $D_n^{(i)}$, for $i = 1, 2, 3$.

Theorem 8.3.2. *Let σ be a unimodular substitution of Pisot type such that $\sigma(1)$ begins with 1. Let $\boldsymbol{f}_1^{(n)}$, $\boldsymbol{f}_2^{(n)}$ and $\boldsymbol{f}_3^{(n)}$ be the column vectors of the incidence matrix $\mathbf{M}_\sigma^{-n}$, i.e., $\mathbf{M}_\sigma^{-n} = \left(\boldsymbol{f}_1^{(n)}, \boldsymbol{f}_2^{(n)}, \boldsymbol{f}_3^{(n)}\right)$.*

Then the following dynamical system W_n on D_n is well-defined (up to a set of measure 0):

$$\begin{aligned} D_n &\xrightarrow{W_n} D_n \\ \boldsymbol{x} &\mapsto \boldsymbol{x} - \pi(\boldsymbol{f}_i^{(n)}), \quad \text{if} \quad \boldsymbol{x} \in D_n^{(i)}, \ i = 1, 2, 3. \end{aligned}$$

All the dynamical systems (D_n, W_n), for $n \geq 0$, are isomorphic.

For a proof of this theorem, see [43].

Consider for instance the **Rauzy substitution**. Fig. 8.20 illustrates the successive exchange of pieces W_n over the sets D_n, for $n = 0, 1, 2, 3$. Each vertical arrow shows the action of Θ: the set D_0 goes onto D_1, then onto D_2, and so on. The result of the action of the domain exchange W_n on the right domain D_n divided into its three parts $D_n^{(i)}$, is shown on the domain D_n which is on the left handside of the arrow. Note that for $i = 1, 2, 3$, $D_n^{(i)}$ is mapped onto $D_n^{(i)'}$.

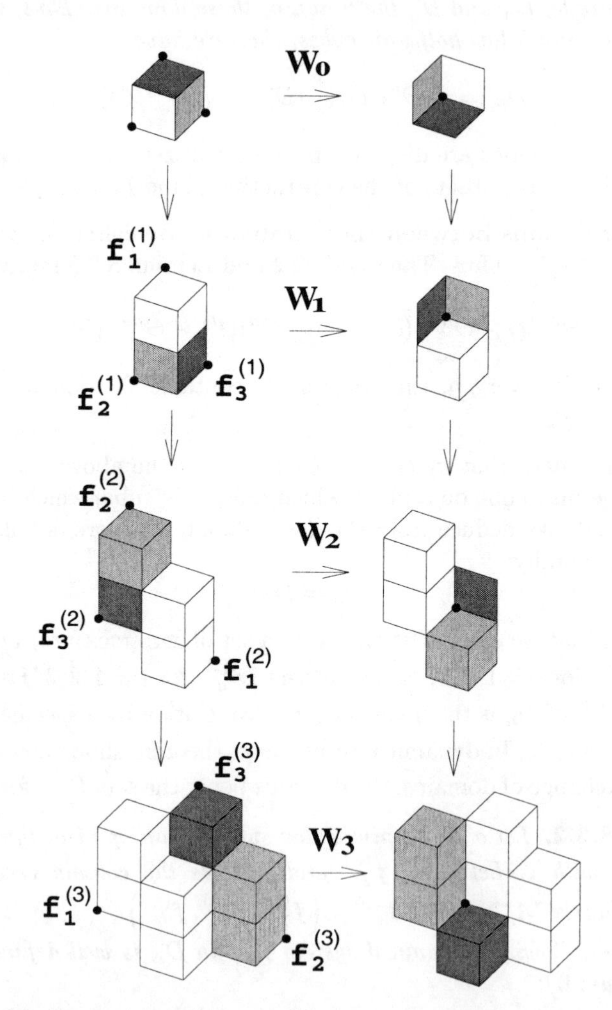

Fig. 8.20. Successive exchanges of domains for the Rauzy substitution.

Figure 8.21 gives the equivalent illustrations for the **modified Jacobi-Perron substitution** $\sigma_{(1,0)}$.

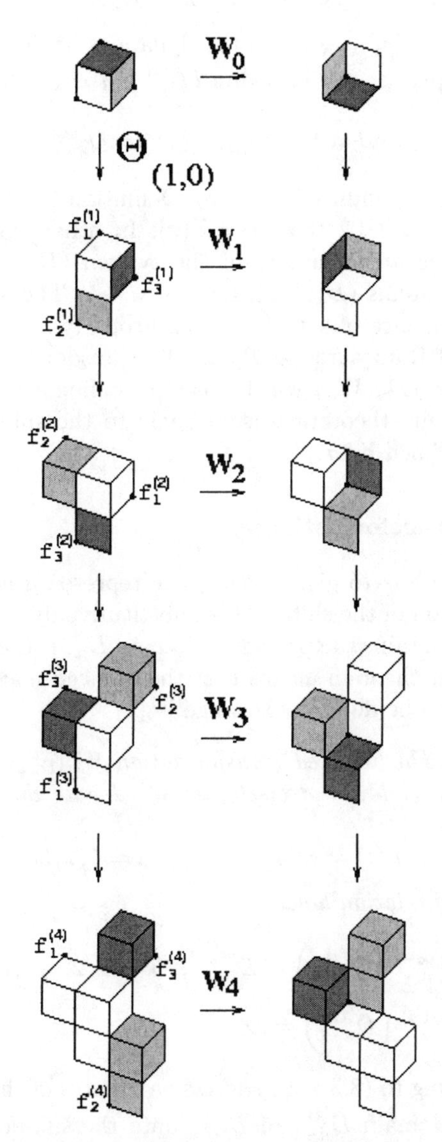

Fig. 8.21. Successive exchanges of domains for a modified Jacobi-Perron substitution.

Since each domain D_n contains the preceding domain D_{n-1}, the map W_n is defined on D_{n-1}. Therefore, with every point x in D_n, one associates two sequences with values in $\{1, 2, 3\}$. The first one $u = (u_k)_{k \in \mathbb{N}}$ codes the dynamics of the map W_n with respect to the partition $\{D_n^{(i)},\ i = 1,2,3\}$ as follows

$$\forall k \in \mathbb{N},\ W_n^k(x) \in D_n^{(u_k)}.$$

The second one $v = (v_k)_{k\in\mathbb{N}}$ codes the dynamics of the induced map of W_n on D_{n-1} with respect to the partition $\{D_{n-1}^{(i)},\ i = 1,2,3\}$:

$$\forall k \in \mathbb{N},\ (W_n|D_{n-1})^k \in D_{n-1}^{(v_k)}.$$

(for a definition of the induced map, see Definition 5.4.7).

We prove in Sec. 8.3.2 that $u = \sigma(v)$. In other words, the dynamical system (D_n, W_n) is an exduction of the system (D_{n-1}, W_{n-1}), and more generally of the systems (D_k, W_k), for $k \le n-1$. The aim of Sec. 8.4 is to prove that the sequence of sets D_n, when properly renormalized, converges to the generalized Rauzy fractal $\mathcal{R}_\sigma$ in the Hausdorff topology. Hence, the dynamical systems (D_n, W_n) will be used to define a dynamical system on $\mathcal{R}_\sigma$ which is measure-theoretical isomorphic to the substitution dynamical system (X_σ, S) defined by σ.

8.3.2 Induced transformations

Our motivation here is to give a geometric representation by means of the sets D_n of the action of the shift on the substitutive dynamical system. Since each domain D_n contains the preceding domain D_{n-1}, the map W_n is defined on D_{n-1}. The next theorem means that the induced map of W_n on D_{n-1} is nothing other than the domain exchange W_{n-1}.

Theorem 8.3.3. *The induced transformation $W_n|_{D_{n-1}}$ of W_n into D_{n-1} coincides with W_{n-1}. More precisely, let $s_k^{(i)}$ denote the letters which occur in $\sigma(i)$:*

$$\sigma(i) = s_1^{(i)} s_2^{(i)} \cdots s_{l(i)}^{(i)},\quad i = 1,2,3.$$

Then the following relation holds:

$$W_n^{k-1}\left(D_{n-1}^{(i)}\right) \subset D_n^{(s_k^{(i)})},\quad k = 1,2,\cdots,l(i), \tag{8.3}$$

$$W_n^{l(i)}\left(D_{n-1}^{(i)}\right) = D_{n-1}^{(i)'}. \tag{8.4}$$

Indeed, according to (8.3), the successive iterates of the domain exchange W_n map the sub-domain $D_{n-1}^{(i)}$ of D_{n-1} onto the sub-domains $D_n^{(j)}$ of D_n, the index j being a letter of $\sigma(i)$ depending on the index of iteration. All the sub-domains $D_n^{(j)}$ are disjoint from D_{n-1} before performing the $l(i)$-th iteration, ($l(i)$ denotes the length of $\sigma(i)$). Concerning this $l(i)$-th iteration, (8.4) means that $D_{n-1}^{(i)}$ is mapped onto D_{n-1}, and that its image is precisely the set $D_{n-1}^{(i)'}$, that is, $W_{n-1}(D_{n-1}^{(i)})$. Thus, the first return map of W_n on D_{n-1} is exactly the domain exchange W_{n-1}.

An illustration of this theorem for $n = 3$ and $i = 1$ in the case of the **Rauzy substitution** is given in Fig. 8.22: the commutative diagram represents the action of the domain exchanges W_2 and W_3, respectively on the domains D_2 and D_3, the set D_3 containing the set D_2. The pictures on the right hand side of the figure represent the successive images of $D_2^{(1)}$ under the action of W_3. The domain $D_2^{(1)}$ is first included in $D_3^{(1)}$ (because $\sigma(1)$ starts with 1). Then W_3 maps $D_2^{(1)}$ inside $D_3^{(2)}$ (because $\sigma(1)$ continues with 2). And since $\sigma(1)$ is of length 2, W_3^2 maps $D_2^{(1)}$ inside D_2 onto $D_2^{(1)'}$.

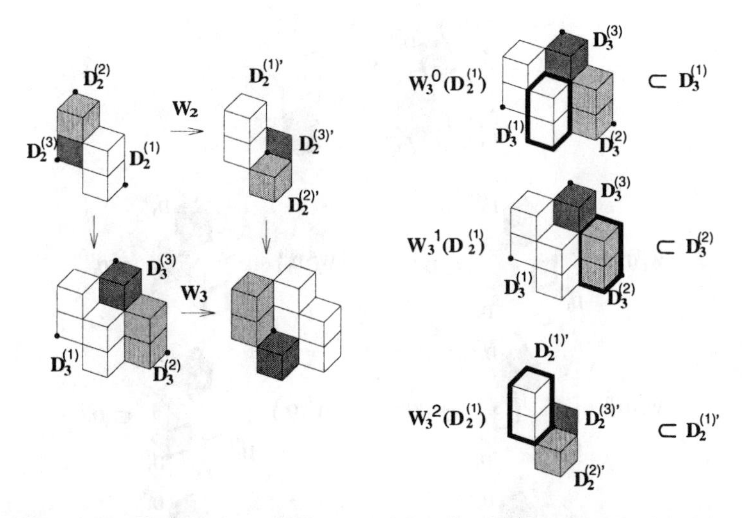

Fig. 8.22. Illustration of Theorem 8.3.3 for the Rauzy substitution.

Since every domain D_n contains the first domain D_0, we deduce from Theorem 8.3.3 the following result.

Corollary 8.3.4. *Let us denote the letters occurring in σ^n by:*

$$\sigma^n(i) = s_1^{(n,i)} s_2^{(n,i)} \cdots s_{l(n,i)}^{(n,i)}, \quad i = 1, 2, 3.$$

Then we have

$$W_n^{k-1}\left(D_0^{(i)}\right) \subset D_n^{\left(s_k^{(n,i)}\right)}, \quad k = 1, 2, \cdots, l(n,i), \tag{8.5}$$

$$W_n^{l(n,i)}\left(D_0^{(i)}\right) = D_0^{(i)'}. \tag{8.6}$$

In particular, we have, where $\mathbf{M}_\sigma^{-n} = \left(\boldsymbol{f}_1^{(n)}, \boldsymbol{f}_2^{(n)}, \boldsymbol{f}_3^{(n)}\right)$,

$$\left\{ W_n^k(\mathbf{0}) \middle| \, k = 1, \cdots, l(n,1) \right\} = \left\{ -\pi \sum_{j=1}^{k} \boldsymbol{f}^{(n)}_{s_j^{(n,1)}} \middle| \; k = 1, \cdots, l(n,1) \right\}. \tag{8.7}$$

Equalities (8.5) and (8.6) are illustrated in Fig. 8.23 for the **Rauzy substitution** in the case $n = 3$ and $i = 1$. Indeed, on the top of this figure one finds the action of the domain exchange W_3 on D_3. Then are drawn the successive images of $D_0^{(1)}$ through the iterations of W_3. These images are successively included in the sets $D_3^{(j)}$, where j follows the sequence $\sigma^3(1) = 1213121$. In this way, these iterations of W_3 on $D_0^{(1)}$ visit successively all the projections of the basic squares of type 1^* that D_3 contains. On the 7-th iteration $(7 = |\sigma^3(1)|)$, $D_0^{(1)}$ comes back into D_0, and more precisely onto $D_0^{(1)'}$.

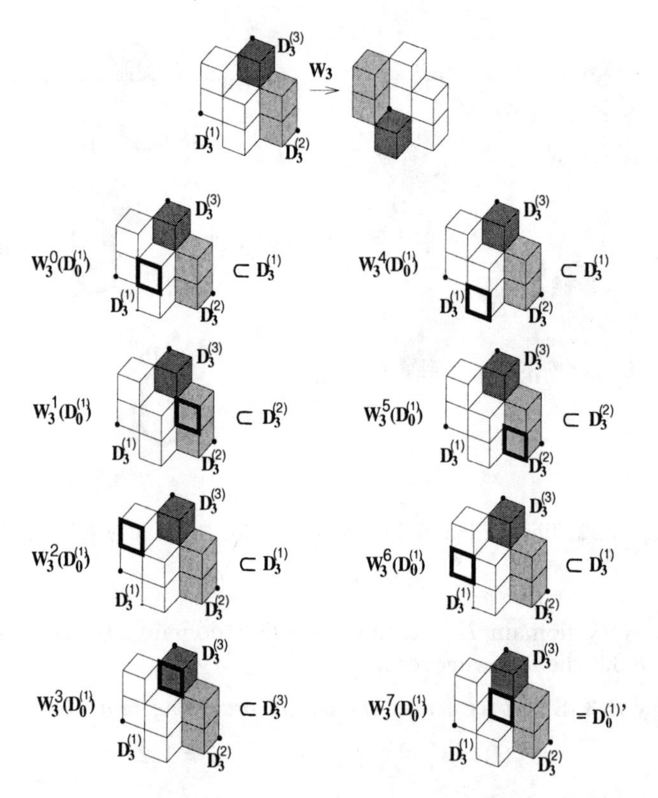

Fig. 8.23. Action of the iterates of W_3 on $D_0^{(1)}$ for the Rauzy substitution.

Equality (8.7) is illustrated in Fig. 8.24 for the Rauzy substitution where

$$\sigma^3(1) = 1213121 \text{ and } \left(\boldsymbol{f}_1^{(3)}, \boldsymbol{f}_2^{(3)}, \boldsymbol{f}_3^{(3)}\right) = \begin{pmatrix} 1 & -1 & -1 \\ -1 & 2 & 0 \\ 0 & -1 & 2 \end{pmatrix}.$$

Rotations. One can prove that the subset D_0 of the contracting plane $\mathcal{P}_{\gamma,\delta}$ tiles periodically this plane under the action of the lattice $\boldsymbol{L}_0 = \{n\pi(\boldsymbol{e}_2 - \boldsymbol{e}_1) + m\pi(\boldsymbol{e}_3 - \boldsymbol{e}_1) \mid m, n \in \mathbb{Z}\}$ (see Fig. 8.26).

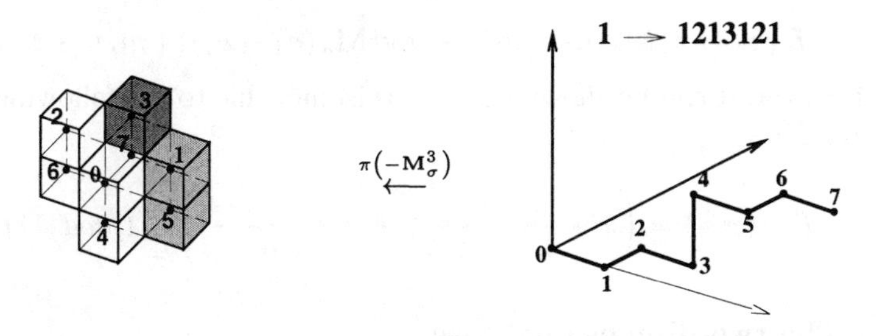

Fig. 8.24. An illustration of (8.7) for the Rauzy substitution.

Furthermore the domain exchange transformation $W_0 : D_0 \to D_0$ (Fig. 8.25) coincides with the following rotation on the fundamental domain D_0 (Fig. 8.26):

$$W_0 : \begin{array}{l} D_0 \longrightarrow D_0 \\ \boldsymbol{x} \mapsto \boldsymbol{x} - \pi(\boldsymbol{e}_1) \quad (\text{mod } \boldsymbol{L}_0). \end{array}$$

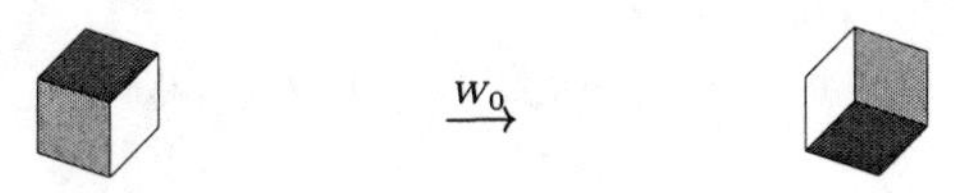

Fig. 8.25. Action of W_0.

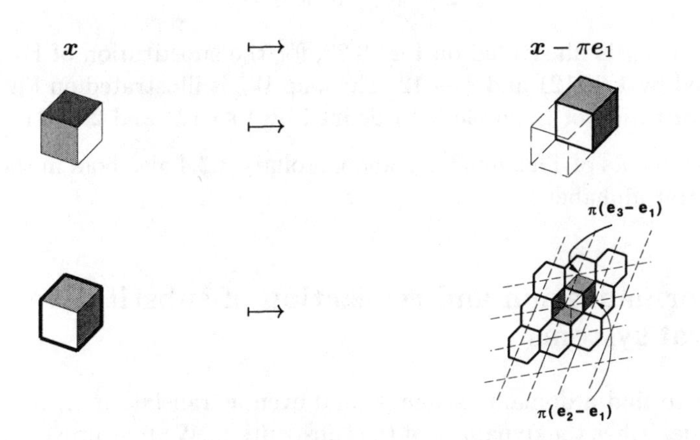

Fig. 8.26. Illustration of the rotation.

The maps W_n similarly perform rotations on the fundamental domains D_n, for every $n \in \mathbb{N}$, the corresponding lattices being easily deduced from $\boldsymbol{L}_0$:

$$\boldsymbol{L}_n = \{n\pi(\mathbf{M}_\sigma(\boldsymbol{e}_2 - \boldsymbol{e}_1)) + m\pi(\mathbf{M}_\sigma(\boldsymbol{e}_3 - \boldsymbol{e}_1)) \mid m, n \in \mathbb{Z}\}.$$

Moreover, it can be shown that W_0 is isomorphic to the following rotation $T_{\alpha,\beta}$ on $\mathbb{T}^2$:

$$T_{\alpha,\beta}(x,y) = \left(x + \frac{\alpha}{1+\alpha+\beta}, x + \frac{\beta}{1+\alpha+\beta}\right) \quad (mod \ \ 1).$$

8.3.3 The two-dimensional case

For two-letter substitutions, let us introduce analogously the unions of intervals on the line l_γ defined by:

$$D_n^{(i)} := \pi\left(\Theta^n(\boldsymbol{e}_i, i^*)\right), \qquad D_n^{(i)'} := \pi\left(\Theta^n(\boldsymbol{0}, i^*)\right), \ \ i = 1, 2.$$

$$D_n := \pi\left(\Theta^n\left(\textstyle\sum_{i=1,2}(\boldsymbol{e}_i, i^*)\right)\right), \qquad D'_n := \pi\left(\Theta^n\left(\textstyle\sum_{i=1,2}(\boldsymbol{0}, i^*)\right)\right).$$

Then the following "union of intervals" exchange on D_n is well defined:

$$\begin{array}{ccl} D_n & \xrightarrow{W_n} & D_n \\ \boldsymbol{x} & \mapsto & \boldsymbol{x} - \pi \boldsymbol{f}_i^{(n)}, \quad \text{if } \boldsymbol{x} \in D_n^{(i)}. \end{array}$$

This map is isomorphic to the interval exchange (where $\mathbf{M}_\sigma^{-n} = \left(\boldsymbol{f}_1^{(n)}, \boldsymbol{f}_2^{(n)}\right)$):

$$\begin{array}{ccl} D_0 & \xrightarrow{W_0} & D_0 \\ \boldsymbol{x} & \mapsto & \boldsymbol{x} - \pi \boldsymbol{e}_i \quad \text{if } \boldsymbol{x} \in D_0^{(i)}. \end{array}$$

The map W_n is illustrated on Fig. 8.27, for the substitution of Example 8.2.6 defined by $1 \mapsto 121$ and $2 \mapsto 12$. The map W_n is illustrated on Fig. 8.28 for the substitution of Example 8.2.7 defined by $1 \mapsto 121$ and $2 \mapsto 21$.

The statements of Theorem 8.3.3 and Corollary 8.3.4 also hold in the case of a two-letter alphabet.

8.4 Renormalization and realization of substitutive dynamical systems

Our aim is to find a domain exchange (and even a translation on the torus $\mathbb{T}^2$) which describes the dynamics of the substitution. We first prove that for unimodular substitutions of Pisot type, and under suitable conditions, the sequence of sets D_n, when properly renormalized, converges to a compact space in the Hausdorff topology, which is exactly the generalized Rauzy fractal $\mathcal{R}_\sigma$ associated with the substitution σ. We then equip this limit set with two transformations producing a realization of the substitutive dynamical system.

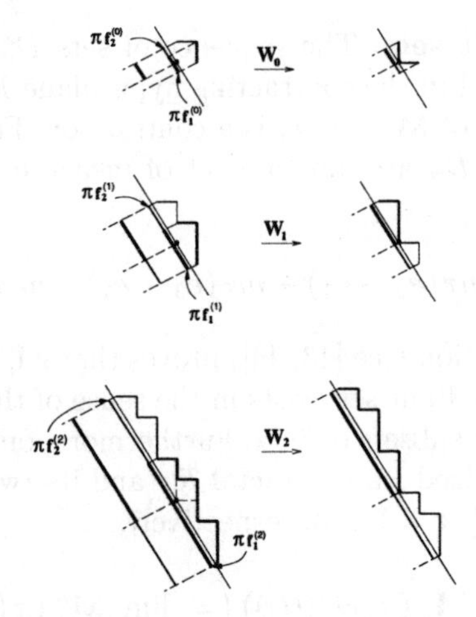

Fig. 8.27. The maps W_0, W_1 and W_2 for the substitution $1 \mapsto 121$, $2 \mapsto 12$.

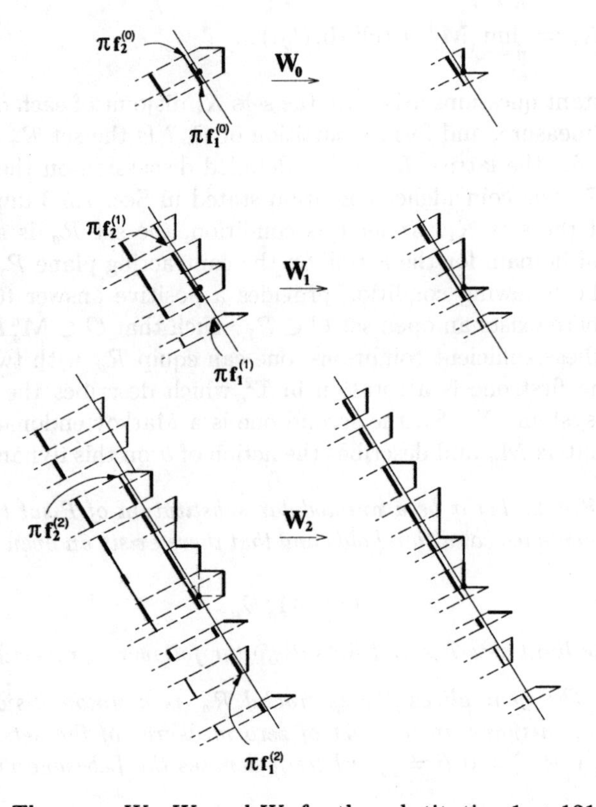

Fig. 8.28. The maps W_0, W_1 and W_2 for the substitution $1 \mapsto 121$, $2 \mapsto 21$.

Renormalized limit sets. The sequence of sets D_n is increasing. But all these sets are included in the contracting hyperplane $\mathcal{P}_{\delta,\gamma}$ of the matrix $\mathbf{M}_\sigma$. Therefore, the action of $\mathbf{M}_\sigma$ on D_n is a contraction. Furthermore, the renormalized domains $\mathbf{M}_\sigma^n D_n$ are, up to a set of measure zero, fundamental domains for the lattice

$$\boldsymbol{L}_0 = \{n\pi(\boldsymbol{e}_2 - \boldsymbol{e}_1) + m\pi(\boldsymbol{e}_3 - \boldsymbol{e}_1) \mid m, n \in \mathbb{Z}\}.$$

A simple computation (see [43, 44]) proves that $\mathbf{M}_\sigma^n$ contracts the sets D_n in such a way that the limit set exists in the sense of the Hausdorff metric on the family of compact subsets of $\mathcal{P}_{\gamma,\delta}$. Furthermore, one can prove that these sets equal the generalized Rauzy fractal $\mathcal{R}_\sigma$ and its two decompositions into $\mathcal{R}_i$, $i = 1, 2, 3$, and $\mathcal{R}'_i$, $i = 1, 2, 3$, respectively.

$$\begin{aligned}
\mathcal{R}_\sigma &:= \lim_{n\to\infty} \mathbf{M}_\sigma^n \left(\pi\left(\Theta^n(\mathcal{U})\right)\right) \left(= \lim_{n\to\infty} \mathbf{M}_\sigma^n \left(\pi\left(\Theta^n(\mathcal{U}')\right)\right)\right), \\
\mathcal{R}_i &:= \lim_{n\to\infty} \mathbf{M}_\sigma^n \left(\pi\left(\Theta^n(\boldsymbol{e}_i, i^*)\right)\right), \\
\mathcal{R}'_i &:= \lim_{n\to\infty} \mathbf{M}_\sigma^n \left(\pi\left(\Theta^n(\mathbf{0}, i^*)\right)\right).
\end{aligned}$$

Two important questions arise: are the sets $\mathcal{R}_i$ disjoint of each other, up to a set of zero measure, and form a partition of $\mathcal{R}_\sigma$? Is the set $\mathcal{R}_\sigma$ a fundamental domain for the lattice $\boldsymbol{L}_0$? For a detailed discussion on these problems, see Chap. 7. The coincidence condition stated in Sec. 7.5.3 implies the disjointness of the sets $\mathcal{R}_i$. Under this condition, the set $\mathcal{R}_\sigma$ is a measurable fundamental domain for the action on the contracting plane $\mathcal{P}_{\gamma,\delta}$ of the lattice $\boldsymbol{L}_0$. The following condition provides a positive answer to the second question: there exists an open set $\mathcal{O} \subset \mathcal{P}_{\gamma,\delta}$ such that $\mathcal{O} \subset \mathbf{M}_\sigma^n D_n$.

Under these sufficient conditions, one can equip $\mathcal{R}_\sigma$ with two dynamical systems: the first one is a rotation in $\mathbb{T}^2$ which describes the shift on the dynamical system (X_σ, S); the second one is a Markov endomorphism with structure matrix $\mathbf{M}_\sigma$ and describes the action of σ on this dynamical system.

Theorem 8.4.1. *Let σ be a unimodular substitution of Pisot type. Suppose that the coincidence condition holds and that there exists an open set $\mathcal{O} \subset \mathcal{P}_{\gamma,\delta}$ such that*

$$\mathcal{O} \subset \mathbf{M}_\sigma^n D_n.$$

Then the limit sets $\mathcal{R}_\sigma$ and $\mathcal{R}_i$ satisfy the following properties:

(1) *The generalized Rauzy fractal $\mathcal{R}_\sigma$ is a union disjoint in measure (i.e., disjoint up to a set of zero measure) of the sets $\mathcal{R}_i$, that is, $\mu(\mathcal{R}_i \bigcap \mathcal{R}_j) = 0$ $(i \neq j)$, where μ denotes the Lebesgue measure:*

$$\mathcal{R}_\sigma = \bigcup_{i=1,2,3} \mathcal{R}_i \quad \textit{(disjoint in measure)}.$$

The set $\mathcal{R}_\sigma$ *generates a periodic tiling of the contracting plane* $\mathcal{P}_{\gamma,\delta}$*, for the lattice* $\boldsymbol{L}_0 = \{n\pi(\boldsymbol{e}_2 - \boldsymbol{e}_1) + m\pi(\boldsymbol{e}_3 - \boldsymbol{e}_1) \mid m, n \in \mathbb{Z}\}$*. More precisely,* $\mu(\ (R_\sigma + \boldsymbol{z}) \bigcap (R_\sigma + \boldsymbol{z}')\ ) = 0$ *for every* $\boldsymbol{z} \neq \boldsymbol{z}' \in \boldsymbol{L}_0$*:*

$$\mathcal{P}_{\gamma,\delta} = \bigcup_{\boldsymbol{z} \in \boldsymbol{L}_0} (\mathcal{R}_\sigma + \boldsymbol{z}) \quad \textit{(disjoint in measure)} .$$

(2) *The following transformation*

$$W : \mathcal{R}_\sigma \to \mathcal{R}_\sigma,\ \boldsymbol{x} \mapsto \boldsymbol{x} - \pi(\boldsymbol{e}_i),\ \textit{if} \quad \boldsymbol{x} \in \mathcal{R}_i,$$

is well-defined and is isomorphic to $W_0 : D_0 \to D_0$.

(3) *The induced transformation* $W|_{\mathbf{M}_\sigma \mathcal{R}_\sigma}$ *is isomorphic to* W *and satisfies*

$$W^{k-1}(\mathcal{R}_i^{(1)}) \subset \mathcal{R}_{s_k^{(i)}}, \quad k = 1, 2, \cdots, l(i),$$
$$W^{l(i)}(\mathcal{R}_i^{(1)}) = \mathcal{R}_i^{(1)'},$$

where $\sigma(i) = s_1^{(i)} \cdots s_{l(i)}^{(i)}$ *and* $\mathcal{R}_i^{(1)} := \mathbf{M}_\sigma \mathcal{R}_i$ $(\mathcal{R}_i^{(1)'} := \mathbf{M}_\sigma \mathcal{R}_i')$.

(4) *The following transformation* $T : \mathcal{R}_\sigma \longrightarrow \mathcal{R}_\sigma$ *is well defined:*

$$T\boldsymbol{x} = \mathbf{M}_\sigma^{-1}\boldsymbol{x} - \mathbf{M}_\sigma^{-1}(f(W)) + \boldsymbol{e}_j,\ \textit{if} \quad \mathbf{M}_\sigma^{-1}\boldsymbol{x} \in \mathbf{M}_\sigma^{-1}(f(W)) + \mathcal{R}_j,$$

where the word W *is given by the following formula:*

$$\Theta(\mathbf{0}, i^*) := \textstyle\sum_{j=1,2,3} \sum_{W:\sigma(j)=Y \cdot i \cdot W} \left(\mathbf{M}_\sigma^{-1}(f(W)), j^*\right).$$

The transformation T *is a Markov endomorphism, the structure matrix of which is* $\mathbf{M}_\sigma$.

Remark. Theorem 8.4.1 requires the assumption that the sequences of renormalized approximations we build for the domain exchange contains a fixed open set, or equivalently, that for n large enough, D_n contains a disk of large diameter. This sufficient condition guarantees that the substitution system (X_σ, S) is isomorphic to a toral translation. See also Corollary 7.5.20.

In all known examples, the substitution system is in fact measure-theoretically isomorphic to a toral translation. We do not know of any example which does not satisfy the previous assumption. We believe that this assumption holds for any unimodular substitution of Pisot type. However, it is still an open problem, except in the case of a two-letter alphabet where this conjecture has been proved proved recently [54]. For more details, see the discussion in Chap. 7.

The transformation W on the set $\mathcal{R}_\sigma$ is illustrated in Fig. 8.29, for the **Rauzy substitution** of Example 8.1.1. The transformation W on the set $\mathcal{R}_\sigma$ is illustrated in Fig. 8.30, for the **modified Perron-Frobenius substitution** $\sigma_{(0,1)}$ of Example 8.1.2. For the substitution of Example 8.2.4, the transformation W on the non-simply connected set $\mathcal{R}_\sigma$ is illustrated in Fig. 8.31.

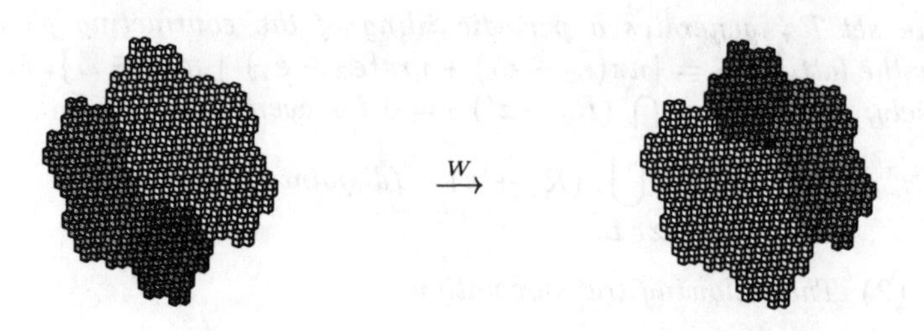

Fig. 8.29. The domain exchange W isomorphic to the dynamical system associated with the Rauzy substitution.

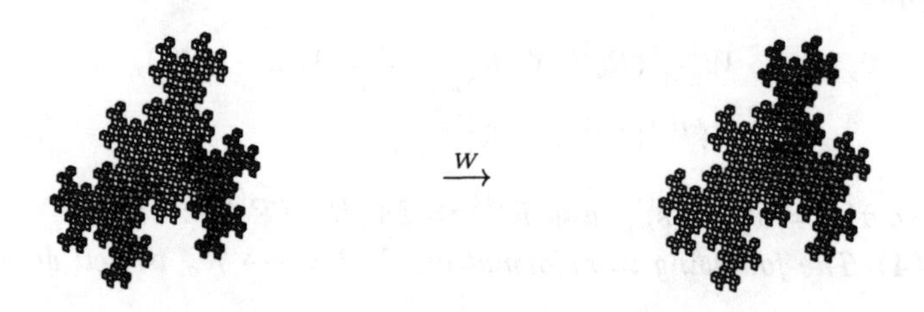

Fig. 8.30. The domain exchange W isomorphic to the dynamical system associated with a modified Jacobi-Perron substitution.

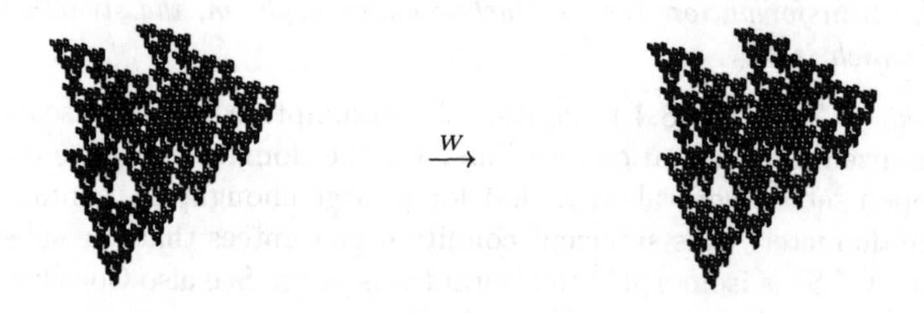

Fig. 8.31. The map W for a non-simply connected domain.

8.5 Fractal boundary

In each example, the boundary of the domain $\mathcal{R}_\sigma$ seems to be fractal. To observe the properties of these boundaries, let us introduce some notation.

Unit intervals. Let us first fix some notation. Let $(x, i) \in \mathbb{Z}^3 \times \{1, 2, 3\}$ be the following intervals, illustrated on Fig. 8.32:

$$(x, i) = \{x + \lambda e_i \mid 0 \leq \lambda \leq 1\}, \quad i = 1, 2, 3;$$

$$y + (x, i) := (y + x, i);$$

$$\forall t, u \in \{1, 2, 3\}, \quad (x, t \wedge u) = \begin{cases} (x, s) & \text{if } (t, u) \in \{(2,3), (3,1), (1,2)\} \\ -(x, s) & \text{if } (t, u) \in \{(3,2), (1,3), (2,1)\} \\ 0 & \text{if } t = u, \end{cases}$$

where $\{s, t, u\} = \{1, 2, 3\}$.

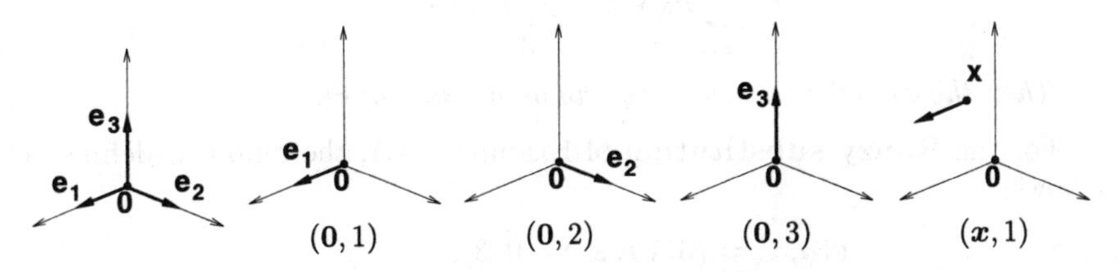

Fig. 8.32. Illustration of $(0, i)$, $i = 1, 2, 3$, and $(x, 1)$.

Boundary map. Let us define the *boundary map* ∂ on $\mathbb{Z}^3 \times \{1, 2, 3\}$ by

$$\partial(x, i^*) := (x, j) + (x + e_j, k) - (x, k) - (x + e_k, j),$$

where $\{i, j, k\} \in \{\{1, 2, 3\}, \{2, 3, 1\}, \{3, 1, 2\}\}$.

This map extends in a natural way to $\mathcal{G}$ as follows: given $\gamma = \sum_{\lambda \in \Lambda} m_\lambda \lambda \in \mathcal{G}$, we set $\partial\gamma = \sum_{\lambda \in \Lambda} m_\lambda \partial\lambda$.

The set of points $\partial\gamma$, for $\gamma \in \mathcal{G}$, is denoted by $\mathcal{G}_1$. Then $\mathcal{G}_1$ is a $\mathbb{Z}$-module.

Boundary endomorphism of $\mathcal{G}_1$.

Definition 8.5.1. *An endomorphism θ of $\mathcal{G}_1$ is called a* boundary endomorphism *associated with Θ if the following commutative relationship is satisfied:*

$$\begin{array}{ccc} \mathcal{G} & \xrightarrow{\Theta} & \mathcal{G} \\ \partial \downarrow & & \downarrow \partial \\ \mathcal{G}_1 & \xrightarrow{\theta} & \mathcal{G}_1 \end{array}.$$

Roughly speaking, the boundary of the image under Θ of a weighted sum of faces is the image of the boundary of this sum under the map θ, which acts on weighted sums of unit intervals.

The following theorem proves the existence of such a boundary endomorphism.

Theorem 8.5.2 ([44]). *Let σ be a Pisot substitution, and write $\sigma(i) = s_1^{(i)} \cdots s_{l(i)}^{(i)}$, $i = 1, 2, 3$.*

Let θ be defined on the unit intervals $(0, i)$ by

$$\theta(0, i) := \sum_{\substack{1 \le t \le 3 \\ 1 \le u \le 3}} \sum_{\substack{s_l^{(t)} = j \\ s_m^{(u)} = k}} \left(\mathbf{M}_\sigma^{-1} \left(f \left(S_l^{(t)} \right) \right) + \mathbf{M}_\sigma^{-1} \left(f \left(S_m^{(u)} \right) \right), t \wedge u \right)$$

where • $\{i,j,k\} \in \{\{1,2,3\},\{2,3,1\},\{3,1,2\}\}$,
• $S_l^{(t)}$ *is the suffix of* $\sigma(t)$ *after the occurrence of* $s_l^{(t)}$, *i.e.,* $\sigma(t) = P_l{}^{(t)} s_l{}^{(t)} S_l{}^{(t)}$.

We extend the definition of the map θ *to* $\mathcal{G}_1$ *as follows:*

$$\theta(\boldsymbol{x}, i) := \mathbf{M}_\sigma^{-1}\boldsymbol{x} + \theta(\mathbf{0}, i),$$
$$\theta(\sum_{\lambda\in\Lambda} n_\lambda \lambda) := \sum_{\lambda\in\Lambda} n_\lambda \theta(\lambda).$$

Then the map θ *is a boundary endomorphism of* Θ.

For the **Rauzy substitution** of Example 8.1.1, the map θ is defined as follows:

$$\begin{aligned}
\theta(\mathbf{0},1) &= (\mathbf{0}, 1\wedge 2) = (\mathbf{0},3),\\
\theta(\mathbf{0},2) &= \left(\mathbf{M}_\sigma{}^{-1}(\boldsymbol{e}_2), 2\wedge 1\right) + (\mathbf{0}, 2\wedge 3)\\
&= -\,([1,0,-1],3) + (\mathbf{0},1),\\
\theta(\mathbf{0},3) &= \left(\mathbf{M}_\sigma{}^{-1}(\boldsymbol{e}_3), 2\wedge 1\right) + (\mathbf{0}, 3\wedge 1)\\
&= -\,([0,1,-1],3) + (\mathbf{0},2).
\end{aligned}$$

The map θ is represented in Fig. 8.33.

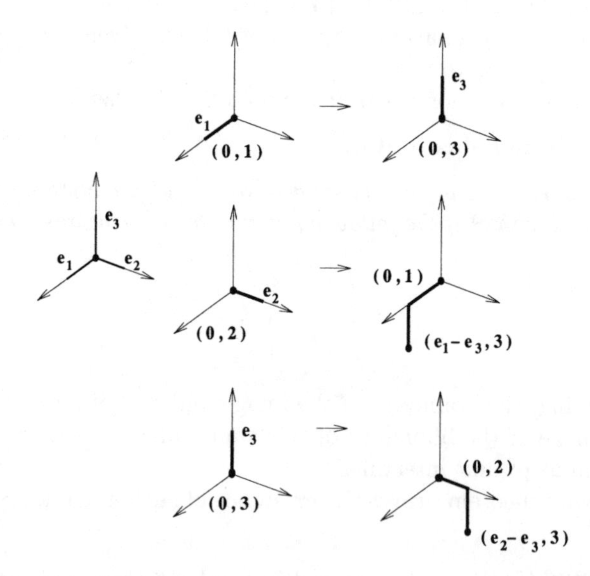

Fig. 8.33. Boundary endomorphism for the Rauzy substitution.

Computation of Haussdorf dimension. If the boundary endomorphism θ satisfies some non-cancellation conditions, then it is possible to compute explicitly the dimension of $\mathcal{R}_\sigma$ (see [221], [223] and [222]).

8.6 Continued fraction expansions and substitutions

The aim of this section is to apply the previous constructions to the modified Jacobi-Perron substitutions and to make explicit the connection with the corresponding simultaneous continued fraction algorithm. We first introduce this algorithm, and then we discuss its natural extension. We conclude by stating the previous results in the one-dimensional case.

More precisely, the idea is to extend what we previously did for one substitution to a family of substitutions (such a composition of substitutions is called S-adic; for more details, see Chap. 12). Instead of iterating always the same substitution, we will iterate modified Jacobi-Perron substitutions according to the natural extension of this algorithm applied to the parameters $(\alpha, \beta, \gamma, \delta)$. Throughout this chapter we have worked with algebraic parameters $(\alpha, \beta, \gamma, \delta)$ associated with a unimodular substitution of Pisot type, to produce a generalized Rauzy fractal $\mathcal{R}_\sigma$ which tiles periodically (under extra assumptions) the contracting plane $\mathcal{P}_{\gamma,\delta}$. We now want to extend these results to any parameters $(\alpha, \beta, \gamma, \delta)$ (not necessarily algebraic), and to associate a set $\mathcal{R}_{\alpha,\beta,\gamma,\delta}$ tiling periodically the plane $\mathcal{P}_{\gamma,\delta}$ with respect to the lattice $\boldsymbol{L}_0$, on which one can perform an exchange of pieces isomorphic to the rotation $T_{\alpha,\beta}$. In other words, we are giving a symbolic interpretation of the natural extension of the modified Jacobi-Perron algorithm. Let us note that we work here with the modified Jacobi-Perron algorithm mainly for two reasons: the first one is that we know a simple realization of its natural extension; the second one is that it is almost everywhere strongly convergent with exponential rate. This will the keypoint of the proof of Theorem 8.6.2 below.

This approach is directly inspired by the Sturmian case (see Chap. 6) and is motivated by the following idea: we thus obtain a combinatorial description via generalized substitutions of toral rotations. Some applications to Diophantine approximation are mentioned in the rest of this chapter.

8.6.1 The modified Jacobi-Perron algorithm

The classical Jacobi-Perron algorithm is an example of a multi-dimensional generalization of the regular continued fraction expansion, as an attempt to characterize cubic irrationals. This algorithm generates a sequence of simultaneous rational approximations of a pair of points with the same denominator (see for instance [64, 90, 378]). The modified Jacobi-Perron algorithm introduced by E. V. Podsypanin in [332] shares this property; this algorithm is a two-point extension of the Brun algorithm. Moreover, both algorithms are shown to be strongly convergent (in the sense of Brentjes [90]) almost everywhere with exponential rate (see also [255]); the Jacobi-Perron case is dealt with in [381] and the modified Jacobi-Perron in [181] (see also for a simpler proof [285]). Both Jacobi-Perron algorithms (the classical one and the modified one) are known to have an invariant ergodic probability measure equivalent to the Lebesgue measure (see for instance [379] and [382]).

However, this measure is not known explicitly in the classical case (the density of the measure is shown to be a piecewise analytical function in [92, 94]), whereas it is known explicitly for the modified case [48, 181].

Let $X = [0,1) \times [0,1)$ and

$$X_0 = \{(\alpha, \beta) \mid \alpha \geq \beta\},$$
$$X_1 = \{(\alpha, \beta) \mid \alpha < \beta\}.$$

Let us define the transformation T on X by

$$T(\alpha,\beta) := \begin{cases} \left(\dfrac{\beta}{\alpha}, \dfrac{1}{\alpha} - \left[\dfrac{1}{\alpha}\right]\right) & \text{if } (\alpha,\beta) \in X_0 - \{(0,0)\}, \\ \left(\dfrac{1}{\beta} - \left[\dfrac{1}{\beta}\right], \dfrac{\alpha}{\beta}\right) & \text{if } (\alpha,\beta) \in X_1, \\ (0,0) & \text{if } \alpha \text{ or } \beta = 0. \end{cases}$$

If $T^n(\alpha, \beta)$ equals zero, for some integer n, then the algorithm stops. By using the integer value functions

$$a(\alpha,\beta) := \begin{cases} \left[\dfrac{1}{\alpha}\right] & \text{if } (\alpha,\beta) \in X_0, \\ \left[\dfrac{1}{\beta}\right] & \text{if } (\alpha,\beta) \in X_1, \end{cases}$$

$$\varepsilon(\alpha,\beta) := \begin{cases} 0 & \text{if } (\alpha,\beta) \in X_0, \\ 1 & \text{if } (\alpha,\beta) \in X_1, \end{cases}$$

we define for each $(\alpha, \beta) \in X - \{(0,0)\}$ a sequence of digits ${}^t(a_n, \varepsilon_n)$ by

$${}^t(a_n, \varepsilon_n) := {}^t\left(a\left(T^{n-1}(\alpha,\beta)\right), \varepsilon\left(T^{n-1}(\alpha,\beta)\right)\right), \text{ if } T^{n-1}(\alpha,\beta) \neq (0,0).$$

We denote

$$(\alpha_n, \beta_n) := T^n(\alpha, \beta).$$

Definition 8.6.1. *The triple* $(X, T; (a(\alpha,\beta), \varepsilon(\alpha,\beta)))$ *is called the* modified Jacobi-Perron algorithm.

8.6.2 Natural extension

Let us introduce a transformation $(\overline{X}, \overline{T})$ called a *natural extension* of the modified Jacobi-Perron algorithm. Roughly speaking, we try to make the map T one-to-one by considering a larger space.

Let $\overline{X} = X \times X$ and $\overline{T}$ be the transformation defined on $\overline{X}$ by

$$\overline{T}(\alpha,\beta,\gamma,\delta) = \begin{cases} \left(\dfrac{\beta}{\alpha}, \dfrac{1}{\alpha} - a_1, \dfrac{\delta}{a_1+\gamma}, \dfrac{1}{a_1+\gamma}\right) & \text{if } (\alpha,\beta) \in X_0 - \{(0,0)\}, \\ \left(\dfrac{1}{\beta} - a_1, \dfrac{\alpha}{\beta}, \dfrac{1}{a_1+\delta}, \dfrac{\gamma}{a_1+\delta}\right) & \text{if } (\alpha,\beta) \in X_1, \\ (0,0,\gamma,\delta) & \text{if } \alpha \text{ or } \beta = 0. \end{cases}$$

The transformation $\overline{T}$ is bijective.

We denote

$$(\alpha_n, \beta_n, \gamma_n, \delta_n) := \overline{T}^n(\alpha, \beta, \gamma, \delta).$$

We will assume $1, \alpha, \beta$ linearly independent, so that the algorithm never stops.

8.6.3 A matricial point of view

For every $a \in \mathbb{N}^+$, let us introduce the following family of matrices:

$$\mathbf{A}_{(a,0)} = \begin{pmatrix} a & 0 & 1 \\ 1 & 0 & 0 \\ 0 & 1 & 0 \end{pmatrix}, \quad \mathbf{A}_{(a,1)} = \begin{pmatrix} a & 1 & 0 \\ 0 & 0 & 1 \\ 1 & 0 & 0 \end{pmatrix}.$$

Then we have the following formulas:

$$\begin{pmatrix} 1 \\ \alpha_n \\ \beta_n \end{pmatrix} = \frac{1}{\theta\theta_1 \cdots \theta_{n-1}} \mathbf{A}^{-1}_{(a_n,\varepsilon_n)} \mathbf{A}^{-1}_{(a_{n-1},\varepsilon_{n-1})} \cdots \mathbf{A}^{-1}_{(a_1,\varepsilon_1)} \begin{pmatrix} 1 \\ \alpha \\ \beta \end{pmatrix},$$

$$\begin{pmatrix} 1 \\ \gamma_n \\ \delta_n \end{pmatrix} = \frac{1}{\eta\eta_1 \cdots \eta_{n-1}} {}^t\mathbf{A}_{(a_n,\varepsilon_n)} {}^t\mathbf{A}_{(a_{n-1},\varepsilon_{n-1})} \cdots {}^t\mathbf{A}_{(a_1,\varepsilon_1)} \begin{pmatrix} 1 \\ \gamma \\ \delta \end{pmatrix},$$

where
$$\theta_k = \max(\alpha_k, \beta_k),$$
$$\eta_k = \begin{cases} a_k + \gamma_{k-1}, & \text{if } (\alpha_{k-1}, \beta_{k-1}) \in X_0 \\ a_k + \delta_{k-1}, & \text{if } (\alpha_{k-1}, \beta_{k-1}) \in X_1 \end{cases}$$

8.6.4 A geometrical point of view

Let $\mathcal{P}_{\gamma_n,\delta_n}$ be the orthogonal plane with respect to ${}^t(1, \gamma_n, \delta_n)$:

$$\mathcal{P}_{\gamma_n,\delta_n} = \{\boldsymbol{x} \mid \langle \boldsymbol{x}, {}^t(1, \gamma_n, \delta_n)\rangle = 0\}$$

and let us introduce the linear map $\varphi_{(a_n,\varepsilon_n)} : \mathbb{R}^3 \to \mathbb{R}^3$ with matrix $\mathbf{A}_{(a_n,\varepsilon_n)}$ in the canonical basis of $\mathbb{R}^3$. Then we see that the following relationship is true:

$$\varphi^{-1}_{(a_n,\varepsilon_n)} \mathcal{P}_{\gamma_{n-1},\delta_{n-1}} = \mathcal{P}_{\gamma_n,\delta_n}.$$

Link with the modified Jacobi-Perron substitutions. The endomorphisms $\Theta_{(a,\varepsilon)}$ associated with the modified Jacobi-Perron substitutions $\sigma_{(a,\varepsilon)}$ satisfy the following property:

$$\forall n \geq 1,\ \Theta_{(a_n,\varepsilon_n)} \cdots \Theta_{(a_2,\varepsilon_2)} \Theta_{(a_1,\varepsilon_1)}(\mathcal{U}) \in \Gamma^+_{\gamma_n,\delta_n}.$$

Let us consider the renormalization of the set

$$\pi_n \Theta_{(a_n,\varepsilon_n)} \cdots \Theta_{(a_2,\varepsilon_2)} \Theta_{(a_1,\varepsilon_1)}(\mathcal{U}) \subset \mathcal{P}_{\gamma_n,\delta_n},$$

that is,

$$\lim_{n\to\infty} \mathbf{A}_{(a_1,\varepsilon_1)} \cdots \mathbf{A}_{(a_n,\varepsilon_n)} \pi_n \Theta_{(a_n,\varepsilon_n)} \cdots \Theta_{(a_1,\varepsilon_1)}(\mathcal{U}) \subset \mathcal{P}_{\gamma,\delta}$$

where $\pi_n : \mathbb{R}^3 \to \mathcal{P}_{\gamma_n,\delta_n}$ is the projection along ${}^t(1, \alpha_n, \beta_n)$. The properties of this set are stated in the following theorem.

Theorem 8.6.2 ([218]). *For almost every $(\gamma, \delta) \in [0,1) \times [0,1)$ the following limit sets exist*

$$\mathcal{R}^{(i)}_{\alpha,\beta,\gamma,\delta} := \lim_{n\to\infty} \mathbf{A}_{(a_1,\varepsilon_1)} \cdots \mathbf{A}_{(a_n,\varepsilon_n)} \pi_n \Theta_{(a_n,\varepsilon_n)} \cdots \Theta_{(a_1,\varepsilon_1)}(\mathbf{e}_i, i^*),$$

$$\mathcal{R}^{(i)'}_{\alpha,\beta,\gamma,\delta} := \lim_{n\to\infty} \mathbf{A}_{(a_1,\varepsilon_1)} \cdots \mathbf{A}_{(a_n,\varepsilon_n)} \pi_n \Theta_{(a_n,\varepsilon_n)} \cdots \Theta_{(a_1,\varepsilon_1)}(\mathbf{0}, i^*).$$

They satisfy the following properties:

(1) $\mathcal{R}_{\alpha,\beta,\gamma,\delta} = \bigcup_{i=1,2,3} \mathcal{R}^{(i)}_{\alpha,\beta,\gamma,\delta}$ is a periodic tiling on $\mathcal{P}_{\gamma,\delta}$, that is,

$$\bigcup_{\mathbf{z}\in \mathbf{L}_0} (\mathcal{R}_{\alpha,\beta,\gamma,\delta} + \mathbf{z}) = \mathcal{P}_{\gamma,\delta};$$

$$int(\mathcal{R}_{\alpha,\beta,\gamma,\delta} + \mathbf{z}) \bigcap int(\mathcal{R}_{\alpha,\beta,\gamma,\delta} + \mathbf{z}') = \phi \quad (\mathbf{z} \neq \mathbf{z}').$$

(2) the following domain exchange transformation $W_{\alpha,\beta,\gamma,\delta}$ is well-defined:

$$W_{\alpha,\beta,\gamma,\delta} : \mathcal{R}_{\alpha,\beta,\gamma,\delta} \to \mathcal{R}_{\alpha,\beta,\gamma,\delta}$$
$$\mathbf{x} \mapsto \mathbf{x} - \pi_0(\mathbf{e}_i), \text{ if } \mathbf{x} \in \mathcal{R}^{(i)}_{\alpha,\beta,\gamma,\delta}.$$

8.6.5 The one-dimensional case and usual continued fractions

All these results also hold in the one-dimensional case. The usual continued fraction algorithm $T : [0,1) \to [0,1)$ is given by:

$$T(\alpha) = \frac{1}{\alpha} - \left[\frac{1}{\alpha}\right] \quad \text{on } [0,1), \text{ if } \alpha \neq 0, \text{ otherwise } T(0) = 0,$$

and the continued fraction expansion:

$$\alpha = \cfrac{1}{a_1 + \cfrac{1}{a_2 + \cfrac{1}{\ddots + \cfrac{1}{\ddots + \cfrac{1}{a_n + T^n(\alpha)}}}}}$$

where $a_k := \left[\frac{1}{T^{k-1}\alpha}\right]$. A realization of its natural extension $\overline{T}$ is given over $[0,1) \times [0,1)$ by (see also Chap. 6):

$$\overline{T}(\alpha,\gamma) := \left(\frac{1}{\alpha} - a_1, \frac{1}{a_1 + \gamma}\right).$$

The linear map $\varphi_a : \mathbb{R}^2 \to \mathbb{R}^2$ with matrix in the canonical basis of $\mathbb{R}^3$:

$$\mathbf{A}_a = \begin{pmatrix} a & 1 \\ 1 & 0 \end{pmatrix}$$

where φ_a satisfies $\varphi_{a_n}^{-1} l_{\gamma_{n-1}} = l_{\gamma_n}$, with $(\alpha_n, \gamma_n) := T^n(\alpha,\gamma)$, and where the line l_{γ_n} (with corresponding stepped curves $\mathcal{S}_{\gamma_n}$ and $\mathcal{S}_{\gamma_n}$) equals

$$l_{\gamma_n} := \left\{ \,{}^t(x,y) \mid \left\langle \,{}^t(x,y),\ {}^t(1,\gamma_n)\right\rangle = 0 \right\}$$

Moreover, one has

$$\Theta_{a_n}\left(\mathcal{S}_{\gamma_{n-1}}\right) = \mathcal{S}_{\gamma_n}$$

where Θ_a was defined in Sec. 8.2.5. Theorem 8.6.2 holds in this context and has to be compared with the study of Sturmian sequences held in Chap. 6.

8.7 Diophantine applications

Let us apply the previous results first to obtain quasi-periodic tilings of a given plane by projection of the associated stepped surface, second to construct Markov partition of group automorphism, third to β-numbers, and fourth, to Diophantine approximation.

8.7.1 Quasi-periodic tiling related to the stepped surface

Let us start with a plane $\mathcal{P}_{\gamma,\delta}$, $0 < \gamma, \delta < 1$, given by

$$\mathcal{P}_{\gamma,\delta} = \left\{ \boldsymbol{x} \mid \left\langle \boldsymbol{x},\ {}^t(1,\gamma,\delta)\right\rangle = 0 \right\},$$

and let $\mathcal{S}_{\gamma,\delta}$ be anyone of the two stepped surfaces associated with $\mathcal{P}_{\gamma,\delta}$, as explained in Sec. 8.2.1.

Let $\pi : \mathbb{R}^3 \longrightarrow \mathcal{P}_{\gamma,\delta}$ be the projection along ${}^t(1,\alpha,\beta)$. Then we have a tiling of $\pi(\mathcal{S}_{\alpha,\beta})$ generated by three parallelograms $\pi(\mathbf{0}, i^*)$ $(i = 1,2,3)$ and their translates. Let us denote the above tiling by $\mathcal{T}_{\alpha,\beta}(= \pi(\mathcal{S}_{\alpha,\beta}))$. In other words, we project the discrete plane (that is, the stepped surface $\mathcal{S}_{\gamma,\delta}$) associated with the plane $\mathcal{P}_{\gamma,\delta}$ onto this plane to obtain a tiling by parallelograms being the projections of the basic squares.

Let Γ_n be the family of patches which are generated by n parallelograms and which are simply connected, that is,

$$\Gamma_n = \{\gamma \mid \gamma \prec \mathcal{T}_{\alpha,\beta},\ \#\gamma = n,\ \gamma \text{ is simply connected}\}.$$

Definition 8.7.1. *A tiling $\mathcal{T}$ of a plane is said to be* quasi-periodic *if for any $n > 0$, there exists $R > 0$ such that any configuration $\gamma \in \Gamma_n$ occurs somewhere in any neighborhood of radius R of any point.*

This definition is the analog of that of uniform recurrence for symbolic sequences (see Chap. 1).

Theorem 8.7.2. *Let $(1, \gamma, \delta)$ be rationally linearly independent, then the tiling $\mathcal{T}_{\gamma,\delta}$ is a quasi-periodic tiling.*

The essential idea for the proof of this result comes from the following fact: for each $(1, \gamma, \delta)$ there exists a sequence

$$\begin{pmatrix} a_1 & a_2 & \cdots & \cdots \\ \varepsilon_1 & \varepsilon_2 & \cdots & \cdots \end{pmatrix}$$

such that the stepped surface of $\mathcal{P}_{\gamma,\delta}$ is given by

$$\lim_{n\to\infty} \Theta_{(a_1,\varepsilon_1)}\Theta_{(a_2,\varepsilon_2)}\cdots\Theta_{(a_n,\varepsilon_n)}(\mathcal{U}),$$

where the sequence (a_n, ε_n) is obtained by the modified Jacobi-Perron algorithm, and $\Theta_{(a,\varepsilon)}$ is the corresponding substitution. For more details, see [224, 223]. Let us note that we have a result that holds for all the parameters (with the assumption of rational independence) contrary to the previous one in Sec. 8.6 on the construction of a fundamental domain $\mathcal{R}_{\alpha,\beta,\gamma,\delta}$ for the rotation $T_{\alpha,\beta}$ in $\mathbb{T}^2$, which only held for almost every parameter. See also [440, 78, 77, 40] for a combinatorial and arithmetic study of two-dimensional sequences coding these tilings over a three-letter alphabet. These sequences are shown to code a $\mathbb{Z}^2$-action on the torus $\mathbb{T}$. We are in a dual situation with respect to that described in Sec. 8.6. On the one hand, we have a $\mathbb{Z}^2$ action by two rotations on the one-dimensional torus, and, on the other hand, we have a two-dimensional rotation on $\mathbb{T}^2$. More generally, quasi-periodic tilings generated by a non-negative integer matrix are discussed following the same ideas in [185]. The tilings we consider here have been obtained via the classical method of cut and projection (see for instance [385]). Some attempts to generalize these ideas towards different tilings have been made. For instance it would be tempting to obtain a description by generalized substitutions of the dynamics of the Penrose tiling described in [360].

8.7.2 The Markov partition of group automorphisms on $\mathbb{T}^3$

Let us consider the incidence matrix of a unimodular substitution of Pisot type satisfying the assumption of Theorem 8.4.1.

Following the ideas of Sec. 7.6.2, it is possible to associate with the generalized Rauzy fractal $\mathcal{R}_\sigma$ and its decomposition into the three parts $\mathcal{R}_i$ $(i = 1, 2, 3)$ a fundamental domain of $\mathbb{R}^3$ realizing a Markov partition for the toral automorphism of matrix $\mathbf{M}_\sigma$ of $\mathbb{T}^3$ (the sets $\mathcal{R}_i$ are thus the bases for the cylinders of the partition, and each cylinder is directed by the expanding direction of the matrix). See Sec. 7.1.2 for the definition of Markov partitions, and see also Sec 7.6.2 in Chap. 7.

Theorem 8.7.3 ([223]). *Over the assumption of Theorem 8.4.1, it is possible to define sets $\overline{\mathcal{R}}_i$ $(i = 1, 2, 3)$ of $\mathbb{R}^3$ such that the set $\Delta := \bigcup_{i=1,2,3} \overline{\mathcal{R}}_i$ is a fundamental domain, which can be identified with $\mathbb{T}^3$. The partition $\{\overline{\mathcal{R}}_i, i = 1, 2, 3\}$ of $\mathbb{T}^3$ is a Markov partition of the group automorphism*

$$\begin{aligned} T_{M_\sigma} : \mathbb{T}^3 &\longrightarrow \mathbb{T}^3 \\ x &\longmapsto \mathbf{M}_\sigma x \quad (\ mod\ 1) \end{aligned}$$

with the structure matrix ${}^t\mathbf{M}_\sigma$.

Remark. The existence of Markov partitions of group automorphisms on $\mathbb{T}^n$ is discussed in [6] and [398]. In [89], Bowen claims that the boundary of Markov partition of 3-dimensional group automorphisms cannot be smooth. Theorem 8.7.3 explains how we can construct (non-smooth) Markov partitions (an analogous discussion can be found in [61]).

The following question is reasonable. For any element of $\mathbf{A} \in SL(3, \mathbb{Z})$, does there exist a substitution σ such that its incidence matrix $\mathbf{M}_\sigma$ satisfies the assumption of Theorem 8.4.1? Is $\mathbf{M}_\sigma$ isomorphic to $\mathbf{A}$? We only know that for any $\mathbf{A}$ there exists $N > 0$ such that $\mathbf{A}^N$ satisfies the assumption of Theorem 8.4.1 (see [186]).

8.7.3 Application to β-numbers

Let us consider *β-expansions*, that is, expansions of real numbers in $[0, 1[$ as powers of a number β: $x = \sum_{k=1}^{\infty} b_k \beta^{-k}$, with some conditions on the non-negative integers b_k. There is a well-known "greedy algorithm" to write such an expansion, and this algorithm is obviously related to the *β-transformation* $x \mapsto \beta x - [\beta x]$ (for more details, see for instance [179]). It is a natural question to investigate those numbers that have an eventually periodic β-expansion; in the case where β is an integer, these are the rational numbers, and one can characterize among them those with purely periodic expansion. In the general case, one can prove easily that these numbers can be expressed as a rational fraction of β. The question is specially interesting when β is an algebraic number and in particular, a Pisot number. It has been studied by several authors, see for instance [226, 411, 410], the references in [179] and Sec. 7.3.3.

In [373] (see also [374, 225]), a complete answer is given in the case of a particular class of Pisot numbers, those that satisfy an equation

$$\beta^d = k_1 \beta^{d-1} + k_2 \beta^{d-2} + \cdots + k_{d-1} \beta + 1,$$

where the k_i are nonnegative integers such that $k_1 \geq k_2 \geq \cdots \geq k_{d-1} \geq 1$; in that case, x has an eventually periodic β-expansion if and only if x belongs to the field $\mathbb{Q}(\beta)$; furthermore, a characterization of those points having an immediately periodic expansion is given by introducing a realization of the natural extension of the β-transformation on a particular domain with a fractal boundary of $\mathbb{R}^d$. This domain can be explicitly described using the previous ideas on generalized substitutions associated with the companion matrix of the previous equation satisfied by β.

8.7.4 Diophantine approximation

Let us end this survey of the Diophantine applications of the generalized substitutions by mentioning the following result.

Theorem 8.7.4. *Let $\langle 1, \alpha, \beta \rangle$ be the basis of the cubic field $\mathbf{Q}(\lambda)$ given by*

$$\mathbf{A} \begin{pmatrix} 1 \\ \alpha \\ \beta \end{pmatrix} = \lambda \begin{pmatrix} 1 \\ \alpha \\ \beta \end{pmatrix}$$

for some $\mathbf{A} \in SL(3, \mathbb{Z})$, and let us assume that λ is a complex Pisot number. The limit set of

$$\left\{ \sqrt{q} \begin{pmatrix} q\alpha - p \\ q\beta - r \end{pmatrix} \mid (q, p, r) \in \mathbb{Z}^3, q > 0 \right\}$$

consist in a family of ellipses.

Theorem 8.7.4 is proved by algebraic geometry in [4]. For a proof of this result using modified Jacobi-Perron substitutions in the case where (α, β) is a purely periodic point with period 1 under this algorithm, see [220, 219]: the nearest ellipses near to the origin are shown to be given by the modified Jacobi-Perron algorithm.

Part III

Extensions to free groups and interval transformations

9. Infinite words generated by invertible substitutions

Combinatorial properties of finite and infinite words are of increasing importance in various fields of science (for a general reference, the reader should consult [270, 271] and the references therein). The combinatorial properties of the Fibonacci infinite word have been studied extensively by many authors (see for examples [65, 129, 175, 451]).

This chapter looks at infinite words generated by invertible substitutions. As we shall see, words of this family generalize the Fibonacci infinite word.

The combinatorial properties of those infinite words are of great interest in various fields of mathematics, such as number theory [10, 95, 163, 231], dynamical systems [330, 339, 349, 399], fractal geometry [43, 102, 221, 273], tiling theory [43, 408], formal languages and computational complexity [65, 67, 201, 200, 232, 294, 383, 384, 448], and also in the study of quasicrystals [82, 99, 256, 257, 273, 455, 453, 454, 458].

The purpose of this chapter is to study the properties of factors of infinite words generated by primitive invertible substitutions over a two-letter alphabet. It is deeply linked to Chap. 6: as we will see, all these words are Sturmian words. The chapter is organized as follows.

Section 9.1 is devoted to the introduction and preliminaries. Section 9.2 is concerned with the structure of the semigroup of invertible substitutions for which we give a set of generators; we derive various consequences from the structure theorem.

Section 9.3 lists properties of factors of the Fibonacci word. After recalling some basic facts, we introduce what we call the singular words of the Fibonacci word and give their properties. We then establish two decompositions of the Fibonacci word involving the singular words. Some applications are derived from those decompositions, describing combinatorial properties of the Fibonacci word related to powers of its factors, for example. We also prove local isomorphisms properties of the Fibonacci word and the overlap properties of the factors.

In Sec. 9.4, we consider more general invertible substitutions. We define singular factors of fixed points of invertible substitutions, and then give applications of decompositions associated with these singular words, after proving some general properties of factors of fixed points of invertible substitutions.

[1] This chapter has been written by Z. -Y. Wen

As we shall see, the general framework we introduce is at a much higher level of complexity than with the Fibonacci word.

9.1 Preliminary

Let us now introduce some definitions and notation (see also Chap. 1).

Free monoid and free group. Let $\mathcal{A}_n = \{a_1, a_2, \cdots, a_n\}$ be an alphabet with n letters. Let $\mathcal{A}_n^*$ and Γ_n respectively denote the free monoid with empty word ε as neutral element and the free group generated by $\mathcal{A}_n$. By a natural embedding, we can regard the set $\mathcal{A}_n^*$ as a subset of Γ_n, and consider any $W \in \mathcal{A}_n^*$ as a reduced element of Γ_n (see [330] and the references therein, see also [274, 277, 415]).

The canonical abelianization map on $\mathcal{A}_n^*$ is denoted by $\mathbf{l} : \mathcal{A}_n^* \to \mathbb{N}^n$, as defined in Chap. 1.

Factors. Let $V, W \in \mathcal{A}_n^*$.

- We write $V \prec W$ when the finite word V is a *factor* of the word W, that is, when there exist words $U, U' \in \mathcal{A}_n^*$ such that $W = UVU'$.
- We say that V is a *left* (respectively *right*) *factor* of a word W, and we write $V \lhd W$ (respectively $V \rhd W$), if there exists a word $U' \in \mathcal{A}_n^*$ such that $W = VU'$ (respectively $W = U'V$).
- We denote by W^{-1} the *inverse word* of W, that is, $W^{-1} = w_p^{-1} \cdots w_2^{-1} w_1^{-1}$ if $W = w_1 w_2 \dots w_p$. This only makes sense in Γ_n; but if V is a right factor of W, we can write $WV^{-1} = U$, with $W = UV$; this makes sense in $\mathcal{A}_n^*$, since the reduced word associated with WV^{-1} belongs to $\mathcal{A}_n^*$. This abuse of language will be very convenient in what follows.

Definition 9.1.1. *Let u and v be two infinite words over $\mathcal{A}_n$. We say that u and v are* locally isomorphic *if any factor of u is also a factor of v and vice versa. If u and v are locally isomorphic, we shall write $u \simeq v$.*

Remarks.

- This means that the dynamical systems associated with u and v under the action of the shift, in the sense of Chap. 1, are the same. If we are only interested in the language of factors of an infinite word, or in the system associated with this infinite word, we do not need to differentiate between locally isomorphic infinite words. We will use this remark several times.
- The notion of local isomorphism was first introduced in connection with quasicrystals, and for physical reasons, the definition was that each factor of u, or its mirror image, should also be a factor of v. Since however all the languages we will consider in this chapter are closed under mirror image, the definition we take here is equivalent in our case, and will allow simpler proofs.

Conjugation of a word. Let $W = w_1 w_2 \dots w_p \in \mathcal{A}_n^*$.

- The *reversed word* (or *mirror image*) of W, denoted $\overline{W}$, is defined as $\overline{W} = w_p \dots w_2 w_1$.
- A word W is called a *palindrome* if $W = \overline{W}$. We denote by $\mathcal{P}$ the set of palindromic words.
- For $1 \leq k \leq |W|$, we define the *k-th conjugation* of W as $C_k(W) = w_{k+1} \dots w_n w_1 \dots w_k$, and we denote by $C(W) = \{C_k(W);\ 1 \leq k \leq |W|\}$. By convention, $C_{-k}(W) = C_{|W|-k}(W)$. We will say that two words are *conjugate* if they belong to the same conjugacy class.

This may seem an abuse of language, since there is another notion of conjugacy, if we consider the two words as an element of the free group; in this sense, U and V are conjugate if there exist an element U' of the free group Γ_n such that $U = U'VU'^{-1}$. These two definitions turn out to be equivalent:

Lemma 9.1.2. *Two words U and V in $\mathcal{A}_n^*$ are conjugate as elements of the free monoid $\mathcal{A}_n^*$ if and only if they are conjugate as elements of the free group Γ_n; in that case, we can always find a conjugating element of length strictly less than $|U| = |V|$.*

Exercise 9.1.3. Give a proof of this lemma, and give the most general form for the conjugating word U' such that $U = U'VU'^{-1}$.

A word $W \in \mathcal{A}_n^*$ is called *primitive* if $W = U^p$, $U \in \mathcal{A}_n^*$, $p > 0$, implies $W = U$ (and thus $p = 1$). That is, W cannot be expressed as a proper power of any other word. Notice that the k-th conjugation C_k is an action from $\mathcal{A}_n^*$ to $\mathcal{A}_n^*$ preserving primitivity. That is, the word $C_k(W)$ is primitive if and only if W is itself a primitive word. We will use the following lemma.

Lemma 9.1.4. *A word $W \in \mathcal{A}_n^*$ is primitive if and only if the cardinality of $C(W)$ is $|W|$.*

Proof. If the cardinality is smaller, it means that we can find k, l distinct, $1 \leq k < l \leq n$, such that $C_k(W) = C_l(W)$. It is immediate to check that we can then find two nonempty words U, V such that $C_k(W) = UV$, $C_l(W) = VU$; since $UV = VU$, it is well known that they must be a power of the same word U', and $C_k(W)$ is not primitive. Hence W is not primitive. The reciprocal is trivial. ∎

There is a useful criterion of primitivity:

Lemma 9.1.5. *If the numbers $|W|_{a_1}, |W|_{a_2}, \dots, |W|_{a_n}$ are relatively prime, the word W is primitive.*

We leave the proof to the reader.

Substitution. Let us recall that a *substitution* (or a *morphism*) over $\mathcal{A}_n$ is a map $\sigma : \mathcal{A}_n^* \to \mathcal{A}_n^*$, such that for any two words U and V, one has $\sigma(UV) = \sigma(U)\sigma(V)$. The substitution σ is uniquely determined by the image of the elements of $\mathcal{A}_n$. We denote by $\sigma = (U_1, U_2, \cdots, U_n)$ the substitution defined by $\sigma(a_i) = U_i$, $1 \leq i \leq n$.

The notions of incidence matrix and primitivity were introduced in Chap. 1. In this chapter, unless stated otherwise, all substitutions we consider are primitive.

If for some letter $a \in \mathcal{A}_n$, the word $\sigma(a)$ begins with a and has length at least 2, then the sequence of words $\sigma^n(a)$ converges to a fixed point $\sigma^\omega(a) \in \mathcal{A}_n^{\mathbb{N}}$. In this chapter, we shall only consider fixed points of a substitution that can be obtained by iterating a morphism in the above manner. We denote by u_σ any one of the fixed points of σ, if it exists.

Invertible substitution. Let $\sigma : \mathcal{A}_n^* \to \mathcal{A}_n^*$ be a substitution over $\mathcal{A}_n^*$. Then σ can be naturally extended to Γ_n by defining $\sigma(a_i^{-1}) = (\sigma(a_i))^{-1}$, $1 \leq i \leq n$. Denote by $\mathrm{Aut}(\Gamma_n)$ the *group of automorphisms* over Γ_n.

Definition 9.1.6. *A substitution σ is called an* invertible substitution *if it can be extended to all of Γ_n by an automorphism (also called σ), that is, if there exists a map $\eta : \Gamma_n \to \Gamma_n$ such that $\sigma\eta = \eta\sigma = Id$ on Γ_n (where Id is the substitution given by $Id = (a_1, a_2, \cdots, a_n)$). The set of invertible substitutions over $\mathcal{A}_n^*$ is denoted by* $\mathrm{Aut}(\mathcal{A}_n^*)$.

Remark. The set $\mathrm{Aut}(\mathcal{A}_n^*)$ is obviously a semigroup. Remark that its elements extend to automorphism of the free group, but they are not automorphisms of the free monoid $\mathcal{A}_n^*$ in the usual sense, since they are not onto, except for the trivial case of the permutations of the alphabet.

Example 9.1.7 (The Fibonacci substitution). Set $\mathcal{A}_2 = \{a, b\}$ for convenience. The *Fibonacci substitution* $\sigma = (ab, a)$ is an invertible substitution over $\mathcal{A}_2$: an easy computation shows that the map $\eta = (b, b^{-1}a)$ is its inverse.

REcall that the fixed point generated by the Fibonacci substitution $\sigma = (ab, a)$ (iterated over a) is called the *infinite Fibonacci word*;it is denoted by u_σ. We shall often simply call it the *Fibonacci word* when no confusion occurs.

General structure of $\mathrm{Aut}(\Gamma_n)$. The structure of $\mathrm{Aut}(\Gamma_n)$ is well known [277, 313, 330]. In particular, we have the following theorem, where, as usual, $\langle S \rangle$ denotes the group generated by a set S of generators.

Theorem 9.1.8. *Let $n > 0$. The number of generators of the group* $\mathrm{Aut}(\Gamma_n)$ *is $3n - 3$ and these generators can be determined explicitly.*

If $n = 2$, then

$$\mathrm{Aut}(\Gamma_2) = \langle \sigma, \quad \pi, \quad \psi \rangle,$$

where $\sigma = (ab, a)$, $\pi = (b, a)$ and $\psi = (a, b^{-1})$

In the following section, we will give a description in terms of generators of the related semigroup $\mathrm{Aut}(\mathcal{A}_2^*)$.

Inner automorphisms and conjugacy. Let $W \in \Gamma_n$. We denote by $i_W : \Gamma_n \to \Gamma_n$ the *inner automorphism* defined by $U \mapsto WUW^{-1}$, for any $U \in \Gamma_n$. The set of all inner automorphisms of Γ_n will be denoted by $\mathrm{Inn}(\Gamma_n)$.

For any substitution τ, the composition $i_W \circ \tau$ results in $U \mapsto W\tau(U)W^{-1}$; it is a morphism of the free group, but not a substitution in general. We have the following lemma:

Lemma 9.1.9. *Let τ be a substitution, and $W \in \Gamma_n$. The composition $i_W \circ \tau$ is a substitution if and only if: either $W \in \mathcal{A}_n^*$ and W is of the form $X_a\tau(a)^k$, where X_a is a suffix of $\tau(a)$, for any letter a , or $W^{-1} \in \mathcal{A}_n^*$ and W^{-1} is of the form $\tau(a)^k X_a$, where X_a is a prefix of $\tau(a)$, for any letter a.*

If $i_W \circ \tau$ is a substitution, then, for any letter a, $i_W \circ \tau(a)$ admits X_a as prefix in the first case, and X_a as suffix in the second.

Proof. This is an immediate consequence of Lemma 9.1.2; $i_W \circ \tau$ is a substitution if and only if, for all a, $W\tau(a)W^{-1}$ is an element of the free monoid; but this gives the necessary and sufficient condition for the form of W. ∎

We will say that two substitutions σ, τ are *conjugate* if there exists a word $W \in \Gamma_n$ such that $\sigma = i_W \circ \tau$, which will be denoted as $\sigma \sim \tau$.

Note that the conjugating word W can be strictly longer than the images of some letters; for example, the two substitutions (aba, ba) and (aba, ab) are conjugate by the word aba. However, if σ and τ are two substitutions conjugate by a word W, that is $\sigma = i_W \circ \tau$, it is easy to check that, for any suffix V of W, $i_V \circ \tau$ is again a substitution. Considering the particular case where V is a letter, we see that the only way to conjugate a substitution is to suppress a common initial (or final) letter from the images of the letters, to replace it as a final (or initial) letter, and to iterate the operation; in particular, a substitution can be conjugate to another substitution if and only if all the images of letters have a common final or initial letter.

We can be more precise; we know that the word W or its inverse belong to $\mathcal{A}_n^*$. We will say that the substitution σ is a *left conjugate* of τ if there exists $W \in \mathcal{A}_n^*$ such that $\sigma = i_W \circ \tau$.

Exercise 9.1.10.
1. Prove that the conjugacy is an equivalence relation on the set of substitutions.
2. Prove that the left conjugacy induces a total order relation on each class.
3. Prove that an element is maximal (respectively minimal) for that order if the images of letters have no common prefix (respectively suffix).

Conjugacy has an interesting consequence:

Proposition 9.1.11. *Let τ, σ be two conjugate primitive substitutions. Let u_τ, u_σ be fixed points of τ and σ. Then u_τ and u_σ are locally isomorphic.*

Proof. We can suppose, by exchanging τ and σ if needed, that $\tau = i_W \circ \sigma$, with $W \in \mathcal{A}^*$. Then by induction we have

$$\tau^k = (i_W \sigma)^k = i_W i_{\sigma(W)} \cdots i_{\sigma^{n-1}(W)} \sigma^k.$$

Let $W_k = W\sigma(W) \cdots \sigma^{k-1}(W)$. We have $\tau^k = i_{W_k}\sigma^k$, that is, $\tau^k(a)W_k = W_k\sigma^k(a)$. Thus there exist U_k and V_k such that $\tau^k(a) = U_k V_k$, and $\sigma^k(a) = V_k U_k$. Since, by primitivity, $|\tau^k(a)|$ tends to infinity, so does $|U_k|$ or $|V_k|$.

Let U be a factor of u_τ, by minimality of the fixed words of primitive substitutions, $U \prec U_k$ or $U \prec V_k$ for k large enough, thus $U \prec u_\sigma$. The same argument proves that any factor of u_σ is a factor of u_τ. Hence $u_\tau \simeq u_\sigma$. ∎

9.2 Structure of invertible substitutions

This section describes a set of generators for the semigroup $\mathrm{Aut}(\mathcal{A}_2^*)$, thus revealing part of its structure.

We will prove the following conclusion: *any invertible substitution over a three-letter alphabet can be written as a finite product of three special invertible substitutions.* This first result is a starting point for studying invertible substitutions, since many basic properties will be derived from it.

The proof we present here is borrowed from [449]. A geometrical proof of the same result is given in [157].

Theorem 9.2.1 ([449]). *Let* $\pi = (b, a)$, $\sigma = (ab, a)$, $\varrho = (ba, a)$, $\gamma = (ab, b)$, *and* $\delta = (ba, b)$. *We have,*

$$\mathrm{Aut}(\mathcal{A}_2^*) = \langle \pi, \varrho, \sigma \rangle = \langle \pi, \gamma, \delta \rangle.$$

The proof follows from the following lemmas.

Lemma 9.2.2. *An automorphism* $\tau \in \mathrm{Aut}(\Gamma_2)$ *is an inner automorphism if and only if its incidence matrix* $\mathbf{M}_\tau$ *is the identity matrix.*

Proof. This means that the kernel of the abelianization map: $\mathrm{Aut}(\Gamma_2) \to GL(2, \mathbb{Z})$ is exactly the group of inner automorphisms; it is clear that the matrix associated with an inner automorphism is the identity. To prove the converse, it suffices to give a presentation of $GL(2, \mathbb{Z})$ by generators and relations, to find generators of $\mathrm{Aut}(\Gamma_2)$ that project to the generators of $GL(2, \mathbb{Z})$, and to check that the combinations of these elements corresponding to the relations are inner automorphisms, see [274]. ∎

Remark. This lemma does not generalize to more letters: the map (bab^{-1}, cbc^{-1}, c) gives an automorphism of the free group on three letters whose matrix is the identity, and which is not an inner automorphism.

Lemma 9.2.3. *Let* $\mathcal{M}$ *be the monoid of square matrices of size* 2 *with non-negative integer coefficients, and determinant* ± 1. *We have*

$$\mathcal{M} = \langle \mathbf{A},\ \mathbf{B} \rangle, \quad \textit{where } \mathbf{A} = \begin{pmatrix} 0 & 1 \\ 1 & 0 \end{pmatrix}, \quad \mathbf{B} = \begin{pmatrix} 1 & 1 \\ 1 & 0 \end{pmatrix}.$$

Proof. Let

$$\mathbf{C} = \begin{pmatrix} 1 & 1 \\ 0 & 1 \end{pmatrix}, \quad \mathbf{D} = \begin{pmatrix} 1 & 0 \\ 1 & 1 \end{pmatrix}.$$

Let also $\mathbf{M} = \begin{pmatrix} p & r \\ q & s \end{pmatrix}$ be an element of $\mathcal{M}$ with at least one entry ≥ 1. Up to multiplication by $\mathbf{A}$, we can suppose $\det \mathbf{M} = 1$. It was proved in Chap. 6, by an easy induction using Euclidean divisions on the columns of the matrix, that the set of matrices with nonnegative coefficients and determinant 1 is the free monoid generated by $\mathbf{C}$ and $\mathbf{D}$ (see Lemma 6.5.14.). But one can easily check the following relations: $\mathbf{C} = \mathbf{BA}$, $\mathbf{D} = \mathbf{AB}$. This ends the proof. ∎

Proof of Theorem 9.2.1. Note that we have the trivial relations: $\sigma = \pi \circ \delta$, $\delta = \pi \circ \sigma$, $\varrho = \pi \circ \gamma$, $\gamma = \pi \circ \varrho$; hence, a substitution can be decomposed on $\{\pi, \sigma, \varrho\}$ if and only if it can be decomposed on $\{\pi, \gamma, \delta\}$. We will prove that this last set is a set of generators for $\mathrm{Aut}(\mathcal{A}_2^*)$.

The idea of the proof is, given an invertible substitution τ, to find, using Lemma 9.2.3, a substitution ϕ, written as a product of π and γ, with the same matrix as τ. Then, using Lemma 9.2.2, these two substitutions are conjugate; but then $\tau = i_W \circ \phi$ can be written as a product of π, γ and δ (this is a generalization of the easy relation $i_b \circ \gamma = \delta$).

Let $\tau \in \mathrm{Aut}(\mathcal{A}_2^*)$. Then $\mathbf{M}_\tau \in \mathcal{M}$. By Lemma 9.2.3, there exist $\mathbf{M}_1, \mathbf{M}_2, \dots \mathbf{M}_k \in \{\mathbf{A}, \mathbf{D}\}$ such that $\mathbf{M}_\tau = \mathbf{M}_1 \cdots \mathbf{M}_k$. Note that $\mathbf{M}_\pi = \mathbf{A}$ and $\mathbf{M}_\gamma = \mathbf{D}$. Take $\phi = \phi_1 \cdots \phi_k$, with $\phi_i = \pi$ or $\phi_i = \gamma$, according to $\mathbf{M}_i = \mathbf{A}$ or $\mathbf{D}$. Then, we have $\mathbf{M}_\tau = \mathbf{M}_\phi$. Therefore Lemma 9.2.2 implies that $\tau \circ \phi^{-1}$ is an inner automorphism i_W. Hence we can write $\tau = i_W \circ \phi$, and the two substitutions τ, ϕ are conjugate; by Lemma 9.1.9, we have $W \in \mathcal{A}_2^*$ or $W^{-1} \in \mathcal{A}_2^*$. But it is clear that $\phi(a)$ and $\phi(b)$ have a different initial letter, since π and γ send words with different initial letters on words with different initial letters. Hence they have no common prefix, and $W \in \mathcal{A}_2^*$.

Thus we only need to prove the following claim: if $\phi \in \langle \pi, \ \gamma, \ \delta \rangle$ is such that $\phi(a) = UW$ and $\phi(b) = VW$ (with $U, V, W \in \mathcal{A}_2^*$), then $i_W \circ \phi \in \langle \pi, \ \gamma, \ \delta \rangle$. By induction, it is enough to prove it in the case where W is reduced to one letter $z \in \{a, b\}$.

First consider $z = b$. Since $i_b \circ \phi$, where $\phi = \phi_1 \phi_2 \dots \phi_t$ ($\phi_j \in \{\pi, \ \gamma, \ \delta\}$), is a substitution, $\phi(a)$ and $\phi(b)$ must have b as last letter. This is only possible if one of the ϕ_i is equal to γ, since it is obvious that, for a product of π and δ, the images of a and b end with different letters. Let i be the first index such that $\phi_i = \gamma$, the proposition will follow if we can prove $i_b \phi_1 \phi_2 \dots \phi_i \in \langle \pi, \ \gamma, \ \delta \rangle$.

Using the properties $\pi^2 = id$, $\gamma\delta = \delta\gamma$, we see that the product $\phi_1 \phi_2 \dots \phi_i$ can be written $\delta^{e_1} \pi \delta^{e_2} \dots \delta^{e_n} \pi \gamma$, with all the e_k strictly positive, except maybe the first, and n even (otherwise, $\tau(a)$ and $\tau(b)$ would end in a). A proof by induction, starting from the remark that $i_b \gamma = \delta$, shows that $i_b \phi_1 \phi_2 \dots \phi_i \in \langle \pi, \ \gamma, \ \delta \rangle$.

A similar proof can be done for $z = a$, in which case n is odd. ∎

Corollary 9.2.4. *Let τ_1, τ_2 be two invertible substitutions and let u_{τ_1}, u_{τ_2} be the fixed points of τ_1 and τ_2 respectively. If there exist $m, n \in \mathbb{N}$ such that $\mathbf{M}_{\tau_1}^n = \mathbf{M}_{\tau_2}^m$, then $u_{\tau_1} \simeq u_{\tau_2}$.*

Proof. It is clear that a fixed point of a substitution is also a fixed point of all its powers; hence, by replacing τ_1, τ_2 by τ_1^n, τ_2^m, we can suppose that $\mathbf{M}_{\tau_1} = \mathbf{M}_{\tau_2}$. But in that case, $\tau_1 \circ \tau_2^{-1}$ is an automorphism of the free group whose matrix is the identity. By Lemma 9.2.2, it is an inner automorphism, hence the substitutions τ_1, τ_2 are conjugate; by Proposition 9.1.11, their fixed points are locally isomorphic. ■

Exercise 9.2.5. The conjugacy class of a substitution is completely explicit: it is obtained by exchanging γ and δ, since this does not change the matrix. We know that left conjugacy defines a total order on this conjugacy class.

Find the maximal and the minimal element of a class, and explain how to find the successor and the predecessor (for this order) of an invertible substitution with given decomposition in the generators.

An interesting consequence of Corollary 9.2.4 is that fixed points of invertible substitutions are dense in the system generated by an invertible substitution:

Corollary 9.2.6. *Let ϕ be an invertible substitution. Let u_ϕ be a fixed point of ϕ, and let U be a factor of u_ϕ. Then there exists a substitution τ such that u_τ is locally isomorphic to u_ϕ, and admits U as prefix.*

Proof. The substitution ϕ is a product of π, ϱ and σ; we can replace all the ϱ by σ, this will change u_σ to a locally isomorphic word, and will not change the factors.

In that case, one of $\phi^n(a)$, $\phi^n(b)$ is a prefix of the other for all n; suppose it is $\phi^n(a)$. Then, by definition of U, it appears in $\phi^n(a)$, for n large enough, and we can find words W, A, B such that: $\phi^n(a) = WUA$, $\phi^n(b) = WUAB$. But then, $\tau = i_{W^{-1}} \circ \phi$ is a substitution, whose fixed point, by the preceding theorem, is locally isomorphic to u_ϕ and admits U as prefix. ■

Another consequence of the proof of Theorem 9.2.1 is that any invertible substitution τ is conjugate to a substitution σ which is a product of π and γ. If we take the square, we can suppose that the number of π in the decomposition of σ is even, and an easy induction shows that σ can be written as a product of the substitutions γ and $\pi \circ \gamma \circ \pi$. But these are the two basic substitutions considered in Chap. 6, and from the results of this chapter, we obtain:

Corollary 9.2.7. *A fixed point of a primitive invertible substitution on two letters is Sturmian.*

Remarks.

- Observe that $\gamma \notin \langle \pi, \delta \rangle$ and $\delta \notin \langle \pi, \gamma \rangle$. Since the length of a substitution can only increase by composition, this proves that the minimal number of generators of $\mathrm{Aut}(\mathcal{A}_2^*)$ is exactly three.
- Brown [95], Kósa [253], Séébold [384], Mignosi and Séébold [295] have considered invertible substitutions in $\langle \pi, \pi_1, \pi_2 \rangle$, with $\pi_1 = (a, ab)$, $\pi_2 = (a, ba)$. They studied combinatorial and arithmetic properties of the infinite words generated by those invertible substitutions.
- Mignosi and Séébold [295] have considered Sturmian substitutions, that is, substitutions preserving all Sturmian sequences. They proved that a fixed point of a substitution over $\mathcal{A}_2$ is Sturmian if and only if the substitution is invertible [295, 449], giving a more precise form of the corollary 9.2.7 above. This motivates us to study the set of invertible substitutions.
- Let $n \geq 2$ and let $\mathcal{A}_n = \{a_1, \cdots, a_n\}$ be a n-letter alphabet. Invertible substitutions over $\mathcal{A}_n$ are defined in complete analogy with invertible substitutions over $\mathcal{A}_2$. The *Rauzy substitution*, $\varrho_n : \mathcal{A}_n^* \to \mathcal{A}_n^*$, is a typical invertible substitution and is defined as follows:

$$\varrho_n(a_1) = a_1 a_2,\ \varrho_n(a_2) = a_2 a_3, \cdots, \varrho_n(a_{n-1}) = a_{n-1} a_n,\ \varrho_n(a_n) = a_1.$$

  From its definition, we see that the Rauzy substitution is a generalization of the Fibonacci substitution. This substitution has many remarkable properties and has some interesting relations with other domains; for more details, see Chap. 7. Other interesting invertible substitutions over $\mathcal{A}_n$ are described in [163].
- Theorems 9.1.8 and 9.2.1, bring us to naturally conjecture that the number of generators of $\mathrm{Aut}(\mathcal{A}_n^*)$ is $3n - 3$. However, authors in [452] have proved that the set of invertible substitutions over an alphabet of more than three letters is not finitely generated (their proof uses the notion of "prime substitution"). Also, they gave examples showing that, in that case, the structure of $\mathrm{Aut}(\mathcal{A}_n^*)$ is much more complex.
- Let τ be a substitution, we denote by Φ_τ the trace mapping induced from τ, see [330] and Chap. 10. Let $\tau_1, \tau_2 \in \mathrm{Aut}(\mathcal{A}_2^*)$, then u_{τ_1} and u_{τ_2} are locally isomorphic if and only $\Phi_{\tau_1} = \Phi_{\tau_2}$ [330].
- In [152], Durand has proved the following result: *If two primitive substitutions over $\mathcal{A}_n$ have the same nonperiodic fixed point, then they have some powers the incidence matrices of which have the same eigenvalues.*

9.3 Singular words of the Fibonacci word and applications

As we have seen in the last section, the Fibonacci substitution plays an important role in the study of the structure of the group of invertible substitutions. Our investigation of the properties of infinite words generated by invertible substitutions will be guided by the study of the factors of the Fibonacci word. Those factors have been studied in relation with mathematics and physics in theories such as number theory, fractal geometry, formal languages, computational complexity, quasicrystals, and so on; see for examples, [10, 82, 231, 339, 383, 448]. Moreover, the properties of the factors of the Fibonacci word have been studied extensively by many authors [65, 129, 232, 294, 383, 384, 451]. We give new properties of the factors of the Fibonacci word based on a new family of factors we call singular words. As we shall see, the most striking of those properties is that two distinct occurrences of a singular word never overlap.

This section is organized as follows. After making preliminary remarks on the Fibonacci word, we introduce the singular words and give elementary properties they satisfy. Then, we establish two decompositions of the Fibonacci word in terms of the singular words (Theorems 9.3.6 and 9.3.8), and describe some of their applications.

Those results (and, in particular, the positively separated property of the singular words) bring us in Sec. 9.3.2 to other combinatorial properties of the Fibonacci infinite word, such as local isomorphism, the overlap properties of the factors and the powers of factors of the Fibonacci word.

9.3.1 Singular words and their properties

As earlier, let $\sigma = (ab, a)$ be the Fibonacci substitution and u_σ denote the Fibonacci infinite word.

Two of the motivations we have are the following.

- The Fibonacci word is closely related to the Fibonacci numbers defined by the recurrence formula $f_{n+2} = f_{n+1} + f_n$, with the initial conditions $f_{-1} = f_0 = 1$, as shown in Chap. 2. Consider the following decomposition of the Fibonacci word

$$\underline{a}\,\underline{b}\,\underline{aa}\,\underline{bab}\,\underline{aabaa}\,\underline{babaabab}\,\underline{aabaababaabaa}\,\underline{babaababaabaababaabab}\cdots$$

  where the n-th block in the decomposition is of length f_n, $n \geq -1$. The question of whether those blocks share special properties naturally comes to mind. As we shall see, Theorem 9.3.6 will answer this question completely.
- As we know, the Fibonacci word is a Sturmian word, hence it has exactly $n + 1$ factors of length n. It is also balanced (see Chap. 6), hence, for any n, there exists p such that each factor of length n has p or $p + 1$ letters a. As we will see, for $n = f_k$ a Fibonacci number, there are n factors with f_{k-1} a's, all in the same conjugacy class, and one factor (the singular word) with $f_{k-1} \pm 1$ a's, depending on the parity. This only happens for special values of n, and is linked to the continued fraction expansion of the golden number.

We first need to state basic facts about the Fibonacci word, for which the reader is referred to [65, 129, 339, 383]; see also Chap. 6.

Proposition 9.3.1. *Let $F_n = \sigma^n(a)$, then*

1. $\mathbf{l}(F_n) = (f_{n-1}, f_{n-2})$;
2. *F_n is primitive;*
3. *$|\sigma^n(a)| = \text{Card } C(F_n) = f_n$, that is, all conjugates of the word F_n are distinct. Moreover, $C(F_n) = \{\overline{W};\ W \in C(F_n)\}$.*
4. $F_{n+1} = F_n F_{n-1}$.
5. *For any $k \geq 1$, $\sigma^k(u_\sigma) = u_\sigma$, that is, $u_\sigma = F_k F_{k-1} F_k F_k F_{k-1} \cdots$.*
6. *The word ab is a suffix of F_n for odd positive n, and the word ba is a suffix of F_n for even strictly positive n.*
7. $b^2 \not\prec u_\sigma$, $a^3 \not\prec u_\sigma$.
8. *Any factor of u_σ appears infinitely many times in u_σ.*
9. *$W \prec u_\sigma$ if and only if $\overline{W} \prec u_\sigma$.*

Proof. The first statement is immediate by recurrence, or by computing the powers of the matrix of σ.

The second comes from the fact that consecutive Fibonacci numbers are relatively prime, which allows us to use Lemma 9.1.5; alternatively, one could use the fact that the matrix of σ^n has determinant 1 (the same argument shows that, for any substitution τ such that the associated matrix has determinant 1, $\tau^n(a)$ is primitive).

The third statement is a consequence of Lemma 9.1.4. The rest is easy to check. ■

Remark. No other results than those cited in Proposition 9.3.1 will be used.

A consequence of this proposition is that for any $n \geq 1$, the set $\mathcal{L}_{f_n}$ splits into two parts. The first part coincides with the conjugacy class of the word F_n, and the second part consists solely of some word W_n. We want to describe more precisely W_n.

Assume $\alpha, \beta \in \mathcal{A}_2$. Then $\alpha\beta \triangleright F_n$ implies $\alpha \neq \beta$, by virtue of Proposition 9.3.1.6. An induction using Proposition 9.3.1 proves the following lemma.

Lemma 9.3.2. *Let $n \geq 2$, and assume that $\alpha\beta$ is a suffix of F_n. Then*

$$F_n = F_{n-2}F_{n-1}\alpha^{-1}\beta^{-1}\alpha\beta,$$

Remark. Observe that $F_n = F_{n-1}F_{n-2}$. Using the preceding lemma, we see that the words $F_{n-1}F_{n-2}$ and $F_{n-2}F_{n-1}$ coincide except on the two last letters which are distinct. This remarkable property will be useful when studying the factors of the Fibonacci word.

Definition 9.3.3. *Let $\alpha\beta$ be a suffix of F_n. We denote by W_n the word $\alpha F_n \beta^{-1}$.*

The word W_n is called the n-th singular word *of the Fibonacci word u_σ. For convenience, we define $W_{-2} = \varepsilon$, $W_{-1} = a$ and $W_0 = b$. We denote by $\mathcal{S}$ the set of singular words of u_σ.*

Lemma 9.3.4. *We have:*

1. $W_n \notin C(F_n)$;
2. $\mathcal{L}_{f_n}(F_{n-1}F_n) = \{W_n\} \bigcup \{C_k(F_n);\ 0 \le k \le f_{n-1} - 2\}$. *In particular, as a factor,* W_n *appears only once in* $F_{n-1}F_n$.

Proof. Statement 1. follows from $\alpha \neq \beta$, so that $\mathbf{l}(W_n) \neq \mathbf{l}(F_n)$.

Lemma 9.3.2 implies that $F_{n-1}F_n = F_nF_{n-1}\alpha^{-1}\beta^{-1}\alpha\beta$ when $\alpha\beta \triangleright F_n$. Since $F_{n-1} \triangleleft F_n$, the first f_{n-1} factors of length f_n of the word $F_{n-1}F_n$ are exactly $C_k(F_n)$, $1 \le k \le f_{n-1} - 2$, and the last factor is $F_n = C_0(F_n)$. The $(f_{n-1} - 1)$-th factor is $\alpha F_n \beta^{-1} = W_n$, which ends the proof of 2. ■

We now describe properties of the singular words:

Proposition 9.3.5. *Let* $\{W_n\}_{n \ge -1}$ *be the singular words.*

1. *If* $n \ge 1$, *then*

$$\text{If } n \ge 0, \text{ then } \mathbf{l}(W_n) = \begin{cases} (f_{n-1} + 1, f_{n-2} - 1) & \text{if } n \text{ is odd;} \\ (f_{n-1} - 1, f_{n-2} + 1) & \text{if } n \text{ is even.} \end{cases}$$

2. $W_n \not\prec W_{n+1}$.
3. *If* $\alpha \triangleright W_{n+1}$, *then* $W_{n+2} = W_nW_{n+1}\alpha^{-1}\beta$.
4. $W_n = W_{n-2}W_{n-3}W_{n-2}$, $n \ge 1$.
5. *For* $n \ge 1$,

$$C_{f_{n-1}-1}(F_n) = W_{n-2}W_{n-1},$$
$$C_{f_n-1}(F_n) = W_{n-1}W_{n-2}.$$

In particular,

$$W_{n-2} \prec C_k(F_n) \text{ if and only if } 0 \le k \le f_{n-1} - 1;$$
$$W_{n-1} \prec C_k(F_n) \text{ if and only if } f_{n-1} - 1 \le k \le f_n - 1.$$

6. *For* $n \ge 1$, $W_{2n-1} = aaUaa$, $W_{2n} = bVb$, *where* $U, V \in \mathcal{A}_2^*$.
7. *For* $n \ge 2$, $1 < k < f_n$, *no proper conjugate of* W_n *is a subword of* F_∞.
8. *For* $n \ge 0$, $W_n^2 \not\prec F_\infty$.
9. *For* $n \ge -1$, W_n *is a palindrome.*
10. W_n *is not the product of two palindromes for* $n \ge 2$.
11. *If* $n \ge 2$, *then* W_n *is primitive.*
12. *For* $n \ge 1$, *let* $\alpha_n = a$ *if* n *is odd, and* $\alpha_n = b$ *if* n *is even. we have:*

$$W_n = \alpha_n \left(\prod_{j=-1}^{n-2} W_j \right) = \left(\prod_{j=-1}^{n-2} W_{n-j-3} \right) \alpha_n.$$

13. $W_n \not\prec \prod_{j=-1}^{n-1} W_j$.

14. Let $k \geq -1$ and $p \geq 1$, and let $U = \prod_{j=k}^{k+p} W_j$. *Then* $U \notin S$.

The proofs are left to the readers as exercises (see also [451]).

The following theorem answers the question raised in the introductory part of this section.

Theorem 9.3.6. *We have* $u_\sigma = \prod_{j=-1}^{\infty} W_j$.

Proof. One easily checks that $F_n = abF_0F_1 \cdots F_{n-3}F_{n-2}$. By definition of the singular words W_n, we have $F_k = \alpha^{-1}W_k\beta$. Then by Proposition 9.3.1.4., we get

$$F_n = W_{-1}W_0 \cdots W_{n-2}\gamma_n,$$

where $\gamma_n = a$ if n is even, and b is n is odd. Letting $n \to \infty$ we obtain the conclusion of the theorem. ■

Some additional work [451] is needed to obtain an even more striking decomposition into singular words. In particular, this decomposition shows the non-overlapping property of the singular words and plays an important role in the study of the factors of the Fibonacci word.

Definition 9.3.7. *For every $n \geq 1$, we consider the set $\Sigma_n = \{W_{n-1}, W_{n+1}\}$ as a new alphabet of two letters, and we define a morphism $\phi_n : \mathcal{A}_2 \to \Sigma_n$ by setting $\phi_n(a) = W_{n+1}, \phi_n(b) = W_{n-1}$. We call the sequence $Z = \phi_n(u_\sigma)$ the* Fibonacci word over Σ_n.

By Proposition 9.3.1.6., the word W_n will appear in u_σ infinitely many times. We can now state the main result of this section:

Theorem 9.3.8. *For any $n \geq 0$, we have*

$$u_\sigma = \left(\prod_{j=-1}^{n-1} W_j\right) W_nZ_1W_nZ_2 \cdots W_nZ_kW_n \cdots$$

where $Z = Z_1Z_2 \cdots Z_n \cdots$ is the Fibonacci word over Σ_n, and W_n has no other occurrences than the occurrences shown in the formula.

For example, when $n = 2$, $W_2 = bab$ and the decomposition predicted by Theorem 9.3.8 is:

$$abaa(bab)\underline{aabaa}(bab)\underline{aa}(bab)\underline{aabaa}(bab)\underline{aabaa}(bab)\underline{aa}(bab)\underline{aabaa}(bab)\underline{aa}\cdots$$

Definition 9.3.9. *Let $v = v_1 v_2 \cdots v_n \cdots$ be an infinite word over $\mathcal{A}_2$. Let $X = v_k v_{k+1} \cdots v_{k+p}$ and $Y = v_l v_{l+1} \cdots v_{l+m}$ (with $l \geq k$) be two factors of v, then the distance between the occurrences of the factors X and Y is defined by*

$$d(X, Y) = \begin{cases} l - k - p & \text{if } l > k + p; \\ 0 & \text{otherwise.} \end{cases}$$

If $d(X, Y) > 0$, we say that the words X and Y are positively separated.

Let $W = v_k v_{k+1} \cdots v_{k+p}$ $(k, p \geq 1)$ be a factor of v. If there is an integer l, $1 \leq l \leq p$, such that $W = v_{k+l} v_{k+l+1} \cdots v_{k+l+p}$, then we say that W has an overlap over $p - l$ letters.

The following statement can be seen as equivalent to the above definition. Let $U \prec u_\sigma$ be a factor of the Fibonacci word. Then, the word U has an overlap if and only if there exist words X, Y and Z such that $U = XY = YZ$ and such that the word $UZ = XYZ$ is a factor of u_σ.

The next corollaries are consequences of Theorem 9.3.8.

Corollary 9.3.10. *Two adjacent occurrences of a singular word W_n, $n \geq 1$, are positively separated. More precisely, for any n and k, we have*

$$d(W_{n,k}, W_{n,k+1}) \in \{f_{n+1}, f_{n-1}\}.$$

Moreover, at least one of $d(W_{n,k}, W_{n,k+1})$ and $d(W_{n,k+1}, W_{n,k+2})$ is f_{n+1}.

In particular, for $n \geq 1$, W_n has no overlap, and the words of length f_{n-2k} adjacent to the word W_{n+1} on the left and on the right are precisely W_{n-2k}.

Corollary 9.3.11. *Let $U \prec u_\sigma$ and assume $f_n < |U| \leq f_{n+1}$. Let W be a singular word of maximal length contained in U. Then W appears only once in U. Moreover, W must be one of the three following singular words: W_{n-1}, W_n or W_{n+1}.*

9.3.2 Some applications of singular words

The results we will now describe follow from properties of singular words, and more particularly from the fact that two occurrences of a singular word are positively separated. Some of the results we give here are known (Example 1 and Example 3), but the proofs we provide are simpler.

Example 1. Power of factors [65, 232, 294, 383].

Theorem 9.3.12. *We have:*

1. *for any n, $W_n^2 \not\prec u_\sigma$;*
2. *for $0 \leq k \leq f_n - 1$, $\left(C_k(F_n)\right)^2 \prec u_\sigma$;*
3. *if $U \prec u_\sigma$, with $f_{n-1} < |U| < f_n$, then $U^2 \not\prec u_\sigma$;*
4. *if $0 \leq k \leq f_{n-1} - 2$, then $\left(C_k(F_n)\right)^3 \prec u_\sigma$;*
5. *if $f_{n-1} - 2 < k < f_n$, then $\left(C_k(F_n)\right)^3 \not\prec u_\sigma$;*
6. *for any $U \prec u_\sigma$, $U^4 \not\prec u_\sigma$.*

Proof.

1. The statement follows from Proposition 9.3.1.5. and 9.3.5.6.
2. Let $C_k(F_n) = UV$ with $F_n = VU$. Then $U \triangleright F_n$ and $V \triangleleft F_n$. The result follows from $F_n^3 \prec u_\sigma$, since $\left(C_k(F_n)\right)^2 = UVUV = UF_nV \prec (F_n)^3$.
3. Suppose that W_k is the largest singular word contained in U (as in Corollary 9.3.11), and let $U = V_1 W_k V_2$. Assume that $U^2 = V_1 W_k V_2 V_1 W_k V_2 \prec u_\sigma$. Then $W_k \not\prec V_2 V_1$, otherwise Theorem 9.3.8 implies either that $W_{k+1} \prec V_1$, or $W_{k+1} \prec V_2$, which would contradict the hypothesis on W_k. But then these are consecutive occurrences, of W_k, so by Theorem 9.3.8 again, $V_2 V_1$ must be either W_{k+1} or W_{k-1}. Hence, $U = V_1 W_k V_2$ is conjugated to $W_k V_2 V_1$, which is equal to $W_k W_{k+1}$ or to $W_k W_{k-1}$, and by Proposition 9.3.5.5, U is conjugated either to F_{k+2}, or to F_{k+1}. But these two cases are impossible, since they violate the hypothesis made on U.
4. We have $F_n F_n F_{n-1} F_n \prec f_\infty$ since $aaba \prec f_\infty$. Let $\alpha\beta \triangleright F_{n-1}$. Then by Lemma 9.3.2, we have
$$F_n^2 F_{n-1} F_n = F_n^2 F_{n-1} F_{n-2} F_{n-1} \alpha^{-1}\beta^{-1}\alpha\beta = F_n^3 F_{n-1}\alpha^{-1}\beta^{-1}\alpha\beta \prec u_\sigma.$$
Note that $F_{n-1} \triangleleft F_n$. Hence, if $0 \le k \le f_{n-1} - 2$, then
$$\left(C_k(F_n)\right)^3 \prec F_n^3 F_{n-1}\alpha^{-1}\beta^{-1} \prec u_\sigma.$$
5. Now suppose that $f_{n-1} - 1 < k < f_n$. Then Proposition 9.3.5.5 implies $W_{n-1} \prec C_k(F_n)$. Let $C_k(F_n) = UW_{n-1}V$, so that $VU = W_{n-2}$. Thus
$$\left(C_k(F_n)\right)^3 = UW_{n-1}W_{n-2}W_{n-1}W_{n-2}W_{n-1}V.$$
Now, if $\left(C_k(F_n)\right)^3 \prec u_\sigma$, then the word $W_{n-1}W_{n-2}W_{n-1} = W_{n+1}$ would overlap itself which is impossible, by Corollary 9.3.11.
6. The statement is proved using an argument similar to that developed in statement 5. ∎

Remark. From Theorem 9.3.12.2., we see that any two occurrences of a conjugate of F_n, $n \ge 0$, are not positively separated. This is a major difference between the conjugates of F_n and W_n.

Example 2. Local isomorphism.

Denote by $S^k = S(S^{k-1})$ the kth iteration of the shift S. Properties of the singular words of the Fibonacci word can be used to obtain the following results on the local isomorphism properties of the Fibonacci word.

Theorem 9.3.13. *1. An infinite word u'_σ obtained by changing a finite number of letters in u_σ is not locally isomorphic to u_σ.*

2. Let $U \in \mathcal{A}_2^*$. Then $u_\sigma \simeq Uu_\sigma$ if and only if there exists an integer $m > -1$ such that $U \triangleright W_m\alpha$, with $\alpha = a$ if m is odd, and $\alpha = b$ if m is even.

Proof.

1. Let $u_\sigma = \prod_{j=-1}^{\infty} W_j$ as in Theorem 9.3.6. Because only a finite number of letters are changed in u_σ, we can find an integer m and words $U, V \in \mathcal{A}_2^*$, such that
$$F'_\sigma = UV \prod_{j=m}^{\sigma} W_j,$$
where $|V| = f_{m-1}$, $V \neq W_{m-1}$. Therefore, $VW_m \not\prec u_\sigma$, that is, $u_\sigma \not\simeq F'_\sigma$.
2. From Theorem 9.3.6 and Proposition 9.3.5.12., for any $k > 0$ and $m \geq 0$,
$$\begin{aligned} W_{2m}au_\sigma &= W_{2m}a \left(\prod_{j=-1}^{2m+2k-1} W_j \right) \left(\prod_{j=2m+2k}^{\infty} W_j \right) \\ &= W_{2m}W_{2m+2k+1} \left(\prod_{j=2m+2k}^{\infty} W_j \right). \end{aligned}$$
Then, $W_{2m}W_{2m+2k+1} \prec u_\sigma$. That is, for any $V \prec W_{2m}au_\sigma$, we can find an integer k, such that $V \prec W_{2m}W_{2m+2k+1}$, so $V \prec u_\sigma$. The case dealing with $W_{2m+1}b$ can be proved similarly. That is, if $U \rhd W_m\alpha$ for some m, then $u_\sigma \simeq Uu_\sigma$. The case where U is not a right factor of any $W_m\alpha$, follows from an argument similar to that developed in the proof of statement 1. ■

Example 3. Overlap of the factors of the Fibonacci word.

In this example, we shall determine the factors which have an overlap. First recall the notation we introduced earlier. We say that the word $U \prec u_\sigma$ *has an overlap* if there exist words X, Y and Z such that $U = XY = YZ$ and $XYZ \prec u_\sigma$. Write $\hat{U}(Y) = UZ = XYZ$. We shall say that the word U has an *overlap with overlap factor* Y (or overlap length $|Y|$). The word $\hat{U}(Y)$ is called the *overlap* of U with overlap factor Y. We denote by $\mathcal{O}(u_\sigma) = \mathcal{O}$ the set of factors which have an overlap. Obviously, if $U \in \mathcal{O}$, then
$$|U| + 1 \leq |\hat{U}(Y)| \leq 2|U| - 1, \tag{9.1}$$
where Y is any overlap factor of U.

Lemma 9.3.14. *Let $f_n < |U| \leq f_{n+1}$, and $U \neq W_{n+1}$. Then $U \in \mathcal{O}$ if and only if $W_n \not\prec U$.*

Proof. Let $W_n \prec U$ and write $U = SW_nT$. Note that $W_n \notin \mathcal{O}$, if $U \in \mathcal{O}$. Thus any overlap of U must be of the form SW_nVW_nT. Hence,
$$|SW_nvW_nT| \geq |S| + |T| + 2f_n + f_{n-1} = |U| + f_{n+1} \geq 2|U|,$$
which is in contradiction with (9.1).

Now suppose that $W_n \not\prec U$. We have

- either $U = SF_nT$, where $S, T \neq \varepsilon$, $|S| + |T| \leq f_{n-1}$, $S \triangleright F_n$, $T \triangleleft F_n$;
- or $U \prec F_n^2$.

In the first case, if $|T| = f_n - 1$, then $U = W_{n+1} \notin \mathcal{O}$. Now assume $|T| < f_n - 1$. We can write $F_n = TXS$, because $|S| + |T| \leq f_{n-1}$, with $S \triangleright F_n$, $T \triangleleft F_n$. Since $|T| < f_n - 1$, Theorem 9.3.12.4. implies

$$\left(C_{|T|}(F_n)\right)^3 = (XST)^3 = XSTXSTXST \prec u_\sigma.$$

That is, $U = SF_nT = STXST$ has an overlap with factor ST.

In the second case, observe that $U \prec F_n^2$ and $|U| > f_n$. Writing $U = ST$, with $|T| = f_n$, we get $T = C_k(F_n)$ for some k, and $S \triangleright T$, so $U = SXS$. On the other hand, since $U = SC_k(F_n) \prec F_n^2$, we have $SXSXS = S\left(C_k(F_n)\right)^2 \prec F_n^3 \prec u_\sigma$. That is, $U = SXS$ has an overlap with overlap factor S. ■

Lemma 9.3.15. *If $U \in \mathcal{O}$, then U has a unique overlap factor.*

Proof. Let $f_n < |U| \leq f_{n+1}$, and let W be a singular word contained in U with maximal length. By Corollary 9.3.16, W is one of W_{n-1}, W_n and W_{n+1}. By virtue of Lemma 9.3.14, W must be W_{n-1} since $U \in \mathcal{O}$. So we can write $U = SW_{n-1}T$. Now suppose that U admits two different overlap factors. Then W_{n-1} will appear three times in one of these two overlap factors. Since $W_{n-1} \notin \mathcal{O}$, this overlap factor must be of the form $SW_{n-1}V_1W_{n-1}V_2W_{n-1}T$. An analogous argument to the one developed in Lemma 9.3.14 brings us to a contradiction with (9.1). ■

The following corollary is immediate from Lemma 9.3.15 and the proof of Lemma 9.3.14.

Corollary 9.3.16. *Let $U \in \mathcal{O}$ be such that $f_n < |U| \leq f_{n+1}$. Then U can be decomposed under the form $U = VV'V$, the length of $|V|$ being the overlap length.*

The following theorem summarizes the results above.

Theorem 9.3.17. *Let U be a word such that $f_n < |U| \leq f_{n+1}$. Moreover assume that $U \neq W_{n+1}$ and $U \prec u_\sigma$. Then $U \in \mathcal{O}$ if and only if $W_n \not\prec U$. If $U \in \mathcal{O}$, then U has a unique overlap factor V satisfying $U = VV'V$ and $|V| = |U| - f_n$. In particular, $C_k(F_n) \in \mathcal{O}$ if and only if $0 \leq k \leq f_n - 2$.*

Remark. Note that:

1. $f_{n+1} < 2f_n < f_{n+2} < 3f_n < f_{n+3}$;
2. for any k, $W_{n+1} \not\prec \left(C_k(F_n)\right)^2$;
3. for any k, $W_{n+2} \not\prec \left(C_k(F_n)\right)^3$.

The following corollary is an immediate consequence of Theorem 9.3.17.

Corollary 9.3.18. *For any k, $\left(C_k(F_n)\right)^2 \in \mathcal{O}$, $\left(C_k(F_n)\right)^3 \in \mathcal{O}$.*

Remark. If $W^2 \not\prec u_\sigma$ and W has no overlap factor, then two consecutive occurrences of W are positively separated. Moreover, we can give for those words a decomposition similar to the one we gave for singular words.

For example, let $W = abab$, by Theorem 9.3.12.3. and Theorem 9.3.17, $W^2 \not\prec u_\sigma$ and $W \notin \mathcal{O}$, so W is positively separated. The following decomposition illustrates the remark above:

$$aba(abab)\underline{aaba}(abab)\underline{a}(abab)\underline{aaba}(abab)\underline{aaba}(abab)\underline{a}(abab)\underline{aaba}(abab)\cdots$$

We totally order $\mathcal{A}_2$ by setting $a < b$ and we extend this order to the set $\mathcal{A}_2^*$ of all words lexicographically. A *Lyndon word* is a word strictly smaller than its proper right factors. Melançon [286] established the factorization of the Fibonacci word into a nonincreasing product of Lyndon words. He also described the intimate links between the Lyndon factorization and the factorization into singular words, and also gave a self-similarity property of the Lyndon factorization. Those results enabled him to give new proofs for Theorems 9.3.6 and 9.3.8. He also introduced generalizations of singular words to the Sturmian words. Using another approach, Cao and Wen [103] generalized singular words and the corresponding factorizations to general Sturmian words and gave some applications to the studies of the factor properties of Sturmian words.

9.4 Properties of factors of the fixed points of invertible substitutions

In Sec. 9.3 we discussed some factor properties of the Fibonacci word. It is natural to seek for more general invertible substitutions based on the knowledge we developed on the Fibonacci substitution.

We will study some particular invertible substitutions; note that the matrices associated with σ and π generate the set of invertible nonnegative integral matrices, hence, for any invertible substitution, there is a conjugate substitution which is a product of σ and π.

Such a substitution can be written $\tau = \sigma^{n_k}\pi\sigma^{n_{k-1}}\pi\ldots\sigma^{n_2}\pi\sigma^{n_1}$, with $n_1 \geq 0$, $n_k \geq 0$, and $n_i > 0$ for $2 \leq i \leq k-1$. It is easy to check that the word appears in the fixed point u_τ if and only if $n_k > 0$, and that $\tau(b)$ is a prefix of $\tau(a)$ if $n_1 > 0$, $\tau(a)$ is a prefix of $\tau(b)$ if $n_1 = 0$. We see that we can define in this way four classes of substitutions. We will only study the substitutions such that $n_k > 0$, $n_1 > 0$, the other three classes have similar properties.

Definition 9.4.1. *We denote by $\mathcal{G}_\sigma$ the set of invertible substitutions that can be written as a product $\sigma^{n_k}\pi\sigma^{n_{k-1}}\pi\ldots\sigma^{n_2}\pi\sigma^{n_1}$, with $n_i > 0$ for $1 \leq i \leq k$.*

Section 9.4.1 is concerned with elementary factor properties for the fixed word of a substitution in $\mathcal{G}_\sigma$. Section 9.4.2 defines singular words for general invertible substitutions, and give some applications. These words are defined by using a substitution in $\mathcal{G}_\sigma$; however, as we will see, they have intrinsic properties which allow to define them just by considering the language associated with an invertible substitution.

9.4.1 Elementary properties of factors

Lemma 9.4.2. *Let τ be an invertible substitution. For all $\alpha \in \mathcal{A}_2$, the word $\tau(\alpha)$ is primitive.*

Proof. Suppose that $\tau(\alpha)$ is not primitive. Then there exist a word $W \in \mathcal{A}_2^*$ and an integer $p \geq 2$, such that $\tau(\alpha) = W^p$. Hence $\mathbf{l}(\tau(\alpha))) = (p|W|_a, p|W|_b)$. Thus $\det M_\tau$ will be divisible by p, which is in contradiction with the fact $\det M_\tau = \pm 1$. ■

We remark that, for any $\tau \in \mathcal{G}_\sigma$, we have $\tau(b) \triangleleft \tau(a)$. We remark also that, if $|\tau(b)| > 1$, the last two letters of $\tau(b)$ are distinct, and if we denote them by $\beta\alpha$, the last two letters of $\tau(a)$ are $\alpha\beta$; the proof is easy by induction.

We are now going to establish an important factorization of $\tau \in \mathcal{G}_\sigma$ which will be used in the sequel. We can write $\tau(a) = \tau(b)U^*$, by virtue of the remark. Let $U \triangleleft \tau(a)$ with $|U| = |U^*|$. We can also write $\tau(a) = UV$ with $|V| = |\tau(b)|$.

Lemma 9.4.3. *Let $\tau \in \mathcal{G}_\sigma$ with $|\tau(b)| \geq 2$ and let $\tau(a) = \tau(b)U^* = UV$ as above. If $\alpha\beta \triangleright \tau(a)$, then*

1. *$U = U^*$;*
2. *$\tau(b)\alpha^{-1}\beta^{-1} = V\beta^{-1}\alpha^{-1}$;*
3. *$\tau(a) = \tau(b)U = U\tau(b)\alpha^{-1}\beta^{-1}\alpha\beta$.*

Proof. Since $\alpha\beta \triangleright \tau(a)$, we have $\beta\alpha \triangleright \tau(b)$.

We prove the lemma by induction with respect to the length of τ in terms of $\mathcal{A}_\sigma := \{\sigma, \pi\}$.

1. The inequality $|\tau(b)| \geq 2$ holds only if τ has length at least 2; if this length is 2, then $\tau = \sigma^2$. In this case, $\tau(a) = \sigma^2(a) = aba = \tau(b)a$, $U^* = U = a$, and the statement is true.
   Suppose that for $\tau = \tau_n \cdots \tau_1$, we have
   $$\tau_n \cdots \tau_1(a) = \tau_n \cdots \tau_1(b)U_n^* = U_n V_n, \tag{9.2}$$
   with $U_n^* = U_n$ (so $|V_n| = |\tau_n \cdots \tau_1(b)|$).
   Let $\tau = \tau_{n+1}\tau_n \cdots \tau_1$. Write
   $$\tau(a) = \tau_{n+1}\tau_n \cdots \tau_1(a) = \tau_{n+1}\tau_n \cdots \tau_1(b)U_{n+1}^* = U_{n+1}V_{n+1},$$

with $|U_{n+1}^*| = |U_{n+1}|$. We will prove that $U_{n+1}^* = U_{n+1}$.
If $\tau_{n+1} = \pi$, the proof is trivial. Suppose now that $\tau_{n+1} = \sigma$. By virtue of (9.2), we have:

$$\sigma(\tau_n \cdots \tau_1(a)) = \sigma(U_n V_n) = \sigma(U_n)\sigma(V_n) = \sigma(\tau_n \cdots \tau_1(b))\sigma(U_n^*).$$

Thus $U_{n+1}^* = \sigma(U_n^*)$, $U_{n+1} = \sigma(U_n)$. So we get $U_{n+1}^* = U_{n+1}$, since $U_n^* = U_n$ by induction.

2. The case of $|\tau(b)| = 2$ can be checked directly. Let $\tau = \tau_n \cdots \tau_1$, then by statement 1. we have

$$\tau_{n+1}\tau_n \cdots \tau_1(a) = \tau_{n+1}\tau_n \cdots \tau_1(b)U_{n+1} = U_{n+1}V_{n+1}.$$

Let $\alpha\beta \triangleright \tau_n \cdots \tau_1(a)$ and suppose that $V_n\beta^{-1}\alpha^{-1} = \tau_n \cdots \tau_1(b)\alpha^{-1}\beta^{-1}$. We shall prove

$$V_{n+1}\alpha^{-1}\beta^{-1} = \tau_{n+1}\tau_n \cdots \tau_1(b)\beta^{-1}\alpha^{-1}.$$

As in the proof of statement 1., it suffices to consider $\tau_{n+1} = \sigma$. Since $\alpha\beta \triangleright \tau_{n+1}\tau_n \cdots \tau_1(b)$, $\beta\alpha \triangleright \sigma(V_{n+1})$, we have

$$\begin{aligned} V_{n+1} = \sigma(V_n) &= \sigma(\tau_n \cdots \tau_1(b)\alpha^{-1}\beta^{-1}\alpha\beta) \\ &= \sigma(\tau_n \cdots \tau_1(b))(\sigma(\beta\alpha))^{-1}\sigma(\alpha\beta). \end{aligned}$$

Notice that $\sigma(\alpha\beta) = a\beta\alpha$ for any $\alpha \neq \beta$, and therefore we obtain

$$V_{n+1}\alpha^{-1}\beta^{-1} = \tau_{n+1}\tau_n \cdots \tau_1(b)\beta^{-1}\alpha^{-1}.$$

3. The conclusion follows immediately from statements 1. and 2. ■

Lemma 9.4.4. *Let $W \in \mathcal{P}$, then $\sigma(W)a$, $a^{-1}\sigma(W) \in \mathcal{P}$.*

Proof. Since $W \in \mathcal{P}$, we can write W in the following form:

$$a^{k_1}b^{l_1}a^{k_2}b^{l_2} \cdots a^{k_m}\zeta a^{k_m} \cdots b^{l_2}a^{k_2}b^{l_1}a^{k_1},$$

where $\zeta \in \{a, b, \varepsilon\}$, $k_j, l_j = 1$, if $2 \leq j \leq m-1$, and $k_1, k_m = 0$.
Thus we have

$$\begin{aligned} \sigma(W)a &= (ab)^{k_1}a^{l_1} \cdots (ab)^{k_m}\sigma(\zeta)(ab)^{k_m} \cdots a^{l_2}(ab)^{k_2}a^{l_1}(ab)^{k_1}a \\ &= (ab)^{k_1}a^{l_1} \cdots (ab)^{k_m}\sigma(\zeta)a(ba)^{k_m} \cdots a^{l_2}(ba)^{k_2}a^{l_1}(ba)^{k_1}. \end{aligned}$$

Since $\sigma(\zeta)a$ is equal to a, aba, aa respectively according to ζ being ε, a, b, $\sigma(W)a$ is a palindrome with center a, b, ε respectively by the formula above. ■

Corollary 9.4.5. *Let $\zeta \in \mathcal{A}_2$. Suppose that $\tau \in \mathcal{G}_\sigma$ with $\beta \triangleright \tau(\zeta)$, then $\alpha\tau(\zeta)\beta^{-1} \in \mathcal{P}$.*

In particular, if $|\tau(\zeta)| \geq 3$, and $\alpha\beta \triangleright \tau(\zeta)$, then $\tau(\zeta)\beta^{-1}\alpha^{-1} \in \mathcal{P}$.

Proof. Let $\tau = \tau_n\tau_{n-1}\cdots\tau_1$.

We prove this corollary by induction as in Lemma 9.4.3. We only prove the case $\zeta = a$, the case of $\zeta = b$ can be proved in a similar way.

The case of $n = 1$ is evident.

Now suppose that $\beta \triangleright \tau_n\tau_{n-1}\cdots\tau_1(a)$ and $\alpha\tau_n\tau_{n-1}\cdots\tau_1(a)\beta^{-1} \in \mathcal{P}$.

If $\tau_{n+1} = \pi$, notice that the word $\pi(\tau_n\tau_{n-1}\cdots\tau_1(a))$ is obtained by exchanging the letters a and b in the word $\tau_n\tau_{n-1}\cdots\tau_1(a)$. So by the induction hypothesis, we get $\beta\pi(\tau_n\tau_{n-1}\cdots\tau_1(a))\alpha^{-1} \in \mathcal{P}$.

Now suppose that $\tau_{n+1} = \sigma$. Notice that $\alpha \triangleright \sigma\tau_n\tau_{n-1}\cdots\tau_1(a)$. Thus

$$\begin{aligned}\beta\sigma\tau_n\tau_{n-1}\cdots\tau_1(a)\alpha^{-1} &= \beta\sigma(\alpha^{-1}\alpha\tau_n\tau_{n-1}\cdots\tau_1(a)\beta^{-1}\beta)\alpha^{-1}\\ &= \beta\sigma(\alpha^{-1})\sigma(\alpha\tau_n\tau_{n-1}\cdots\tau_1(a)\beta^{-1})\sigma(\beta)\alpha^{-1}.\end{aligned}$$

By the induction hypothesis, $\alpha\tau_n\tau_{n-1}\cdots\tau_1(a)\beta^{-1} \in \mathcal{P}$. On the other hand, by a simple calculation, we have either $\beta\sigma(\alpha^{-1}) = a^{-1}$, $\sigma(\beta)\alpha^{-1} = \varepsilon$, or $\beta\sigma(\alpha^{-1}) = \varepsilon$, $\sigma(\beta)\alpha^{-1} = a$. Hence by Lemma 9.4.4, we get

$$\beta\sigma\tau_n\tau_{n-1}\cdots\tau_1(a)\alpha^{-1} \in \mathcal{P}.$$

■

Let $\tau(a) = \tau(b)U$ be the factorization of Lemma 9.4.3. Let $\tau = \sigma^{n_k}\pi\cdots\sigma^{n_2}\pi\sigma^{n_1}$, then $U = \sigma^{n_k}\pi\cdots\sigma^{n_2}\pi\sigma^{n_1-1}(b)$. Thus by Lemma 9.4.3 and Corollary 9.4.5, we get:

Corollary 9.4.6. *Let $\tau \in \mathcal{G}_\sigma$ and let $\alpha\beta \triangleright \tau(a)$.*

1. *If $|\tau(b)| = 1$, then $\tau(a) = a^n b$ for some n, i.e., $\tau(a)$ is the product of palindromes a^n and b;*
2. *If $|\tau(b)| = 2$, then $\tau(a)$ is a palindrome;*
3. *If $|\tau(b)| > 2$, then $\tau(a)$ is the product of palindromes $\tau(b)\alpha^{-1}\beta^{-1}$ and $\beta\alpha u$.*

Lemma 9.4.7. *Suppose that $W \in \mathcal{A}_2^*$ is a primitive word and suppose that $W = U_1U_2$ is a product of two palindromes U_1 and U_2.*

1. *If $|U_1|, |U_2| \in 2\mathbb{N}+1$, then for any k, $C_k(W) \notin \mathcal{P}$.*
2. *If $|U_1|, U_2| \in 2\mathbb{N}$, then $C_{|U_1|/2}(W) \in \mathcal{P}$, $C_{|U_1|+|U_2|/2}(W) \in \mathcal{P}$. For other k, $C_k(W) \notin \mathcal{P}$;*
3. *If $|U_1| \in 2\mathbb{N}$, $|U_2| \in 2\mathbb{N}+1$ (or $|U_1| \in 2\mathbb{N}+1$, $|U_2| \in 2\mathbb{N}$), then $C_{|U_1|/2}(W) \in \mathcal{P}$ (or $C_{|U_1|+|U_2|/2}(W) \in \mathcal{P}$). For other k, $C_k(W) \notin \mathcal{P}$.*

Proof. We only prove statement 1., the other cases can be discussed in the same way.

Notice that at least one of $|U_1|$, $|U_2|$ is larger than 2 (otherwise U_1, U_2 will be a^2 or b^2). Suppose that without loosing generality $|U_1| = 3$. Then we can write $U_1 = xU_1'x$, where $x \in \mathcal{A}_2$, $U_1' \in \mathcal{P}$. Thus $C_1(W) = C_1(U_1U_2) = (U_1')(xU_2x)$ is a product of two palindromes, and $C_1(U_1U_2) \notin \mathcal{P}$. Since $|U_1'|, |xU_2x| \in 2\mathbb{N}+1$ and $U_1', xU_2x \in \mathcal{P}$, we can repeat the discussion above which follows the proof. ■

Lemma 9.4.8. *Let $\tau \in \mathcal{G}_\sigma$. Suppose that $|\tau(a)| \in 2\mathbb{N}$. If $\tau(a) = U_1U_2$ is a product of two palindromes, then $|U_1|, |U_2| \in 2\mathbb{N}+1$.*

Proof. If $|U_1|, |U_2| \in 2\mathbb{N}$, then $|U_1|_a + |U_1|_b$, $|U_2|_a + |U_2|_b \in 2\mathbb{N}$. Thus both components of $\mathbf{l}(\tau(a))$ are even which follows that $|\det M_\tau| \in 2\mathbb{N}$. But we know that $\det M_\tau$ must be ± 1. ∎

From Corollary 9.4.6, Lemma 9.4.7, Lemma 9.4.8 and Propositions 9, 10, 11 of [130], we obtain

Proposition 9.4.9. *Let $\tau \in \mathcal{G}_\sigma$.*

1. *Any conjugate of $\tau(a)$ is primitive.*
2. *$|C(\tau(a))| = |\tau(a)|$. That is, all conjugates of $\tau(a)$ are distinct.*
3. *Any conjugation of $\tau(a)$ (containing $\tau(a)$ itself) is either a palindrome, or a product of two palindromes. In the later case, the factorization is unique.*
4. *$C(\tau(a)) = \overline{C(\tau(a))}$, where $\overline{C(\tau(a))} = \{\overline{W};\ W \in C(\tau(a))\}$.*
5. *If $|\tau(a)| \in 2\mathbb{N}$, there is no palindrome in $C(\tau(a))$; if $|\tau(a)| \in 2\mathbb{N}+1$, there is only one palindrome in $C(\tau(a))$.*

9.4.2 Decomposition into singular words and its application

Similarly as in the case of the Fibonacci infinite word, we will introduce singular words for the invertible substitutions in $\mathcal{G}_\sigma$ and study their properties.

Let $\tau \in \mathcal{G}_\sigma$ and let $u_\tau = \tau(u_\tau) = w_1w_2\cdots w_k\cdots$ be the fixed point of τ. Then for any $n \in \mathbb{N}$, we have

$$u_\tau = \tau^n(u_\tau) = \tau^n(w_1)\tau^n(w_2)\cdots\tau^n(w_k)\ldots \tag{9.3}$$

We now discuss the properties of the factors of the infinite word u_τ.

Definition 9.4.10. *We denote $\tau^n(a)$ by A_n, and $\tau^n(b)$ by B_n.*

Notice that if $\tau \in \mathcal{G}_\sigma$, then for any $n \geq 2$, $\tau^n \in \mathcal{G}_\sigma$. Therefore the conclusions about τ which we have obtained above hold also for τ^n. In particular, A_n and B_n are primitive. We also remark that A_n and B_n always have a different final letter.

Lemma 9.4.11. *Any factor of length $|A_n|$ of u_τ is contained either in A_nA_n or B_nA_n.*

Proof. Let W be a factor of u_τ of length $|A_n|$. Since the decomposition of τ in product of σ, π begins with σ (from the definition of $\mathcal{G}_\sigma$, we have $n_k > 0$) it follows that the fixed point contains only isolated occurrences of b, hence it has a canonical decomposition in A_n and B_n where no two consecutive B_n occur. Hence W will be contained in one of the following four words: A_nA_n, A_nB_n, B_nA_n, $A_nB_nA_n$. We have $B_n \lhd A_n$, so $A_nB_n \prec A_nA_n$.

Let $W \prec A_n B_n A_n$, write $W = W_1 B_n W_2$. Then $W_1 \triangleright A_n, W_2 \triangleleft A_n$. Let $A_n = B_n U_n$ be the factorization as in Lemma 9.4.3. Then $U_n \triangleleft A_n$ by Lemma 9.4.3.3. Since $|W| = |A_n| = |B_n| + |U_n|$, $|W_2| < |U_n|$. Consequently $W_2 \triangleleft U_n$, and hence $W = W_1 B_n W_2 \prec A_n B_n U_n = A_n A_n$. ∎

We are going to study the two cases.

Lemma 9.4.12. *Let $W \prec A_n A_n$ with $|W| = |A_n|$, then $W \in C(A_n)$.*

Proof. Since $W \prec A_n A_n$, and $|W| = |A_n|$, we can write $W = UV$ with $U \triangleright A_n, V \triangleleft A_n$, and $|U| + |V| = |A_n|$, which follows that $A_n = VU$. That is, $W = C_{|V|}(A_n)$. ∎

Since all conjugates of A_n are distinct by Primitivity of A_n, we get

Corollary 9.4.13. *The set of the factors of length $|A_n|$ of A_n^k, $k = 2$ is exactly $C(A_n)$.*

Now we study the factors of the word $B_n A_n$. For this aim, we introduce *singular words* of u_τ as follows:

Definition 9.4.14. *The n-th singular words of u_τ with respect to a or b are defined respectively as:*

$$W_{n,a} = \beta A_n \alpha^{-1}, \quad W_{n,b} = \alpha B_n \beta^{-1},$$

where α (respectively β) is the last letter of A_n (respectively B_n).

From the definition of the singular words, we get immediately:

Lemma 9.4.15. *With the notation above, we have*

$$\begin{pmatrix} \mathbf{l}(W_{n,a}) \\ \mathbf{l}(W_{n,b}) \end{pmatrix} = \begin{cases} \begin{pmatrix} |A_n|_a - 1 & |A_n|_b + 1 \\ |B_n|_a + 1 & |B_n|_b - 1 \end{pmatrix} & \text{if } \alpha = a; \\ \begin{pmatrix} |A_n|_a + 1 & |A_n|_b - 1 \\ |B_n|_a - 1 & |B_n|_b + 1 \end{pmatrix} & \text{if } \alpha = b. \end{cases}$$

Hence the abelianizations of $W_{n,a}$ and A_n differ, and we get:

Corollary 9.4.16. $\forall n \in \mathbb{N},\ W_{n,a} \notin C(A_n),\ W_{n,b} \notin C(B_n)$.

Let $\alpha\beta \triangleright A_n$ (so $\beta\alpha \triangleright B_n$). By Lemma 9.4.3,

$$B_n A_n = B_n u_n B_n \alpha^{-1} \beta^{-1} \alpha\beta = A_n B_n \alpha^{-1} \beta^{-1} \alpha\beta.$$

Since $A_n B_n \alpha^{-1} \beta^{-1} \prec A_n B_n \prec A_n A_n$, we see that the first $|B_n| - 2$ factors of length $|A_n|$ of $B_n A_n$ are those of $A_n A_n$ which are distinct from each other by Corollary 9.4.13, and the $(|B_n| - 1)$-th factor is exactly $W_{n,a}$ which is not in $C(A_n)$ by Corollary 9.4.16. We get from the analysis above

Lemma 9.4.17. *Let $W \prec B_nA_n$ with $|W| = |A_n|$. Then W is either in $C(A_n)$ or equal to $W_{n,a}$. In particular, as a factor, $W_{n,a}$ appears only once in B_nA_n.*

Since we know that the fixed point is Sturmian, we deduce that $|\mathcal{L}_{|A_n|}| = |A_n| + 1$, and from the previous lemma we obtain:

Proposition 9.4.18. *Let $\tau \in \mathcal{G}$, $n \in \mathbb{N}$. Then $\mathcal{L}_{|A_n|} = C(A_n(a)) \bigcup \{W_{n,a}\}$, $\mathcal{L}_{|B_n|} = C(B_n(b)) \bigcup \{W_{n,b}\}$.*

We now discuss the properties of the singular words. The above proposition can be rephrased: since the fixed word is Sturmian, it is balanced. Hence, words of the same length as $\tau(a)$ can only have the same number of a as $\tau(a)$, or one more or less a. We have shown that all words of same length as $\tau(a)$, except one, the singular word, have same number of a as $\tau(a)$. We will see that this difference appears in a very conspicuous way: either $W_{n,a}$ has b as initial and final letter, or $a^{p(\tau)+1}$ as initial and final prefix, where $p(\tau)$ is the minimum of the number of a's that separate two occurrences of b. ($p(\tau)$ can also be defined in the following way: let $(\sigma\pi)^k$ the highest power of $\sigma\pi$ that occurs as prefix in the decomposition of τ in the generators σ, π; then $p(\tau) = k + 1$.

Proposition 9.4.19. *$\forall n, m \in \mathbb{N}$, $W_{n,a}, W_{n,b}$ and $\beta A_n^m B_n \alpha^{-1} \in \mathcal{P}$.*

Proof. The fact that $W_{n,a}, W_{n,b} \in \mathcal{P}$ follows immediately from the definition of the singular words and Proposition 9.4.5.

Let $A_n = B_nU_n$ be the factorization as in Lemma 9.4.3, then for any $m \geq 1$,

$$\beta A_n^m B_n \alpha^{-1} = ((\beta B_n \alpha^{-1})(\alpha U_n \beta^{-1}))^m (\beta B_n \alpha^{-1}),$$

by Corollary 9.4.6, $\beta B_n \alpha^{-1}, \alpha u_n \beta^{-1} \in \mathcal{P}$. Thus $\beta A_n^m B_n \alpha^{-1} \in \mathcal{P}$. ■

Lemma 9.4.20. *$W_{n,a} \triangleright W_{n+2,a}$, $W_{n,b} \triangleright W_{n+2,b}$.*

Proof. Let $\alpha\beta \triangleright \tau^n(a)$, then $\alpha\beta \triangleright \tau^{n+2}(a)$, thus

$$W_{n+2,a} = \alpha\tau^{n+2}(a)\beta^{-1} = \alpha\tau^n(\tau^2(a))\beta^{-1}.$$

We have $a \triangleleft \tau^2(a)$, so $\tau^2(a) = ua$ for some $u \in \mathcal{A}_2$. Thus $W_{n+2,a} = \alpha\tau^n(u)\tau^n(a)\beta^{-1}$ which yields the conclusion of the lemma. ■

Lemma 9.4.21. *Either $a^{p(\tau)+1} \triangleright W_{n,a}$, $a^{p(\tau)+1} \triangleleft W_{n,a}$, or $b \triangleright W_{n,a}$, $b \triangleleft W_{n,a}$.*

Proof. This follows from Proposition 9.4.18 and the definition of $W_{n,a}$. ■

Proposition 9.4.22. *$\forall n \in \mathbb{N}$, if $W_{n,a}$ is not a power of a, then $W_{n,a}$ is primitive.*

Proof. Suppose that $W_{n,a}$ is not a power of a. If $W_{n,a} = w^p$ for some $w \in \mathcal{A}_2^*$ and $p \geq 2$, then $W_{n,a}$ will be equal to one of its conjugates. But we will then have either $b^2 \prec u_\tau$ or $a^{2p(\tau)+2} \prec u_\tau$. This is impossible. ∎

Let $u_\infty = w_1 \cdots w_n \cdots$ be an infinite word over $\mathcal{A}_2$. Let $W \prec u_\infty$ be a word which appears infinitely many times in u_∞. If any two adjacent W's (i.e., two successive occurrences) appearing in u_∞ are separated by a factor of u_∞ (this assertion is equivalent to $\nexists U \prec u_\infty$ such that $U = XYZ$ with $W = XY = YZ$), we say that the word W possesses the *positive separation property*, and the factor is called a *separate factor* (with respect to W). If any separate factor is not equal to ε, we say that W is *strongly separated* (otherwise, we say that W is *weakly separated*).

Definition 9.4.23. *A word $W \in \mathcal{A}_2^*$ of the form $W = (UV)^k U$, where $k > 0$, UV is a primitive word, is called a* sesquipower. *The positive integer k is said to be the* order *of the sesquipower. A sesquipower of order larger than* 1 *is called a* strong sesquipower.

From Lemma 9.4.12, Lemma 9.4.17, and the definition of $W_{n,a}$, we obtain the following

Theorem 9.4.24. *Let $\tau \in \mathcal{G}_\sigma$ and let $\alpha\beta \triangleright A_n$. For $n \geq 1$, we have*

$$u_\tau = W_0 W_{n,a} z_1^{(n)} W_{n,a} z_2^{(n)} \cdots z_k^{(n)} W_{n,a} z_{k+1}^{(n)} \cdots ,$$

where $W_0 = A_n^{p(\tau)} B_n \alpha^{-1}$, $\forall i$, $z_i^{(n)} \in \{\beta A_n^{p(\tau)-1} B_n \alpha^{-1}, \beta A_n^{p(\tau)} B_n \alpha^{-1}\}$. We call the above decomposition of u_τ the $n-$decomposition of u_τ,

Denote by $u_\tau(z^{(n)})$ the infinite word $z_1^{(n)} z_2^{(n)} \ldots z_k^{(n)} \ldots$.

From the discussions above and Proposition 9.4.19, we obtain

Theorem 9.4.25. *Let $\tau \in \mathcal{G}_\sigma$, then for any n,*

1. *$W_{n,a}$ is strongly separated.*
2. *The separate factor is either $\beta A_n^{p(\tau)-1} B_n \alpha^{-1}$, or $\beta A_n^{p(\tau)} B_n \alpha^{-1}$.*
3. *The separate factors are sesquipower palindromes of order $p(\tau)$ and $p(\tau)-$ 1 respectively.*

Example 9.4.26. Let $\tau = (\sigma\pi)^{n-1}\sigma$, then

$$\tau = (a^n b, a), \quad p(\tau) = n.$$

Consider 1−decomposition of u_τ, we have

$$ab \triangleright A_1^n = (a^n b)^n, \quad W_0 = (A_1)^n B_1 a^{-1} = (a^n b)^n, \quad W_{1,a} = a(a^n b) b^{-1} = a^{n+1}.$$

Put

$$A = bA_1^n B_1 a^{-1} = b(a^n b)^n a a^{-1} = b(a^n b)^{n-1}, \quad B = b(a^n b)^n.$$

When $n = 1$, $\tau = \sigma$, then $\forall m \in \mathbb{N}$, $u_\sigma(z^{(m)}) = u_\sigma$ over $\{A, B\}$;
When $n = 2$, $\tau = \sigma\pi\sigma$, then

$$u_{\sigma\pi\sigma}(z^{(1)}) = ABABAABABAABABABAABABAABABABAABA\ldots$$

We see that $u_{\sigma\pi\sigma}(z) \neq u_{\sigma\pi\sigma}$ over $\{A, B\}$. A calculation on computer shows that

$$u_{\sigma\pi\sigma}(z^{(1)}) = u_\tau = u_\tau(z^{(1)})$$

where $\tau = (ab, aba) = \sigma\sigma\pi$. ■

As said in the introduction, the motivation of this section is to generalize the properties of the factors of the Fibonacci word to the case of all invertible substitutions. By comparing with the results of Sec. 9.3, we see that one main difference is the structure of the singular words and the singular decomposition of the infinite word according to the singular words. In fact, the structure of the singular words of a general invertible substitution is much more complicated than that of the Fibonacci substitution: for example, all singular words and separate words of the Fibonacci word are sesquipowers of order 1, i.e, weak sesquipowers; on the other hand, those of a general invertible substitution are strong sesquipowers. But we have seen that the singular words still play an important role in the studies of the factors, such as power of factors, overlap of factors and local isomorphism. On the other hand, singular words and their applications can be generalized to the case of Sturmian words [103].

Remark. By introducing *patterns* in high dimensional space (see for example [24, 326]), we can generalize the definition of local isomorphism to the colored lattice, that are applicable in the studies of quasicrystals [458]. But it is a difficult problem to characterize local isomorphism for the general case. In fact, in Theorem 9.2.4, we only deal with the invertible substitutions over the alphabet of two letters. We do not know any results for the general substitutions (which are non-invertible) over $\mathcal{A}_2$. For the alphabet over more than two letters, the question is still open even for invertible substitutions.

10. Polynomial dynamical systems associated with substitutions

It is natural, in many situations related to physics or geometry, to study representations of the free group with values in $SL(2,\mathbb{C})$. Conjugate representations just differ by a change of coordinates, so one should be interested in quantities invariant by conjugacy; the trace is one such quantity, so that, for a representation ϕ, one is interested in studying $\operatorname{tr}\phi(W)$, for an element W of the free group.

For instance, consider the following problem. Given two complex 2×2-matrices $\mathbf{A}_0$ and $\mathbf{B}_0$ with determinant 1, define, for $n \geq 0$, $\mathbf{A}_{n+1} = \mathbf{A}_n\mathbf{B}_n$ and $\mathbf{B}_{n+1} = \mathbf{B}_n\mathbf{A}_n$ (in other words, $\mathbf{A}_n$ is a product of 2^{n+1} matrices, the factors being chosen according to the beginning of the Thue-Morse sequence). How to compute the traces of $\mathbf{A}_n$ and $\mathbf{B}_n$?

One can obtain $\big(\operatorname{tr}\mathbf{A}_n, \operatorname{tr}\mathbf{B}_n, \operatorname{tr}\mathbf{A}_n\mathbf{B}_n\big)$ by iterating the function Φ : $(x,y,z) \longmapsto (z,z,xyz-x^2-y^2+2)$. To be more precise, one has

$$\big(\operatorname{tr}\mathbf{A}_{n+1}, \operatorname{tr}\mathbf{B}_{n+1}, \operatorname{tr}\mathbf{A}_{n+1}\mathbf{B}_{n+1}\big) = \Phi\big(\operatorname{tr}\mathbf{A}_n, \operatorname{tr}\mathbf{B}_n, \operatorname{tr}\mathbf{A}_n\mathbf{B}_n\big).$$

Had we defined $\mathbf{A}_{n+1} = \mathbf{A}_n\mathbf{B}_n$ and $\mathbf{B}_n = \mathbf{A}_n$ (this time using the Fibonacci substitution), we would have obtained

$$\big(\operatorname{tr}\mathbf{A}_{n+1}, \operatorname{tr}\mathbf{B}_{n+1}, \operatorname{tr}\mathbf{A}_{n+1}\mathbf{B}_{n+1}\big) = \big(\operatorname{tr}\mathbf{A}_n\mathbf{B}_n, \operatorname{tr}\mathbf{A}_n, \operatorname{tr}\mathbf{A}_n \operatorname{tr}\mathbf{A}_n\mathbf{B}_n - \operatorname{tr}\mathbf{B}_n\big).$$

This behavior is general: given a substitution σ on the two-letter alphabet $\{\mathbf{A}_0, \mathbf{B}_0\}$, there exists a polynomial map Φ from $\mathbb{C}^3$ into itself such that, if $\mathbf{A}_n = \sigma^n(\mathbf{A}_0)$ and $\mathbf{B}_n = \sigma^n(\mathbf{B}_0)$, one has $\big(\operatorname{tr}\mathbf{A}_{n+1}, \operatorname{tr}\mathbf{B}_{n+1}, \operatorname{tr}\mathbf{A}_{n+1}\mathbf{B}_{n+1}\big) = \Phi\big(\operatorname{tr}\mathbf{A}_n, \operatorname{tr}\mathbf{B}_n, \operatorname{tr}\mathbf{A}_n\mathbf{B}_n\big)$.

Of course, to find such a recursion relation, one could think of expressing the eight entries of $\mathbf{A}_{j+1}$ and $\mathbf{B}_{j+1}$ in terms of those of $\mathbf{A}_j$ and $\mathbf{B}_j$, and then getting, by elimination, a recursion relation linking nine successive values of $\operatorname{tr}\mathbf{A}_j$. As a matter of fact, on the one hand, this method is not so bad: had we considered a recursion involving n matrices, we should have obtained a recursion relation, the length of which grows linearly in n, for the traces. On the other hand, eliminating variables could be an untractable operation, even when using computer algebra software. Besides, this method gives no idea of the algebraic properties of these recurrence formulae.

[1] This chapter has been written by J. Peyrière

Another way of operating, which is developed here, is to exploit polynomial identities in rings of matrices. This will provide an effective algorithm for constructing such recursion relations for traces, the so-called trace maps. Besides, these trace maps exhibit very interesting algebraic and geometric properties.

More generally, given a representation ϕ in the case of the free group Γ_2 on two elements with generators a, b, for any element $W \in \Gamma_2$, there exists a unique polynomial $P_W(x, y, z)$ such that

$$\operatorname{tr} \phi(W) = P_W(\operatorname{tr} \phi(a) \operatorname{tr} \phi(b), \operatorname{tr} \phi(ab));$$

in other words, the trace of any product of 2 matrices $\mathbf{A}, \mathbf{B}$ can be computed by using only $\operatorname{tr} \mathbf{A}, \operatorname{tr} \mathbf{B}, \operatorname{tr} \mathbf{AB}$. Hence, the traces of the representation is completely determined by $[T]_\phi = (\operatorname{tr} \phi(a), \operatorname{tr} \phi(b), \operatorname{tr} \phi(ab)) \in \mathbb{C}^3$.

An object of particular interest are free subgroups of $SL(2, \mathbb{C})$ whose generators $\mathbf{A}, \mathbf{B}$ have a parabolic commutator (that is, $\mathbf{ABA}^{-1}\mathbf{B}^{-1}$ has trace 2, or $\mathbf{A}$ and $\mathbf{B}$ have a common eigenvector). A computation shows that the polynomial P_W associated with the word $W = aba^{-1}b^{-1}$, as defined above, is $P_W(x, y, z) = x^2 + y^2 + z^2 - xyz - 2 = \lambda(x, y, z) + 2$, where $\lambda(x, y, z) = x^2 + y^2 + z^2 - xyz - 4$. Hence, these free subgroups are given by representations ϕ such that $\lambda([T]_\phi) = 0$.

In Chap. 9, we studied endomorphisms σ of the free group; for any representation ϕ whose image is a free group, such an endomorphism gives rise to a new representation $\phi \circ \sigma$, and, by the above, one can find a polynomial map $\Phi_\sigma : \mathbb{C}^3 \to \mathbb{C}^3$ such that $[T]_{\phi \circ \sigma} = \Phi_\sigma([T_\phi])$. We will show that $\lambda \circ \Phi_\sigma$ is always divisible by λ, and equal to λ if σ is an automorphism. Thus, we recover a dynamical system on the surface $\lambda(x, y, z) = 0$ associated with the automorphisms of the free group Γ_2.

More is true: in Chap. 9, we saw that the inner automorphisms are exactly those whose abelianization is the identity. We will prove here that they are also exactly the automorphisms σ such that $\Phi_\sigma = Id$.

It is of course tempting to try to generalize, by increasing the number of letters, or the dimension. Indeed, when dealing with more than two matrices, the situation is more complex. This time, one gets for Φ a polynomial map from a certain affine algebraic variety into itself. We will show some results in this direction, but everything here becomes more difficult, as we saw already in Chap. 9.

The case of representations ϕ with value in $SL(2, \mathbb{R})$ such that $\lambda([T]_\phi) = 0$ is of particular geometric interest. Indeed, consider a once-punctured torus with an hyperbolic metric; such a torus is obtained as a quotient of the hyperbolic plane under the action of a free subgroup of $SL(2, \mathbb{R})$ of rank 2 whose generators have a parabolic commutator, and it is classical that the conjugacy classes of such groups, with given generators, parametrize the Teichmüller space of the once-punctured torus. In fact, the elements $[T]_\phi$ offer an explicitation of this parametrization, as will be proved below.

One is also interested in the modular space: one would like to forget the particular set of generators of the group. A change of generators is nothing but an automorphism σ of the free group, and this is done, at the level of the parametrization, by the polynomial map Φ_σ. Hence the modular space is the quotient of the surface $\lambda(x, y, z) = 0$ by the action of the automorphism group, via Φ_σ. Since the action of the inner automorphism group is trivial, one obtains an action of the outer automorphism group, which is isomorphic to $SL(2\mathbb{Z})$.

We recover, in a completely different way, something very similar to the modular surface discussed at the end of Chap. 6. This is just the beginning of a long story: it may be proved that the modular surface of the compact torus and of the once-punctured torus are isomorphic in a canonical way; the once-punctured torus (whose fundamental group is Γ_2) can be seen as a non-commutative version of the compact torus (whose fundamental group is $\mathbb{Z}^2$); the group of automorphisms of the free group in the first case plays the role of $SL(2, \mathbb{Z})$ in the second, and the linear representations we consider enter naturally by considering geometric structures, instead of considering lattices as we did in Chap. 6. It turns out that continued fractions and Sturmian sequences also enter naturally in the hyperbolic version of the theory (for example, by way of the parametrization of geodesics without self-intersection on the hyperbolic once-punctured torus).

The trace maps are also useful in studying certain physical problems, namely the heat or electric conduction in one-dimensional quasicrystals, modeled as chains of atoms disposed according to a substitutive sequence.

These trace maps have been widely used and studied from the point of view of iteration. But applications as well as the dynamical properties of trace maps are not within the scope of this chapter, which only aims at defining these dynamical systems.

Part of the material of this chapter comes from [328]. In the same volume, which is the proceedings of a school on quasicrystals and deterministic disorder, one can find, besides mathematical developments, many courses showing the importance of finite automata and substitutive sequences for modeling and describing certain situations in condensed matter physics.

The main additions to the lecture given at Les Houches School in Condensed Matter Physics [328] are the following: the proof that $Q_\sigma = 1$ characterizes automorphisms, and the trace maps for 3×3-matrices.

10.1 Polynomial identities in the algebra of 2 × 2-matrices

10.1.1 Some identities for 2 × 2-matrices

In this section upper case letters will stand for 2×2-matrices the entries of which are complex numbers. The basic identity is given by the Cayley-

Hamilton theorem:

$$\mathbf{A}^2 - (\operatorname{tr}\mathbf{A})\mathbf{A} + \mathbf{I}\det\mathbf{A} = 0, \tag{10.1}$$

where $\mathbf{I}$ is the identity matrix of order 2, and $\operatorname{tr}\mathbf{A}$ stands for the trace of $\mathbf{A}$.

As a consequence, one has

$$\mathbf{A}^n = p_n(\operatorname{tr}\mathbf{A}, \det\mathbf{A})\mathbf{A} - p_{n-1}(\operatorname{tr}\mathbf{A}, \det\mathbf{A})\,(\det\mathbf{A})\,\mathbf{I},$$

where p_n's are polynomials in two variables, independent of $\mathbf{A}$, with integer coefficients. If $\mathbf{A}$ is invertible, such a formula is also valid for negative n. The polynomials p_n's are closely related to the Chebyschev polynomials of the second kind. Indeed, if variables are denoted by x and u, we have the following recursion formula: $p_{n+1} = xp_n - up_{n-1}$, from which it results that $p_n(2\cos\varphi, 1) = \sin n\varphi / \sin\varphi$.

One has $\det\mathbf{A} = \lambda\mu = \left[(\lambda+\mu)^2 - (\lambda^2+\mu^2)\right]/2$, if λ and μ are the eigenvalues of $\mathbf{A}$. Therefore the Cayley-Hamilton relation can be rewritten as

$$\mathbf{A}^2 - \mathbf{A}\operatorname{tr}\mathbf{A} + \frac{1}{2}\left[(\operatorname{tr}\mathbf{A})^2 - \operatorname{tr}\mathbf{A}^2\right]\mathbf{I} = 0\,.$$

This form allows bilinearization: writing this formula for $\mathbf{A}$, $\mathbf{B}$, and $\mathbf{A}+\mathbf{B}$, one gets

$$\mathbf{AB} + \mathbf{BA} = \operatorname{tr}\mathbf{AB} - (\operatorname{tr}\mathbf{A})(\operatorname{tr}\mathbf{B}) + \mathbf{A}\operatorname{tr}\mathbf{B} + \mathbf{B}\operatorname{tr}\mathbf{A} \tag{10.2}$$

(we dropped the identity matrix $\mathbf{I}$ as, from now on, we identify scalars and scalar matrices). As this identity is a polynomial identity with integral coefficients linking the entries of matrices $\mathbf{A}$ and $\mathbf{B}$, it is valid for matrices with entries in any commutative ring.

By using (10.2), one gets $\mathbf{A}(\mathbf{AB}+\mathbf{BA}) = (\operatorname{tr}\mathbf{AB} - tr\mathbf{A}\operatorname{tr}\mathbf{B})\,\mathbf{A} + \mathbf{A}^2\operatorname{tr}\mathbf{B} + \mathbf{AB}\operatorname{tr}\mathbf{A}$. Then $\mathbf{ABA} = (\operatorname{tr}\mathbf{AB} - tr\mathbf{A}\operatorname{tr}\mathbf{B})\,\mathbf{A} + \mathbf{A}^2\operatorname{tr}\mathbf{B} + AB\operatorname{tr}\mathbf{A} - \mathbf{A}^2\mathbf{B} = \mathbf{A}\operatorname{tr}\mathbf{AB} + (\mathbf{A}^2 - \mathbf{A}\operatorname{tr}\mathbf{A})\operatorname{tr}\mathbf{B} - (\mathbf{A}^2 - \mathbf{A}\operatorname{tr}\mathbf{A})\mathbf{B}$. Finally, we get the formula

$$\mathbf{ABA} = \mathbf{A}\operatorname{tr}\mathbf{AB} + \mathbf{B}\det\mathbf{A} - \det\mathbf{A}\operatorname{tr}\mathbf{B} \tag{10.3}$$

which will be useful later.

For the sake of simplicity, we shall mostly deal with complex matrices having determinant 1, i.e., elements of $SL(2,\mathbb{C})$).

Proposition 10.1.1. *If $m_1, n_1, m_2, n_2, \cdots, m_k, n_k$ is a sequence of integers, there exist four polynomials p, q, r, and s in three variables with integer coefficients such that, for any pair of matrices $\mathbf{A}$ and $\mathbf{B}$ in $SL(2,\mathbb{C})$, one has*

$$\begin{aligned}\mathbf{A}^{m_1}\mathbf{B}^{n_1}\mathbf{A}^{m_2}\mathbf{B}^{n_2}\cdots\mathbf{A}^{m_k}\mathbf{B}^{n_k} = \;& p(\operatorname{tr}\mathbf{A}, \operatorname{tr}\mathbf{B}, \operatorname{tr}\mathbf{AB}) + \\ & q(\operatorname{tr}\mathbf{A}, \operatorname{tr}\mathbf{B}, \operatorname{tr}\mathbf{AB})\,\mathbf{A} + \\ & r(\operatorname{tr}\mathbf{A}, \operatorname{tr}\mathbf{B}, \operatorname{tr}\mathbf{AB})\,\mathbf{B} + \\ & s(\operatorname{tr}\mathbf{A}, \operatorname{tr}\mathbf{B}, \operatorname{tr}\mathbf{AB})\,\mathbf{AB}.\end{aligned}$$

Proof. By using repeatedly the Cayley-Hamilton theorem for $\mathbf{A}$ and $\mathbf{B}$, one is left with a linear combination with polynomial coefficients of $\mathbf{I}$, $\mathbf{A}$, $\mathbf{B}$, $\mathbf{AB}$, $\mathbf{BA}$, and of products of the form $\mathbf{ABA}\cdots$ or $\mathbf{BAB}\cdots$. Then by using the Cayley-Hamilton theorem for $\mathbf{AB}$ and $\mathbf{BA}$, one is left with a linear combination of I, $\mathbf{A}$, $\mathbf{B}$, $\mathbf{AB}$, $\mathbf{BA}$, $\mathbf{ABA}$, and $\mathbf{BAB}$. We conclude by using (10.2) and (10.3). ∎

Corollary 10.1.2. *Given m's and n's as in Proposition 10.1.1, there exists a unique polynomial P in three variables with integer coefficients such that, for any pair* $(\mathbf{A}, \mathbf{B})$ *of unimodular* 2×2*-matrices, one has*

$$\operatorname{tr} \mathbf{A}^{m_1}\mathbf{B}^{n_1}\cdots\mathbf{A}^{m_k}\mathbf{B}^{m_k} = P(\operatorname{tr}\mathbf{A}, \operatorname{tr}\mathbf{B}, \operatorname{tr}\mathbf{AB})\,.$$

Proof. The existence results from the above proposition. The uniqueness follows from the fact that $(\operatorname{tr}\mathbf{A}, \operatorname{tr}\mathbf{B}, \operatorname{tr}\mathbf{AB})$ can assume any value (x, y, z). To see this, just take

$$\mathbf{A} = \begin{pmatrix} x & 1 \\ -1 & 0 \end{pmatrix} \quad \text{and} \quad \mathbf{B} = \begin{pmatrix} 0 & t \\ z+t & y \end{pmatrix}$$

with $t(z+t)+1 = 0$. ∎

Remark. Had we not restricted the determinants of $\mathbf{A}$ and $\mathbf{B}$ to be 1, the trace of the product above would have been expressed as a polynomial in the five variables $\operatorname{tr}\mathbf{A}$, $\operatorname{tr}\mathbf{B}$, $\operatorname{tr}\mathbf{AB}$, $\det\mathbf{A}$, and $\det\mathbf{B}$.

Examples. Let $\mathbf{A}$ and $\mathbf{B}$ be two unimodular 2×2-matrices, $x = \operatorname{tr}\mathbf{A}$, $y = \operatorname{tr}\mathbf{B}$, $z = \operatorname{tr}\mathbf{AB}$.

1.

$$\begin{aligned}(\mathbf{AB}-\mathbf{BA})^2 &= (\mathbf{AB})^2 + (\mathbf{BA})^2 - \mathbf{AB}^2\mathbf{A} - \mathbf{BA}^2\mathbf{B} \\ &= z(\mathbf{AB}+\mathbf{BA}) - 2 - y\mathbf{ABA} - x\mathbf{BAB} + \mathbf{A}^2 + \mathbf{B}^2 \\ &= z(z - xy + y\mathbf{A} + x\mathbf{B}) - 2 - y(z\mathbf{A} + \mathbf{B} - y) \\ &\qquad - x(z\mathbf{B} + \mathbf{A} - x) + x\mathbf{A} + y\mathbf{B} - 2 \\ &= x^2 + y^2 + z^2 - xyz - 4.\end{aligned}$$

This result is not surprising because $\operatorname{tr}(\mathbf{AB}-\mathbf{BA}) = 0$. The polynomial

$$\lambda(x, y, z) = x^2 + y^2 + z^2 - xyz - 4 \tag{10.4}$$

will play an important role in the sequel. The above formula says that the determinant of $\mathbf{AB}-\mathbf{BA}$ is $-\lambda(x, y, z)$. This is an easy exercise in linear algebra to show that $\det(\mathbf{AB}-\mathbf{BA}) = 0$ if and only if the matrices $\mathbf{A}$ and $\mathbf{B}$ have a common eigendirection. Indeed, let $\mathbf{e}$ be a non-zero element in $\ker(\mathbf{AB}-\mathbf{BA})$. If $\mathbf{e}$ and $\mathbf{Ae}$ are independent, then $(\mathbf{AB}-\mathbf{BA})\mathbf{Ae} = \mathbf{AABe} - \mathbf{BAAe} = x(\mathbf{AB}-\mathbf{BA})\mathbf{e} = 0$, which means that $\mathbf{AB} = \mathbf{BA}$ (recall that we are here in dimension 2). If $\mathbf{Ae} = \rho\mathbf{e}$, then $\rho\mathbf{Be} = \mathbf{ABe}$, which means that, if $\mathbf{A} \neq \rho\mathbf{I}$, there exists ρ' such that $\mathbf{Be} = \rho'\mathbf{e}$.

Thus $\mathbf{A}$ *and* $\mathbf{B}$ *have a common eigendirection if and only if* $\lambda(x, y, z) = 0$.

2.

$$\begin{aligned}\mathbf{ABA}^{-1}\mathbf{B}^{-1} &= \mathbf{AB}(x - \mathbf{A})(y - \mathbf{B}) \\ &= (\mathbf{AB})^2 + xy\mathbf{AB} - y\mathbf{ABA} - x\mathbf{AB}^2 \\ &= (\mathbf{AB})^2 + x\mathbf{A} - y\mathbf{ABA} \\ &= z\mathbf{AB} - 1 + (x - yz)\mathbf{A} - y(\mathbf{B} - y) \\ &= z\mathbf{AB} + (x - yz)\mathbf{A} - y\mathbf{B} + y^2 - 1.\end{aligned}$$

Therefore $\operatorname{tr} \mathbf{ABA}^{-1}\mathbf{B}^{-1} = \lambda(x, y, z) + 2$. ■

We now turn our attention to formulae involving more than two elements of $SL(2, \mathbb{C})$. As a consequence of (10.2), we have the following proposition.

Proposition 10.1.3. *If* $\{\mathbf{A}_j\}_{1 \leq j \leq n}$ *are elements of* $SL(2, \mathbb{C})$, *then*

1. *any product constructed from these matrices or their inverses, can be written as a linear combination of the* 2^n *matrices* $\mathbf{A}_{i_1}\mathbf{A}_{i_2} \cdots \mathbf{A}_{i_k}$ $(0 \leq k \leq n,\ i_1 < i_2 < \cdots < i_k)$ [1] *the coefficients of which are polynomials in the* $2^n - 1$ *variables* $\operatorname{tr} \mathbf{A}_{i_1}\mathbf{A}_{i_2} \cdots \mathbf{A}_{i_k}$ $(1 \leq k \leq n,\ i_1 < i_2 < \cdots < i_k)$,
2. *the trace of such a product can be expressed as a polynomial with integer coefficients in the* $2^n - 1$ *traces defined above.*

We now turn to formulae which involve three matrices $\mathbf{A}_1$, $\mathbf{A}_2$, and $\mathbf{A}_3$ in $SL(2, \mathbb{C})$. Let x_1, x_2, x_3, y_1, y_2, and y_3 denote the traces of $\mathbf{A}_1$, $\mathbf{A}_2$, $\mathbf{A}_3$, $\mathbf{A}_2\mathbf{A}_3$, $\mathbf{A}_3\mathbf{A}_1$, and $\mathbf{A}_1\mathbf{A}_2$.

Define the following polynomials:

$$p(X, Y) = x_1y_1 + x_2y_2 + x_3y_3 - x_1x_2x_3 \tag{10.5}$$

$$\begin{aligned}q(X, Y) = x_1^2 + x_2^2 + x_3^2 + y_1^2 + y_2^2 + y_3^2 \\ - x_1x_2y_3 - x_2x_3y_1 - x_3x_1y_2 + y_1y_2y_3 - 4\end{aligned} \tag{10.6}$$

where X stands for the collection of x's and similarly for Y.

Proposition 10.1.4 (Fricke lemma). *One has*

1. $\operatorname{tr}(\mathbf{A}_1\mathbf{A}_2\mathbf{A}_3) + \operatorname{tr}(\mathbf{A}_1\mathbf{A}_3\mathbf{A}_2) = p(X, Y)$
2. $\operatorname{tr}(\mathbf{A}_1\mathbf{A}_2\mathbf{A}_3) \operatorname{tr}(\mathbf{A}_1\mathbf{A}_3\mathbf{A}_2) = q(X, Y)$.

[1] Of course, for $k = 0$, this product should be interpreted as the identity matrix.

Proof. To prove assertion 1, write $\mathbf{A}_1\mathbf{A}_2\mathbf{A}_3 + \mathbf{A}_1\mathbf{A}_3\mathbf{A}_2 = \mathbf{A}_1(\mathbf{A}_2\mathbf{A}_3 + \mathbf{A}_3\mathbf{A}_2)$ and use (10.2).

To prove assertion 2, write $[\mathbf{A}_1(\mathbf{A}_2\mathbf{A}_3)\mathbf{A}_1]\mathbf{A}_3\mathbf{A}_2 = \mathbf{A}_1\mathbf{A}_2(\mathbf{A}_3\mathbf{A}_1\mathbf{A}_3)\mathbf{A}_2$ and use (10.3) twice. We obtain

$$\mathbf{A}_1\mathbf{A}_3\mathbf{A}_2\operatorname{tr}(\mathbf{A}_1\mathbf{A}_2\mathbf{A}_3) + \mathbf{A}_2\mathbf{A}_3^2\mathbf{A}_2 - y_1\mathbf{A}_3\mathbf{A}_2 = \\ y_2\mathbf{A}_1\mathbf{A}_2\mathbf{A}_3\mathbf{A}_2 + (\mathbf{A}_1\mathbf{A}_2)^2 - x_1\mathbf{A}_1\mathbf{A}_2^2.$$

By reducing further, we get

$$\mathbf{A}_1\mathbf{A}_3\mathbf{A}_2\operatorname{tr}(\mathbf{A}_1\mathbf{A}_2\mathbf{A}_3) = -2 + x_3^2 + y_1^2 - y_1x_2x_3 + (x_1 - y_2x_3)\mathbf{A}_1 + x_2\mathbf{A}_2 \\ -(x_3 - y_1x_2)\mathbf{A}_3 + (y_1y_2 + y_3 - x_1x_2)\mathbf{A}_1\mathbf{A}_2 + y_2\mathbf{A}_1\mathbf{A}_3 - y_1\mathbf{A}_2\mathbf{A}_3\,,$$

from which assertion 2 follows by taking the trace. ■

Corollary 10.1.5. $\operatorname{tr}(\mathbf{A}_1\mathbf{A}_2\mathbf{A}_3)$ *and* $\operatorname{tr}(\mathbf{A}_1\mathbf{A}_3\mathbf{A}_2)$ *are the roots of the equation*

$$z^2 - p(X,Y)z + q(X,Y) = 0.$$

This leads to define a polynomial in seven variables

$$\Lambda(X,Y,z) = z^2 - p(X,Y)z + q(X,Y) \tag{10.7}$$

This corollary means that variables x's, y's, and $z = \operatorname{tr}(\mathbf{A}_1\mathbf{A}_2\mathbf{A}_3)$ are not independent. Indeed the set of polynomials P, in seven variables with integer coefficients such that, for any triple $\{\mathbf{A}_j\}_{1\le j\le 3}$ of elements of $SL(2,\mathbb{C})$, one has

$$P(\operatorname{tr}\mathbf{A}_1, \operatorname{tr}\mathbf{A}_2, \operatorname{tr}\mathbf{A}_3, \operatorname{tr}\mathbf{A}_2\mathbf{A}_3, \operatorname{tr}\mathbf{A}_3\mathbf{A}_1, \operatorname{tr}\mathbf{A}_1\mathbf{A}_2, \operatorname{tr}\mathbf{A}_1\mathbf{A}_2\mathbf{A}_3) = 0,$$

is an ideal containing Λ. It can be shown that this ideal is generated by Λ. Therefore, the polynomial the existence of which is asserted in Proposition 10.1.3-2 is not unique when $n > 2$. In the case $n = 3$ it is defined up to a multiple of Λ.

Proposition 10.1.6. *Let z stand for* $\operatorname{tr}\mathbf{A}_1\mathbf{A}_2\mathbf{A}_3$. *One has*

$$2\mathbf{A}_1\mathbf{A}_2\mathbf{A}_3 = z - x_1y_1 - x_3y_3 + x_1x_2x_3 + (y_1 - x_3x_3)\mathbf{A}_1 - y_2\mathbf{A}_2 \\ +(y_3 - x_1x_2)\mathbf{A}_3 + x_3\mathbf{A}_1\mathbf{A}_2 + x_2\mathbf{A}_1\mathbf{A}_3 + x_1\mathbf{A}_2\mathbf{A}_3\,.$$

Proof. In $\mathbf{A}_1(\mathbf{A}_2\mathbf{A}_3)$, commute $\mathbf{A}_1$ and $\mathbf{A}_2\mathbf{A}_3$ by using (10.2). We get, among other terms, $-\mathbf{A}_2\mathbf{A}_3\mathbf{A}_1$. By using (10.2) twice, on can make $\mathbf{A}_1$ to jump over $\mathbf{A}_3$ and $\mathbf{A}_2$. So $\mathbf{A}_1\mathbf{A}_2\mathbf{A}_3$ can be written as $-\mathbf{A}_1\mathbf{A}_2\mathbf{A}_3$ plus a linear combination of I, $\mathbf{A}_1$, $\mathbf{A}_2$, $\mathbf{A}_3$, and products of two such matrices. ■

Corollary 10.1.7. *If n is larger than 3 and $\{\mathbf{A}_j\}_{1\le j\le n}$ are elements of $SL(2,\mathbb{C})$, then*

1. *any product constructed from these matrices or their inverses, can be written as a linear combination of the matrices* $\mathbf{I}$, $\mathbf{A}_i$ $(1 \le i \le n)$, *and* $\mathbf{A}_{i_1}\mathbf{A}_{i_2}$ $(1 \le i_1 < i_2 \le n)$, *the coefficients of which are polynomials in the variables* $\operatorname{tr}\mathbf{A}_i$ $(1 \le i \le n)$, $\operatorname{tr}\mathbf{A}_{i_1}\mathbf{A}_{i_2}$ $(1 \le i_1 < i_2 \le n)$, *and* $\operatorname{tr}\mathbf{A}_{i_1}\mathbf{A}_{i_2}\mathbf{A}_{i_3}$ $(1 \le i - 1 < i_2 < i_3 \le n)$.
2. *the trace of such a product can be expressed as a polynomial with rational coefficients in the* $n(n^2+5)/6$ *traces defined above.*

This last corollary is a significant improvement on Proposition 10.1.3 when n is larger than 3.

As a matter of fact, one can go further reducing the number of traces needed. It results from [276] that the trace of a product of the kind considered above can be expressed as a rational fraction in the variables $\operatorname{tr}\mathbf{A}_i$ $(1 \le i \le n)$, $\operatorname{tr}\mathbf{A}_i\mathbf{A}_j$ $(i < j,\ 1 \le i \le 3,\ 1 < j \le n)$, and $\operatorname{tr}\mathbf{A}_1\mathbf{A}_2\mathbf{A}_3$ (see also [329]).

10.1.2 Free groups and monoids

Let $\mathcal{A} = \{A_1, A_2, \cdots, A_n\}$ be a finite set called alphabet.

Free semi-group generated by $\mathcal{A}$. Let $\mathcal{A}^*$ be the set of words over the alphabet $\mathcal{A}$. Recall that the product W_1W_2 of two of its elements is just the word obtained by putting the word W_2 after the word W_1. This operation is called concatenation. It is associative and has a unit element, the empty word, denoted by ε. The set $\mathcal{A}^*$ endowed with this structure is called the free semi-group or free monoid generated by $\mathcal{A}$.

Free group generated by $\mathcal{A}$. We perform the same construction as above with the alphabet $\{A_1, A_2, \cdots, A_n,\ A_1^{-1}, \cdots, A_n^{-1}\}$, but we introduce the following simplification rules:

$$A_jA_j^{-1} = A_j^{-1}A_j = \varepsilon \ \ (\text{for } 1 \le j \le n).$$

We obtain a group[2] which we will denote by $\Gamma_{\mathcal{A}}$ and call the free group generated by $\mathcal{A}$.

In the case of a two-letter alphabet, $\Gamma_{\{a,b\}}$ will be simply denoted by Γ.

Abelianization map. The notations defined for the free monoid extend to the free group: if W is an element of $\Gamma_{\mathcal{A}}$ and $a \in \mathcal{A}$, $|W|_a$ stands for the sums of exponents of a in W (one can easily be convinced that $|W|_a$ only depends on W and not of the particular word used to represent it). Moreover, $|W|_a$ is independent of the order of the factors in W. Recall that the mapping $W \longmapsto \mathbf{l}(W) = (|W|_{A_1}, \cdots, |W|_{A_n})$ (where $A_1, A_2, \cdots, A_n$ are the elements of $\mathcal{A}$) is a group homomorphism of $\Gamma_{\mathcal{A}}$ onto $\mathbb{Z}^n$ known as the abelianization map.

[2] Strictly speaking, elements of this group are not words, but equivalence classes of words.

For instance,

$$|aba^{-1}|_a = 0 \qquad |aba^{-1}|_b = 1$$
$$|ab^{-1}a^{-2}|_a = -1 \quad |ab^{-1}a^{-2}|_b = -1.$$

Representations in $SL(2,\mathbb{C})$. A representation of $\mathcal{A}^*$ or $\Gamma_\mathcal{A}$ in $SL(2,\mathbb{C})$ is a mapping φ from $\mathcal{A}^*$ or $\Gamma_\mathcal{A}$ into $SL(2,\mathbb{C})$ such that

$$\varphi(W_1W_2) = \varphi(W_1)\varphi(W_2)$$

for any W_1 and W_2 in $\mathcal{A}^*$ or $\Gamma_\mathcal{A}$.

Such a representation is determined by the values $\mathbf{A}_j$ of $\varphi(A_j)$ for $j = 1, 2, \cdots, n$. Computing $\varphi(W)$ simply consists in replacing each letter in W by the corresponding matrix.

10.1.3 Reformulation in terms of PI-algebras

The following proposition is mainly a reformulation of the corollary to Proposition 10.1.1.

Proposition 10.1.8. *For any $W \in \Gamma$, there exists a unique polynomial P_W with integer coefficients such that, for any representation φ of Γ in $SL(2,\mathbb{C})$, one has*

$$\operatorname{tr}\varphi(W) = P_W(\operatorname{tr}\varphi(a), \operatorname{tr}\varphi(b), \operatorname{tr}\varphi(ab))\,.$$

Moreover, if $\mathbf{l}(W_1) = \mathbf{l}(W_2)$, the polynomial $P_{W_1} - P_{W_2}$ is divisible by λ.

Proof. We only have to prove the second assertion. Consider a representation φ such that the matrices $\mathbf{A} = \varphi(a)$ and $\mathbf{B} = \varphi(b)$ share an eigenvector. As these matrices are simultaneously trigonalizable, the trace of a product of $\mathbf{A}$'s and $\mathbf{B}$'s does not depend on the order of factors. This means that $\operatorname{tr}\varphi(W_1) = \operatorname{tr}\varphi(W_2)$. In other terms, we have $P_{W_1}(x,y,z) = P_{W_2}(x,y,z)$ as soon as $\lambda(x,y,z) = 0$. The conclusion then follows from the irreducibility of λ. ■

According to Horowitz [213], polynomials P_W are called Fricke characters of Γ.

The following notation will prove to be convenient: if φ is a representation of Γ in $SL(2,\mathbb{C})$, set

$$[T]\varphi = \big(\operatorname{tr}\varphi(a), \operatorname{tr}\varphi(b), \operatorname{tr}\varphi(ab)\big). \tag{10.8}$$

With this notation, the equation of definition of P_W is $\operatorname{tr}\varphi(W) = P_W([T]\varphi)$.

The reader may wonder whether it was necessary to replace lower case letters by upper case ones (i.e., to replace a letter by its image under a representation) in the previous calculations. Indeed, this is not compulsory. If we analyze what we have done, we have just considered a and b to be generators of an algebra on the ring $\mathbb{Z}[x, y, z]$, subject to the following relations:

$$a^2 - xa + 1 = 0,\ b^2 - yb + 1 = 0, \text{and } ab + ba = z - xy + ya + xb\,.$$

It is an algebra with polynomial identities (a PI algebra) which we call the Procesi-Razmyslov algebra on a two-letter alphabet. Then Proposition 10.1.1 can be interpreted as giving a homomorphism of the group algebra[3] of Γ to the Procesi-Razmyslov algebra.

We denote again by tr the $\mathbb{Z}[x, y, z]$-linear form on this algebra which maps 1, a, b, and ab respectively on 2, x, y, and z. Then, for any $W \in \Gamma$, one has $\operatorname{tr} W = P_W$. Moreover, it is easy to show that, if u and v are two elements of this algebra, one has $\operatorname{tr} uv = \operatorname{tr} vu$.

10.2 Trace maps

10.2.1 Endomorphisms of free groups

A map σ from $\Gamma_\mathcal{A}$ to $\Gamma_\mathcal{A}$ is an endomorphism of $\Gamma_\mathcal{A}$ if, for any W_1 and W_2 in $\Gamma_\mathcal{A}$, one has

$$\sigma(W_1 W_2) = \sigma(W_1)\sigma(W_2)\ .$$

In the case where none of the words $\sigma(A_j)$ contains negative powers, σ is an endomorphism of $\mathcal{A}^*$ and we recover the notion of substitution on the alphabet $\mathcal{A}$.

Obviously, an endomorphism σ is determined by $\sigma(A_j)$ $(j = 1, 2, \cdots, n)$. Hereafter we shall identify an endomorphism σ and the collection of words $(\sigma(A_1), \sigma(A_2), \cdots, \sigma(A_n))$.

For instance[4], $\sigma = (ab, a)$ means that σ is the endomorphism, so called the Fibonacci substitution, such that $\sigma(a) = ab$ and $\sigma(b) = a$. In this case, as an example, let us compute $\sigma(aba^{-1})$:

$$\begin{aligned}\sigma(aba^{-1}) &= \sigma(a)\sigma(b)\sigma(a)^{-1} \\ &= aba(ab)^{-1} = abab^{-1}a^{-1}\ .\end{aligned}$$

The composition of endomorphisms is simply the composition of maps. This is illustrated by the following examples:

- $(ab, ba) \circ (ab, a) = (abba, ab)$,
- $(ab, a) \circ (b, b^{-1}a) = (b, b^{-1}a) \circ (ab, a) = (a, b)$ (this means that, as an endomorphism of $\Gamma_\mathcal{A}$, the Fibonacci substitution is invertible).

[3] This algebra is the set of finite formal linear combinations of elements of Γ endowed with the bilinear multiplication which extends the product in Γ.

[4] In the case of a two-letter alphabet, we prefer to denote by a and b the generators instead of A_1 and A_2.

Recall that the $n \times n$-matrix $\mathbf{M}_\sigma$ whose entry of indices (i,j) is $|\sigma(A_j)|_{A_i}$ is, by definition, the matrix of the endomorphism σ.

For instance, the Fibonacci substitution matrix is $\begin{pmatrix} 1 & 1 \\ 1 & 0 \end{pmatrix}$, and the one of the Morse substitution, (ab, ba), is $\begin{pmatrix} 1 & 1 \\ 1 & 1 \end{pmatrix}$.

One has, for any σ and W,

$$\mathbf{l}(\sigma(W)) = \mathbf{M}_\sigma \mathbf{l}(W)$$

and, for any pair of endomorphisms,

$$\mathbf{M}_{\sigma_1 \circ \sigma_2} = \mathbf{M}_{\sigma_1} \times \mathbf{M}_{\sigma_2} .$$

10.2.2 Trace maps (two-letter alphabet)

Recall that Γ stands for the free group on the two generators a and b.

Definition of trace maps. Let σ be an endomorphism of Γ. We define the trace map associated with σ to be

$$\Phi_\sigma = \left(P_{\sigma(a)}, P_{\sigma(b)}, P_{\sigma(ab)}\right) . \tag{10.9}$$

It can be considered as well as a map from $\mathbb{C}^3$ to $\mathbb{C}^3$.

Let us compute the trace map for the Morse substitution $\sigma = (ab, ba)$. We operate in the Procesi-Razmyslov algebra. We wish to compute $(\operatorname{tr} ab, \operatorname{tr} ba, \operatorname{tr} ab^2a)$. We have $\operatorname{tr} ab = \operatorname{tr} ba = z$ and $ab^2a = a(yb-1)a = y(za+b-y) - (xa-1)$, so $\operatorname{tr} ab^2a = xyz - y^2 - x^2 + 2$. At last, $\Phi_\sigma = \left(z, z, xyz - x^2 - y^2 + 2\right)$.

Here are a few examples of trace maps.

σ	Φ_σ
inner automorphism	(x, y, z)
(a^{-1}, b^{-1})	(x, y, z)
(b, a)	(y, x, z)
(ab, b^{-1})	(z, y, x)
(b, a^{-1})	$(y, x, xy - z)$
(ab, a)	$(z, x, xz - y)$
$(b, b^{-1}a)$	$(y, xy - z, x)$
(ab, ba)	$(z, z, xyz - x^2 - y^2 + 2)$
(aba, b)	$(xz - y, y, z^2 - 2)$
(a^2b, ba)	$(xz - y, z, x^2yz - x^3 - xy^2 - yz + 3x)$
(aab, bab)	$(xz - y, yz - x, xyz^2 - (x^2 + y^2 - 1)z)$

First properties of trace maps.

Proposition 10.2.1. *For any endomorphisms σ and τ of Γ, we have $\Phi_{\sigma \circ \tau} = \Phi_\tau \circ \Phi_\sigma$.*

Proof. We have the following characterization of Φ_τ: for any representation φ,

$$[T](\varphi \circ \sigma) = \Phi_\sigma([T]\varphi).$$

The proposition then results from

$$\Phi_{\sigma\circ\tau}([T]\varphi) = [T](\varphi \circ \sigma \circ \tau) = \Phi_\tau([T](\varphi \circ \sigma)) = \Phi_\tau \circ \Phi_\sigma([T]\varphi)\,.$$

■

Corollary 10.2.2. *For any endomorphism σ of Γ, and for any $W \in \Gamma$, one has*

$$P_{\sigma(W)} = P_W \circ \Phi_\sigma.$$

Proof. Let τ be the endomorphism (W, b). Then $P_{\sigma(W)}$ is the first component of $\Phi_{\sigma\circ\tau}$, i.e., the first component of Φ_τ composed with Φ_σ.

As a consequence, if σ has a fixed point W, the corresponding trace map Φ_σ leaves the surfaces of $P_W(x, y, z) = \text{constant}$ globally invariant. ■

If σ is invertible, then $\Phi_\sigma \circ \Phi_{\sigma^{-1}} = \text{id}$. Taking the Jacobian, we get

$$\det(\Phi'_\sigma \circ \Phi_{\sigma^{-1}}) \det(\Phi'_{\sigma^{-1}}) = 1\,.$$

As these determinants are polynomials with integer coefficients, we must have $\det \Phi'_\sigma \equiv 1$ or $\det \Phi'_\sigma \equiv -1$.

As an example, consider the Morse substitution for which we have

$$\Phi'_\sigma = \begin{pmatrix} 0 & 0 & 1 \\ 0 & 0 & 1 \\ yz - 2x & xz - 2y & xy \end{pmatrix}.$$

The corresponding determinant is 0, so the Morse substitution is not invertible.

Let us consider another example: $\sigma = (aba, b)$. Then

$$\Phi'_\sigma = \begin{pmatrix} z & -1 & x \\ 0 & 1 & 0 \\ 0 & 0 & 2z \end{pmatrix}.$$

The determinant equals $2z^2$, therefore σ is not invertible.

Proposition 10.2.3. *For any endomorphism σ of Γ, there exists a polynomial Q_σ with integer coefficients such that*

$$\lambda \circ \Phi_\sigma = \lambda \cdot Q_\sigma.$$

Proof. Let x, y, and z be such that $\lambda(x,y,z) = 0$ and $z \neq 0$. Choose unimodular matrices $\mathbf{A}$ and $\mathbf{B}$ such that $\operatorname{tr}\mathbf{A} = x$, $\operatorname{tr}\mathbf{B} = y$, and $\operatorname{tr}\mathbf{AB} = z$, and consider the representation φ defined by $\varphi(a) = \mathbf{A}$ and $\varphi(b) = \mathbf{B}$. As $\mathbf{A}$ and $\mathbf{B}$ share an eigenvector, so do $\varphi(\sigma(a))$ and $\varphi(\sigma(b))$. So we have $\lambda(\mathrm{T}(\varphi \circ \sigma)) = 0$. Therefore $\lambda(x,y,z) = 0$ implies $\lambda\left(\Phi_\sigma(x,y,z)\right) = 0$. Since λ is irreducible, it divides $\lambda \circ \Phi_\sigma$. ■

As a consequence, any Φ_σ leaves globally invariant the surface Ω the equation of which is $\lambda(x,y,z) = 0$. Moreover, the restriction of Φ_σ to Ω only depends on $\mathbf{M}_\sigma$.

Lemma 10.2.4. *For any σ, we have $Q_\sigma(0,0,0) = 0$ or 1.*

Proof. This is checked by testing on matrices $\begin{pmatrix} 0 & 1 \\ -1 & 0 \end{pmatrix}$ and $\begin{pmatrix} 0 & i \\ i & 0 \end{pmatrix}$. ■

Proposition 10.2.5. *If σ and τ are endomorphisms, we have*

$$Q_{\sigma\circ\tau} = Q_\sigma \cdot Q_\tau \circ \Phi_\sigma .$$

Proof. We have

$$\lambda \cdot Q_{\sigma\circ\tau} = \lambda \circ \Phi_{\sigma\circ\tau} = (\lambda \circ \Phi_\tau) \circ \Phi_\sigma = \lambda \circ \Phi_\sigma \cdot Q_\tau \circ \Phi_\sigma = \lambda \cdot Q_\sigma \cdot Q_\tau \circ \Phi_\sigma .$$

■

Corollary 10.2.6. *If σ is invertible, then $Q_\sigma \equiv 1$ and Φ_σ leaves globally invariant each surface $\lambda(x,y,z) =$ constant.*

Characterization of automorphisms of Γ in terms of $\mathbf{Q}_\sigma$. In this section, if $W \in \Gamma$, we shall also denote the polynomial P_W by $\operatorname{tr} W$. Let us set $\operatorname{tr} a = x$, $\operatorname{tr} b = y$, and $\operatorname{tr} ab = z$. As we have seen in Sec. 10.1.1, we have $\operatorname{tr} a^n = u_n(x)a - u_{n-1}(x)$ for $n \in \mathbb{Z}$, where the polynomials u_n satisfy $u_0 = 0$, $u_1(x) = 1$, and $u_{n+1}(x) + u_{n-1}(x) = x\, u_n(x)$.

Two elements W and W' of Γ are conjugate if there exists $V \in \Gamma$ such that $W' = VWV^{-1}$. In this case, we have $\operatorname{tr} W = \operatorname{tr} W'$.

Any $W \in \Gamma$ is conjugate to ε, a^m, b^n, or to $a^{m_1}b^{n_1}a^{m_2}b^{n_2}\cdots a^{m_k}b^{m_k}$ with $\prod_{j=1}^k m_j n_j \neq 0$ (such a form will be called a cyclic reduction of W). In the latter case, we shall say that k is the width of W, otherwise the width of W is 0.

Lemma 10.2.7. *If $W \in \Gamma$, the degree, $\mathrm{d}_z^\circ P_W$, of P_W with respect to the variable z equals the width of W.*

Proof. It goes by induction on the width k of W. This is true for $k = 0$.

Suppose that $k \geq 1$ and the lemma is true for any word of width less than k. Consider $W = a^{m_1}b^{n_1}a^{m_2}b^{n_2}\cdots a^{m_k}b^{m_k}$, and write $W = a^{m_1}b^{n_1}W'$. One has $W = a^{m_1}b^{n_1}a^{m_1}a^{-m_1}W'$. Equality (10.3) shows that

$$\begin{aligned} W &= \left[a^{m_1}\operatorname{tr} a^{m_1}b^{n_1} - b^{-n_1}\right] a^{-m_1}W' \\ &= W' \operatorname{tr} a^{m_1}b^{n_1} - b^{-n_1}a^{-m_1}W'. \end{aligned}$$

But, the second term of the last equality has a width less than k. One has $a^{m_1}b^{n_1} = \left(u_{m_1}(x)a - u_{m_1-1}(x)\right)\left(u_{n_1}(y)b - u_{n_1-1}(y)\right)$, from which it results that $\mathrm{d}_z^\circ \operatorname{tr} a^{m_1}b^{n_1} = 1$. Therefore, $\mathrm{d}_z^\circ \operatorname{tr} W = \mathrm{d}_z^\circ \operatorname{tr} W' + 1$. This proves the lemma. ■

Lemma 10.2.8. *If $W \in \Gamma$ is such that $\operatorname{tr} W = \alpha z$, with $\alpha \in \mathbb{Z}$, then $\alpha = 1$ and the cyclic reduction of W is either ab or $a^{-1}b^{-1}$.*

Proof. If we had $\alpha = 0$, the cyclic reduction of W would be ε, a^m, or b^n. But, the trace is nonzero in any of these cases. Therefore $\alpha \neq 0$.

So, by the preceding lemma, a cyclic reduction of W is of the form $a^m b^n$ with $mn \neq 0$. Therefore we have

$$\alpha z = u_m(x)u_n(y)\, z - y\, u_{m-1}(x)u_n(y) - x\, u_m(x)u_{n-1}(y) + 2u_{m-1}(x)u_{n-1}(y).$$

By looking at the coefficients of z in both sides, we get $|m| = |n| = 1$. It is then easy to show that we must have $mn = 1$. ■

Lemma 10.2.9. *For $W \in \Gamma$, if $\operatorname{tr} W = \alpha x$, then $\alpha = 1$ and W is conjugate either to a or to a^{-1}.*

Proof. Consider the following automorphism of Γ: $\sigma = (ab, b^{-1})$. As one has $\Phi_\sigma(x, y, z) = (z, y, x)$, it results from the corollary to Proposition 10.2.1 that $\operatorname{tr} \sigma(W) = \alpha z$. Then, by virtue of the preceding lemma, $\alpha = 1$ and $\sigma(W)$ writes $V(ab)^{\pm 1}V^{-1}$. Then, $W = \sigma^{-1}(V)a^{\pm 1}\sigma^{-1}(V^{-1})$. ■

Of course, a similar result holds if $\operatorname{tr} W = \alpha y$.

Proposition 10.2.10. *Let σ be an endomorphism of Γ. Then $\Phi_\sigma = \mathrm{Id}$ if and only if σ is either an inner automorphism of Γ or an inner automorphism composed with the involution (a^{-1}, b^{-1}).*

Proof. Suppose $\Phi_\sigma = \mathrm{Id}$. It results from the preceding lemma that $\sigma(a) = Ua^\varepsilon U^{-1}$ and $\sigma(b) = Vb^\eta V^{-1}$, with $|\varepsilon| = |\eta| = 1$. By Proposition 10.1.8, λ divides $\operatorname{tr} \sigma(ab) - \operatorname{tr} a^\varepsilon b^\eta$. This implies $\varepsilon = \eta$. By composing, if necessary, with (a^{-1}, b^{-1}), we may suppose that $\varepsilon = \eta = 1$.

We assume that the words UaU^{-1} and VbV^{-1} are reduced (i.e., there are no cancellations). If $U = V = \varepsilon$, there is nothing to be proved. Suppose that $|U| > 1$ and write $U = Wb^n$, with either $W = \varepsilon$ or W ending with an a. Then, $\sigma(ab) = Wb^n ab^{-n}W^{-1}VbV^{-1}$; so, $z = \operatorname{tr} \sigma(ab) =$

$\mathrm{tr}(ab^{-n}W^{-1}VbV^{-1}Wb^n)$. This means that, once reduced, $W^{-1}V$ does not contain a. Therefore $W^{-1}V = b^k$. This shows that σ is an inner automorphism.

Now we turn to the study of polynomial maps from $\mathbb{C}^3$ to $\mathbb{C}^3$ which leave λ invariant. Let us set

$$\mathcal{G} = \left\{\psi \in \mathbb{C}[x,y,z]^3 \;\middle|\; \lambda \circ \psi = \lambda\right\}.$$

Of course $\mathcal{G}$ contains $\left\{\Phi_\sigma \;\middle|\; \sigma \in \mathrm{Aut}\,\Gamma\right\}$ (by the corollary to Proposition 10.2.5).

It will be convenient to name some elements of $\mathrm{Aut}\,\Gamma$:

$$\alpha = (b,a), \quad \beta = (ab, b^{-1}), \quad \gamma = (a, b^{-1}).$$

The corresponding trace maps are (y,x,z), (z,y,x), and $(x,y,xy-z)$.

We also consider the following elements of $\mathcal{G}$: $\rho(x,y,z) = (-x,-y,z)$ and $\theta(x,y,z) = (-x,y,-z)$.

We shall use the following notations: d° stands for the total degree of a polynomial in three variables, and, if $\psi = (\psi_1,\psi_2,\psi_3) \in \mathbb{C}[x,y,z]^3$, $\deg\psi = \sum_{j=1}^{3} \mathrm{d}^\circ\,\psi_j$. ■

Lemma 10.2.11. *If $\psi = (\psi_1,\psi_2,\psi_3) \in \mathcal{G}$, then, for $j = 1, 2, 3$, we have $\mathrm{d}^\circ\,\psi_j \geq 1$*

Proof. If, for instance, we had $\psi_3 = c \in \mathbb{C}$, then we would have $\psi_1^2 + \psi_2^2 - c\psi_1\psi_2 = x^2 + y^2 + z^2 - xyz - c^2$, which is impossible, for the left-hand side is reducible whereas the right-hand side is not. ■

Lemma 10.2.12. *The set $\mathcal{L} = \left\{\psi \in \mathcal{G} \;\middle|\; \deg\psi = 3\right\}$ is the group generated by Φ_α, Φ_β, and ρ.*

Proof. Let us call the variables x_1, x_2, and x_3 instead of x, y, and z. We have $\psi_j = \ell_j + h_j$, where ℓ_j is linear and $h_j \in \mathbb{C}$. We have

$$\sum_{j=1}^{3}(\ell_j + h_j)^2 - \prod_{j=1}^{3}(\ell_j + h_j) = x_1^2 + x_2^2 + x_3^2 - x_1x_2x_3.$$

Looking at terms of degree 3 gives $\ell_j = k_j x_{\tau(j)}$, where τ is a permutation, $k_j \in \mathbb{C}$, and $k_1k_2k_3 = 1$. Looking at quadratic terms gives $h_1 = h_2 = h_3 = 0$ and $k_1^2 = k_2^2 = k_3^2 = 1$. The result follows easily. ■

Lemma 10.2.13. *If $\psi \in \mathcal{G}$ is such that $\deg\psi > 3$, there exists σ in $\langle\alpha,\beta,\gamma\rangle$, the group generated by α, β, and γ, such that $\deg\Phi_\sigma \circ \psi < \deg\psi$.*

Proof. By replacing ψ by $\Phi_\sigma \circ \psi$, where σ is a suitable element of $\langle \alpha, \beta \rangle$, we may suppose $\mathrm{d}^\circ \psi_3 \geq \mathrm{d}^\circ \psi_2 \geq \mathrm{d}^\circ \psi_1$. Moreover, since $\deg \psi > 3$, we have $\mathrm{d}^\circ \psi_3 \geq 2$.

Since $\psi \in \mathcal{G}$, we have

$$\psi_3(\psi_3 - \psi_1\psi_2) + \psi_2^2 + \psi_1^2 = x^2 + y^2 + z^2 - xyz. \tag{10.10}$$

If we had $\mathrm{d}^\circ \psi_3 \neq \mathrm{d}^\circ \psi_1\psi_2$, the degree of the left-hand side of (10.10) would be

$$\max(2\,\mathrm{d}^\circ \psi_3, \mathrm{d}^\circ \psi_1 + \mathrm{d}^\circ \psi_2 + \mathrm{d}^\circ \psi_3) \geq 4,$$

which is impossible.

We have $\mathrm{d}^\circ \psi_3 = \mathrm{d}^\circ \psi_1\psi_2 > \mathrm{d}^\circ \psi_2$. If we had $\mathrm{d}^\circ(\psi_3 - \psi_1\psi_2) \geq \mathrm{d}^\circ \psi_3$, then we would have $2\,\mathrm{d}^\circ \psi_3 = 3$, which is absurd. Therefore $\mathrm{d}^\circ(\psi_3 - \psi_1\psi_2) < \mathrm{d}^\circ \psi_3$, and $\deg \Phi_\gamma \circ \psi < \deg \psi$. ■

Proposition 10.2.14. *$\mathcal{G}$ is the group generated by Φ_α, Φ_β, Φ_γ, and ρ.*

Proof. Apply Lemma 10.2.13 repeatedly and conclude by using Lemma 10.2.12. ■

Proposition 10.2.15. *For an endomorphism σ of Γ, $Q_\sigma = 1$ if and only if σ is an automorphism.*

Proof. $Q_\sigma = 1$ is equivalent to $\Phi_\sigma \in \mathcal{G}$. Due to Proposition 10.2.14 and to commutation relations

$$\Phi_\alpha \rho \Phi_\alpha = \rho, \quad \Phi_\beta \theta \Phi_\beta = \theta, \quad \Phi_\alpha \theta \Phi_\alpha = \Phi_\beta \rho \Phi_\beta = \rho\theta,$$

there exists $\tau \in \langle \alpha, \beta, \gamma \rangle$ such that $\Phi_\tau \circ \Phi_\sigma \in \langle \rho, \theta \rangle$. Then Lemma 10.2.8 and 10.2.9 and Proposition 10.2.10 show that $\sigma \circ \tau$ is an automorphism. ■

Corollary 10.2.16. $\mathrm{Aut}\, F = \langle \alpha, \beta, \gamma \rangle$.

Proof. If $\sigma \in \mathrm{Aut}\, F$, then $\Phi_\sigma \in \mathcal{G}$. So, there exists $\tau \in \langle \alpha, \beta, \gamma \rangle$ such that $\tau\sigma$ is an inner automorphism or an inner automorphism composed with (a^{-1}, b^{-1}). But, as $(a^{-1}, b^{-1}) = (\alpha\gamma)^2$, the corollary will be proved once we have shown that an inner automorphism is in $\langle \alpha, \beta, \gamma \rangle$. It is easily checked that, if i_W stands for the inner automorphism $V \mapsto WVW^{-1}$, we have $i_a = \alpha\gamma\beta\alpha\gamma\alpha\gamma\beta\alpha\gamma$ and $i_b = \alpha i_a \alpha$. ■

For further properties of trace maps, see [330].

10.2.3 Trace maps (n-letter alphabet)

Three-letter alphabet. If φ is a representation of Γ_A in $SL(2, \mathbb{C})$, we define $[T]\varphi$ to be the following collection of traces:

$$\big(\operatorname{tr}\varphi(A_1), \operatorname{tr}\varphi(A_2), \operatorname{tr}\varphi(A_3), \operatorname{tr}\varphi(A_2A_3), \operatorname{tr}\varphi(A_3A_1), \\ \operatorname{tr}\varphi(A_1A_2), \operatorname{tr}\varphi(A_1A_2A_3)\big)$$

and recall the definitions of several polynomials

$$\Lambda(X,Y,z) = z^2 - p(X,Y)z + q(X,Y)$$

where

$$p(X,Y) = x_1y_1 + x_2y_2 + x_3y_3 - x_1x_2x_3$$

and

$$q(X,Y) = x_1^2 + x_2^2 + x_3^2 + y_1^2 + y_2^2 + y_3^2 - x_1x_2y_3 - x_2x_3y_1 - x_3x_1y_2 + y_1y_2y_3 - 4$$

(as previously, X stands for the collection of x's, and similarly for Y).

Let $\mathcal{V}$ be the hyper-surface in $\mathbb{C}^7$ the equation of which is $\Lambda(X,Y,z) = 0$. It can be seen that any point of $\mathcal{V}$ is of the form $[T]\varphi$ (see [330, 329]).

Proposition 10.1.3 (or the corollary to Proposition 10.1.6), shows that, for any $W \in \Gamma_A$, there exits a polynomial P_W such that $\operatorname{tr}\varphi(W) = P_W([T]\varphi)$ for any representation φ. As we have already observed, this polynomial is no longer unique. It is indeed defined up to the addition of a multiple of polynomial Λ (i.e., modulo the ideal $\mathcal{I}$ generated by Λ).

Now, if we have an endomorphism σ of Γ_A, we choose a collection of polynomials

$$\Phi_\sigma = \left(P_{\sigma(A_1)}, P_{\sigma(A_2)}, P_{\sigma(A_3)}, P_{\sigma(A_2A_3)}, P_{\sigma(A_3A_1)}, P_{\sigma(A_1A_2)}, P_{\sigma(A_1A_2A_3)}\right) .$$

This Φ_σ defines a map from $\mathbb{C}^7$ to $\mathcal{V}$, the restriction of which to $\mathcal{V}$ does not depend on the different choices. Indeed, this is this map from $\mathcal{V}$ to $\mathcal{V}$ which is the trace map and which we call Φ_σ. As previously, $[T](\varphi \circ \sigma) = \Phi_\sigma([T]\varphi)$ for any φ, and $\Phi_{\sigma\circ\tau} = \Phi_\tau \circ \Phi_\sigma$.

In order to show that, as previously, there exists an algebraic sub-manifold Ω of $\mathcal{V}$ which is globally invariant under any Φ_σ, we need the following lemma of which we omit the proof.

Lemma 10.2.17. *Three matrices* $\mathbf{A}_1$, $\mathbf{A}_2$, *and* $\mathbf{A}_3$ *in* $SL(2,\mathbb{C})$ *have a common eigendirection if and only if* $\Lambda(X,Y,z) = \lambda(x_1,x_2,y_3) = \lambda(x_2,x_3,y_1) = \lambda(x_3,x_1,y_2) = 0$ *and* $p(X,Y)^2 - 4q(X,Y) = 0$, *where*

$$z = \operatorname{tr}\mathbf{A}_1\mathbf{A}_2\mathbf{A}_3,\ X = (x_1,x_2,x_3) = (\operatorname{tr}\mathbf{A}_1, \operatorname{tr}\mathbf{A}_2, \operatorname{tr}\mathbf{A}_3), \\ \text{and}\ \ Y = (y_1,y_2,y_3) = (\operatorname{tr}\mathbf{A}_2\mathbf{A}_3, \operatorname{tr}\mathbf{A}_3\mathbf{A}_1, \operatorname{tr}\mathbf{A}_1\mathbf{A}_2).$$

Let Ω be the manifold associated with the ideal $\mathcal{J}$ generated by the polynomials Λ, $\lambda(x_1,x_2,y_3)$, $\lambda(x_2,x_3,y_1)$, $\lambda(x_3,x_1,y_2)$, and $p^2 - 4q$.

Then an argument similar to the one used in the proof of Proposition 10.2.3 shows that Ω is invariant under any Φ_σ.

n-letter alphabet. When n is larger than 3, some complications occur and the situation is less easy to describe. In view of the corollary to Proposition 10.1.6, we need $n(n^2+5)/6$ variables. The ideal $\mathcal{I}$ of relations between these variables is no longer principal. The trace maps take the variety $\mathcal{V}$ of $\mathcal{I}$ to itself. They still leave globally invariant a sub-variety Ω of $\mathcal{V}$ defined by an ideal $\mathcal{J}$ the definition of which comes from expressing that n elements of $SL(2,\mathbb{C})$ have a common eigendirection.

10.3 The case of 3 × 3-matrices

If $\mathbf{M}$ is a 3 × 3-matrix, the Cayley-Hamilton identity can be written as

$$\mathbf{N}^3 - (\operatorname{tr}\mathbf{N})\mathbf{N}^2 + \frac{1}{2}\left((\operatorname{tr}\mathbf{N})^2 - \operatorname{tr}\mathbf{N}^2\right)\mathbf{N} - \frac{1}{6}(\operatorname{tr}\mathbf{N})^3 + \frac{1}{2}(\operatorname{tr}\mathbf{N})(\operatorname{tr}\mathbf{N}^2) - \frac{1}{3}\operatorname{tr}\mathbf{N}^3 = 0. \tag{10.11}$$

So, by trilinearization, ones get the formula

$$\begin{aligned}\sum_{\eta} \mathbf{N}_{\eta(1)}\mathbf{N}_{\eta(2)}\mathbf{N}_{\eta(3)} &- (\operatorname{tr}\mathbf{N}_{\eta(1)})\mathbf{N}_{\eta(2)}\mathbf{N}_{\eta(3)} \\ &+ \frac{1}{2}(\operatorname{tr}\mathbf{N}_{\eta(1)})(\operatorname{tr}\mathbf{N}_{\eta(2)})\mathbf{N}_{\eta(3)} \\ &- \frac{1}{2}(\operatorname{tr}\mathbf{N}_{\eta(1)}\mathbf{N}_{\eta(2)})\mathbf{N}_{\eta(3)} - \frac{1}{6}\operatorname{tr}\mathbf{N}_{\eta(1)}\mathbf{N}_{\eta(2)}\mathbf{N}_{\eta(3)} \\ &+ \frac{1}{2}(\operatorname{tr}\mathbf{N}_{\eta(1)})\operatorname{tr}\mathbf{N}_{\eta(2)}\mathbf{N}_{\eta(3)} \\ &- \frac{1}{3}(\operatorname{tr}\mathbf{N}_{\eta(1)})(\operatorname{tr}\mathbf{N}_{\eta(2)})(\operatorname{tr}\mathbf{N}_{\eta(3)}) = 0,\end{aligned}$$

where the summation runs over the permutations η of $\{1,2,3\}$ and where $\mathbf{N}_1$, $\mathbf{N}_2$, and $\mathbf{N}_3$ are arbitrary 3 × 3-matrices. If in this formula one takes $\mathbf{N}_1 = \mathbf{N}_2 = \mathbf{M}$ and $\mathbf{N}_3 = \mathbf{N}$, one gets

$$\begin{aligned}\mathbf{MNM} + \mathbf{M}^2\mathbf{N} + \mathbf{NM}^2 &= (\operatorname{tr}\mathbf{N})\mathbf{M}^2 + (\operatorname{tr}\mathbf{M})(\mathbf{NM} + \mathbf{MN}) \\ &\quad - (\operatorname{tr}\mathbf{M}\operatorname{tr}\mathbf{N} - \operatorname{tr}\mathbf{MN})\mathbf{M} \\ &\quad - \frac{1}{2}\left((\operatorname{tr}\mathbf{M})^2 - \operatorname{tr}\mathbf{M}^2\right)(\mathbf{N} - \operatorname{tr}\mathbf{N}) \\ &\quad + \operatorname{tr}\mathbf{M}^2\mathbf{N} - \operatorname{tr}\mathbf{M}\operatorname{tr}\mathbf{MN}.\end{aligned} \tag{10.12}$$

Let $\mathcal{A} = \{a, b\}$ be a two-letter alphabet. Consider the subset $\mathcal{S}_0 = \{\varepsilon\}$ of the free monoid $\mathcal{A}^*$, of which the unit ε is the only element. We are going to construct by induction a sequence $\mathcal{S}_n$ of subsets of $\mathcal{A}^n$. Suppose that we know $\mathcal{S}_n$. Then, $\mathcal{S}_{n+1}$ will be the set $\mathcal{S}_n\, a \cup \mathcal{S}_n\, b$ from which the elements ending exactly by a^3, aba, bab, b^3, ab^2a, ba^2b, a^2ba^2, or b^2ab^2 have been removed (by "ending exactly", we mean, for instance, that a^2ba is *not* removed). We get

$$\begin{aligned} \mathcal{S}_1 &= \{a, b\} \\ \mathcal{S}_2 &= \{a^2, ab, ba, b^2\} \\ \mathcal{S}_3 &= \{ba^2, b^2a, a^2b, ab^2\} \\ \mathcal{S}_4 &= \{b^2a^2, a^2ba, b^2ab, a^2b^2\} \\ \mathcal{S}_5 &= \{a^2b^2a, b^2a^2b\} \\ \mathcal{S}_6 &= \{b^2a^2ba, a^2b^2ab\} \\ \mathcal{S}_7 &= \emptyset \end{aligned} \tag{10.13}$$

Define

$$\mathcal{S} = \bigcup_{n=0}^{6} \mathcal{S}_n. \tag{10.14}$$

The property of the $\mathcal{S}_n$'s which matters to us is the following. Suppose we are given a representation φ from $\mathcal{A}^*$ in $\mathcal{M}_3(\mathbb{C})$, the ring of 3 × 3-matrices with complex entries. Then, for any $W \in \mathcal{S}_n$, the matrices $\varphi(Wa)$ and $\varphi(Wb)$ can be expressed as a linear combination of the matrices $\{\varphi(V)\}_{V \in \bigcup_{i=0}^{n+1} \mathcal{S}_i}$, of which the coefficients are polynomials, which can be chosen independent of φ, in the variables $\operatorname{tr}\varphi(a^3)$, $\operatorname{tr}\varphi(b^3)$, and $\{\operatorname{tr}\varphi(V)\}_{V \in \bigcup_{i=0}^{n+1} \mathcal{S}_i}$. The verification of this property is left to the reader. It involves repeated use of (10.11) and (10.12). Also, it is important to notice that the traces of $\varphi(V)$, for $V \in \mathcal{S}_6$, are not involved.

This can be summarized in the following proposition.

Proposition 10.3.1. *Given a word W in $\{a, b\}^*$, there exists polynomials $\{p_V\}_{V \in \mathcal{S}}$ in ten variables with rational coefficients such that, for any representation φ of $\{a, b\}^*$ in $\mathcal{M}_3(\mathbb{C})$, one has*

$$\varphi(W) = \sum_{V \in \mathcal{S}} p_V(T_\varphi)\varphi(V),$$

where

$$T_\varphi = \Big(\operatorname{tr}\varphi(a), \operatorname{tr}\varphi(a^2), \operatorname{tr}\varphi(a^3),$$
$$\operatorname{tr}\varphi(b), \operatorname{tr}\varphi(b^2), \operatorname{tr}\varphi(b^3), \operatorname{tr}\varphi(ab), \operatorname{tr}\varphi(ab^2), \operatorname{tr}\varphi(a^2b), \operatorname{tr}\varphi(a^2b^2)\Big).$$

Lemma 10.3.2. *There exists a polynomial p in ten variables with rational coefficients such that, for any representation φ of $\mathcal{A}^*$ in $\mathcal{M}_3(\mathbb{C})$, one has*

$$\operatorname{tr}\varphi(b^2a^2ba) + \operatorname{tr}\varphi(a^2b^2ab) = p(T_\varphi).$$

Proof. Multiply on the left identity (10.12) by $\mathbf{N}^2\mathbf{M}$, then use (10.11) and (10.12). ∎

Lemma 10.3.3. *There exists a polynomial q such that, for any representation of $\mathcal{A}^*$ in $\mathcal{M}_3(\mathbb{C})$, we have*

$$\operatorname{tr}\varphi(a^2b^2ab)\,\operatorname{tr}\varphi(b^2a^2ba) = q(T_\varphi).$$

Proof. By multiplying (10.12) on the right by $\mathbf{L}$, one gets

$$\begin{aligned}\operatorname{tr}\mathbf{MN}\operatorname{tr}\mathbf{ML} = {} & \operatorname{tr}\mathbf{MNML} + \operatorname{tr}\mathbf{M}^2\mathbf{NL} + \operatorname{tr}\mathbf{M}^2\mathbf{NL} - \operatorname{tr}\mathbf{N}\operatorname{tr}\mathbf{M}^2\mathbf{L} \\ & - \operatorname{tr}\mathbf{L}\operatorname{tr}\mathbf{M}^2\mathbf{N} - (\operatorname{tr}\mathbf{M})(\operatorname{tr}\mathbf{MLN} + \operatorname{tr}\mathbf{MNL}) \\ & + \operatorname{tr}\mathbf{M}\operatorname{tr}\mathbf{N}\operatorname{tr}\mathbf{ML} + \operatorname{tr}\mathbf{M}\operatorname{tr}\mathbf{L}\operatorname{tr}\mathbf{MN} \\ & + \frac{1}{2}\Big((\operatorname{tr}\mathbf{M})^2 - \operatorname{tr}\mathbf{M}^2\Big)(\operatorname{tr}\mathbf{NL} - \operatorname{tr}\mathbf{N}\operatorname{tr}\mathbf{L}).\end{aligned}$$

By putting $\mathbf{M} = \varphi(ab)$, $\mathbf{N} = \varphi(a^2b^2)$, and $\mathbf{L} = \varphi(ba^2b)$ in the preceding identity, one gets

$$\begin{aligned}&\operatorname{tr}\varphi(b^2a^2ba) \times \operatorname{tr}\varphi(a^2b^2ab) = \\ &\quad p_1(T_\varphi)\,\operatorname{tr}\varphi(b^2a^2ba) + p_2(T_\varphi)\,\operatorname{tr}\varphi(a^2b^2ab) + p_3(T_\varphi),\end{aligned} \tag{10.15}$$

where p_1, p_2, and p_3 are polynomials in ten variables.

Then, by replacing in (10.15) the matrices $\varphi(a)$ and $\varphi(b)$ by their transpose, one gets

$$\begin{aligned}&\operatorname{tr}\varphi(b^2a^2ba) \times \operatorname{tr}\varphi(a^2b^2ab) = \\ &\quad p_2(T_\varphi)\,\operatorname{tr}\varphi(b^2a^2ba) + p_1(T_\varphi)\,\operatorname{tr}\varphi(a^2b^2ab) + p_3(T_\varphi).\end{aligned} \tag{10.16}$$

By adding (10.15) and (10.16) and taking Lemma 10.3.2 into account, one gets

$$\operatorname{tr}\varphi(b^2a^2ba) \times \operatorname{tr}\varphi(a^2b^2ab) = \frac{1}{2}\Big(p_1(T_\varphi) + p_2(T_\varphi)\Big) + p_3(T_\varphi).$$

Let us consider the following polynomial in eleven variables with rational coefficients

$$\Lambda = \tau^2 - p\tau + q,$$

where p and q are defined in Lemma 10.3.2 and 10.3.3. It results from these lemmas that, for any homomorphism φ of $\Gamma_{\langle a,b\rangle}$ in $\mathcal{M}_3(\mathbb{C})$, the roots of $\Lambda(T_\varphi, \tau)$ are $\operatorname{tr} a^2b^2ab$ and $\operatorname{tr} b^2a^2ba$. ∎

Proposition 10.3.4. *The polynomial Λ is irreducible on $\mathbb{C}$.*

Proof. Consider the homomorphism φ of $\Gamma_{\langle a,b\rangle}$ in $\mathcal{M}_3(\mathbb{C})$ so defined

$$\varphi(a) = \begin{pmatrix} 0 & t & 0 \\ 0 & 0 & t^{-1} \\ 1 & 0 & 0 \end{pmatrix} \text{ and } \varphi(b) = \begin{pmatrix} 0 & 1 & 0 \\ 0 & 0 & 1 \\ 1 & 0 & 0 \end{pmatrix}.$$

It is straightforward to check that $T\varphi = (0,0,3,0,0,3,0,x,x,0)$, $p(T_\varphi) = x^2 - 3$, and $q(T_\varphi) = 2x^3 - 6x^2 + 9$, where $x = 1 + t + t^{-1}$. This gives $(p^2 - 4q)(0,0,3,0,0,3,0,x,x,0) = (x-3)^3(x+1)$.

If Λ were not irreducible, the polynomial $p^2 - 4q$ would be a square and so would be the polynomial $(p^2 - 4q)(0,0,3,0,0,3,0,x,x,0)$, which it is not. ∎

Proposition 10.3.5. *For any $W \in \Gamma_{\langle a,b\rangle}$ there exists a polynomial P_W with rational coefficients in eleven variables such that, for any homomorphism φ from $\Gamma_{\langle a,b\rangle}$ to $\mathcal{M}_3(\mathbb{C})$, one has*

$$\operatorname{tr}\varphi(W) = P_W(T_\varphi, \operatorname{tr}\varphi(a^2b^2ab)).$$

This polynomial is unique modulo the principal ideal generated by Λ.

Proof. The existence of P_W results from Proposition 10.3.1 and Lemma 10.3.2. Its uniqueness modulo λ comes from the fact that the derivative of the mapping $\varphi \longmapsto (T_\varphi, \operatorname{tr}(\varphi(a^2b^2ab)))$ is of rank 10 at some φ, for instance for φ such that

$$\varphi(a) = \begin{pmatrix} 1 & 1 & 0 \\ 0 & 1 & 2 \\ 0 & 0 & 1 \end{pmatrix} \text{ and } \varphi(b) = \begin{pmatrix} 1 & 0 & 0 \\ 3 & 1 & 0 \\ 0 & 4 & 1 \end{pmatrix}.$$

As in the case of 2×2-matrices, one can define a trace map associated with an endomorphism of $\Gamma_{\langle a,b\rangle}$: it is a polynomial map of the variety of $\langle \Lambda \rangle$ into itself. ∎

10.4 Comments

Fricke formula and the corollary to Proposition 10.1.4 (Fricke lemma) appear in [178], but were also stated by Vogt in 1889.

Proposition 10.1.3-1 has been stated by Fricke [178] and proved by Horowitz [212]. Since then, it has been rediscovered several times: Allouche and Peyrière [23] for $n = 2$, for general n by Kolář and Nori [251] (although they gave a formula involving a number of traces much larger than $2^n - 1$), and Peyrière et al. [330].

Traina [432, 433] gave an efficient algorithm for computing P_W in the case of a two-letter alphabet; also in this case Wen Z.-X. and Wen Z.-Y. [447] determined the leading term of P_W. Procesi [337] and Razmyslov [353], instead of considering relations between traces only, used more general polynomial identities. This gives simple algorithms for computing polynomials P_W with an arbitrary alphabet. This is this method which is exposed here.

Proposition 10.1.6 and its corollary appear in Avishai, Berend and Glaubman [51].

The trace map appears in Horowitz [213]. It has also been rediscovered a number of times: by Kohmoto et al. [317] in the case of Fibonacci, by Allouche and Peyrière [23] and Peyrière [327] for $n = 2$, by Peyrière et al. [330] for $n > 2$.

Proposition 10.2.3 essentially appears in Horowitz [213]. It has also be rediscovered. Kolář and Ali [250] conjectured it after having used a computer algebra software. The proof given here appears in Peyrière [327]. For a generalization, see [331]

Results in Sec. 10.2.2 can be found in Horowitz [213] and Peyrière et al. [330]. For recent developments, see Wen &Wen [450, 449]: they prove that $Q_\sigma(2,2,z) \equiv 1$ implies that σ is an automorphism; they show that, on a two-letter alphabet, invertible substitutions (i.e., morphisms of the free monoid which extend as automorphisms of the free group) are generated by three substitutions.

The structure of the ideal $\mathcal{I}$, for a four-letter alphabet, is studied by Whittemore [456] and completely elucidated by Magnus [276] for an arbitrary alphabet. It also results from Magnus [276] that, for a n-letter alphabet, one can use $3n - 3$ variables only in trace maps with the counterpart that Φ_σ is a rational map instead of being a polynomial one. See also [329]

For a study of the quotient ring modulo $\mathcal{I}$ (the ring of Fricke characters) see Magnus [276].

Polynomial identities for $p \times p$-matrices are studied by Procesi [337], Razmyslov [353], and Leron [263]. Wen Z.-X. [446, 447] gives some algorithms for getting such identities. He also constructs a trace map for 3×3-matrices and a two-letter alphabet. This is his derivation which is given in this course.

For basic references on free groups, see [277, 312, 313, 314].

11. Piecewise linear transformations of the unit interval and Cantor sets

We discussed in Chap. 7 the relationship between substitutive dynamical systems and shifts of finite type, in terms of Markov expanding maps on the interval. As an example, the Morse substitution and the expanding map $F(x) = 2x \pmod 1$ on $[0,1]$ have the same matrix $\begin{pmatrix} 1 & 1 \\ 1 & 1 \end{pmatrix}$ (see Sec. 7.1.2).

Substitutions and expanding maps have other similarities. For instance, G. Rauzy gave a geometric realization of the Tribonacci substitution as an expanding dynamical system on a compact subset of the Euclidean plane, namely the Rauzy fractal (see Secs. 7.5 and 8.1.2). If β denotes the golden ratio, the map $F(x) = \beta x \pmod 1$ has the same symbolic dynamics on the Rauzy fractal as the action of the matrix $\begin{pmatrix} 1 & 1 \\ 1 & 0 \end{pmatrix}$.

The aim of this chapter is to study ergodicity of expanding maps from a symbolic point of view, through the study of a few examples. Ergodic properties of dynamical systems were obtained in the preceding chapters by using the spectrum of the unitary operator U (let us recall that U is defined on (X, T, μ) by $U(f) = f \circ T$; see Sec. 1.4). In this chapter, we will focus on another operator, that is, the Perron-Frobenius operator, which has the same spectrum on the unit circle as the unitary operator. Thus, the spectrum of the Perron-Frobenius operator determines the ergodicity of the dynamical system. We will give an algorithm to compute this spectrum concretely in a few situations.

In [259], Lasota and Yorke focus their attention on expanding maps of the interval: they state a relationship between the ergodicity of these dynamical systems and the spectrum of their Perron–Frobenius operator. In Sec. 11.2.1 are summarized the definitions of the unitary operator and the Perron–Frobenius operator, and the relationship between the spectrum of these operators and the ergodicity of the dynamical systems. We do not claim here to give an exhaustive presentation of these topics. We rather chose to introduce these notions through the study of examples, in order to give an idea of these classical methods.

[1] This chapter has been written by M. Mori

Hofbauer and Keller [204] stated a relationship between the singularities of the Ruelle's zeta function and the spectrum of the Perron–Frobenius operator. Baladi and Ruelle [53] and Mori [299] proved similar results. However, their methods are quite different. In Sec. 11.2.2, along [299], is defined a *Fredholm determinant*, though the Perron–Frobenius operator is not compact. This definition involves generating functions and renewal equations. This Fredholm determinant allows us in Sec. 11.2.3 to determine explicitly the spectrum of the Perron–Frobenius operator as well as the Ruelle's zeta function. We deduce ergodic properties of the dynamical system such as ergodicity, mixing and rate of decay of correlations.

As an application, Sec. 11.4 is devoted to the computation of the Hausdorff dimension of some Cantor sets naturally associated with piecewise linear maps. For that purpose, we will define α–Fredholm matrices.

11.1 Definitions

Let F be a map from the unit interval $[0,1]$ into itself. Our main interest is to study the asymptotic behavior of orbits x, $F(x)$, $F^2(x)$, $\cdots$. Note that $F^n(x)$ denotes the n–th iteration and not the n–th power $(F(x))^n$ of F.

One–dimensional dynamical system. We focus on maps for which there exists a probability measure μ on $[0,1]$ which satisfies the following conditions:

1. μ is absolutely continuous with respect to the Lebesgue measure, i.e., there exists a *density function* $\varrho(x)$, denoted by $\varrho(x) = \frac{d\mu}{dx}(x)$, such that $\int_0^1 \varrho\, dx = 1$ and for every measurable set A, $\mu(A) = \int_A \varrho(x)\, dx$;
2. for any measurable set A, $\mu(A) = \mu(F^{-1}(A))$ holds, that is, μ is an *invariant measure* with respect to the map F.

From a physical view point, we are studying a dynamical system in an equilibrium state: F is a *time evolution*, and μ is an *equilibrium state*. To be more precise we should mention the σ–algebra which is generated by the intervals. We shall omit it since it does not play an essential role in our discussion.

Class of transformations. We call F a *piecewise monotonic transformation* if there exists a finite set $\mathcal{A}$ and a finite partition into intervals $\{\langle a\rangle\}_{a\in\mathcal{A}}$ of $[0,1]$ such that on each subinterval $\langle a\rangle$ the map F is monotone (increasing or decreasing).

We call a transformation F *piecewise linear* if the derivative F' is constant on each interval of monotonicity $\langle a\rangle$.

We call a transformation F *Markov* if the following holds:

$$\text{if } F(\langle a\rangle) \cap \langle b\rangle^o \neq \emptyset, \text{ then } \overline{F(\langle a\rangle)} \supset \langle b\rangle,$$

where J^o and $\overline{J}$ denote respectively the interior and the closure of a set J. Namely, for any $a \in \mathcal{A}$, $F(\langle a\rangle)$ is essentially a union of several $\langle b\rangle$ $(b \in \mathcal{A})$.

We call a map F *expanding* if

$$\liminf_{n\to\infty} \frac{1}{n} \inf_{x\in[0,1]} \frac{1}{n} \log |(F^n)'(x)| > 0.$$

This coefficient of expansivity will be denoted in all that follows ξ.

Symbolic Dynamics. Let $\mathcal{A}$ be a finite set and S denotes the shift map on $\mathcal{A}^{\mathbb{N}}$. As defined in Chap. 1, a triple (X, S, ν) is called a *symbolic dynamical system* if X is a closed invariant subset of $\mathcal{A}^{\mathbb{N}}$ (i.e., $TX = X$) and ν an invariant probability measure.

The one–dimensional dynamical system considered in the former subsection can be expressed in terms of symbolic dynamics. Namely, we code in a natural way the orbits of the points of the unit interval with respect to the partition into intervals of monotonicity. More precisely, let F denote a piecewise monotonic transformation. Let $\{\langle a\rangle\}_{a\in\mathcal{A}}$ be a partition of $[0,1]$ into subintervals of monotonicity, that is, F is monotone on each $\langle a\rangle$, $\cup_{a\in\mathcal{A}}\langle a\rangle = [0,1]$ and $\langle a\rangle \cap \langle b\rangle = \emptyset$ for $a \neq b$. Let $x \in [0,1]$. The *expansion* $w^x = (w_n^x)_{n\in\mathbb{N}}$ of x is defined by: $\forall n \in \mathbb{N}^+$, $F^n(x) \in \langle w_{n+1}^x\rangle$.

We denote by X the closure of all the expansions of the points x: $X := \overline{\{w^x | x \in [0,1]\}}$. Then the shift map S on X corresponds to F on $[0,1]$, that is, for a sequence $w^x \in X$ which is the expansion of x, the expansion of $F(x)$ equals $S(w^x)$.

Admissible words. Let $W = w_1 \cdots w_n$ be a finite word in $\mathcal{A}^\star$ and $|W| = n$ be the length of W. Let

$$\langle W\rangle = \cap_{k=1}^n F^{-k+1}(\langle w_k\rangle) \subset [0,1]$$

be defined as the interval coded by the word W, i.e., $\forall x \in \langle W\rangle$, $x \in \langle w_1\rangle, \ldots, F^{n-1}(x) \in \langle w_n\rangle$. We say that a word W is *admissible* if $\langle W\rangle \neq \emptyset$.

Structure matrix. For a Markov map F, the following $\mathcal{A}\times\mathcal{A}$ matrix $\mathbf{M}$ plays an essential role, particularly when F is piecewise linear. Let for $a, b \in \mathcal{A}$, $\mathbf{M}_{a,b} = 1$ if $\overline{F(\langle a\rangle)} \supset \langle b\rangle$, $\mathbf{M}_{a,b} = 0$ otherwise. We call this matrix the *structure matrix* of F.

Functions of bounded variations. We denote by $\mathcal{L}^1$ and $\mathcal{L}^\infty$ the set of functions f such that $\int_0^1 |f|\,dx < \infty$ and $\sup_{x\in[0,1]} |f(x)| < \infty$, respectively. To study the ergodic properties of the dynamical systems we consider here, the space of functions of *bounded variation* is one of the most important notions. We will denote it by BV.

Definition 11.1.1. *A function $f : [0,1] \to [0,1]$ is of bounded variation if its total variation $Var(f) := \sup \sum_{k=1}^n |f(x_k) - f(x_{k+1})|$, is finite. The supremum is taken over all finite subdivisions $(x_k)_{1\leq k\leq n}$ of $[0,1]$.*

A function $f \in \mathcal{L}^1$ is of bounded variation if $v(f) = \inf Var(g)$ is finite, the infimum being taken over the class of those g such that $f = g$ in $\mathcal{L}^1$.

11.2 Ergodic properties of piecewise linear Markov transformations

The aim of this section is to illustrate, through the study of the "simple" example of a piecewise linear Markov transformation, how to deduce from the spectral study of the Perron–Frobenius operator ergodic properties of the dynamical systems arising from piecewise monotonic transformations.

Let

$$F(x) = \begin{cases} x/\eta_a & 0 \leq x < \eta_a \\ (x - \eta_a)/\eta_b & \eta_a \leq x \leq 1, \end{cases}$$

where $1 - \eta_a = \eta_a \eta_b$, that is, $\lim_{x \uparrow 1} F(x) = \eta_a$. Let us take as alphabet $\mathcal{A} = \{a, b\}$, and corresponding intervals $\langle a \rangle = [0, \eta_a)$ and $\langle b \rangle = [\eta_a, 1]$.

We are interested first in ergodic properties such as the existence of an invariant measure; second in statistical properties, for instance mixing. Recall that mixing deals with the convergence of $\int f(x) g(F^n(x))\, dx$ towards $\int f\, dx \int g\, d\mu$. For more details, see Chap. 5 or [444]. The rate of convergence is called the *decay of correlations.*

Let us introduce some tools and notation, as the Perron–Frobenius operator, the generating functions or the Fredholm matrix.

11.2.1 The Perron–Frobenius operator

Definition 11.2.1. *The operator defined on* $\mathcal{L}^1$ *by:*

$$Pf(x) = \sum_{y:\ F(y)=x} f(y) |F'(y)|^{-1},$$

is called the Perron–Frobenius operator *associated with* F.

The name "Perron–Frobenius operator" is related to the fact that it has similar properties to matrices with positive entries. For more details on this operator, see for instance [93, 233, 334].

This operator corresponds to the change of variables $F(x) \to x$ in the following integration:

$$\int f(x) g(F(x))\, dx = \int Pf(x) g(x)\, dx \quad f \in \mathcal{L}^1,\ g \in \mathcal{L}^\infty.$$

However, restricting its domain to the set BV of bounded variation functions, the eigenvalues of this operator allow one to determine ergodic properties of the dynamical system [259]. There is no restriction here since eigenfunctions of expanding maps of the interval are shown to be BV [259, 300]. This implies that we can equivalently work here with $\mathcal{L}^1$ or with $\mathcal{L}^2$, as in the rest of this book. Indeed, the following properties are satisfied by the Perron–Frobenius operator

1. $P\colon \mathcal{L}^1 \to \mathcal{L}^1$ is positive,
2. the spectrum of P on the unit circle coincides with the spectrum of the unitary operator defined by $Uf = f \circ F$.

This implies the following:

1. The system admits 1 as an eigenvalue, and the eigenspace associated with the eigenvalue 1 is contained in BV. Moreover, there exists a basis with nonnegative functions $\varrho \in BV$ such that $\int \varrho \, dx = 1$, and these ϱ are the density functions of invariant probability measures [259];
2. if 1 is a simple eigenvalue, then the dynamical system is ergodic;
3. if there exists no other eigenvalue except 1 on the unit circle, then the dynamical system is strongly mixing [88], i.e., for $f \in \mathcal{L}^1$ and $g \in \mathcal{L}^\infty$

$$\int f(x) g(F^n(x)) \, dx \to \int f \, dx \int g \, d\mu,$$

where μ denotes the invariant probability measure absolutely continuous with respect to the Lebesgue measure. For more details on mixing, see Chap. 5.

As an operator from $\mathcal{L}^1$ into itself, every $|z| < 1$ is an eigenvalue of P with infinite multiplicity [239]. Thus the rate of convergence of $\int f(x) g(F^n(x)) \, dx$ to $\int f \, dx \int g \, d\mu$ has no meaning. However, restricting P to BV, every $|z| < e^{-\xi}$, (where ξ corresponds to the expansivity coefficient, i.e., $\xi = \liminf_{n\to\infty} \inf_{x\in[0,1]} \frac{1}{n} \log |(F^n)'(x)|$) is an eigenvalue with infinite multiplicity. Hence, the eigenvalues of P restricted to BV in the annulus $e^{-\xi} < |z| \le 1$ play an essential role to determine ergodic properties such as ergodicity, mixing, the rate of decay of convergence and so on.

Remark. Let 1_J denote the indicator function of a set J. Note that

$$P1_{\langle a \rangle}(x) = \sum_{y\colon F(y)=x} 1_{\langle a \rangle}(y) |F'(y)|^{-1} = \eta_a, \quad P1_{\langle b \rangle}(x) = \begin{cases} \eta_b & \text{if } x \in \langle a \rangle \\ 0 & \text{otherwise.} \end{cases}$$

11.2.2 Generating functions and Fredholm matrix

To study ergodic properties of a dynamical system, we need to determine the eigenvalues of the Perron-Frobenius operator. Similarly as in the case of matrices, the Fredholm determinant of a nuclear operator P is defined to be $\det(I - zP)$. It is an entire function, the zeros of which are the reciprocals of the eigenvalues of P. Though the Perron-Frobenius operator is not nuclear, the Fredholm determinant can be defined by constructing renewal equations. However, in general, it is not an entire function and has for natural boundary $|z| = e^{\xi}$.

The *generating functions* of any $g \in \mathcal{L}^\infty$, are the formal power series defined by

$$s_g(z) = \begin{pmatrix} s_g^{\langle a \rangle}(z) \\ s_g^{\langle b \rangle}(z) \end{pmatrix} \quad \text{with} \quad \begin{cases} s_g^{\langle a \rangle}(z) = \sum_{n=0}^{\infty} z^n \int 1_{\langle a \rangle}(x) g(F^n(x))\, dx \\ s_g^{\langle b \rangle}(z) = \sum_{n=0}^{\infty} z^n \int 1_{\langle b \rangle}(x) g(F^n(x))\, dx. \end{cases}$$

Since

$$s_g^{\langle a \rangle}(z) = \sum_{n=0}^{\infty} z^n \int \left[P^n 1_{\langle a \rangle}\right](x)\, g(x)\, dx,$$

we get as a formal expression

$$s_g^{\langle a \rangle}(z) = \int \left[(I - zP)^{-1} 1_{\langle a \rangle}\right](x)\, g(x)\, dx.$$

This asserts, very roughly, that the reciprocals of eigenvalues of the Perron–Frobenius operator become singular points of $s_g^{\langle a \rangle}(z)$.

The aim of this section is thus to use the above generating functions, in order to study the ergodic properties of the map F. We are interested in particular in the eigenvalues of the Perron–Frobenius operator in order to prove the existence of an invariant probability measure μ, absolutely continuous with respect to the Lebesgue measure such that the dynamical system $([0,1], \mu, F)$ is strongly mixing.

Lemma 11.2.2. *Let* $\Phi(z) = \begin{pmatrix} z\eta_a & z\eta_a \\ z\eta_b & 0 \end{pmatrix}$ *(called a* Fredholm matrix*). Then taking* $s_g(z) = \begin{pmatrix} s_g^{\langle a \rangle}(z) \\ s_g^{\langle b \rangle}(z) \end{pmatrix}$, *we get* $s_g(z) = \begin{pmatrix} \int_{\langle a \rangle} g\, dx \\ \int_{\langle b \rangle} g\, dx \end{pmatrix} + \Phi(z) s_g(z)$.

Proof. Dividing the sum of the generating functions into the term corresponding to $n = 0$ and the remaining part, we get:

$$\begin{aligned} s_g^{\langle a \rangle}(z) &= \int 1_{\langle a \rangle}(x) g(x)\, dx + \sum_{n=1}^{\infty} z^n \int 1_{\langle a \rangle}(x) g(F^n(x))\, dx \\ &= \int 1_{\langle a \rangle}(x) g(x)\, dx + \sum_{n=1}^{\infty} z^n \int P 1_{\langle a \rangle}(x) g(F^{n-1}(x))\, dx \\ &= \int_{\langle a \rangle} g(x)\, dx + \eta_a \sum_{n=1}^{\infty} z^n \int g(F^{n-1}(x))\, dx \\ &= \int_{\langle a \rangle} g(x)\, dx + z\eta_a \sum_{n=0}^{\infty} z^n \int g(F^n(x))\, dx \\ &= \int_{\langle a \rangle} g(x)\, dx + z\eta_a \left(s_g^{\langle a \rangle}(z) + s_g^{\langle b \rangle}(z)\right). \end{aligned}$$

Similarly, we get

$$\begin{aligned} s_g^{\langle b\rangle}(z) &= \int 1_{\langle b\rangle}(x) g(x)\, dx + \sum_{n=1}^{\infty} z^n \int 1_{\langle b\rangle}(x) g(F^n(x))\, dx \\ &= \int_{\langle b\rangle} g(x)\, dx + \sum_{n=1}^{\infty} z^n \int P1_{\langle b\rangle}(x) g(F^{n-1}(x))\, dx \\ &= \int_{\langle b\rangle} g(x)\, dx + \eta_b \sum_{n=1}^{\infty} z^n \eta_b \int_{\langle a\rangle} g(F^{n-1}(x))\, dx \\ &= \int_{\langle b\rangle} g(x)\, dx + z\eta_b s_g^{\langle a\rangle}(z). \end{aligned}$$

■

We thus get

$$s_g(z) = (I - \Phi(z))^{-1} \begin{pmatrix} \int_{\langle a\rangle} g\, dx \\ \int_{\langle b\rangle} g\, dx \end{pmatrix}.$$

We call this equation a *renewal equation.* In our case,

$$s_g(z) = \frac{1}{(1-z)(1-(\eta_a - 1)z)} \begin{pmatrix} 1 & z\eta_a \\ z\frac{1-\eta_a}{\eta_a} & 1 - z\eta_a \end{pmatrix} \begin{pmatrix} \int_{\langle a\rangle} g\, dx \\ \int_{\langle b\rangle} g\, dx \end{pmatrix}.$$

11.2.3 Ergodic properties

An invariant measure for the map F. Let us introduce now an invariant measure for the map F absolutely continuous with respect to the Lebesgue measure. Let us recall that the density of this measure is an eigenvector for the Perron–Frobenius operator associated with the eigenvalue 1, that is to say, it corresponds to the singularity 1 for the generating functions.

Taking

$$\varrho(x) = \begin{cases} \frac{1}{(2-\eta_a)\eta_a} & x \in \langle a\rangle, \\ \frac{1}{2-\eta_a} & x \in \langle b\rangle, \end{cases}$$

we get

$$\begin{aligned} s_g^{\langle a\rangle}(z) &= \frac{\int_{\langle a\rangle} g\, dx + z\eta_a \int_{\langle b\rangle} g\, dx}{(1-z)(1-(\eta_a-1)z)} \\ &= \frac{\eta_a \int g(x)\varrho(x)\, dx}{1-z} + \frac{(1-\eta_a)\int_{\langle a\rangle} g\, dx - \eta_a \int_{\langle b\rangle} g\, dx}{(1-(\eta_a-1)z)(2-\eta_a)} \end{aligned}$$

and

$$s_g^{\langle b\rangle}(z) = \frac{(1-\eta_a)\int g(x)\varrho(x)\, dx}{1-z} + \frac{-\frac{1-\eta_a}{\eta_a}\int_{\langle a\rangle} g\, dx + (2-\eta_a)\int_{\langle b\rangle} g\, dx}{(1-(\eta_a-1)z)(2-\eta_a)}.$$

We thus can imagine that $z = 1$ and $\eta_a - 1$ may be the eigenvalues of the Perron–Frobenius operator. As 1 is a singularity of order 1, 1 will be a simple eigenvalue of P, and there may be no other eigenvalue in the unit circle. This is what we hope to prove now in a more formal way.

Noticing $\int \varrho(x)\,dx = 1$, we denote by μ a probability measure of density function ϱ. The measure μ is an invariant probability measure for the transformation F.

Mixing. Returning to the definition of $s_g(z)$, we get by comparing the coefficient of z^n

$$\int 1_{\langle a\rangle}(x)g(F^n(x))\,dx = \eta_a \int g\,d\mu + C_a(\eta_a - 1)^n,$$
$$\int 1_{\langle b\rangle}(x)g(F^n(x))\,dx = (1-\eta_a) \int g\,d\mu + C_b(\eta_a - 1)^n,$$

where

$$C_a = \frac{1-\eta_a}{2-\eta_a}\int_{\langle a\rangle} g\,dx - \frac{\eta_a}{2-\eta_a}\int_{\langle b\rangle} g\,dx,\quad C_b = \frac{-(1-\eta_a)}{\eta_a(2-\eta_a)}\int_{\langle a\rangle} g\,dx + \int_{\langle b\rangle} g\,dx.$$

Therefore, for example, $\int 1_{\langle a\rangle}(x)g(F^n(x))\,dx$ converges to $\eta_a \int g\,d\mu$ exponentially as $n \to \infty$ with the rate $(\eta_a - 1)$. Moreover, since $\eta_a - 1 < 0$, it is oscillating. Note that the Lebesgue measure of $\langle a\rangle$ and $\langle b\rangle$ are η_a and $1-\eta_a$, respectively.

More generally, for a word $W = w_1 \cdots w_m$, let η_W denote the product $\eta_W = \eta_{w_1} \cdots \eta_{w_m}$. Consider

$$\begin{aligned}
&\sum_{n=0}^{\infty} z^n \int 1_{\langle W\rangle}(x)g(F^n(x))\,dx \\
&= \sum_{n=0}^{m-1} z^n \int 1_{\langle W\rangle}(x)g(F^n(x))\,dx + \sum_{n=m}^{\infty} z^n \int 1_{\langle W\rangle}(x)g(F^n(x))\,dx \\
&= \sum_{n=0}^{m-1} z^n \int 1_{\langle W\rangle}(x)g(F^n(x))\,dx + \begin{cases} z^m\eta_W(s_g^{\langle a\rangle}(z) + s_g^{\langle b\rangle}(z)) & \text{if } w_m = a, \\ z^m\eta_W s_g^{\langle a\rangle}(z) & \text{if } w_m = b. \end{cases}
\end{aligned}$$

Take

$$\chi_g^W(z) = \sum_{n=0}^{m-1} z^n \int 1_{\langle W\rangle}(x)g(F^n(x))\,dx,$$

$$\Phi^W(z) = \begin{cases} \left(z^m\eta_W, z^m\eta_W\right) & \text{if } w_m = a, \\ \left(z^m\eta_W, 0\right), & \text{if } w_m = b. \end{cases}$$

By dividing f into the sum $f = \sum_W C_W 1_{\langle W\rangle}$, we get

$$\begin{aligned}\sum_{n=0}^{\infty} z^n \int f(x)g(F^n(x))\,dx &= \sum_W C_W \sum_{n=0}^{\infty} z^n \int 1_{\langle W\rangle}(x)g(F^n(x))\,dx \\ &= \sum_W C_W \chi_g^W(z) + \sum_W C_W \Phi^W(z) s_g(z).\end{aligned}$$

Since $f \in BV$, there exists a decomposition $f = \sum_W C_W 1_{\langle W\rangle}$ such that $\sum_{n=1}^{\infty} r^n \sum_{|W|=n} |C_W| < \infty$ for any $0 < r < 1$. This ensures the convergence of $\sum_W C_W \chi_g^W(z)$ and $\sum_W C_W \Phi^W(z)$ in $|z| < e^{\xi}$.

Suppose now that $|z| < e^{\xi}$. Consider the eigenvalues of the Perron–Frobenius operator in BV which are greater than $e^{-\xi}$ in modulus. Let us prove that they coincide with the reciprocals of the solutions of $\det(I - \Phi(z)) = 0$.

Note that for any $\varepsilon > 0$, there exists a constant C such that

$$Lebes(\langle W\rangle) \le Ce^{-(\xi-\varepsilon)n} \quad \text{and} \quad \eta_W = \eta_{w_1}\cdots\eta_{w_n} \le Ce^{-(\xi-\varepsilon)n},$$

where *Lebes* is the Lebesgue measure. Hence,

$$\begin{aligned}|\sum_W C_W \chi_g^W(z)| &= \left|\sum_{n=1}^{\infty} \sum_{|W|=n} |C_W| \chi_g^W(z)\right| \\ &\le \sum_{n=1}^{\infty} \sum_{|W|=n} |C_W| \sum_{m=0}^{n-1} |z|^m \|g\|_\infty Ce^{-(\xi-\varepsilon)n},\end{aligned}$$

$$|\text{each component of } \sum_W \Phi^W(z)| \le \sum_{n=1}^{\infty} \sum_{|W|=n} |C_W||z|^n Ce^{-(\xi-\varepsilon)n}.$$

Therefore for $|z| < e^{\xi}$, $\sum_W C_W \chi_g^W(z)$ and all the components of $\sum_W \Phi^W(z)$ are analytic.

On the other hand, $s_g(z)$ has singularities at $z = 1$ and $z = 1/(\eta_a - 1)$. Therefore, $\sum_{n=0}^{\infty} z^n \int f(x)g(F^n(x))\,dx$ can be extended meromorphically outside of the unit disk and it has at most two singularities in $|z| < e^{\xi}$. This shows that the reciprocals of the solutions of $\det(I - \Phi(z)) = 0$ coincide with the eigenvalues of the Perron–Frobenius operator which is grater than $e^{-\xi}$ in modulus. Namely, $\det(I - \Phi(z))$ plays the role of the Fredholm determinant for nuclear operators. This is the reason why we call $\Phi(z)$ the Fredholm matrix.

Comparing the coefficients as before,

$$\int f(x)g(F^n(x))\,dx \to \sum_W \begin{cases} C_W\eta_W \int g\,d\mu & \text{if } w_m = a \\ C_W\eta_W\eta_a \int g\,d\mu & \text{if } w_m = b \end{cases} = \int f\,dx \int g\,d\mu.$$

The order of convergence is proved to be $\min\{(\eta_a - 1), e^{-\xi}\}$. Hence taking $f \in BV$ and $g \in \mathcal{L}^{\infty}$, we get

$$\int f(x)g(F^n(x))d\mu = \int f \cdot \varrho(x)g(F^n(x))dx \to \int f \cdot \varrho dx \int g d\mu = \int f d\mu \int g d\mu.$$

We call this property of the dynamical system *strong mixing*. The *decay rate of correlations* is $\min\{(\eta_a - 1), e^{-\xi}\}$.

Remark. We can note now that the Fredholm matrix $\Phi(z)$ is essentially a 'weighted structure matrix'. Here the structure matrix of this dynamical system is $\mathbf{M} = \begin{pmatrix} 1 & 1 \\ 1 & 0 \end{pmatrix}$. Namely, the first row expresses that $F(\langle a \rangle)$ contains both $\langle a \rangle$ and $\langle b \rangle$, and the second row expresses that $F(\langle b \rangle)$ contains only $\langle b \rangle$. The trace $\operatorname{tr} \mathbf{M} = 1$ means $F(\langle a \rangle) \supset \langle a \rangle$, and this corresponds to the fixed point $x = 0$. At the same time, since $\mathbf{M}^2 = \begin{pmatrix} 2 & 1 \\ 1 & 1 \end{pmatrix}$, its trace equals 3. This means there exists a fixed point $a \to a \to a$ and another two-periodic orbit $a \to b \to a$ $(b \to a \to b)$. In this way, the trace of the product of the structure matrix expresses a number of periodic orbits.

Ruelle's zeta function. Similarly, the trace of $\Phi^n(z)$ also corresponds to periodic orbits with period n. Let us now define *Ruelle's zeta function*:

$$\zeta(z) = \exp\left[\sum_{n=1}^{\infty} \frac{z^n}{n} \sum_{F^n(p)=p} |F^{n\prime}(p)^{-1}|\right].$$

Then

$$\zeta(z) = \exp\left[\sum_{n=1}^{\infty} \frac{1}{n} \operatorname{tr} \Phi^n(z)\right] = \exp[-\operatorname{tr}\log(I - \Phi(z))] = (\det(I - \Phi(z))^{-1}.$$

This shows that the reciprocals of the singularities of the zeta function become eigenvalues of the Perron–Frobenius operator.

Final result. The above discussions can be easily generalized to general piecewise linear Markov transformations. We can construct a Fredholm matrix $\Phi(z)$ by considering generating functions, and we get

Theorem 11.2.3. *In* $|z| < e^{\xi}$, *the reciprocals of the solutions of* $\det(I - \Phi(z)) = 0$ *coincide with the eigenvalues of the Perron–Frobenius operator restricted to BV which are greater than* $e^{-\xi}$ *in modulus. Moreover,* $\det(I - \Phi(z)) = 1/\zeta(z)$.

Thus, if the solutions of the equation $\det(I - \Phi(z)) = 0$ satisfy

1. 1 is simple,
2. there exist no others eigenvalues on the unit circle,

then there exists an invariant probability measure μ absolutely continuous with respect to the Lebesgue measure and the dynamical system $([0,1], \mu, F)$ is strongly mixing.

11.3 Non-Markov transformation: β–expansion

As an example of a non-Markov transformation, let us study the β–transformations.

Let $1 < \eta_a + \eta_b \leq 2$, $\mathcal{A} = \{a, b\}$, $\langle a \rangle = [0, \eta_a)$ and $\langle b \rangle = [\eta_a, 1]$. Let

$$F(x) = \begin{cases} x/\eta_a & x \in \langle a \rangle \\ (x - \eta_a)/\eta_b & x \in \langle b \rangle. \end{cases}$$

The map F is called a *β-transformation* if $1/\eta_a = 1/\eta_b = \beta$. The above case is a generalization of a β–transformation.

The generating function $s_g^{\langle a \rangle}(z)$ satisfies the same renewal equation as before. If $F(\langle b \rangle) \subset \langle a \rangle$, we put $J_1 = F(\langle b \rangle)$. Next if $F(J_1) \supset \langle a \rangle$, we put $J_2 = F(J_1) \cap \langle b \rangle$. Then

$$\begin{aligned} s_g^{\langle b \rangle}(z) &= \int 1_{\langle b \rangle}(x) g(x)\, dx + \eta_b z s_g^{J_1}(z) \\ &= \int 1_{\langle b \rangle}(x) g(x)\, dx + \eta_b z \Big(\int 1_{J_1}(x) g(x)\, dx + z\eta_a (s_g^{\langle a \rangle}(z) + s_g^{J_2}(z) \Big). \end{aligned}$$

Similarly, taking

$$J_n = \begin{cases} F(J_{n-1}) & \text{if } F(J_{n-1}) \subset \langle a \rangle, \\ F(J_{n-1}) \cap \langle b \rangle & \text{if } F(J_{n-1}) \supset \langle a \rangle, \end{cases}$$

and

$$\phi(n) = \begin{cases} 0 & \text{if } F(J_n) \subset \langle a \rangle \ (F^n 1 < \eta_a), \\ \prod_{k=1}^{n} \eta_{w_k^1} & \text{if } F(J_n) \supset \langle a \rangle \ (F^n 1 \geq \eta_a), \end{cases}$$

we get

$$s_g^{\langle b \rangle}(z) = \chi_g^b(z) + \sum_{n=1}^{\infty} z^n \phi(n) s_g^{\langle a \rangle}(z),$$

where $w^1 = (w_n^1)_{n \in \mathbb{N}}$ is the expansion of 1 (following the notation of Sec. 11.1) and

$$\chi_g^b(z) = \sum_{n=0}^{\infty} z^n \prod_{k=1}^{n} \eta_{w_k^1} \int_{J_n} g\, dx,$$

with $J_0 = \langle b \rangle$. Note that all the components of $\chi_g(z)$ and $\Phi(z)$ are analytic in $|z| < e^{\xi}$. Therefore, putting

$$\Phi(z) = \begin{pmatrix} z\eta_a & z\eta_a \\ \sum_{n=1}^{\infty} z^n \phi(n) & 0 \end{pmatrix},$$

we get a renewal equation

$$s_g(z) = (I - \Phi(z))^{-1}\chi_g(z)$$

as before. Here $\chi_g(z) = \begin{pmatrix} \int_{\langle a\rangle} g\,dx \\ \chi_g^b(z) \end{pmatrix}$.

Through a similar discussion to that of the Markov case, we obtain that the reciprocals of the solutions of $\det(I - \Phi(z)) = 0$ coincide with the eigenvalues of the Perron–Frobenius operator which are greater than $e^{-\xi}$ in modulus. Using this expression, we can prove that the dynamical system is strongly mixing, and we can also calculate the density function ϱ of the invariant probability measure μ absolutely continuous with respect to the Lebesgue measure, using

$$\lim_{z\uparrow 1}(1-z)(I - \Phi(z))^{-1}\begin{pmatrix} \int_{\langle a\rangle} g\,dx \\ \chi_g^b(1) \end{pmatrix} = \begin{pmatrix} \eta_a \int g\,d\mu \\ (1-\eta_a)\int g\,d\mu \end{pmatrix}.$$

Therefore,

$$\begin{pmatrix} \eta_a \int g\,d\mu \\ (1-\eta_a)\int g\,d\mu \end{pmatrix} = \lim_{z\uparrow 1}\frac{1-z}{\det(I-\Phi(z))}\begin{pmatrix} 1 & \eta_a \\ \sum_{n=1}^{\infty} z^n\phi(n) & 1-\eta_a \end{pmatrix}\begin{pmatrix} \int_{\langle a\rangle} g\,dx \\ \chi_g^b(1) \end{pmatrix}.$$

Hence,

$$\varrho(x) = C^{-1}\left[1_{\langle a\rangle}(x) + \eta_a \sum_{n=0}^{\infty}\prod_{k=1}^{n}\eta_{w_k^1} 1_{J_n}(x)\right],$$

where C is the normalizing constant.

In general, we can also prove that

$$\zeta(z) = \frac{1}{\det(I - \Phi(z))}.$$

We need unessential detailed discussion to prove this result (see [299]). Anyway, we get

$$\frac{1}{\zeta(z)} = 1 - z\eta_a \sum_{n=0}^{\infty} z^n\phi(n),$$

where $\phi(0) = 1$. Using the above expression, we can calculate the spectrum of the Perron–Frobenius operator P. Thus by Theorem 11.2.3, we can determine the ergodicity of the dynamical system.

11.4 Cantor sets

The usual Cantor set is generated by the map $F(x) = 3x \pmod 1$ by discarding the inverse images of $[\frac{1}{3}, \frac{2}{3}]$. Namely, first we discard the interval $[\frac{1}{3}, \frac{2}{3}]$. Next we discard its inverse images $[\frac{1}{9}, \frac{2}{9}]$ and $[\frac{7}{9}, \frac{8}{9}]$. Then we discard their inverse images $[\frac{1}{27}, \frac{2}{27}]$ $[\frac{7}{27}, \frac{8}{27}]$, $[\frac{19}{27}, \frac{20}{27}]$ and $[\frac{25}{27}, \frac{26}{27}]$, and continue this procedure.

We consider here sets obtained by an analogous device of construction and show how to compute their Hausdorff dimension by introducing the α–Fredholm matrix.

11.4.1 Hausdorff Dimension

The aim of this section is to measure the size of a Cantor set by computing its Hausdorff dimension.

Definition 11.4.1. *Let us consider a set $\mathcal{C} \subset [0,1]$. Let $\delta > 0$. A covering of $\mathcal{C}$ by a family of intervals which is at most countable $\{I_i\}$ ($\cup_i I_i \supset \mathcal{C}$), where the length of each I_i is less than δ, is called a δ*–covering.

Now for $\alpha \geq 0$, put

$$\nu_\alpha(\mathcal{C}, \delta) = \inf \sum_i |I_i|^\alpha,$$

where the infimum is taken over all δ–coverings. Then, since $\nu_\alpha(\mathcal{C}, \delta)$ is monotone increasing as $\delta \downarrow 0$, the limit

$$\nu_\alpha(\mathcal{C}) := \lim_{\delta \downarrow 0} \nu_\alpha(\mathcal{C}, \delta)$$

exists.

Lemma 11.4.2. *Let us assume that $\nu_\alpha(\mathcal{C}) < \infty$ for $\alpha \geq 0$. Then for any $\alpha' > \alpha$, $\nu_{\alpha'}(\mathcal{C}) = 0$.*

Proof. Let $\{I_i\}$ be a δ–covering such that $\sum_i |I_i|^\alpha < \nu_\alpha(\mathcal{C}) + 1$. Then

$$\nu_{\alpha'}(\mathcal{C}) \leq \sum_i |I_i|^{\alpha'} \leq \delta^{\alpha'-\alpha} \sum_i |I_i|^\alpha \leq \delta^{\alpha'-\alpha}(\nu_\alpha(\mathcal{C}) + 1).$$

Take $\delta \downarrow 0$, then we get $\nu_{\alpha'}(\mathcal{C}) = 0$. ■

The Hausdorff measure. We can construct a measure space over $(\mathcal{C}, \nu_\alpha)$ with a suitable σ–algebra. We will call ν_α a *Hausdorff measure* with coefficient α. However, as one can see from Lemma 11.4.2, this Hausdorff measure usually has no meaning. Namely, there exists $\alpha_0 \geq 0$ such that for any $\alpha < \alpha_0$ every set has measure infinity with respect to ν_α, and for any $\alpha > \alpha_0$ they have measure 0. We call this critical value α_0 the *Hausdorff dimension* of a set $\mathcal{C}$.

11.4.2 A Cantor set associated with a piecewise linear Markov transformation

Let

$$F(x) = \begin{cases} x/\eta_a & x \in [0, \eta_a), \\ (x - \eta_a)/\eta_b & x \in [\eta_a, \eta_a + \eta_b), \\ (x - \eta_a - \eta_b)/\eta_c & x \in [\eta_a + \eta_b, 1], \end{cases}$$

and $0 < \eta_a, \eta_b, \eta_c < 1$ and

$$\eta_a < (1 - \eta_a - \eta_b)/\eta_c < \eta_a + \eta_b.$$

Put $\lambda_a = \eta_a$, $\lambda_c = 1-\eta_a-\eta_b$. These are the lengths of $\langle a \rangle$ and $\langle c \rangle$, respectively. Now, let us define the α–Fredholm matrix by

$$\Phi_\alpha(z) = \begin{matrix} & a & c \\ a & z\eta_a^\alpha & z\eta_a^\alpha \\ c & z\eta_c^\alpha & 0 \end{matrix}, \quad \text{i.e.} \quad \Phi_\alpha(z) = \begin{pmatrix} z\eta_a^\alpha & z\eta_a^\alpha \\ z\eta_c^\alpha & 0 \end{pmatrix}.$$

Let $\mathcal{C}$ be the Cantor set of points such that their positive orbit under the action of F never enters the interval $\langle b \rangle$.

As a rough discussion, we will calculate the Hausdorff dimension by covering the set $\mathcal{C}$ only by words with the same length.

As a first approximation, let us consider the covering by intervals $\langle a \rangle$ and $\langle c \rangle$. This can be expressed by

$$\eta_a^\alpha + \eta_c^\alpha = (1,1)\begin{pmatrix} \lambda_a^\alpha \\ \lambda_c^\alpha \end{pmatrix}.$$

The second approximation is the covering by intervals $\langle aa \rangle$, $\langle ac \rangle$ and $\langle ca \rangle$. The subintervals corresponding to the word cc do not exist. Therefore, the second approximation equals

$$(\eta_a\lambda_a)^\alpha + (\eta_a\lambda_c)^\alpha + (\eta_c\lambda_a)^\alpha = (1,1)\Phi_\alpha(1)\begin{pmatrix} \lambda_a^\alpha \\ \lambda_c^\alpha \end{pmatrix}.$$

Generally, the n–th approximation becomes

$$(1,1)\Phi_\alpha(1)^{n-1}\begin{pmatrix} \lambda_a^\alpha \\ \lambda_c^\alpha \end{pmatrix}.$$

If all the eigenvalues of the α–Fredholm matrix are less than 1 in modulus, $(1,1)\Phi_\alpha(1)^{n-1}\begin{pmatrix} \lambda_a^\alpha \\ \lambda_c^\alpha \end{pmatrix}$ converges to 0 as n tends to ∞. This says that α is greater than the Hausdorff dimension of $\mathcal{C}$. On the contrary, if one of the eigenvalue is greater than 1 in modulus, then $(1,1)\Phi_\alpha(1)^{n-1}\begin{pmatrix} \lambda_a^\alpha \\ \lambda_c^\alpha \end{pmatrix}$ will diverge.

Note that by Perron–Frobenius' theorem the maximal eigenvalue of $\Phi_\alpha(1)$ is positive. Thus, when α equals the Hausdorff dimension α_0 of $\mathcal{C}$, $\Phi_{\alpha_0}(1)$ has eigenvalue 1. Therefore α_0 will be the solution of $\det(I - \Phi_\alpha(1)) = 0$. Namely, α_0 will satisfy

$$\eta_a^{\alpha_0} + (\eta_a \eta_c)^{\alpha_0} = 1.$$

Take $\begin{pmatrix} \lambda_a^* \\ \lambda_c^* \end{pmatrix}$ as an eigenvector of $\Phi_{\alpha_0}(1)$ associated with the eigenvalue 1 with $\lambda_a^* + \lambda_c^* = 1$. To be more precise, $\lambda_a^* = \eta_a^{\alpha_0}$ and $\lambda_c^* = (\eta_a\eta_c)^{\alpha_0} = 1 - \eta_a^{\alpha_0}$. The other eigenvalue equals $(\eta_a^{\alpha_0} - 1)$ and $\begin{pmatrix} \eta_a^{\alpha_0} \\ -1 \end{pmatrix}$ is an eigenvector associated with it. Let us denote

$$\begin{pmatrix} \lambda_a^{\alpha_0} \\ \lambda_c^{\alpha_0} \end{pmatrix} = C_1 \begin{pmatrix} \lambda_a^* \\ \lambda_c^* \end{pmatrix} + C_2 \begin{pmatrix} \eta_a^{\alpha_0} \\ -1 \end{pmatrix},$$

where

$$C_1 = \frac{(1-\eta_a-\eta_c)^{\alpha_0}+1}{(1-\eta_a)^{\alpha_0}+1} \quad \text{and} \quad C_2 = \frac{(1-\eta_a)^{\alpha_0} - (1-\eta_a-\eta_c)^{\alpha_0}}{(1-\eta_a)^{\alpha_0}+1}.$$

This shows $0 < C_1 < 1$.

For a word $W = w_1 \cdots w_n$, $(w_i \in \{a, c\})$, its Hausdorff measure will be

$$\nu(\langle W \rangle) = \lim_{m\to\infty} \eta_{w_1}^{\alpha_0} \cdots \eta_{w_{n-1}}^{\alpha_0} v(w_n) \Phi_{\alpha_0}(1)^m \begin{pmatrix} \lambda_a^{\alpha_0} \\ \lambda_c^{\alpha_0} \end{pmatrix} = C_1 \eta_{w_1}^{\alpha_0} \cdots \eta_{w_{n-1}}^{\alpha_0} \lambda_{w_n}^*, \tag{11.1}$$

where $v(a) = (1, 0)$ and $v(c) = (0, 1)$. Hence, the total measure $0 < \nu(\mathcal{C}) = C_1 < 1$. The corresponding generating function is

$$s_g(z) = (I - \Phi_{\alpha_0}(z))^{-1} \chi_g = \frac{1}{1 - \eta_a^{\alpha_0} z - (\eta_a\eta_c)^{\alpha_0} z^2} \begin{pmatrix} 1 & z\eta_a^{\alpha_0} \\ z\eta_c^{\alpha_0} & 1 - z\eta_a^{\alpha_0} \end{pmatrix} \chi_g,$$

where $\chi_g = \begin{pmatrix} \chi_g^a \\ \chi_g^c \end{pmatrix}$ and $\chi_g^d = \int_{\langle d \rangle} g \, d\nu$ $(d = a, c)$. Therefore,

$$\begin{aligned} s_g(z) &= \frac{1}{(1-z)(1-(1-\eta_a^{\alpha_0})z)} \begin{pmatrix} \int_{\langle a \rangle} g\, d\nu + \eta_a^{\alpha_0} z \int_{\langle c \rangle} g \, d\nu \\ \eta_c^{\alpha_0} z \int_{\langle a \rangle} g\, d\nu + (1 - \eta_a^{\alpha_0} z) \int_{\langle c \rangle} g\, d\nu \end{pmatrix} \\ &= \frac{1}{1-z} \frac{1}{2 - \eta_a^{\alpha_0}} \left[\frac{1}{\eta_a^{\alpha_0}} \int_{\langle a \rangle} g\, d\nu + \int_{\langle c \rangle} g\, d\nu \right] \begin{pmatrix} \eta_a^{\alpha_0} + \text{small order} \\ 1 - \eta_a^{\alpha_0} + \text{small order} \end{pmatrix}. \end{aligned}$$

Now put

$$\frac{d\mu}{d\nu}(x) = \begin{cases} \dfrac{1}{C_1(2-\eta_a^{\alpha_0})\eta_a^{\alpha_0}} & \text{if } x \in \langle a \rangle, \\ \dfrac{1}{C_1(2-\eta_a^{\alpha_0})} & \text{if } x \in \langle c \rangle. \end{cases}$$

Then, μ becomes an invariant probability measure absolutely continuous with respect to the Hausdorff measure. By Equation (11.2),

$$\int 1_{\langle a\rangle}(x)g(F^n(x))\,d\nu - \nu(\langle a\rangle)\int g\,d\mu \sim (1-\eta_a^{\alpha_0})^n.$$

A similar equation also holds for the letter c. Therefore, this shows that the dynamical system is strongly mixing.

Let us construct another map $G : [0,1] \to [0,1]$. Let the lengths of $\langle a^*\rangle$ equal $\lambda_a^* = \eta_a^{\alpha_0}$, and the length of $\langle c^*\rangle$ equal $\lambda_c^* = 1 - \eta_a^{\alpha_0}$, corresponding to the eigenvector of $\Phi_{\alpha_0}(1)$ associated with the eigenvalue 1, respectively. Take

$$G(x) = \begin{cases} \eta_a^{-\alpha_0}x & \text{if } x \in \langle a^*\rangle, \\ \eta_c^{-\alpha_0}(x-\lambda_a^*) & \text{if } x \in \langle c^*\rangle. \end{cases}$$

Note that $G(1) = \lambda_a^*$. Also in this case, the generating function equals

$$s_g(z\colon G) = (I - \Phi_{\alpha_0}(z))^{-1}\chi_g(G).$$

One may easily understand that the invariant probability measure derived from the above equation becomes Markov, and at the same time, the Fredholm matrix associated with G equals $\Phi_{\alpha_0}(z)$.

11.4.3 Rigorous Results

Let us state now the above results more precisely. We can extend these results even to general piecewise linear cases by using signed symbolic dynamics. We restrict ourselves to the Markov case in this chapter.

Let $F : [0,1] \to [0,1]$ be a piecewise linear Markov transformation. We are going to consider a subset $\mathcal{A}_1 \subset \mathcal{A}$ and put

$$\mathcal{C} = \{x \in [0,1]\colon w_i^x \in \mathcal{A}_1 \text{ for all } i\}.$$

The aim of this section is to study the Hausdorff dimension of $\mathcal{C}$ and the ergodic properties of this dynamical system.

The α–Fredholm matrix is defined by

$$\Phi_\alpha(z)_{a,b} = \begin{cases} z\eta_a^\alpha & \text{if } \overline{F(\langle a\rangle)} \supset \langle b\rangle, \\ 0 & \text{otherwise.} \end{cases}$$

Moreover we assume that F is expanding and the α–Fredholm matrix irreducible ($z \neq 0$).

Theorem 11.4.3. *Let α_0 be the maximal solution of* $\det(I - \Phi_\alpha(1)) = 0$. *Then α_0 is the Hausdorff dimension of $\mathcal{C}$.*

First, let us state elementary lemmas.

Lemma 11.4.4. *Let $0 < \alpha < 1$, and $x_i > 0$ $(1 \le i \le k)$. Then,*

$$k^{\alpha-1}(\sum_{i=1}^{k} x_i^\alpha) \le (\sum_{i=1}^{k} x_i)^\alpha \le \sum_{i=1}^{k} x_i^\alpha.$$

Lemma 11.4.5. *There exists a constant $K > 1$ such that*

$$\frac{1}{K} < \frac{Lebes(\langle Wa\rangle)}{Lebes(\langle Wb\rangle)} < K,$$

for any word W and any symbols $a, b \in \mathcal{A}$ such that $\langle Wa\rangle, \langle Wb\rangle \neq \emptyset$.

To prove Theorem 11.4.3, we need to define another Hausdorff dimension. Let μ be a probability measure on $[0, 1]$. Define

$$\mu_\alpha(\mathcal{C}, \delta) = \inf \sum_i \mu(\langle W_i\rangle)^\alpha,$$

where the infimum is taken over all coverings by words $\{W_i\}$ such that $\mu(\langle W_i\rangle) < \delta$. The difference between ν_α and μ_α lies in two points:

1. μ_α uses the probability measure μ, and ν_α uses the Lebesgue measure,
2. to define μ, we only consider coverings by words, and to define ν_α, we consider any covering by intervals

Put

$$\mu_\alpha(\mathcal{C}) = \lim_{\delta\to 0} \mu_\alpha(\mathcal{C}, \delta),$$

and denote by $\dim_\mu(\mathcal{C})$ the critical point whether $\mu_\alpha(\mathcal{C})$ converges or diverges.

Lemma 11.4.6. *The Hausdorff dimension of $\mathcal{C}$ equals $\dim_{Lebes}(\mathcal{C})$.*

Proof. It is clear that the Hausdorff dimension is less than or equal to $\dim_{Lebes}(\mathcal{C})$. We will prove another inequality. Let $\{J_i\}$ be a covering by words such that $\sum (Lebes(J_i))^\alpha < M < \infty$. For each J_i, put

$$n_i = \min\{n \colon |W| = n, \langle W\rangle \subset J_i\}.$$

If there exist words $W_1, \ldots, W_k$ such that $|W_j| = n_i - 1$ and $\langle W_j\rangle$ intersects J_i, we will divide J_i into k subintervals $J_i \cap \langle W_1\rangle, \ldots, J_i \cap \langle W_k\rangle$. We denote this new covering by words again by $\{J_i\}$. Therefore we can assume that J_i is contained in some $\langle W_i\rangle$ with $|W_i| = n_i - 1$, and contains at least one $\langle W_i'\rangle$ with $|W_i'| = n_i$. Then by Lemma 11.4.4 and the assumption,

$$\sum_i (Lebes(J_i))^\alpha \le \text{Card } \mathcal{A}^{1-\alpha} M < \text{Card } \mathcal{A} \cdot M.$$

Now we will choose a family of words $\{W_{i,j}\}$ such that $\cup_j \langle W_{i,j}\rangle = J_i$ as follows. First we will choose words $w_{i,1}, \ldots, W_{i,j_1}$ with length n_i such that $\langle W_{i,j}\rangle \subset J_i$. Note that $j_1 \le \text{Card } \mathcal{A}$. Next we will choose words $W_{i,j_1+1}, \ldots,$ W_{i,j_2+1} with length $n_i + 1$ and $\langle w_{i,j}\rangle \subset J_i \backslash \cup_{j=1}^{j_1} \langle W_{i,j}\rangle$, and repeat this procedure. It can be noted that the number of words $W_{i,j}$ with length $n_i + k$ is at most $2\,\text{Card } \mathcal{A}$, and by Lemma 11.4.5,

$$\begin{aligned}
Lebes(\langle W_{i,j}\rangle) &\le e^{-k\xi} Lebes(\langle V_i\rangle) \\
&\le e^{-k\xi} K \text{Card } \mathcal{A} \cdot \max\{Lebes(\langle W_{i,j}\rangle) \colon 1 \le j \le j_1\} \\
&\le e^{-k\xi} K \text{Card } \mathcal{A} \cdot Lebes(J_i).
\end{aligned}$$

Then

$$\begin{aligned}
\sum_i \sum_j (Lebes(\langle W_{i,j}\rangle))^\alpha &\le \sum_i \sum_k \sum_{j \colon |W_{i,j}| = n_i + k} (Lebes(\langle W_{i,j}\rangle))^\alpha \\
&\le \sum_i \sum_k 2\text{Card } \mathcal{A}\,(e^{-k\xi} K \text{Card } \mathcal{A} \cdot Lebes(J_i))^\alpha \\
&\le \frac{2K^\alpha (\text{Card } \mathcal{A})^{1+\alpha}}{1 - e^{-\xi\alpha}} \sum_i (Lebes(J_i))^\alpha \\
&\le K' M,
\end{aligned}$$

where

$$K' = \frac{2K^\alpha (\text{Card } \mathcal{A})^{2+\alpha}}{1 - e^{-\xi}}.$$

Now let us take any α which is greater than the Hausdorff dimension of $\mathcal{C}$. Then for any $\varepsilon > 0$, there exists a covering by intervals $\{J_i\}$ such that $\sum_i (Lebes(J_i))^\alpha < \varepsilon$. Then we can choose a covering by words $\{\langle W_{i,j}\rangle\}$ such that

$$\sum_i \sum_j (Lebes(\langle W_{i,j}\rangle)))^\alpha < K'\varepsilon.$$

This shows that $\dim_{Lebes}(\mathcal{C}) < \alpha$. Thus $\dim_{Lebes}(\mathcal{C})$ is less than or equal to the Hausdorff dimension of $\mathcal{C}$. Namely, the Hausdorff dimension of $\mathcal{C}$ equals $\dim_{Lebes}(\mathcal{C})$. ∎

A proof of the following theorem can be found in [80] (Theorem 14.1).

Theorem 11.4.7 (Billingsley). *Let μ_1, μ_2 be probability measures. Assume that*

$$\mathcal{C} \subset \left\{ x \colon \lim_{n\to\infty} \frac{\log \mu_1(\langle w^x[1,n]\rangle)}{\log \mu_2(\langle w^x[1,n]\rangle)} = \alpha \right\}$$

holds for some $0 \le \alpha \le \infty$, where $w^x[1,n]$ is a word with length n and $x \in \langle w^x[1,n]\rangle$. Then

$$\dim_{\mu_2}(\mathcal{C}) = \alpha \dim_{\mu_1}(\mathcal{C}).$$

In the examples studied in the former sections, the mapping $G\colon [0,1] \to [0,1]$ was built from an eigenvector of $\Phi_{\alpha_0}(1)$. We are going to construct it for general piecewise linear Markov transformations.

Since $\Phi_{\alpha}(1)$ is irreducible, and $\Phi_{\alpha_0}(1)$ is a non–negative matrix, by Perron–Frobenius' theorem, the maximal eigenvalue is non–negative (according to the assumption, this equals 1), and we can choose a non–negative eigenvector associated with the maximal eigenvalue 1 for which the sum of its components equals 1. Let the measure of a^* $(a \in \mathcal{A}_1)$ be the component of the eigenvector. We arrange a^* as in the natural order of a, and the slope of G for $x \in \langle a^* \rangle$

$$G'(x) = \begin{cases} +\eta_a^{-\alpha_0} & \text{if } F'(y) > 0 \\ -\eta_a^{-\alpha_0} & \text{if } F'(y) < 0 \end{cases} \quad \text{for } y \in \langle a \rangle.$$

Then we can construct a mapping G which has the same symbolic dynamics as that of F restricted to $\mathcal{C}$. Hence the Fredholm matrix of G equals $\Phi_{\alpha_0}(z)$.

We denote by μ_G the induced measure from the Lebesgue measure over $[0,1]$ where G acts over $[0,1]$, where F acts. Then

$$\lim_{n\to\infty} \frac{\log Lebes(\langle w^x[1,n]\rangle)}{\log \mu_G(\langle w^x[1,n]\rangle)} = \frac{1}{\alpha_0}$$

holds for every $x \in \mathcal{C}$. Of course, $\mu_G(\mathcal{C}) = Lebes([0,1]) = 1$, i.e., $\dim_{\mu_G}(\mathcal{C}) = 1$. Therefore,

$$\dim_{Lebes}(\mathcal{C}) = \alpha_0 \dim_{\mu_G}(\mathcal{C}) = \alpha_0.$$

This shows that the Hausdorff dimension of $\mathcal{C}$ equals α_0.

The measure μ_G is called a *conformal measure* and using this we can prove that the Hausdorff measure on $\mathcal{C}$ is absolutely continuous with respect to μ_G, and is nonzero and finite [203]. Moreover, according to the the assumption made, G is expanding. Therefore it has an invariant probability measure μ absolutely continuous with respect to the Lebesgue measure, and the dynamical system is strongly mixing. We denote by $\varrho(x) = \frac{d\mu}{dx}(x)$ the density of μ with respect to the Lebesgue measure. The induced measure $\hat{\mu}$ on $\mathcal{C}$ (deduced from this measure μ) is absolutely continuous with respect to the Hausdorff measure, and the density function $\hat{\varrho}(y)$ of $\hat{\mu}$ with respect to the Hausdorff measure ν equals $\varrho(x)/\nu(\mathcal{C})$, where x and y have respectively the same expansion with respect to G and F. This shows that the dynamical system equipped with this measure is also strongly mixing.

12. Some open problems

As a conclusion, let us emphasize the underlying arithmetic structure and the interaction between the geometric and symbolic nature of the dynamical systems we have considered throughout this book. We are mainly interested in the two following problems: first, finding geometric interpretations of various symbolic dynamical systems including those generated by substitutions, and secondly, developing multidimensional continued fraction algorithms reflecting the dynamics of the systems.

12.1 The S-adic conjecture

Let us start with an algorithmic approach. For more details on the subject, the reader is referred to [167, 169].

A sequence is said to have an *at most linear complexity* if there exists a constant C such that for every positive integer n, $p(n) \leq Cn$, or in other words, if $p(n) = O(n)$. Many sequences that we have encountered in this book have at most linear complexity, including fixed points of primitive substitutions, automatic sequences, Sturmian sequences, Arnoux-Rauzy sequences, and so on. On the other hand, numerous combinatorial, ergodic or arithmetic properties can be deduced from this indication on the growth-order of the complexity function.

Let us start with a purely combinatorial result.

Theorem 12.1.1 (Cassaigne [106]). *A sequence has at most linear complexity if and only if the first difference of the complexity $p(n+1) - p(n)$ is bounded.*

Furthermore, an upper bound on the first difference can be explicitly given. Let us note that the equivalent result does not hold in the case of sequences with at most quadratic complexity: in [167], one can find an example of a sequence with a quadratic complexity function and unbounded second differences $p(n+2) + p(n) - 2p(n+1)$.

[1] This chapter has been written by P. Arnoux and V. Berthé

12.1.1 S-adic expansions

The following result [167] can be deduced from Theorem 12.1.1 and considerations on the graph of words. The complexity function of a symbolic dynamical system is defined as the function which counts the number of distinct factors of the *language* of this system, i.e., the union of the sets of factors of the sequences of the system.

Theorem 12.1.2 (Ferenczi [167]). *Let X be a minimal symbolic system on a finite alphabet $\mathcal{A}$ such that its complexity function $p_X(n)$ is at most linear, or, equivalently (following Theorem 12.1.1), such that $p_X(n+1) - p_X(n)$ is bounded; then there exists a finite set of substitutions $\mathcal{S}$ over an alphabet $\mathcal{D} = \{0, ..., d-1\}$, a substitution φ from $\mathcal{D}^\star$ to $\mathcal{A}^\star$, and an infinite sequence of substitutions $(\sigma_n)_{n\geq 1}$ with values in $\mathcal{S}$ such that $|\sigma_1\sigma_2...\sigma_n(r)| \to +\infty$ when $n \to +\infty$, for any letter $r \in \mathcal{D}$, and any word of the language of the system is a factor of $\varphi(\sigma_1\sigma_2...\sigma_n)(0)$ for some n.*

The above proposition can be read as follows: *minimal systems with at most linear complexity are generated by a finite number of substitutions.* Using a variation of the Vershik terminology, we propose to call such systems *S-adic systems*, and the pair $(\varphi, (\sigma_n))_{n\geq 1})$ is called an *S-adic expansion* of X. We similarly call a sequence *S-adic* if the symbolic dynamical system generated by the sequence is itself S-adic, and the sequence $(\varphi(\sigma_n))_{n\geq 1}$ is again called an *S-adic expansion* of the sequence x. Such expansions appear for instance in Chap. 6.

The fact that the lengths of the words tend to infinity, which generalizes the notion of everywhere growing substitutions (i.e., substitutions such that $\forall r$, $\exists n \in \mathbb{N}$, $|\sigma^n(r)| \geq 2$), is necessary to make Theorem 12.1.2 nonempty. Furthermore, it can be seen in the proof of [167] that this prevents us from getting a universal bound on the number d of letters of the alphabet $\mathcal{D}$. However, in an important particular case generalizing the Arnoux-Rauzy sequences [49], we do have a universal upper bound:

Proposition 12.1.3. *For minimal systems over a three-letter alphabet such that $p_X(n+1) - p_X(n) \leq 2$ for every n large enough, Theorem 12.1.2 is satisfied with $d \leq 3$ and $\operatorname{Card} S \leq 3^{27}$.*

Such universal upper bounds exist (but are very large) as soon as one has an upper bound on $p(n+1) - p(n)$.

A measure-theoretic consequence of Theorem 12.1.2 is that a minimal and uniquely ergodic system of at most linear complexity cannot be strongly mixing [167].

The S-adic expansion is known in an explicit way for the symbolic dynamical systems generated by Sturmian sequences, by the Arnoux-Rauzy sequences [49], for the systems generated by some binary codings of rotations [3, 146] (i.e., for codings with respect to a partition of the unit circle into intervals the lengths of which are larger than or equal to the angle of the rotation), and for systems generated by irrational interval exchanges (see in particular [2, 171, 270]). In the Sturmian case (see Sec. 6.4.4 in Chap. 6), one gets a more precise result: one can expand a given Sturmian sequence as an infinite composition of two substitutions; the rules for the iteration of these substitutions follow the Ostrowski expansion with respect to the angle of the initial point whose orbit is coded (for more details, see [41]). In some other cases (Arnoux-Rauzy sequences, binary codings, interval exchanges) one uses multidimensional continued fraction expansions. See also [1, 201] for a detailed sudy of the behaviour of sequences for which $p(n+1) - p(n) \in \{1, 2\}$ through the use of graphs of words. See also for more results on S-adicity [446].

12.1.2 The conjecture

The converse of Theorem 12.1.2 is clearly false. To produce a counter-example, it is sufficient to consider a non-primitive substitution [320] with a fixed point of complexity function satisfying $\Theta(n \log\log n)$ as $a \mapsto aba$, $b \mapsto bb$, or $\Theta(n \log n)$ as $a \mapsto aaba$, $b \mapsto bb$. Nevertheless, in the case where the initial sequence u is minimal, the fact that it is a fixed point of a substitution guarantees that the complexity is at most linear. This is not true any more if the sequence u is generated by two substitutions, even if u is minimal. Such an example was proposed by Cassaigne: one uses a positive substitution which appears infinitely often in the iteration, and a substitution having a fixed point of quadratic complexity which appears in long ranges in the iteration.

Question. Let u be a sequence generated by the iteration of a finite number of substitutions; which restrictions should one add to these substitutions, so that the sequence u has at most linear complexity?

We still have to find a *stronger* form of S-adicity which would be equivalent to at most linear complexity. This is the *S-adic conjecture* stating that minimal systems have at most linear complexity if and only if they are *strongly* S-adic.

12.1.3 Linear recurrence

Durand gives a sufficient condition in [155] for a sequence to have at most linear complexity. Namely, let u be a given recurrent sequence and let W be a factor of the sequence u. Let us recall that a *return word* over W is a word V such that VW is a factor of the sequence u, W is a prefix of VW and W has exactly two occurrences in VW. A sequence is *linearly recurrent* if there exists a constant $C > 0$ such that for every factor W, the length of every return word V of W satisfies $|V| \leq C|W|$. Such a sequence always has at most linear complexity [156]. Unfortunately, this condition is strictly stronger than having at most linear complexity. Durand shows that a sequence is linearly recurrent if and only if it is proper S-adic with bounded partial quotients, i.e., every substitution comes back with bounded gaps in the S-adic expansion [155]. In particular, a Sturmian sequence is linearly recurrent if and only if the partial quotients in the continued fraction expansion of its angle are bounded, which means that not every sequence with at most linear complexity is linearly recurrent.

12.1.4 Periodic case and substitutive sequences

Given an S-adic sequence, one can ask whether this sequence is substitutive, that is, whether it is a letter-to-letter projection of a fixed point of a substitution. Substitutive Sturmian sequences correspond to quadratic angles (for more details, see [126, 465] and Chap. 9). This can be deduced from Durand's characterization of substitutive sequences based on return words (the notion of derived sequence can be geometrically seen using the induction). Hence, this result can be considered as a generalization of Galois' theorem for continued fraction expansions (see Sec. 6.5). See also [86] for a connected result: if all the parameters of an interval exchange belong to the same quadratic extension, the sequence of induced interval exchanges (by performing always the same induction process) is ultimately periodic.

12.2 Multidimensional continued fraction expansions

12.2.1 Arithmetics and S-adicity

One of the main interests of the S-adic expansion lies in the fact that it provides an arithmetic description of the sequences we consider. For instance, the sequence (σ_n) (in Theorem 12.1.2) is governed by the continued fraction expansion of the angle in the Sturmian case. More generally, in numerous cases, the "partial quotients" (i.e., the gaps between successive occurrences of runs of the same substitution which appear in the iteration in Theorem 12.1.2) provide a generalized continued fraction expansion which describes the combinatorial properties of the sequence: one can find such examples of continued fraction expansions in [2, 114, 146, 170, 171, 172, 270, 349, 358, 467]. The techniques which are usually used in these problems are, on the one hand, the use of the graphs of words (see for instance [73]), and, on the other hand, the induction process (see Chaps. 5 and 6).

The usual continued fraction algorithm provides the best rational approximations of a real number α; it is fundamentally connected to the toral rotation of $\mathbb{T}$, $x \mapsto x+\alpha$. It describes in a natural way the combinatorial properties of Sturmian sequences, as the properties of their geometric representation as a coding of a rotation, as well as those of their associated symbolic dynamical system. What can be said now when considering the two-dimensional torus $\mathbb{T}^2$, or when trying to get simultaneous approximation results?

There exist several unidimensional or multidimensional symbolic objects which generalize Sturmian sequences in a natural way and provide multidimensional continued fraction algorithms. One can for instance code a rotation over $\mathbb{T}$ with respect to a partition into two intervals or more, the lengths of which do not depend on the angle α of the rotation. Such sequences have been studied for instance in [8, 75, 146, 363].

The following three particular generalizations have been the subject of recent studies: interval exchanges, the Arnoux-Rauzy sequences and codings of the $\mathbb{Z}^2$-action of two rotations on the unit circle.

12.2.2 Interval exchanges

Three-interval exchanges are fundamentally connected (via an induction process) to binary codings of rotations, that is, to codings of irrational rotations on $\mathbb{T}$ with respect to a two-interval partition. Ferenczi, Holton, and Zamboni develop in [170, 171, 172] a multidimensional algorithm which generates the orbits of the discontinuity points and opens new ways in the study of the ergodic and spectral properties of interval exchanges. This algorithm is produced in a combinatorial way by catenative rules of production of bispecial factors, or in a geometric way, by the observation of the evolution of the intervals corresponding to the bispecial factors. It can also be expressed in terms of a new induction process which does not correspond to the usual ones. See [2, 270] and also [269] for an S-adic expansion.

12.2.3 Arnoux-Rauzy sequences

Let us consider now the *Arnoux-Rauzy sequences.* It will be recalled that these are recurrent sequences defined over a three-letter alphabet with the following extra combinatorial property [49]: for every n, there is exactly one right special factor and one left special factor of length n, and these special factors can be extended in three different way. It will be recalled that a factor W of the sequence u is called right special (respectively left special) if W is a prefix (respectively suffix) of at least two words of length $|W|+1$ which are factors of the sequence u (see also Chap. 6). Let us note that they can be similarly defined over any alphabet of larger size, say d; one thus obtains sequences of complexity $(d-1)n+1$. Arnoux-Rauzy sequences are a natural generalization of Sturmian sequences, since they share with Sturmian sequences the fundamental combinatorial property of the unicity of the right (and left) special factors of given length. Contrary to the Sturmian case, these sequences are not characterized by their complexity function any more.

One knows perfectly well the S-adic expansion of the Arnoux-Rauzy sequences [49], and more precisely, the S-adic expansion of the associated dynamical system and the S-adic expansion of the sequence which has as prefixes the left special factors; following the terminology in the Sturmian case, this sequence is called the *left special sequence* or the *characteristic sequence.* The combinatorial properties of the Arnoux-Rauzy sequences are well-understood and are perfectly described by a two-dimensional continued fraction algorithm defined over a subset of zero measure of the simplex introduced in [358, 467] and in [116]. See also [114] for the connections with a generalization of the Fine and Wilf's theorem for three periods. By using this algorithm, one can express in an explicit way the frequencies of factors of given length [461], one can count the number of all the factors of the Arnoux-Rauzy sequences [297], or prove that the associated dynamical system has always simple spectrum [116].

These sequences can also be described as an exchange of six intervals of the unit circle [49]. Let us recall that the *Tribonacci sequence* (i.e., the fixed point beginning with a of the *Rauzy substitution*: $a \mapsto ab$, $b \mapsto ac$, $c \mapsto a$) is an Arnoux-Rauzy sequence. In this case, one obtains good approximation results [117]. The dynamical system generated by this sequence is isomorphic to a rotation of the torus $\mathbb{T}^2$ (for more details, see Chaps. 7 and 8). More generally, in the periodic case, i.e., in the case where the S-adic expansion of the special sequence is purely periodic, this sequence is a fixed point of a unimodular substitution of Pisot type (and conversely). All the results of Chaps. 7 and 8 apply. In particular, the dynamical systems generated by such sequences are obtained as toral rotations (the sufficient condition of Theorem 8.4.1 holds).

It was believed that all Arnoux-Rauzy sequences originated from toral rotations, and more precisely, that they were natural codings of rotations over $\mathbb{T}^2$. We say that a sequence u is a *natural coding* of a rotation if there exists a measurable dynamical system (X, T, μ), which is itself measure-theoretically isomorphic to a rotation, a generating partition $\mathcal{P} = \{P_1, P_2, P_3\}$ of X, and a point $x \in X$ such that

$$\forall n \in \mathbb{N},\ u_n = i \text{ if and only if } T^n(x) \in P_i;$$

furthermore, the map T is supposed to be a piecewise translation, the translation vector taking exactly one value on each P_i. Namely, in all the examples, we know that the rotation is constructed as an exchange of domains in $\mathbb{R}^2$, the pieces of the exchange being isomorphic to the cylinders (see Chaps. 7 and 8).

This conjecture was disproved in [110], where an example of a totally unbalanced Arnoux-Rauzy sequence is constructed (a sequence is said to be *totally unbalanced* if there exists a letter a such that for each positive integer n, there exist two factors of the sequence with equal length, with one having at least n more occurrences of the letter a than the other). Namely, a natural coding of a rotation cannot be totally unbalanced following [165]. Let us note that the partial quotients (in the sense of the S-adic expansion) are unbounded in the above example of totally unbalanced Arnoux-Rauzy sequence.

The following questions are thus natural.

Questions. How to characterize the Arnoux-Rauzy sequences which are natural codings of rotations? In a weaker form, how to characterize the Arnoux-Rauzy sequences which are measure-theoretically isomorphic to rotations? Is linear recurrence a sufficient condition for either question? Are there Arnoux-Rauzy sequences which are weakly mixing? What are the eigenvalues of the Arnoux-Rauzy sequences? Do they admit a rotation as a factor?

12.2.4 Codings of two rotations

A third way of generalizing the Sturmian case consists in introducing a second parameter. One thus gets two dual approaches: one can either code a rotation of angle (α, β) in the two-dimensional torus $\mathbb{T}^2$, or a $\mathbb{Z}^2$-action by two irrational rotations of angle α and β on $\mathbb{T}$. In the first case, one gets a unidimensional sequence, in the second, a two-dimensional sequence. We discussed the first case by considering Arnoux-Rauzy sequences and their connections with rotations over $\mathbb{T}^2$ in the previous section. Consider now the second approach.

Let $(\alpha, \beta) \in \mathbb{R}^2$, such that $1, \alpha, \beta$ are rationally independent, and let $\rho \in \mathbb{R}$. Let u be the two-dimensional sequence defined over $\mathbb{Z}^2$ with values in $\{0, 1\}$ by

$$\forall (m,n) \in \mathbb{Z}^2, \; (u(m,n) = 0 \Longleftrightarrow m\alpha + n\beta + \rho \in [0, \alpha[\text{ modulo } 1).$$

Such a two-dimensional sequence has $mn+n$ rectangular factors of size (m, n) and is *uniformly recurrent* (i.e., for every positive integer n, there exists an integer N such that every square factor of size (N, N) contains every square factor of size (n, n)). The function which counts the number of rectangular factors of given size is called the *rectangle complexity function* (for more details, see Sec. 12.3). Conversely, every two-dimensional uniformly recurrent sequence with complexity $mn + n$ admits such a geometric description [77].

Let us see why these sequences can be considered as a generalization of Sturmian sequences.

They have the smallest complexity function known among two-dimensional sequences which are uniformly recurrent and not periodic. They are obtained as a letter-to-letter projection of two-dimensional sequences defined over a three-letter alphabet which code a plane approximation (let us recall that Sturmian sequences code and describe discrete lines [66, 69]): one can approximate a plane with irrational normal by square faces oriented along the three coordinates planes; this approximation is called a *discrete plane* or a *stepped surface* as in Chap. 8; after projection on the plane $x+y+z=0$, along $(1, 1, 1)$, one obtains a tiling of the plane with three kinds of diamonds, namely the projections of the three possible faces (for more details, see [441, 78], see also [356, 177] and Chap. 8). One can code this projection over $\mathbb{Z}^2$ by associating with each diamond the name of the projected face. One thus gets a sequence with values in a three-letter alphabet. These sequences are shown to code a $\mathbb{Z}^2$-action over the unit circle $\mathbb{T}$ [78].

These latter sequences are generated by two-dimensional substitutions governed by the Jacobi-Perron algorithm. Namely, a geometric interpretation of the Jacobi–Perron algorithm is given in [40]: in the same way as classical continued fractions can be interpreted in terms of induction of rotations, this algorithm can be considered as an induction algorithm for a group of rotations operating on the unidimensional torus; it is not trivial that this group can be induced on a subinterval to obtain a new group of rotations, but by inducing on a suitable interval, the induction of the $\mathbb{Z}^2$-action we consider is again generated by a $\mathbb{Z}^2$-action through a pair of rotations. However, unlike the classical $\mathbb{Z}$-action, the generators of the $\mathbb{Z}^2$-action are not canonically defined (since we can find an infinite number of bases for the lattice $\mathbb{Z}^2$). This can be related to the fact that there seems to be no way of defining a "best" two-dimensional continued fraction algorithm. Using this induction process, a sequence of two-dimensional substitutions is defined in [40] associated with the Jacobi–Perron algorithm (see also Chap. 8) which generates the two-dimensional Sturmian sequences mentioned above over a three-letter alphabet. We shall allude again to this notion of two-dimensional substitution in Sec. 12.3.

12.3 Combinatorics on two-dimensional words

When one works with multidimensional sequences, fundamental problems in the definition of the objects appear; for instance, how can one define a multidimensional complexity function? Consider two-dimensional sequences, i.e., sequences defined over $\mathbb{Z}^2$ with values in a finite alphabet. A possible notion of complexity consists in counting the rectangular factors of given size; we thus define the *rectangle complexity function*: $(m,n) \mapsto P(m,n)$. This notion is not completely satisfactory since it depends on the choice of a basis of the lattice $\mathbb{Z}^2$. For a more general definition of complexity, see [371]. The following questions are thus natural:

Questions. Can one characterize a two-dimensional sequence with respect to its rectangle complexity function? Which functions do exist as rectangle complexity functions?

A two-dimensional sequence is *periodic* if it is invariant under translation, i.e., if it admits a nonzero vector of periodicity. Note that the fact that the lattice of periodicity vectors has rank 2 is characterized by a bounded rectangle complexity function. There is no characterization of periodic sequences by means of the complexity function: one can construct two-dimensional sequences with a nonzero periodicity vector of very large complexity function;

namely, consider the sequence $(u_{m+n})_{(m,n)\in\mathbb{Z}^2}$, where the unidimensional sequence $(u_n)_{n\in\mathbb{Z}}$ has maximal complexity. Conversely Nivat has conjectured the following:

Conjecture 12.3.1. If there exists (m_0, n_0) such that $P(m_0, n_0) \leq m_0 n_0$, then the two-dimensional sequence u is periodic.

Let us note that the conjecture was proved for factors of size $(2, n)$ or $(n, 2)$ in [373]. A more general conjecture is given in [372, 373, 371]. Let us remark that in higher-dimensional cases, counter-examples to the conjecture can be produced [373].

It is interesting, with respect to this conjecture, to consider the "limit" case of sequences of rectangle complexity function $mn+1$. Such sequences are fully described in [109] and are all proved to be non-uniformly recurrent. The two-dimensional Sturmian sequences mentioned in Sec. 12.2.4 of complexity $mn+n$ are conjectured to be the uniformly recurrent sequences with smallest complexity function (it remains to give a more precise meaning to the term "smallest" for parameters belonging to $\mathbb{N}^2$).

Let us end this section with the problem of the definition of two-dimensional substitutions. Unlike the classical one-dimensional case, the notion of a two-dimensional substitution is not trivial. Consider a map which associates with a letter a pointed two-dimensional pattern. We first need more information to know where to place the image of a letter. Two natural problems then arise. First, it is not immediate to prove the consistency if one wants to apply the substitution to a finite pattern or to a double sequence. Secondly, how can one iterate such a process to generate a double sequence? Two notions of substitutions are introduced in [40]. *Pointed substitutions* are first defined: given the value of the initial sequence at the point x, one can deduce the value of the image sequence on a pointed pattern situated at a point y that can be computed from x and its value. This is however inconvenient for explicit computation, since one needs at each step global information. In particular it is difficult to iterate it in order to generate a double sequence. It is much more convenient to be able to use a local information, i.e., *local rules* (this is exactly what is done when one computes one-dimensional substitutions: one does not compute the exact position of a given pattern, but one only uses the fact that patterns follow each other). Roughly speaking, a local rule says how to place the image of a pointed letter with respect to the images of the letters belonging to a finite neighborhood. If we know the image of the initial point, we can compute the values of adjacent points by using a finite number of patterns, and in this way, compute the image of the complete sequence. See also for a study of rectangle substitutions [195].

Questions. Can one build other examples of substitutions endowed with local rules as the ones produced in [40]? Given a substitution with local rules, does the limit of the iterates cover $\mathbb{Z}^2$? Can one introduce a different notion of substitution which can be used for the generation of two-dimensional sequences?

12.4 Substitutions, rotations and toral automorphisms

The general question of the isomorphism between a substitutive dynamical system and a rotation brings many open questions. For more details, see Chap. 7.

The dynamical system generated by the Tribonacci substitution is measure-theoretically isomorphic to a translation on the torus $\mathbb{T}^2$. The isomorphism is a continuous onto map from the orbit closure of the fixed point of the substitution to the torus. The images of the three basic cylinders corresponding to the letters are connected, and even simply connected domains of the torus. The three basic sets are bases for the three cylinders of a Markov partition for the automorphism of the torus $\mathbb{T}^3$ associated with the substitution. Do these properties extend to some other symbolic dynamical systems?

Let us recall (Theorem 7.5.18) that the symbolic dynamical system associated with a unimodular substitution of Pisot type over a two-letter alphabet $\{0, 1\}$ is measure-theoretically isomorphic to a rotation on $\mathbb{T}$ (for more details, see Chap. 7, and in particular Sec. 7.5.3). This is obtained by proving [54] that this class of substitution satisfies the following coincidence condition: there exist two integers k, n such that $\sigma^n(0)$ and $\sigma^n(1)$ have the same k-th letter, and the prefixes of length $k-1$ of $\sigma^n(0)$ and $\sigma^n(1)$ have the same number of occurrences of the letter 0.

It is conjectured that every unimodular substitution of Pisot type over three letters satisfies a coincidence condition on three letters (see Chap. 7). It is not known whether there exists a unimodular substitution of Pisot type that does not satisfy the coincidence condition.

Conjecture 12.4.1. If the Perron–Frobenius eigenvalue of the incidence matrix of a unimodular substitution is a Pisot number, then the dynamics of the substitution is measure-theoretically isomorphic to a rotation on the torus.

A connected problem is the following, addressed by Liardet in [188]: ***is the odometer associated with the numeration scale related to a real number $\beta > 1$ isomorphic in measure to a translation on a compact group if β is a canonical Pisot number (i.e., β has a finite β-expansion and the associated companion polynomial is irreducible over $\mathbb{Q}$)?***

Other natural questions arise concerning the topological properties of the fractal domain X_σ associated with a substitution, such as the question of connectedness and simple connectedness : Canterini gives in [101] a sufficient condition for the connectivity of the exchange of domains associated with unimodular substitutions of Pisot type.

Question. Can one give a characterization of those substitutions for which the associated domains are connected or simply connected?

12.5 Arithmetics in $SL(d,\mathbb{N})$ and $SL(d,\mathbb{Z})$

We have considered here products of substitutions via the S-adic conjecture. This can also be expressed in terms of products of matrices. It is thus natural from an algebraic point of view to consider the properties of the monoid $SL(d,\mathbb{N})$ and of the group $SL(d,\mathbb{Z})$, for $d \geq 2$.

Definitions. For $d \geq 2$, let $SL(d,\mathbb{N})$ (respectively $SL(d,\mathbb{Z})$) denote the set of matrices of determinant 1 with nonnegative (respectively integer) coefficients. This set endowed with the multiplication is a monoid (respectively a group), whose identity element is the identity matrix $\mathbf{I}_d$.

A matrix $\mathbf{M}$ in $SL(d,\mathbb{N})$ is said to be a *unit* if it admits an inverse in $SL(d,\mathbb{N})$, i.e., if there exists a matrix $\mathbf{N} \in SL(d,\mathbb{N})$, such that $\mathbf{MN} = \mathbf{I}_d$. It is easy to check that the unit matrices are exactly the even permutation matrices; there is no nontrivial unit for $d = 2$, and there are 2 units different from the identity if $d = 3$, that is, the matrices:

$$\mathbf{P} = \begin{pmatrix} 0&1&0 \\ 0&0&1 \\ 1&0&1 \end{pmatrix}, \ \mathbf{P}^2 = \begin{pmatrix} 0&0&1 \\ 1&0&0 \\ 0&1&0 \end{pmatrix}.$$

A matrix $\mathbf{M}$ in $SL(d,\mathbb{N})$ is said to be *undecomposable* if it is not a unit matrix, and if, for any pair of matrices $\mathbf{A}, \mathbf{B}$ in $SL(d,\mathbb{N})$ such that $\mathbf{M} = \mathbf{AB}$, $\mathbf{A}$ or $\mathbf{B}$ is a unit.

Structure of $SL(2,\mathbb{N})$. The structure of the monoid $SL(2,\mathbb{N})$ is very simple; this is a free monoid with two generators (for more details, see Chap. 6)

$$\mathbf{A} = \begin{pmatrix} 1&0 \\ 1&1 \end{pmatrix}, \ \mathbf{B} = \begin{pmatrix} 1&1 \\ 0&1 \end{pmatrix}.$$

These two matrices $\mathbf{A}$ and $\mathbf{B}$ are the only undecomposable matrices in $SL(2,\mathbb{N})$, and any matrix admits a unique decomposition in terms of these two undecomposable elements. This decomposition is a matricial translation of the Euclidean algorithm and it corresponds to the continued fraction expansion.

Structure of $SL(3,\mathbb{N})$. The situation in $SL(3,\mathbb{N})$ is completely different. Note that this monoid is not free. Furthermore, we have the following theorem. For a proof of this result (by J. Rivat), see Appendix A.

Theorem 12.5.1 (Rivat). *There exist infinitely many undecomposable matrices in $SL(3,\mathbb{N})$.*

Questions. Can one give a simple characterization of the undecomposable matrices in $SL(3,\mathbb{N})$? Can one find an explicit family of undecomposable generators of $SL(3,\mathbb{N})$?

Some motivations for this question are the following: first, a decomposition of a matrix as a product of undecomposable matrices can be seen as a multidimensional continued fraction algorithm; secondly, it can also provide a better understanding of the automorphisms of the torus.

Invertible substitutions. A substitution over the alphabet $\mathcal{A}$ is said to be *invertible* if it extends to a morphism of the free group generated by $\mathcal{A}$.

Invertible substitutions over a two-letter alphabet are completely characterized (see Sec. 9.2 and Chap. 6). In particular they are Sturmian, i.e., they preserve Sturmian words. In particular, the monoid of invertible substitutions is finitely generated. One is far from understanding the structure of the monoid of invertible substitutions over a larger size alphabet: indeed the monoid of invertible substitutions over a d-letter alphabet is no more finitely generated, for $d > 2$ (see [453]). Let us recall that $SL(3,\mathbb{N})$ is not finitely generated (Theorem 12.5.1). However, these two results, which could be thought nearly equivalent, are in fact completely unrelated. First, the substitutions of [453] have matrices which are decomposable as products of elementary matrices, and second, the abelianization map is not onto: there are positive matrices of determinant 1 that are not matrices of invertible substitutions for $d = 3$.

Question. Can one characterize invertible substitutions over a larger size alphabet?

Some decision problems. Some related decision problems can be considered, and most of them turn out to be undecidable in dimension at least 3, such as the mortality problem, that is, the presence of the zero matrix in a finitely generated subsemigorup.

In particular, is it possible to find an algorithm which decides whether the semigroup generated by two or more square matrices of dimension two over the nonnegative integers is free? The answer is negative in the three-dimensional case: namely, this problem is shown to be undecidable by reducing the Post correspondence problem to it [248]. It is shown in [111] that the above problem is still undecidable for square matrices of dimension 3 that are upper-triangular. Furthermore, in the case of two upper-triangular 2×2 matrices, sufficient conditions for freeness of the semigroup generated by these two matrices are given. Related undecidable problems for 2-generator matrix semigroups are the following (see [112]): given two square matrices, decide whether the semigroup that they generate contains the zero matrix, and whether it contains a matrix having a zero in the right upper corner.

A. Undecomposable matrices in dimension 3 (by J. Rivat)

Let us recall that $SL(d, \mathbb{N})$ denotes the monoid of square matrices of dimension d with nonnegative integer coefficients and determinant 1. A matrix $\mathbf{M}$ is a *unit* if there exists a matrix $\mathbf{N}$ such that $\mathbf{MN} = \mathbf{I}_d$. A matrix $\mathbf{M}$ of $SL(d, \mathbb{N})$ is *undecomposable* if it is not a unit, and if, for all pair of matrices $\mathbf{A}, \mathbf{B}$ in $SL(d, \mathbb{N})$ such that $\mathbf{M} = \mathbf{AB}$, $\mathbf{A}$ or $\mathbf{B}$ is a unit. The aim of this appendix is to answer to the natural question: does there exist an infinite number of undecomposable matrices in $SL(3, \mathbb{N})$? The following theorem gives a positive answer:

Theorem A.0.2. *For any integer $n \geq 3$, the following matrix is undecomposable:*

$$\mathbf{M}_n = \begin{pmatrix} 1 & 0 & n \\ 1 & n-1 & 0 \\ 1 & 1 & n-1 \end{pmatrix}.$$

Remark. The matrices $\mathbf{M}_1$ and $\mathbf{M}_2$ are not undecomposable, since

$$\mathbf{M}_1 = \begin{pmatrix} 1&0&1 \\ 1&0&0 \\ 1&1&0 \end{pmatrix} = \begin{pmatrix} 1&0&0 \\ 0&1&0 \\ 0&1&1 \end{pmatrix} \begin{pmatrix} 1&0&1 \\ 1&0&0 \\ 0&1&0 \end{pmatrix}, \quad \mathbf{M}_2 = \begin{pmatrix} 1&0&0 \\ 0&1&0 \\ 0&1&1 \end{pmatrix} \begin{pmatrix} 1&0&2 \\ 1&1&0 \\ 0&0&1 \end{pmatrix}.$$

Proof. We fix $n \geq 3$. We consider $\mathbf{A}, \mathbf{B} \in SL(3, \mathbb{N})$ such that $\mathbf{AB} = \mathbf{M}_n$, and we denote:

$$\mathbf{A} = \begin{pmatrix} a_1 & a_2 & a_3 \\ b_1 & b_2 & b_3 \\ c_1 & c_2 & c_3 \end{pmatrix}, \quad \mathbf{B} = \begin{pmatrix} x_1 & x_2 & x_3 \\ y_1 & y_2 & y_3 \\ z_1 & z_2 & z_3 \end{pmatrix}.$$

If $a_1a_2a_3 \neq 0$, using the relation $a_1x_2 + a_2y_2 + a_3z_2 = 0$, we deduce that $x_2 = y_2 = z_2 = 0$, which is impossible for a matrix of determinant 1. Hence $a_1a_2a_3 = 0$, and, up to a cyclic permutation of the columns (multiplication by $\mathbf{P}$ or $\mathbf{P}^2$), one has either $a_1 \neq 0$, $a_2 = a_3 = 0$ or $a_1 \neq 0$, $a_2 = 0$, $a_3 \neq 0$ (the case $a_1 = a_2 = a_3 = 0$ is impossible since $\det(\mathbf{A}) = 1$).

[1] This appendix has been written by J. Rivat; Institut Elie Cartan de Nancy; Faculté des Sciences; B.P. 239; 54506 Vandoeuvre-lès-Nancy Cedex; FRANCE; rivat@iecn.u-nancy.fr

From the two null coefficients of $\mathbf{M}_n$, one obtains:

$$a_1x_2 = a_2y_2 = a_3z_2 = b_1x_3 = b_2y_3 = b_3z_3 = 0.$$

First case: Suppose $a_1 \neq 0$, $a_2 = a_3 = 0$, hence $x_2 = 0$.

Since $\det(\mathbf{A}) = a_1(b_2c_3 - b_3c_2) = 1$, one has $a_1 = 1$. Hence the first line of $\mathbf{A}$ is $(1\ 0\ 0)$, which implies that the first line of $\mathbf{B}$ is the same as the first line of $\mathbf{M}_n$.

From the equation above $b_1x_3 = 0$, and $x_3 = n$, one obtains $b_1 = 0$; from the equation $c_1x_3 + c_2y_3 + c_3z_3 = n - 1$ and $x_3 = n$, one obtains $c_1 = 0$, hence:

$$\mathbf{A} = \begin{pmatrix} 1 & 0 & 0 \\ 0 & b_2 & b_3 \\ 0 & c_2 & c_3 \end{pmatrix}, \quad \mathbf{B} = \begin{pmatrix} 1 & 0 & n \\ y_1 & y_2 & y_3 \\ z_1 & z_2 & z_3 \end{pmatrix}.$$

Since $\det(\mathbf{A}) = b_2c_3 - b_3c_2 = 1$, b_2 must be strictly positive; hence, from the equation above $b_2y_3 = 0$, one obtains $y_3 = 0$. From the equation $c_3z_3 = n - 1$, one obtains that z_3 is strictly positive, hence, since $b_3z_3 = 0$, one has $b_3 = 0$; using again $\det \mathbf{A} = 1$, we get $b_2 = c_3 = 1$. Since the second line of $\mathbf{A}$ is $(0\ 1\ 0)$, the second line of $\mathbf{B}$ must be that of $\mathbf{M}_n$, hence:

$$\mathbf{A} = \begin{pmatrix} 1 & 0 & 0 \\ 0 & 1 & 0 \\ 0 & c_2 & 1 \end{pmatrix}, \quad \mathbf{B} = \begin{pmatrix} 1 & 0 & n \\ 1 & n-1 & 0 \\ z_1 & z_2 & z_3 \end{pmatrix}.$$

This implies that $(n-1)c_2 + z_2 = 1$, hence $c_2 = 0$, and $\mathbf{A}$ is the identity matrix.

Second case: Suppose $a_1 \neq 0$, $a_2 = 0$, $a_3 \neq 0$.

We have then $x_2 = z_2 = 0$. Since $\det(\mathbf{B}) = y_2(x_1z_3 - x_3z_1) = 1$, one obtains $y_2 = 1$; this implies as above that the second column of $\mathbf{A}$ is that of $\mathbf{M}_n$. Hence:

$$\mathbf{A} = \begin{pmatrix} a_1 & 0 & a_3 \\ b_1 & n-1 & b_3 \\ c_1 & 1 & c_3 \end{pmatrix}, \quad \mathbf{B} = \begin{pmatrix} x_1 & 0 & x_3 \\ y_1 & 1 & y_3 \\ z_1 & 0 & z_3 \end{pmatrix}.$$

From the determinant of $\mathbf{B}$, one obtains that x_1 and z_3 are strictly positive, and from the equation $a_1x_1 + a_3z_1 = 1$, one obtains that $a_1 = x_1 = 1$, $z_1 = 0$, hence, again from the determinant of $\mathbf{B}$, $z_3 = 1$. Hence:

$$\mathbf{A} = \begin{pmatrix} 1 & 0 & a_3 \\ b_1 & n-1 & b_3 \\ c_1 & 1 & c_3 \end{pmatrix}, \quad \mathbf{B} = \begin{pmatrix} 1 & 0 & x_3 \\ y_1 & 1 & y_3 \\ 0 & 0 & 1 \end{pmatrix}.$$

From the equation $b_1 + (n-1)y_1 = 1$, one obtains $y_1 = 0$, $b_1 = 1$, and from the equation $b_1x_3 + (n-1)y_3 + b_3 = 0$, one obtains $x_3 = y_3 = 0$, hence $\mathbf{B}$ is the identity, which ends the proof. ∎

References

1. A. ABERKANE, Suites de complexité inférieure à $2n$, *Bull. Belg. Math. Soc.* 8 (2001), 161–180.
2. B. ADAMCZEWSKI, Codages de rotations et phénomènes d'autosimilarité. To appear in *J. Théor. Nombres Bordeaux.*
3. B. ADAMCZEWSKI, Répartitions des suites $(n\alpha)_{n\in GN}$ et substitutions. To appear in *Acta Arith.*
4. W. W. ADAMS, Simultaneous asymptotic diophantine approximations to a basis of a real cubic number field, *J. Number Theory* 1 (1969), 179–194.
5. S. I. ADIAN, *The Burnside problem and identities in groups*, Ergebnisse der Mathematik und ihrer Grenzgebiete, 95. Springer-Verlag, Berlin, 1979. Translated from the Russian by J. Lennox and J. Wiegold.
6. R. L. ADLER AND B. WEISS, Entropy, a complete metric invariant for automorphisms of the torus, *Proc. Nat. Acad. Sci. U.S.A.* 57 (1967), 1573–1576.
7. R. L. ADLER AND B. WEISS, *Similarity of automorphisms of the torus*, American Mathematical Society, Providence, R.I., 1970. Memoirs of the American Mathematical Society, No. 98.
8. P. ALESSANDRI AND V. BERTHÉ, Three distance theorems and combinatorics on words, *Enseign. Math.* 44 (1998), 103–132.
9. J.-P. ALLOUCHE, Somme des chiffres et transcendance, *Bull. Soc. Math. France* 110 (1982), 279–285.
10. J.-P. ALLOUCHE, Arithmétique et automates finis, *Astérisque* 147-148 (1987), 13–26, 343. Journées arithmétiques de Besançon (Besançon, 1985).
11. J.-P. ALLOUCHE, Automates finis en théorie des nombres, *Exposition. Math.* 5 (1987), 239–266.
12. J.-P. ALLOUCHE, Sur le développement en fraction continue de certaines séries formelles, *C. R. Acad. Sci. Paris , Série I* 307 (1988), 631–633.
13. J.-P. ALLOUCHE, Note on an article of H. Sharif and C. F. Woodcock: "Algebraic functions over a field of positive characteristic and Hadamard products", *Sém. Théor. Nombres Bordeaux* 1 (1989), 163–187.
14. J.-P. ALLOUCHE, Sur la transcendance de la série formelle π, *Sém. Théor. Nombres Bordeaux* 2 (1990), 103–117.
15. J.-P. ALLOUCHE, Finite automata and arithmetic, in *Séminaire Lotharingien de Combinatoire (Gerolfingen, 1993)*, pp. 1–18, Univ. Louis Pasteur, Strasbourg, 1993.
16. J.-P. ALLOUCHE, Sur la complexité des suites infinies, *Bull. Belg. Math. Soc. Simon Stevin* 1 (1994), 133–143. Journées Montoises (Mons, 1992).
17. J.-P. ALLOUCHE, Transcendence of the Carlitz-Goss gamma function at rational arguments, *J. Number Theory* 60 (1996), 318–328.
18. J.-P. ALLOUCHE, Nouveaux résultats de transcendance de réels à développement non aléatoire, *Gaz. Math.* 84 (2000), 19–34.

19. J.-P. Allouche and V. Berthé, Triangle de Pascal, complexité et automates, *Bull. Belg. Math. Soc. Simon Stevin* 4 (1997), 1–23.
20. J.-P. Allouche, J. Bétréma and J. Shallit, Sur des points fixes de morphismes d'un monoïde libre, *Theoret. Inform. Appl.* 23 (1989), 235–249.
21. J.-P. Allouche, J. L. Davison, M. Queffélec and L. Q. Zamboni, Transcendence of Sturmian or morphic continued fractions, *J. Number Theory* 91 (2001), 33–66.
22. J.-P. Allouche and M. Mendès France, Automata and automatic sequences, in *Beyond quasicrystals (Les Houches, 1994)*, pp. 293–367, Springer, Berlin, 1995.
23. J.-P. Allouche and J. Peyrière, Sur une formule de récurrence sur les traces de produits de matrices associés à certaines substitutions, *C. R. Acad. Sci. Paris Sér. II Méc. Phys. Chim. Sci. Univers Sci. Terre* 302 (1986), 1135–1136.
24. J.-P. Allouche and O. Salon, Finite automata, quasicrystals, and Robinson tilings, in *Quasicrystals, networks, and molecules of fivefold symmetry*, pp. 97–105, VCH, Weinheim, 1990.
25. J.-P. Allouche and J. O. Shallit, The ring of k-regular sequences, *Theoret. Comput. Sci.* 98 (1992), 163–197.
26. J.-P. Allouche and J. O. Shallit, Complexité des suites de Rudin-Shapiro généralisées, *J. Théor. Nombres Bordeaux* 5 (1993), 283–302.
27. J.-P. Allouche and J. O. Shallit, The ubiquitous Prouhet-Thue-Morse sequence, in *Sequences and their applications, Proceedings of SETA'98*, C. Ding, T. Helleseth and H. Niederreiter (eds), pp. 1–16, Springer Verlag, 1999.
28. J.-P. Allouche and J. O. Shallit, *Automatic sequences: Theory and Applications*, Cambridge University Press, 2002.
29. J.-P. Allouche and D. S. Thakur, Automata and transcendence of the Tate period in finite characteristic, *Proc. Amer. Math. Soc.* 127 (1999), 1309–1312.
30. J.-P. Allouche, F. von Haeseler, E. Lange, A. Petersen and G. Skordev, Linear cellular automata and automatic sequences, *Parallel Comput.* 23 (1997), 1577-1592. Cellular automata (Gießen, 1996).
31. J.-P. Allouche, F. von Haeseler, H.-O. Peitgen, A. Petersen and G. Skordev, Automaticity of double sequences generated by one-dimensional linear cellular automata, *Theoret. Comput. Sci.* 186 (1997), 195–209.
32. J.-P. Allouche, F. von Haeseler, H.-O. Peitgen and G. Skordev, Linear cellular automata, finite automata and Pascal's triangle, *Discrete Appl. Math.* 66 (1996), 1–22.
33. J.-P. Allouche and L. Q. Zamboni, Algebraic irrational binary numbers cannot be fixed points of non-trivial constant length or primitive morphisms, *J. Number Theory* 69 (1998), 119–124.
34. Y. Amice, *Les nombres p-adiques*, Presses Universitaires de France, Paris, 1975. Préface de Ch. Pisot, Collection SUP: Le Mathématicien, No. 14.
35. C. Apparicio, Reconnaissabilité des substitutions de longueur constante. Stage de Maitrise de l'ENS Lyon, 1999.
36. P. Arnoux, Un exemple de semi-conjugaison entre un échange d'intervalles et une translation sur le tore, *Bull. Soc. Math. France* 116 (1988), 489–500.
37. P. Arnoux, Complexité de suites à valeurs dans un ensemble fini: quelques exemples, in *Aspects des systèmes dynamiques, Journées X-UPS 1994*, École Polytechnique, Palaiseau, 1994.
38. P. Arnoux, Le codage du flot géodésique sur la surface modulaire, *Enseign. Math.* 40 (1994), 29–48.

39. P. ARNOUX, Recoding Sturmian sequences on a subshift of finite type. Chaos from order: a worked out example, in *Complex Systems*, E. Goles and S. Martinez (eds), pp. 1–67, Kluwer Academic Publ., 2001.
40. P. ARNOUX, V. BERTHÉ AND S. ITO, Discrete planes, $\mathbb{Z}^2$-actions, Jacobi-Perron algorithm and substitutions, Ann. Inst. Fourier 52 (2002), 1001–1045.
41. P. ARNOUX, S. FERENCZI AND P. HUBERT, Trajectories of rotations, *Acta Arith.* 87 (1999), 209–217.
42. P. ARNOUX AND A. M. FISCHER, The scenery flow for geometric structures on the torus: the linear setting, *Chin. Ann. of Math.* 22B (2001), 1–44.
43. P. ARNOUX AND S. ITO, Pisot substitutions and Rauzy fractals, *Bull. Belg. Math. Soc. Simon Stevin* 8 (2001), 181–207. Journées Montoises (Marne-la-Vallée, 2000).
44. P. ARNOUX, S. ITO AND Y. SANO, Higher dimensional extensions of substitutions and their dual maps, *J. Anal. Math.* 83 (2001), 183–206.
45. P. ARNOUX AND C. MAUDUIT, Complexité de suites engendrées par des récurrences unipotentes, *Acta Arith.* 76 (1996), 85–97.
46. P. ARNOUX, C. MAUDUIT, I. SHIOKAWA AND J.-I. TAMURA, Complexity of sequences defined by billiard in the cube, *Bull. Soc. Math. France* 122 (1994), 1–12.
47. P. ARNOUX, C. MAUDUIT, I. SHIOKAWA AND J.-I. TAMURA, Rauzy's conjecture on billiards in the cube, *Tokyo J. Math.* 17 (1994), 211–218.
48. P. ARNOUX AND A. NOGUEIRA, Mesures de Gauss pour des algorithmes de fractions continues multidimensionnelles, *Ann. Sci. École Norm. Sup.* 26 (1993), 645–664.
49. P. ARNOUX AND G. RAUZY, Représentation géométrique de suites de complexité $2n+1$, *Bull. Soc. Math. France* 119 (1991), 199–215.
50. J. AUSLANDER, *Minimal flows and their extensions*, North-Holland Publishing Co., Amsterdam, 1988. Mathematical Notes, Vol. 122.
51. Y. AVISHAI, D. BEREND AND D. GLAUBMAN, Minimum-dimension trace maps for substitution sequences, *Phys. Rev. Lett.* 72 (1994), 1842–1845.
52. F. AXEL AND D. GRATIAS (eds), *Beyond quasicrystals*, Springer-Verlag, Berlin, 1995. Papers from the Winter School held in Les Houches, March 7–18, 1994.
53. V. BALADI AND D. RUELLE, An extension of the theorem of Milnor and Thurston on the zeta functions of interval maps, *Ergodic Theory Dynam. Systems* 14 (1994), 621–632.
54. M. BARGE AND B. DIAMOND, Coincidence for substitutions of Pisot type. To appear in *Bull. Soc. Math. France.*
55. M. F. BARNSLEY, *Fractals everywhere*, Academic Press Professional, Boston, MA, 1993. Second ed., revised with the assistance of and with a foreword by H. Rising.
56. Y. BARYSHNIKOV, Complexity of trajectories in rectangular billiards, *Comm. Math. Phys.* 174 (1995), 43–56.
57. L. E. BAUM AND M. M. SWEET, Continued fractions of algebraic power series in characteristic 2, *Ann. of Math.* 103 (1976), 593–610.
58. L. E. BAUM AND M. M. SWEET, Badly approximable power series in characteristic 2, *Ann. of Math.* 105 (1977), 573–580.
59. M.-P. BÉAL AND D. PERRIN, Symbolic dynamics and finite automata, in *Handbook of formal languages, Vol. 2*, pp. 463–505, Springer, Berlin, 1997.
60. P.-G. BECKER, k-regular power series and Mahler-type functional equations, *J. Number Theory* 49 (1994), 269–286.
61. T. BEDFORD, Generating special Markov partitions for hyperbolic toral automorphisms using fractals, *Ergodic Theory Dynam. Systems* 6 (1986), 325–333.

62. T. BEDFORD, M. KEANE AND C. SERIES (eds), *Ergodic theory, symbolic dynamics, and hyperbolic spaces*, The Clarendon Press Oxford University Press, New York, 1991. Papers from the Workshop on Hyperbolic Geometry and Ergodic Theory held in Trieste, April 17–28, 1989.
63. A. BERMAN AND R. J. PLEMMONS, *Nonnegative matrices in the mathematical sciences*, Society for Industrial and Applied Mathematics (SIAM), Philadelphia, PA, 1979.
64. L. BERNSTEIN, *The Jacobi-Perron algorithm—Its theory and application*, Springer-Verlag, Berlin, 1971. Lecture Notes in Mathematics, Vol. 207.
65. J. BERSTEL, Mot de Fibonacci, in *Séminaire d'informatique théorique du L.I.T.P.*, Paris, 1980/1981, pp. 57–78.
66. J. BERSTEL, Tracé de droites, fractions continues et morphismes itérés, in *Mots*, pp. 298–309, Hermès, Paris, 1990.
67. J. BERSTEL, Recent results in Sturmian words, in *Developments in Language Theory II (DLT'95), Magdeburg (Allemagne)*, pp. 13–24, World Sci., Singapore, 1996.
68. J. BERSTEL AND P. SÉÉBOLD, Morphismes de Sturm, *Bull. Belg. Math. Soc. Simon Stevin* 1 (1994), 175–189. Journées Montoises (Mons, 1992).
69. J. BERSTEL AND P. SÉÉBOLD, Chapter 2: Sturmian words, in *Algebraic combinatorics on words*, M. Lothaire, Cambridge University Press, 2002.
70. V. BERTHÉ, Automates et valeurs de transcendance du logarithme de Carlitz, *Acta Arith.* 66 (1994), 369–390.
71. V. BERTHÉ, Combinaisons linéaires de $\zeta(s)/\pi^s$ sur $\mathbf{F}_q(x)$, pour $1 \leq s \leq q-2$, *J. Number Theory* 53 (1995), 272–299.
72. V. BERTHÉ, Fréquences des facteurs des suites sturmiennes, *Theoret. Comput. Sci.* 165 (1996), 295–309.
73. V. BERTHÉ, Sequences of low complexity: automatic and Sturmian sequences, in *Topics in symbolic dynamics and applications*, F. Blanchard, A. Nogueira and A. Maas (eds), pp. 1–34, *London Mathematical Society Lecture Note Series* vol. 279, Cambridge University Press, 2000.
74. V. BERTHÉ, Autour du système de numération d'Ostrowski, *Bull. Belg. Math. Soc. Simon Stevin* 8 (2001), 209–239.
75. V. BERTHÉ, N. CHEKHOVA AND S. FERENCZI, Covering numbers: arithmetics and dynamics for rotations and interval exchanges, *J. Anal. Math.* 79 (1999), 1–31.
76. V. BERTHÉ AND H. NAKADA, On continued fraction expansions in positive characteristic: equivalence relations and some metric properties, *Expo. Math.* 18 (2000), 257-284.
77. V. BERTHÉ AND L. VUILLON, Suites doubles de basse complexité, *J. Théor. Nombres Bordeaux* 12 (2000), 179–208.
78. V. BERTHÉ AND L. VUILLON, Tilings and rotations on the torus: a two-dimensional generalization of Sturmian sequences, *Discrete Math.* 223 (2000), 27–53.
79. J.-P. BERTRANDIAS, J. COUOT, J. DHOMBRES, M. MENDÈS FRANCE, P. P. HIEN AND K. VO-KHAC, *Espaces de Marcinkiewicz: corrélations, mesures, systèmes dynamiques*, Masson, Paris, 1987. With an introduction by J. Bass.
80. P. BILLINGSLEY, *Ergodic theory and information*, John Wiley & Sons Inc., New York, 1965.
81. F. BLANCHARD, A. NOGUEIRA AND A. MAAS (eds), *Topics in symbolic dynamics and applications*, Cambridge University Press, 2000. London Mathematical Society Lecture Note Series, Vol. 279.

82. E. Bombieri and J. E. Taylor, Which distributions of matter diffract? An initial investigation, *J. Physique* 47 (1986), C3–19–C3–28. International workshop on aperiodic crystals (Les Houches, 1986).
83. Z. I. Borevich and I. R. Shafarevich, *Number theory*, Academic Press, New York, 1966. Translated from the Russian by N. Greenleaf. Pure and Applied Mathematics, Vol. 20.
84. P. Borwein and C. Ingalls, The Prouhet-Tarry-Escott problem revisited, *Enseign. Math.* 40 (1994), 3–27.
85. M. Boshernitzan and I. Kornfeld, Interval translation mappings, *Ergodic Theory Dynam. Systems* 15 (1995), 821–832.
86. M. D. Boshernitzan and C. R. Carroll, An extension of Lagrange's theorem to interval exchange transformations over quadratic fields, *J. Anal. Math.* 72 (1997), 21–44.
87. R. Bowen, *Equilibrium states and the ergodic theory of Anosov diffeomorphisms*, Springer-Verlag, Berlin, 1975. Lecture Notes in Mathematics, Vol. 470.
88. R. Bowen, Bernoulli maps of the interval, *Israel J. Math.* 28 (1977), 161–168.
89. R. Bowen, Markov partitions are not smooth, *Proc. Amer. Math. Soc.* 71 (1978), 130–132.
90. A. J. Brentjes, *Multidimensional continued fraction algorithms*, Mathematisch Centrum, Amsterdam, 1981.
91. S. Brlek, Enumeration of factors in the Thue-Morse word, *Discrete Appl. Math.* 24 (1989), 83–96. First Montreal Conference on Combinatorics and Computer Science, 1987.
92. A. Broise, Fractions continues multidimensionnelles et lois stables, *Bull. Soc. Math. France* 124 (1996), 97–139.
93. A. Broise, Transformations dilatantes de l'intervalle et théorèmes limites. études spectrales d'opérateurs de transfert et applications, *Astérisque* 238 (1996), 1–109.
94. A. Broise and Y. Guivarch'h, Exposants caractéristiques de l'algorithme de Jacobi-Perron et de la transformation associée, *Ann. Inst. Fourier* 51 (2001), 565–686.
95. T. C. Brown, A characterization of the quadratic irrationals, *Canad. Math. Bull.* 34 (1991), 36–41.
96. H. Bruin, G. Keller and M. St. Pierre, Adding machines and wild attractors, *Ergodic Theory Dynam. Systems* 17 (1997), 1267–1287.
97. V. Bruyère, Automata and numeration systems, *Sém. Lothar. Combin.* 35 (1995), 19 pp. (electronic).
98. V. Bruyère, G. Hansel, C. Michaux and R. Villemaire, Logic and p-recognizable sets of integers, *Bull. Belg. Math. Soc. Simon Stevin* 1 (1994), 191–238. Journées Montoises (Mons, 1992).
99. J. W. Cahn and J. E. Taylor, An introduction to quasicrystals, in *The legacy of Sonya Kovalevskaya (Cambridge, Mass., and Amherst, Mass., 1985)*, pp. 265–286, Amer. Math. Soc., Providence, RI, 1987.
100. V. Canterini, Connectedness of geometric representation of substitutions of pisot type. To appear in *Bull. Belg. Math. Soc. Simon Stevin.*
101. V. Canterini, *Géométrie des substitutions Pisot unitaires*, PhD thesis, Université de la Méditerranée, 2000.
102. V. Canterini and A. Siegel, Automate des préfixes-suffixes associé à une substitution primitive, *J. Théor. Nombres Bordeaux* 13 (2001), 353–369.
103. V. Canterini and A. Siegel, Geometric representation of substitutions of Pisot type, *Trans. Amer. Math. Soc.* 353 (2001), 5121–5144.

104. W.-T. CAO AND Z.-Y. WEN, Some properties of the Sturmian sequences. Preprint, 2001.
105. L. CARLITZ, On certain functions connected with polynomials in a Galois field, *Duke Math. J.* 1 (1935), 137–168.
106. J. CASSAIGNE, Special factors of sequences with linear subword complexity, in *Developments in Language Theory II (DLT'95), Magdeburg (Allemagne)*, pp. 25–34, World Sci., Singapore, 1996.
107. J. CASSAIGNE, Complexité et facteurs spéciaux, *Bull. Belg. Math. Soc. Simon Stevin* 4 (1997), 67–88.
108. J. CASSAIGNE, Sequences with grouped factors, in *Developments in Language Theoy III (DLT'97)*, pp. 211–222, Aristotle University of Thessaloniki, 1998.
109. J. CASSAIGNE, Two dimensional sequences with complexity $mn+1$, *J. Auto. Lang. Comb.* 4 (1999), 153–170.
110. J. CASSAIGNE, S. FERENCZI AND L. Q. ZAMBONI, Imbalances in Arnoux-Rauzy sequences, *Ann. Inst. Fourier* 50 (2000), 1265–1276.
111. J. CASSAIGNE, T. HARJU AND J. KARHUMÄKI, On the undecidability of freeness of matrix semigroups, *Internat. J. Algebra Comput.* 9 (1999), 295–305.
112. J. CASSAIGNE AND J. KARHUMÄKI, Examples of undecidable problems for 2-generator matrix semigroups, *Theoret. Comput. Sci.* 204 (1998), 29–34.
113. J. W. S. CASSELS, *An introduction to Diophantine approximation*, Cambridge University Press, New York, 1957.
114. M. G. CASTELLI, F. MIGNOSI AND A. RESTIVO, Fine and Wilf's theorem for three periods and a genereralization of Sturmian words, *Theoret. Comput. Sci.* 218 (1999), 83–94.
115. R. V. CHACON, Weakly mixing transformations which are not strongly mixing, *Proc. Amer. Math. Soc.* 22 (1969), 559–562.
116. N. CHEKHOVA, Covering numbers of rotations, *Theoret. Comput. Sci.* 230 (2000), 97–116.
117. N. CHEKHOVA, P. HUBERT AND A. MESSAOUDI, Propriétés combinatoires, ergodiques et arithmétiques de la substitution de Tribonacci, *J. Théor. Nombres Bordeaux* 13 (2001), 371–394.
118. G. CHRISTOL, Ensembles presque periodiques k-reconnaissables, *Theoret. Comput. Sci.* 9 (1979), 141–145.
119. G. CHRISTOL, T. KAMAE, M. MENDÈS FRANCE AND G. RAUZY, Suites algébriques, automates et substitutions, *Bull. Soc. Math. France* 108 (1980), 401–419.
120. A. COBHAM, On the base-dependence of sets of numbers recognizable by finite automata, *Math. Systems Theory* 3 (1969), 186–192.
121. A. COBHAM, Uniform tag sequences, *Math. Systems Theory* 6 (1972), 164–192.
122. I. P. CORNFELD, S. V. FOMIN AND Y. G. SINAĬ, *Ergodic theory*, Springer-Verlag, New York, 1982. Translated from the Russian by A. B. Sosinskiĭ.
123. E. M. COVEN, Sequences with minimal block growth. II, *Math. Systems Theory* 8 (1974/75), 376–382.
124. E. M. COVEN AND G. A. HEDLUND, Sequences with minimal block growth, *Math. Systems Theory* 7 (1973), 138–153.
125. E. M. COVEN AND M. S. KEANE, The structure of substitution minimal sets, *Trans. Amer. Math. Soc.* 162 (1971), 89–102.
126. D. CRISP, W. MORAN, A. POLLINGTON AND P. SHIUE, Substitution invariant cutting sequences, *J. Théor. Nombres Bordeaux* 5 (1993), 123–137.
127. K. DADJANI AND C. KRAAIKAMP, *Ergodic theory of numbers*, Preprint, 2002.
128. G. DAMAMME AND Y. HELLEGOUARCH, Transcendence of the values of the Carlitz zeta function by Wade's method, *J. Number Theory* 39 (1991), 257–278.

129. J. L. DAVISON, A class of transcendental numbers with bounded partial quotients, in *Number theory and applications (Banff, AB, 1988)*, pp. 365–371, Kluwer Acad. Publ., Dordrecht, 1989.
130. A. DE LUCA, A combinatorial property of the Fibonacci words, *Inform. Process. Lett.* 12 (1981), 193–195.
131. A. DE LUCA, On some combinatorial problems in free monoids, *Discrete Math.* 38 (1982), 207–225.
132. A. DE LUCA AND S. VARRICCHIO, Some combinatorial properties of the Thue-Morse sequence and a problem in semigroups, *Theoret. Comput. Sci.* 63 (1989), 333–348.
133. B. DE MATHAN, Irrationality measures and transcendence in positive characteristic, *J. Number Theory* 54 (1995), 93–112.
134. J. DE VRIES, *Elements of topological dynamics*, Kluwer Academic Publishers Group, Dordrecht, 1993.
135. F. M. DEKKING, The spectrum of dynamical systems arising from substitutions of constant length, *Z. Wahrscheinlichkeitstheorie und Verw. Gebiete* 41 (1977/78), 221–239.
136. F. M. DEKKING, On the structure of self-generating sequences, in *Séminaire de Théorie des Nombres (Talence, 1980–1981)*, Univ. Bordeaux I, 1981. Exp. No. 31.
137. F. M. DEKKING, Recurrent sets, *Adv. in Math.* 44 (1982), 78–104.
138. F. M. DEKKING, What is the long range order in the Kolakoski sequence?, in *The mathematics of long-range aperiodic order (Waterloo, ON, 1995)*, pp. 115–125, Kluwer Acad. Publ., Dordrecht, 1997.
139. F. M. DEKKING AND M. KEANE, Mixing properties of substitutions, *Z. Wahrscheinlichkeitstheorie und Verw. Gebiete* 42 (1978), 23–33.
140. A. DEL JUNCO, A transformation with simple spectrum which is not rank one, *Canad. J. Math.* 29 (1977), 655–663.
141. A. DEL JUNCO AND M. KEANE, On generic points in the Cartesian square of Chacón's transformation, *Ergodic Theory Dynam. Systems* 5 (1985), 59–69.
142. A. DEL JUNCO, M. RAHE AND L. SWANSON, Chacon's automorphism has minimal self-joinings, *J. Analyse Math.* 37 (1980), 276–284.
143. P. DELIGNE, Intégration sur un cycle évanescent, *Invent. Math.* 76 (1984), 129–143.
144. L. DENIS, Indépendance algébrique des dérivées d'une période du module de Carlitz, *J. Austral. Math. Soc. Ser. A* 69 (2000), 8–18.
145. R. L. DEVANEY, *A first course in chaotic dynamical systems*, Addison-Wesley Publishing Company Advanced Book Program, Reading, MA, 1992.
146. G. DIDIER, Codages de rotations et fractions continues, *J. Number Theory* 71 (1998), 275–306.
147. G. DIDIER, Caractérisation des n-écritures et applications à l'étude des suites de complexité ultimement $n + c^{ste}$, *Theoret. Comput. Sci.* 215 (1999), 31–49.
148. V. G. DRINFELD, Elliptic modules, *Mat. Sb. (N.S.)* 94 (136) (1974), 594–627, 656.
149. X. DROUBAY, J. JUSTIN AND G. PIRILLO, Epi-Sturmian words and some constructions of de Luca and Rauzy, *Theoret. Comput. Sci.* 255 (2001), 539–553.
150. P. DUMAS, *Récurrence mahlérienne, suites automatiques, études asymptotiques*, PhD thesis, Université Bordeaux I, 1993.
151. J.-M. DUMONT AND A. THOMAS, Systèmes de numération et fonctions fractales relatifs aux substitutions, *Theoret. Comput. Sci.* 65 (1989), 153–169.
152. F. DURAND, *Contribution à l'étude des suites et systèmes dynamiques substitutifs*, PhD thesis, Université de la Méditerranée, 1996.

153. F. DURAND, A characterization of substitutive sequences using return words, *Discrete Math.* 179 (1998), 89–101.
154. F. DURAND, A generalization of Cobham's thoerem, *Theory Comput. Syst.* 31 (1998), 169–185.
155. F. DURAND, Linearly recurrent subshifts have a finite number of non-periodic subshift factors, *Ergodic Theory Dynam. Systems* 20 (2000), 1061–1078; corrigendum and addendum: to appear.
156. F. DURAND, B. HOST AND C. SKAU, Substitutional dynamical systems, Bratteli diagrams and dimension groups, *Ergodic Theory Dynam. Systems* 19 (1999), 953–993.
157. A. EHRENFEUCHT, K. P. LEE AND G. ROZENBERG, Subword complexities of various classes of deterministic developmental languages without interactions, *Theoret. Comput. Sci.* 1 (1975), 59–75.
158. H. EI AND S. ITO, Decomposition theorem on invertible substitutions, *Osaka J. Math.* 35 (1998), 821–834.
159. S. EILENBERG, *Automata, languages, and machines. Vol. A*, Academic Press, New York, 1974. Pure and Applied Mathematics, Vol. 58.
160. R. ELLIS, *Lectures on topological dynamics*, W. A. Benjamin, Inc., New York, 1969.
161. R. ELLIS AND W. H. GOTTSCHALK, Homomorphisms of transformation groups, *Trans. Amer. Math. Soc.* 94 (1960), 258–271.
162. K. FALCONER, *Fractal geometry*, John Wiley & Sons Ltd., Chichester, 1990. Mathematical foundations and applications.
163. D.-J. FENG, M. FURUKADO, S. ITO AND J. WU, Pisot substitutions and the Haussdorf dimension of boundaries of atomic surfaces. Preprint, 2000.
164. D.-J. FENG AND Z.-Y. WEN, A property of Pisot numbers. To appear in *J. Number Theory*.
165. S. FERENCZI, Bounded remainder sets, *Acta Arith.* 61 (1992), 319–326.
166. S. FERENCZI, Les transformations de Chacon: combinatoire, structure géométrique, lien avec les systèmes de complexité $2n+1$, *Bull. Soc. Math. France* 123 (1995), 271–292.
167. S. FERENCZI, Rank and symbolic complexity, *Ergodic Theory Dynam. Systems* 16 (1996), 663–682.
168. S. FERENCZI, Systems of finite rank, *Colloq. Math.* 73 (1997), 35–65.
169. S. FERENCZI, Complexity of sequences and dynamical systems, *Discrete Math.* 206 (1999), 145–154. Combinatorics and number theory (Tiruchirappalli, 1996).
170. S. FERENCZI, C. HOLTON AND L. Q. ZAMBONI, The structure of 3-interval exchange transformations I: An arithmetic study, *Ann. Inst. Fourier* 51 (2001), 861–901.
171. S. FERENCZI, C. HOLTON AND L. Q. ZAMBONI, The structure of 3-interval exchange transformations II: Combinatorial properties, Preprint n° 01-25, Institut de Mathématiques de Luminy, 2001.
172. S. FERENCZI, C. HOLTON AND L. Q. ZAMBONI, The structure of 3-interval exchange transformations III: Ergodic and spectral properties, Preprint n° 01-26, Institut de Mathématiques de Luminy, 2001.
173. S. FERENCZI AND C. MAUDUIT, Transcendence of numbers with a low complexity expansion, *J. Number Theory* 67 (1997), 146–161.
174. S. FERENCZI, C. MAUDUIT AND A. NOGUEIRA, Substitution dynamical systems: algebraic characterization of eigenvalues, *Ann. Sci. École Norm. Sup.* 29 (1996), 519–533.
175. A. H. FORREST, K-groups associated with substitution minimal systems, *Israel J. Math.* 98 (1997), 101–139.

176. A. S. Fraenkel and J. Simpson, The exact number of squares in Fibonacci words, *Theoret. Comput. Sci.* 218 (1999), 95–106. Words (Rouen, 1997).
177. J. Françon, Sur la topologie d'un plan arithmétique, *Theoret. Comput. Sci.* 156 (1996), 159–176.
178. J. Fresnel, M. Koskas and B. de Mathan, Automata and transcendence in positive characteristic, *J. Number Theory* 80 (2000), 1–24.
179. R. Fricke and F. Klein, *Vorlesungen über die Theorie der automorphen Funktionen. Band I: Die gruppentheoretischen Grundlagen. Band II: Die funktionentheoretischen Ausführungen und die Anwendungen*, Johnson Reprint Corp., New York, 1965. Bibliotheca Mathematica Teubneriana, Bände 3, 4. Originally published by B. G. Teubner, Leipzig, 1897.
180. C. Frougny, Number representation and finite automata, in *Topics in symbolic dynamics and applications*, F. Blanchard, A. Nogueira and A. Maas (eds), pp. 207–228, Cambridge University Press, 2000. London Mathematical Society Lecture Note Series, Vol. 279.
181. C. Frougny and B. Solomyak, Finite beta-expansions, *Ergodic Theory Dynam. Systems* 12 (1992), 713–723.
182. T. Fujita, S. Ito, M. Keane and M. Ohtsuki, On almost everywhere exponential convergence of the modified Jacobi-Perron algorithm: a corrected proof, *Ergodic Theory Dynam. Systems* 16 (1996), 1345–1352.
183. H. Furstenberg, The structure of distal flows, *Amer. J. Math.* 85 (1963), 477–515.
184. H. Furstenberg, Algebraic functions over finite fields, *J. Algebra* 7 (1967), 271–277.
185. H. Furstenberg, *Recurrence in ergodic theory and combinatorial number theory*, Princeton University Press, Princeton, N.J., 1981. M. B. Porter Lectures.
186. M. Furukado and S. Ito, The quasi-periodic tiling of the plane and Markov subshifts, *Japan. J. Math. (N.S.)* 24 (1998), 1–42.
187. M. Furukado and S. Ito, Connected Markov partitions of group automorphisms and Rauzy fractals. Preprint, 2000.
188. J.-M. Gambaudo, P. Hubert, P. Tisseur and S. Vaienti (eds), *Dynamical systems: from crystal to chaos*, World Scientific, 2000. Proceedings in honor of G. Rauzy on his 60th birthday.
189. F. R. Gantmacher, *The theory of matrices. Vol. 1, 2*, AMS Chelsea Publishing, Providence, RI, 1998. Translated from the Russian by K. A. Hirsch, Reprint of the 1959 translation.
190. D. Goss, *Basic structures of function field arithmetic*, Springer-Verlag, Berlin, 1996.
191. W. H. Gottschalk, Substitution minimal sets, *Trans. Amer. Math. Soc.* 109 (1963), 467–491.
192. F. Q. Gouvêa, *p-adic numbers*, Springer-Verlag, Berlin, 1997. Second ed.
193. P. J. Grabner, P. Liardet and R. F. Tichy, Odometers and systems of numeration, *Acta Arith.* 70 (1995), 103–123.
194. P. R. Halmos, *Lectures on ergodic theory*, Chelsea Publishing Co., New York, 1960.
195. C. W. Hansen, *Dynamics of multi-dimensional substitutions*, PhD thesis, George Washington University, 2000.
196. Y. Hara and S. Ito, On real quadratic fields and periodic expansions, *Tokyo J. Math.* 12 (1989), 357–370.
197. T. Harase, Algebraic elements in formal power series rings, *Israel J. Math.* 63 (1988), 281–288.

198. G. H. HARDY AND E. M. WRIGHT, *An introduction to the theory of numbers*, Oxford Science Publications, 1979.
199. T. HARJU AND M. LINNA, On the periodicity of morphisms on free monoids, *RAIRO Inform. Théor. Appl.* 20 (1986), 47–54.
200. S. HASHIMOTO, *On Phonemics in Ancient Japanese Language*, Meiseidô-shoten, 1942. (Iwanami-shoten, Tokyo, 1980), in Japanese.
201. A. HEINIS, *Arithmetics and combinatorics of words of low complexity*, PhD thesis, University of Leiden, 2001.
202. A. HEINIS AND R. TIJDEMAN, Characterisation of asymptotically Sturmian sequences, *Publ. Math. Debrecen* 56 (2000), 415–430.
203. Y. HELLEGOUARCH, Un analogue d'un théorème d'Euler, *C. R. Acad. Sci. Paris Sér. I Math.* 313 (1991), 155–158.
204. F. HOFBAUER, Hausdorff and conformal measures for expanding piecewise monotonic maps of the interval, *Studia Math.* 103 (1992), 191–206.
205. F. HOFBAUER AND G. KELLER, Zeta-functions and transfer-operators for piecewise linear transformations, *J. Reine Angew. Math.* 352 (1984), 100–113.
206. M. HOLLANDER AND B. SOLOMYAK, Two-symbol Pisot substitutions have pure discrete spectrum. Preprint, 2001.
207. C. HOLTON AND L. ZAMBONI, Initial powers of Sturmian words. Preprint, 2001.
208. C. HOLTON AND L. Q. ZAMBONI, Geometric realizations of substitutions, *Bull. Soc. Math. France* 126 (1998), 149–179.
209. C. HOLTON AND L. Q. ZAMBONI, Descendants of primitive substitutions, *Theory Comput. Syst.* 32 (1999), 133–157.
210. C. HOLTON AND L. Q. ZAMBONI, Substitutions, partial isometries of $\mathbf{R}$, and actions on trees, *Bull. Belg. Math. Soc. Simon Stevin* 6 (1999), 395–411.
211. C. HOLTON AND L. Q. ZAMBONI, Directed graphs and substitutions, 2001.
212. J. E. HOPCROFT AND J. D. ULLMAN, *Introduction to automata theory, languages, and computation*, Addison-Wesley Publishing Co., Reading, Mass., 1979. Addison-Wesley Series in Computer Science.
213. R. D. HOROWITZ, Characters of free groups represented in the two-dimensional special linear group, *Comm. Pure Appl. Math.* 25 (1972), 635–649.
214. R. D. HOROWITZ, Induced automorphisms on Fricke characters of free groups, *Trans. Amer. Math. Soc.* 208 (1975), 41–50.
215. B. HOST, Valeurs propres des systèmes dynamiques définis par des substitutions de longueur variable, *Ergodic Theory Dynam. Systems* 6 (1986), 529–540.
216. B. HOST, Représentation géométrique des substitutions sur 2 lettres. Unpublished manuscript, 1992.
217. B. HOST, Substitution subshifts and Bratteli diagrams, in *Topics in symbolic dynamics and applications*, pp. 35–56, *London Mathematical Society Lecture Note Series* vol. 279, Cambridge University Press, 2000.
218. S. ITO, Some skew product transformations associated with continued fractions and their invariant measures, *Tokyo J. Math.* 9 (1986), 115–133.
219. S. ITO, Fractal domains of quasi-periodic motions on $\mathbb{T}^2$, in *Algorithms, fractals, and dynamics (Okayama/Kyoto, 1992)*, pp. 95–99, Plenum, New York, 1995.
220. S. ITO, Limit set of $\{\sqrt{q}\binom{q\alpha-p}{q\beta-r} \mid (q,p,r) \in \mathbb{Z}^3\}$ and domain exchange transformations, in *Dynamical systems and chaos, Vol. 1 (Hachioji, 1994)*, pp. 101–102, World Sci. Publishing, River Edge, NJ, 1995.
221. S. ITO, Simultaneous approximations and dynamical systems (on the simultaneous approximation of (α,α^2) satisfying $\alpha^3 + k\alpha - 1 = 0$), *Sūrikaisekikenkyūsho Kōkyūroku*, 1996, 59–61. Analytic number theory (Japanese) (Kyoto, 1994).

222. S. ITO AND M. KIMURA, On Rauzy fractal, *Japan J. Indust. Appl. Math.* 8 (1991), 461–486.
223. S. ITO AND M. OHTSUKI, On the fractal curves induced from endomorphisms on a free group of rank 2, *Tokyo J. Math.* 14 (1991), 277–304.
224. S. ITO AND M. OHTSUKI, Modified Jacobi-Perron algorithm and generating Markov partitions for special hyperbolic toral automorphisms, *Tokyo J. Math.* 16 (1993), 441–472.
225. S. ITO AND M. OHTSUKI, Parallelogram tilings and Jacobi-Perron algorithm, *Tokyo J. Math.* 17 (1994), 33–58.
226. S. ITO AND Y. SANO, On periodic β-expansions of Pisot numbers and Rauzy fractals, *Osaka J. Math.* 38 (2001), 349–368.
227. S. ITO AND Y. TAKAHASHI, Markov subshifts and realization of β-expansions, *J. Math. Soc. Japan* 26 (1974), 33–55.
228. J.-P. KAHANE AND R. SALEM, *Ensembles parfaits et séries trigonométriques*, Hermann, Paris, 1994. Second ed., with notes by J.-P. Kahane, T. W. Körner, R. Lyons and S. W. Drury.
229. T. KAMAE, Spectrum of a substitution minimal set, *J. Math. Soc. Japan* 22 (1970), 567–578.
230. T. KAMAE, A topological invariant of substitution minimal sets, *J. Math. Soc. Japan* 24 (1972), 285–306.
231. T. KAMAE, Linear expansions, strictly ergodic homogeneous cocycles and fractals, *Israel J. Math.* 106 (1998), 313–337.
232. T. KAMAE, J.-I. TAMURA AND Z.-Y. WEN, Hankel determinants for the Fibonacci word and Padé approximation, *Acta Arith.* 89 (1999), 123–161.
233. J. KARHUMÄKI, On cube-free ω-words generated by binary morphisms, *Discrete Appl. Math.* 5 (1983), 279–297.
234. A. KATOK AND B. HASSELBLATT, *Introduction to the modern theory of dynamical systems*, Cambridge University Press, Cambridge, 1995.
235. Y. KATZNELSON, *An introduction to harmonic analysis*, Dover Publications Inc., New York, 1976. corrected ed.
236. M. KEANE, Generalized Morse sequences, *Zeit Wahr.* 10 (1968), 335–353.
237. M. KEANE, Irrational rotations and quasi-ergodic measures, in *Séminaires de Mathématiques de l'Université de Rennes*, pp. 17–26, 1970.
238. M. KEANE, Sur les mesures quasi-ergodiques des translations irrationnelles, *C. R. Acad. Sci. Paris Sér. A-B* 272 (1971), A54–A55.
239. M. S. KEANE, Ergodic theory and subshifts of finite type, in *Ergodic theory, symbolic dynamics, and hyperbolic spaces (Trieste, 1989)*, pp. 35–70, Oxford Univ. Press, New York, 1991.
240. G. KELLER, On the rate of convergence to equilibrium in one-dimensional systems, *Comm. Math. Phys.* 96 (1984), 181–193.
241. G. KELLER, *Equilibrium states in ergodic theory*, Cambridge University Press, Cambridge, 1998.
242. R. KENYON, *Self-similar tilings*, PhD thesis, Princeton University, 1990.
243. R. KENYON, The construction of self-similar tilings, *Geom. Funct. Anal.* 6 (1996), 471–488.
244. R. KENYON, *Sur la combinatoire, la dynamique et la statistique des pavages.* Habilitation à diriger des recherches, Université de Paris-Sud, 1999.
245. R. KENYON AND A. VERSHIK, Arithmetic construction of sofic partitions of hyperbolic toral automorphisms, *Ergodic Theory Dynam. Systems* 18 (1998), 357–372.
246. A. Y. KHINTCHINE, *Continued fractions*, P. Noordhoff, Ltd., Groningen, 1963.
247. B. P. KITCHENS, *Symbolic dynamics*, Springer-Verlag, Berlin, 1998.

248. D. A. KLARNER, J.-C. BIRGET AND W. SATTERFIELD, On the undecidability of the freeness of integer matrix semigroups, *Internat. J. Algebra Comput.* 1 (1991), 223–226.
249. B. G. KLEIN, Homomorphisms of symbolic dynamical systems, *Math. Systems Theory* 6 (1972), 107–122.
250. W. KOLAKOSKI, Problem 5304, *Amer. Math. Monthly* 72 (1965), 674.
251. M. KOLÁŘ AND M. K. ALI, Trace maps associated with general two-letter substitution rules, *Phys. Rev. A* 42 (1990), 7112–7124.
252. M. KOLÁŘ AND F. NORI, Trace maps of general substitutional sequences, *Phys. Rev. B* 42 (1990), 1062–1065.
253. T. KOMATSU, A certain power series and the inhomogeneous continued fraction expansions, *J. Number Theory* 59 (1996), 291–312.
254. M. KÓSA, Problems and solutions, *Bull. European Assoc. Theor. Comput. Sci. (EATCS)* 32 (1987), 331–333.
255. M. KOSKAS, *Complexité de suites, fonctions de Carlitz*, PhD thesis, Université de Bordeaux I, 1995.
256. J. C. LAGARIAS, The quality of the Diophantine approximations found by the Jacobi-peron algorithm and related algorithms, *Monatsh. Math.* 115 (1993), 179–194.
257. J. C. LAGARIAS, Geometric models for quasicrystals I. Delone sets of finite type, *Discrete Comput. Geom.* 21 (1999), 161–191.
258. J. C. LAGARIAS, Geometric models for quasicrystals. II. Local rules under isometries, *Discrete Comput. Geom.* 21 (1999), 345–372.
259. A. LASJAUNIAS, A survey of Diophantine approximation in fields of power series, *Monatsh. Math.* 130 (2000), 211–229.
260. A. LASOTA AND J. A. YORKE, On the existence of invariant measures for piecewise monotonic transformations, *Trans. Amer. Math. Soc.* 186 (1973), 481–488 (1974).
261. S. LE BORGNE, Un codage sofique des automorphismes hyperboliques du tore, *C. R. Acad. Sci. Paris Sér. I Math.* 323 (1996), 1123–1128.
262. S. LE BORGNE, Algébricité des partitions markoviennes des automorphismes hyperboliques du tore, *C. R. Acad. Sci. Paris Sér. I Math.* 328 (1999), 1045–1048.
263. M. LEMAŃCZYK, Toeplitz Z_2-extensions, *Ann. Inst. H. Poincaré Probab. Statist.* 24 (1988), 1–43.
264. U. LERON, Trace identities and polynomial identities of $n \times n$ matrices, *J. Algebra* 42 (1976), 369–377.
265. D. LIND AND B. MARCUS, *An introduction to symbolic dynamics and coding*, Cambridge University Press, Cambridge, 1995.
266. A. N. LIVSHITS, On the spectra of adic transformations of Markov compact sets, *Uspekhi Mat. Nauk* 42 (1987), 189–190. English translation: *Russian Math. Surveys* 42(3): 222–223, 1987.
267. A. N. LIVSHITS, Sufficient conditions for weak mixing of substitutions and of stationary adic transformations, *Mat. Zametki* 44 (1988), 785–793, 862. English translation: *Math. Notes* 44: 920–925, 1988.
268. A. N. LIVSHITS, Some examples of adic transformations and automorphisms of substitutions, *Selecta Math. Soviet.* 11 (1992), 83–104. Selected translations.
269. L.-M. LOPEZ AND P. NARBEL, D0L-systems and surface automorphisms, in *Mathematical foundations of computer science, 1998 (Brno)*, pp. 522–532, Springer, Berlin, 1998.
270. L.-M. LOPEZ AND P. NARBEL, Substitutions and interval exchange transformations of rotation class, *Theoret. Comput. Sci.* 255 (2001), 323–344.

271. M. LOTHAIRE, *Combinatorics on words*, Cambridge University Press, Cambridge, 1997. Second ed.
272. M. LOTHAIRE, *Algebraic combinatorics on words*, Cambridge University Press, 2002.
273. J. H. LOXTON AND A. J. VAN DER POORTEN, Arithmetic properties of automata: regular sequences, *J. Reine Angew. Math.* 392 (1988), 57–69.
274. J. M. LUCK, C. GODRÈCHE, A. JANNER AND T. JANSSEN, The nature of the atomic surfaces of quasiperiodic self-similar structures, *J. Phys. A* 26 (1993), 1951–1999.
275. R. C. LYNDON AND P. E. SCHUPP, *Combinatorial group theory*, Springer-Verlag, Berlin, 2001. Reprint of the 1977 edition.
276. A. MAES, *Morphic predicates and applications to the decidability of arithmetic theories*, PhD thesis, Université de Mons-Hainault, 1999.
277. W. MAGNUS, Rings of Fricke characters and automorphism groups of free groups, *Math. Z.* 170 (1980), 91–103.
278. W. MAGNUS, A. KARRASS AND D. SOLITAR, *Combinatorial group theory*, Dover Publications Inc., New York, 1976. Revised ed.
279. K. MAHLER, On the translation properties of a simple class of arithmetical functions, *J. Math. Massachusetts* 6 (1927), 158–163.
280. K. MAHLER, *Lectures on diophantine approximations. Part I: g-adic numbers and Roth's theorem*, University of Notre Dame Press, Notre Dame, Ind., 1961.
281. J. C. MARTIN, Substitution minimal flows, *Amer. J. Math.* 93 (1971), 503–526.
282. J. C. MARTIN, Minimal flows arising from substitutions of non-constant length, *Math. Systems Theory* 7 (1973), 72–82.
283. C. MAUDUIT, Caractérisation des ensembles normaux substitutifs, *Invent. Math.* 95 (1989), 133–147.
284. C. MAUDUIT, Multiplicative properties of the Thue-Morse sequence, *Period. Math. Hungar.* 43 (2001), 137–153.
285. C. MAUDUIT AND A. SÁRKÖZY, On finite pseudorandom binary sequences. II. The Champernowne, Rudin-Shapiro, and Thue-Morse sequences, a further construction, *J. Number Theory* 73 (1998), 256–276.
286. R. MEESTER, A simple proof of the exponential convergence of the modified Jacobi-Perron algorithm, *Ergodic Theory Dynam. Systems* 19 (1999), 1077–1083.
287. G. MELANÇON, Lyndon words and singular factors of Sturmian words, *Theoret. Comput. Sci.* 218 (1999), 41–59. Words (Rouen, 1997).
288. M. MENDÈS FRANCE AND J.-Y. YAO, Transcendence and the Carlitz-Goss gamma function, *J. Number Theory* 63 (1997), 396–402.
289. A. MESSAOUDI, *Autour du fractal de Rauzy*, PhD thesis, Université de la Méditerranée, 1996.
290. A. MESSAOUDI, Propriétés arithmétiques et dynamiques du fractal de Rauzy, *J. Théor. Nombres Bordeaux* 10 (1998), 135–162.
291. A. MESSAOUDI, Frontière du fractal de Rauzy et système de numération complexe, *Acta Arith.* 95 (2000), 195–224.
292. P. MICHEL, Coincidence values and spectra of substitutions, *Z. Wahrscheinlichkeitstheorie und Verw. Gebiete* 42 (1978), 205–227.
293. F. MIGNOSI, Infinite words with linear subword complexity, *Theoret. Comput. Sci.* 65 (1989), 221–242.
294. F. MIGNOSI, On the number of factors for Sturmian words, *Theoret. Comput. Sci.* 82 (1991), 71–84.
295. F. MIGNOSI AND G. PIRILLO, Repetitions in the Fibonacci infinite word, *RAIRO Inform. Théor. Appl.* 26 (1992), 199–204.

296. F. Mignosi and P. Séébold, Morphismes sturmiens et règles de Rauzy, *J. Théor. Nombres Bordeaux* 5 (1993), 221–233.
297. F. Mignosi and L. Q. Zamboni, On the number of Arnoux-Rauzy words, *Acta Arith.* 101 (2002), 121–129.
298. M. H. Mills and D. P. Robbins, Continued fractions for certain algebraic power series, *J. Number Theory* 23 (1986), 388–404.
299. M. Mkaouar, Sur le développement en fraction continue de la série de Baum et Sweet, *Bull. Soc. Math. France* 123 (1995), 361–374.
300. M. Mori, Fredholm determinant for piecewise linear transformations, *Osaka J. Math.* 27 (1990), 81–116.
301. T. Morita, Random iteration of one-dimensional transformations, *Osaka J. Math.* 22 (1985), 489–518.
302. H. M. Morse, Recurrent geodesics on a surface of negative curvature, *Trans. Amer. Math. Soc.* 22 (1921), 84–100.
303. M. Morse and G. A. Hedlund, Symbolic dynamics, *Amer. J. Math.* 60 (1938), 815–866.
304. M. Morse and G. A. Hedlund, Symbolic dynamics II. Sturmian trajectories, *Amer. J. Math.* 62 (1940), 1–42.
305. M. Morse and G. A. Hedlund, Unending chess, symbolic dynamics and a problem in semigroups, *Duke Math. J.* 11 (1944), 1–7.
306. B. Mossé, Puissances de mots et reconnaissabilité des points fixes d'une substitution, *Theoret. Comput. Sci.* 99 (1992), 327–334.
307. B. Mossé, Reconnaissabilité des substitutions et complexité des suites automatiques, *Bull. Soc. Math. France* 124 (1996), 329–346.
308. M. G. Nadkarni, *Basic ergodic theory*, Birkhäuser Verlag, Basel, 1998. Second ed.
309. M. G. Nadkarni, *Spectral theory of dynamical systems*, Birkhäuser Verlag, Basel, 1998.
310. H. Nakada, Metrical theory for a class of continued fraction transformations and their natural extensions, *Tokyo Journal of Mathematics* 4 (1981), 399–426.
311. P. Narbel, The boundary of iterated morphisms on free semi-groups, *Internat. J. Algebra Comput.* 6 (1996), 229–260.
312. J. Néraud (ed.), WORDS (Rouen, september 22–26, 1997), *Theoret. Comput. Sci.* 218 (1999). Elsevier Science Publishers B.V., Amsterdam.
313. B. Neumann, Die Automorphismengruppe der freien Gruppen, *Math. Ann.* 107 (1933), 367–386.
314. J. Nielsen, Die Isomorphismen der allgemeinen unendlichen Gruppen mit zwei Erzengenden, *Math. Ann.* 78 (1918), 385–397.
315. J. Nielsen, Die Isomorphismen der freien Gruppen, *Math. Ann.* 91 (1924), 169–209.
316. K. Nishioka, J.-I. Tamura and I. Shiokawa, Arithmetical properties of a certain power series, *J. Number Theory* 42 (1992), 61–87.
317. D. S. Ornstein, D. J. Rudolph and B. Weiss, *Equivalence of measure preserving transformations*, American Mathematical Society, Providence, R.I., 1982. Memoirs of the American Mathematical Society, No. 262.
318. S. Ostlund, R. Pandit, D. Rand, H. J. Schellnhuber and E. D. Siggia, One-dimensional Schrödinger equation with an almost periodic potential, *Phys. Rev. Lett.* 50 (1983), 1873–1876.
319. A. Ostrowsky, Bemerkungen zur Theorie der Diophantischen Approximationen I, II, *Abh. Math. Sem. Hamburg* 1 (1922), 77–98 and 250–251.
320. J.-J. Pansiot, Complexité des facteurs des mots infinis engendrés par morphismes itérés, in *Automata, languages and programming (Antwerp, 1984)*, pp. 380–389, Springer, Berlin, 1984.

321. J.-J. PANSIOT, Decidability of periodicity for infinite words, *RAIRO Inform. Théor. Appl.* 20 (1986), 43–46.
322. D. P. PARENT, *Exercices de théorie des nombres*, Gauthier-Villars, Paris, 1978.
323. W. PARRY, Intrinsic Markov chains, *Trans. Amer. Math. Soc.* 112 (1964), 55–66.
324. W. PARRY, *Topics in ergodic theory*, Cambridge University Press, Cambridge, 1981.
325. D. PERRIN, Recent results on automata and infinite words, in *Mathematical foundations of computer science (Prague, 1984)*, pp. 134–148, Springer, Berlin, 1984.
326. K. PETERSEN, *Ergodic theory*, Cambridge University Press, Cambridge, 1989. Corrected reprint of the 1983 original.
327. J. PEYRIÈRE, Frequency of patterns in certain graphs and in Penrose tilings, *J. Physique* 47 (1986), C3–41–C3–62. International workshop on aperiodic crystals (Les Houches, 1986).
328. J. PEYRIÈRE, On the trace map for products of matrices associated with substitutive sequences, *J. Statist. Phys.* 62 (1991), 411–414.
329. J. PEYRIÈRE, Trace maps, in *Beyond quasicrystals (Les Houches, 1994)*, pp. 465–480, Springer, Berlin, 1995.
330. J. PEYRIÈRE, On an article by W. Magnus on the Fricke characters of free groups, *J. Algebra* 228 (2000), 659–673.
331. J. PEYRIÈRE, Z.-X. WEN AND Z.-Y. WEN, Polynômes associés aux endomorphismes de groupes libres, *Enseign. Math.* 39 (1993), 153–175.
332. J. PEYRIÈRE, Z.-X. WEN AND Z.-Y. WEN, Endomorphismes de certaines algèbres à identités polynomiales, *C. R. Acad. Sci. Paris* 331 (2000), 111–114.
333. E. V. PODSYPANIN, A generalization of the continued fraction algorithm that is related to the Viggo Brun algorithm, *Zap. Naučn. Sem. Leningrad. Otdel. Mat. Inst. Steklov. (LOMI)* 67 (1977), 184–194.
334. F. POINT AND V. BRUYÈRE, On the Cobham-Semenov theorem, *Theory Comput. Syst.* 30 (1997), 197–220.
335. M. POLLICOTT AND M. YURI, *Dynamical systems and ergodic theory*, Cambridge University Press, Cambridge, 1998.
336. B. PRAGGASTIS, Numeration systems and Markov partitions from self-similar tilings, *Trans. Amer. Math. Soc.* 351 (1999), 3315–3349.
337. N. M. PRIEBE, Towards a characterization of self-similar tilings in terms of derived Voronoi tessellations, *Geom. Dedicata* 79 (2000), 239–265.
338. C. PROCESI, The invariant theory of $n \times n$ matrices, *Advances in Math.* 19 (1976), 306–381.
339. E. PROUHET, Mémoire sur quelques relations entre les puissances des nombres, *C. R. Acad. Sci. Paris* 33 (1851), 31.
340. M. QUEFFÉLEC, *Substitution dynamical systems—spectral analysis*, Springer-Verlag, Berlin, 1987. Lecture Notes in Mathematics, Vol. 1294.
341. M. QUEFFÉLEC, Une nouvelle propriété des suites de Rudin-Shapiro, *Ann. Inst. Fourier* 37 (1987), 115–138.
342. M. QUEFFÉLEC, Spectral study of automatic and substitutive sequences, in *Beyond quasicrystals (Les Houches, 1994)*, pp. 369–414, Springer, Berlin, 1995.
343. M. QUEFFÉLEC, Transcendance des fractions continues de Thue-Morse, *J. Number Theory* 73 (1998), 201–211.
344. M. QUEFFÉLEC, Irrational numbers with automaton-generated continued fraction expansion, in *Dynamical systems: from crystal to chaos*, J.-M. Gambaudo, P. Hubert, P. Tisseur and S. Vaienti (eds), World Scientific, 2000. Proceedings in honor of G. Rauzy on his 60th birthday.

345. B. RANDÉ, *Équations fonctionnelles de Mahler et applications aux suites p-régulières*, PhD thesis, Université Bordeaux I, 1992.
346. M. RATNER, Horocycle flows, joinings and rigidity of products, *Ann. of Math.* 118 (1983), 277–313.
347. M. RATNER, On measure rigidity of unipotent subgroups of semisimple groups, *Acta Math.* 165 (1990), 229–309.
348. G. RAUZY, *Propriétés statistiques de suites arithmétiques*, Presses Universitaires de France, Paris, 1976. Le Mathématicien, No. 15, Collection SUP.
349. G. RAUZY, Une généralisation du développement en fraction continue, in *Séminaire Delange-Pisot-Poitou, 1976/77, Théorie des nombres, Fasc. 1*, pp. Exp. No. 15, 16, Paris, 1977.
350. G. RAUZY, Nombres algébriques et substitutions, *Bull. Soc. Math. France* 110 (1982), 147–178.
351. G. RAUZY, Suites à termes dans un alphabet fini, in *Séminaire de Théorie des Nombres (Talence, 1982–1983)*, Univ. Bordeaux I, 1983. Exp. No. 25.
352. G. RAUZY, Mots infinis en arithmétique, in *Automata on infinite words (Le Mont-Dore, 1984)*, pp. 165–171, Springer, Berlin, 1985.
353. G. RAUZY, Rotations sur les groupes, nombres algébriques, et substitutions, in *Séminaire de Théorie des Nombres (Talence, 1987–1988)*, Univ. Bordeaux I, 1988. Exp. No. 21.
354. J. P. RAZMYSLOV, Trace identities of full matrix algebras over a field of characteristic zero, *Izv. Akad. Nauk SSSR ser. Mat.* 38 (1974). English translation in *Math. USSR Izvestija*, 8(1974): 727–760.
355. F. RECHER, Propriétés de transcendance de séries formelles provenant de l'exponentielle de Carlitz, *C. R. Acad. Sci. Paris Sér. I Math.* 315 (1992), 245–250.
356. J.-P. REVEILLÈS, Combinatorial pieces in digital lines and planes, in *Vision geometry IV (San Diego, CA), Proc. SPIE, 2573*, pp. 23–24, 1995.
357. D. RIDOUT, Rational approximations to algebraic numbers, *Mathematika* 4 (1957), 125–131.
358. R. N. RISLEY AND L. Q. ZAMBONI, A generalization of Sturmian sequences; combinatorial structure and transcendence, *Acta Arith.* 95 (2000), 167–184.
359. R. W. RITCHIE, Finite automata and the set of squares, *J. Assoc. Comput. Mach.* 10 (1963), 528–531.
360. A. M. ROBERT, *A course in p-adic analysis*, Springer-Verlag, New York, 2000.
361. E. A. J. ROBINSON, The dynamical properties of Penrose tilings, *Trans. Amer. Math. Soc.* 348 (1996), 4447–4464.
362. V. A. ROHLIN, Exact endomorphisms of a Lebesgue space, *Izv. Akad. Nauk SSSR Ser. Mat.* 25 (1961), 499–530.
363. G. ROTE, Sequences with subword complexity $2n$, *J. Number Theory* 46 (1994), 196–213.
364. K. F. ROTH, Rational approximations to algebraic numbers, *Mathematika* 2 (1955), 1–20; corrigendum, 168.
365. G. ROZENBERG AND A. LINDENMAYER, Developmental systems with locally catenative formula, *Acta Inf.* 2 (1973), 214–248.
366. W. RUDIN, Some theorems on Fourier coefficients, *Proc. Amer. Math. Soc.* 10 (1959), 855–859.
367. D. J. RUDOLPH, An example of a measure preserving map with minimal self-joinings, and applications, *J. Analyse Math.* 35 (1979), 97–122.
368. A. SALOMAA, *Jewels of formal language theory*, Computer Science Press, Rockville, Md., 1981.
369. O. SALON, Suites automatiques à multi-indices, in *Séminaire de Théorie des Nombres (Talence, 1986–1987)*, Univ. Bordeaux I, 1987. Exp. No. 4.

370. O. SALON, Suites automatiques à multi-indices et algébricité, *C. R. Acad. Sci. Paris Sér. I Math.* 305 (1987), 501–504.
371. J. W. SANDER AND R. TIJDEMAN, The complexity of functions on lattices, *Theoret. Comput. Sci.* 246 (2000), 195–225.
372. J. W. SANDER AND R. TIJDEMAN, Low complexity function and convex sets in $\mathbb{Z}^k$, *Math. Z.* 233 (2000), 205–218.
373. J. W. SANDER AND R. TIJDEMAN, The rectangle complexity of functions on two-dimensional lattices, *Theoret Comput. Sci.* 270 (2002), 857–863.
374. Y. SANO, On purely periodic β-expansions of pisot numbers. To appear in *J. Anal. Math.*
375. Y. SANO AND S. ITO, On periodic β-expansions of Pisot numbers and Rauzy fractals, *Sūrikaisekikenkyūsho Kōkyūroku*, 2000, 186–193. Analytic number theory and related topics (Japanese) (Kyoto, 1999).
376. K. SCHMIDT, *Dynamical systems of algebraic origin*, Birkhäuser Verlag, Basel, 1995.
377. W. M. SCHMIDT, On simultaneous approximations of two algebraic numbers by rationals, *Acta Math.* 119 (1967), 27–50.
378. W. M. SCHMIDT, On continued fractions and Diophantine approximation in power series fields, *Acta Arith.* 95 (2000), 139–166.
379. F. SCHWEIGER, *The metrical theory of Jacobi-Perron algorithm*, Springer-Verlag, Berlin, 1973. Lecture Notes in Mathematics, Vol. 334.
380. F. SCHWEIGER, On the invariant measure for Jacobi-Perron algorithm, *Math. Pannon.* 1 (1990), 91–106.
381. F. SCHWEIGER, *Ergodic theory of fibred systems and metric number theory*, The Clarendon Press Oxford University Press, New York, 1995.
382. F. SCHWEIGER, The exponent of convergence for the 2-dimensional Jacobi-Perron algorithm, in *Proceedings of the Conference on Analytic and Elementary Number Theory (Vienna)*, W. G. Nowak and J. Schoissengeier (eds), pp. 207–213, 1996.
383. F. SCHWEIGER, *Multidimensinal Continued Fraction*, Oxford Univ. Press, New York, 2000.
384. P. SÉÉBOLD, *Propriétés combinatoire des mots infinis engendré par certains morphismes*, PhD thesis, L. I. T. P., 1985.
385. P. SÉÉBOLD, Fibonacci morphisms and Sturmian words, *Theoret. Comput. Sci.* 88 (1991), 365–384.
386. M. SENECHAL, *Quasicrystals and geometry*, Cambridge University Press, Cambridge, 1995.
387. E. SENETA, *Nonnegative matrices and Markov chains*, Springer-Verlag, New York, 1981. Second ed.
388. C. SERIES, The geometry of Markoff numbers, *Math. Intelligencer* 7 (1985), 20–29.
389. C. SERIES, The modular surface and continued fractions, *J. London Math. Soc.* 31 (1985), 69–80.
390. J. SHALLIT, A generalization of automatic sequences, *Theoret. Comput. Sci.* 61 (1988), 1–16.
391. J. SHALLIT, Real numbers with bounded partial quotients: a survey, *Enseign. Math.* 38 (1992), 151–187.
392. J. SHALLIT, Automaticity IV: sequences, sets, and diversity, *J. Théor. Nombres Bordeaux* 8 (1996), 347–367. Errat. 9 (1997), 247.
393. C. E. SHANNON, A mathematical theory of communication, *Bell System Tech. J.* 27 (1948), 379–423, 623–656.
394. H. SHAPIRO, *Extremal problems for polynomials and power series*, PhD thesis, Massachusetts Institute of Technology, 1951.

395. M. SHUB, A. FATHI AND R. LANGEVIN, *Global stability of dynamical systems*, Springer-Verlag, New York, 1987. Translated from the French by J. Christy.
396. A. SIEGEL, Représentation des systèmes dynamiques substitutifs non unimodulaires. To appear in *Ergodic Theory Dynam. Systems.*
397. A. SIEGEL, *Représentation géométrique, combinatoire et arithmétique des substitutions de type Pisot*, PhD thesis, Université de la Méditerranée, 2000.
398. A. SIEGEL, Pure discrete spectrum dynamical system and periodic tiling associated with a substitution, 2002. Preprint.
399. J. G. SINAI, Construction of Markov partitionings, *Funkcional. Anal. i Priložen.* 2 (1968), 70–80.
400. V. F. SIRVENT, Relationships between the dynamical systems associated to the Rauzy substitutions, *Theoret. Comput. Sci.* 164 (1996), 41–57.
401. V. F. SIRVENT, Identifications and dimension of the Rauzy fractal, *Fractals* 5 (1997), 281–294.
402. V. F. SIRVENT, On some dynamical subsets of the Rauzy fractal, *Theoret. Comput. Sci.* 180 (1997), 363–370.
403. V. F. SIRVENT, A semigroup associated with the k-bonacci numbers with dynamic interpretation, *Fibonacci Quart.* 35 (1997), 335–340.
404. V. F. SIRVENT, Modelos geométricos asociados a substituciones. Trabajo de ascenso, Universidad Simon Bolivar, 1998.
405. V. F. SIRVENT, Semigroups and the self-similar structure of the flipped Tribonacci substitution, *Appl. Math. Lett.* 12 (1999), 25–29.
406. V. F. SIRVENT, The common dynamics of the Tribonacci substitutions, *Bull. Belg. Math. Soc. Simon Stevin* 7 (2000), 571–582.
407. V. F. SIRVENT, Geodesic laminations as geometric realizations of sequences of pisot substitutions, *Ergodic Theory Dynamical Systems* 20 (2000), 1253–1266.
408. V. F. SIRVENT, The Arnoux semi-conjugacy is Hölder continuous, *J. Math. Anal. Appl.* 259 (2001), 357–367.
409. V. F. SIRVENT AND Y. WANG, Self-affine tiling via substitution dynamical systems and Rauzy fractals. To appear in *Pacific J. Math*, 2002.
410. S. SMALE, Differentiable dynamical systems, *Bull. Amer. Math. Soc.* 73 (1967), 747–817.
411. B. SOLOMYAK, On the spectral theory of adic transformations, in *Representation theory and dynamical systems*, pp. 217–230, Amer. Math. Soc., Providence, RI, 1992.
412. B. SOLOMYAK, Substitutions, adic transformations, and beta-expansions, in *Symbolic dynamics and its applications (New Haven, CT, 1991)*, pp. 361–372, Amer. Math. Soc., Providence, RI, 1992.
413. M. SOLOMYAK, On simultaneous action of Markov shift and adic transformation, in *Representation theory and dynamical systems*, pp. 231–239, Amer. Math. Soc., Providence, RI, 1992.
414. V. SÒS, On the distribution mod. 1 of the sequences $n\alpha$, *Ann. Univ. Sci. Budapest Eötvös Sect. Math.* 1 (1958), 127–134.
415. J. SOTOMAYOR, *Licoes de equacoes diferenciais ordinarias*, Graphica Editora Hamburg Ltda, Sao Paolo, 1979. Edited by IMPA, Projeto Euclides.
416. J. STILLWELL, *Classical topology and combinatorial group theory*, Springer-Verlag, New-York, 1993.
417. J.-I. TAMURA, Some problems and results having their origin in the power series $\sum_{n=1}^{\infty} z^{[\alpha n]}$, in *Proc. of the colloquium "Analytic Number Theory and its Circumference" held at Gakushûin Univ., 11-13 Nov. 1991*, pp. 190–212, 1992.
418. J.-I. TAMURA, A class of transcendental numbers having explicit g-adic and Jacobi-Perron expansions of arbitrary dimension, *Acta Arith.* 71 (1995), 301–329.

419. J.-I. TAMURA, Certain sequences making a partition of the set of positive integers, *Acta Math. Hungar.* 70 (1996), 207–215.
420. J.-I. TAMURA, Certain partitions of a lattice, in *Dynamical systems: from crystal to chaos*, J.-M. Gambaudo, P. Hubert, P. Tisseur and S. Vaienti (eds), World Scientific, 2000. Proceedings in honor of G. Rauzy on his 60th birthday.
421. J.-I. TAMURA, Certain words, tilings, their non-periodicity, and substitutions of high dimension, in *Analytic number theory*, Kluwer Academic Publishers, Dordrecht, To be published, 2002.
422. K. TANAKA, *A Doubling Rule in "hi-fu-mi"*, pp. 251–254, Chikuma-shobô, Tokyo, 1989. In Japanese.
423. D. THAKUR, Continued fraction for the exponential for $\mathbb{F}_q[T]$, *J. Number Theory* 41 (1992), 150–155.
424. D. THAKUR, Exponential and continued fractions, *J. Number Theory* 59 (1996), 248–261.
425. D. THAKUR, Patterns of continued fractions for the analogues of e and related numbers in the function field case, *J. Number Theory* 66 (1997), 129–147.
426. D. S. THAKUR, Number fields and function fields (zeta and gamma functions at all primes), in *Proceedings of the conference on p-adic analysis (Houthanlen, 1987)*, Brussels, 1986, pp. 149–157, Vrije Univ. Brussel.
427. D. S. THAKUR, Automata-style proof of Voloch's result on transcendence, *J. Number Theory* 58 (1996), 60–63.
428. D. S. THAKUR, Automata and transcendence, in *Number theory (Tiruchirapalli, 1996)*, pp. 387–399, Amer. Math. Soc., Providence, RI, 1998.
429. A. THUE, Über die gegenseitige Lage gleicher Teile gewisser Zeichenreihen, in *Selected mathematical papers*, pp. 413–477, Universitetsforlaget, Oslo, 1977. Originaly published in *Christiania Vidensk. Selsk. Skr.* 1912, no. 1; Jbuch 43, 162.
430. A. THUE, Über unendliche Zeichenreihen, in *Selected mathematical papers*, pp. 139–158, Universitetsforlaget, Oslo, 1977. Originally published in *Christiania Vidensk. Selsk. Skr.* 1906, no. 7; Jbuch 37, 66.
431. W. P. THURSTON, *Groups, tilings and finite state automata.* Lectures notes distributed in conjunction with the Colloquium Series, in *AMS Colloquium lectures*, 1989.
432. M. TOBE, *Introduction to Hungarian*, Tairyûsha, Tokyo, 1988. In Japanese.
433. C. TRAINA, *Representation of the Trace Polynomial of Cyclically Reduced Words in a Free Group on Two Generators*, PhD thesis, Polytechnic Institute of New York, 1978.
434. C. R. TRAINA, Trace polynomial for two-generator subgroups of SL(2, $\mathbb{C}$), *Proc. Amer. Math. Soc.* 79 (1980), 369–372.
435. N. ÜÇOLUK, Solution for problem 5304, *Amer. Math. Monthly* 73 (1996), 681–682.
436. J. V. USPENSKY, On a problem arising out of a certain game, *Amer. Math. Monthly* 34 (1927), 516–521.
437. A. VERSHIK AND N. SIDOROV, Arithmetic expansions associated with the rotation of a circle and continued fractions, *Algebra i Analiz* 5 (1993), 97–115.
438. A. M. VERSHIK, Uniform algebraic approximation of shift and multiplication operators, *Dokl. Akad. Nauk SSSR* 259 (1981), 526–529. English translation: *Soviet Math. Dokl.* 24 (1981), 97–100.
439. A. M. VERSHIK AND A. N. LIVSHITS, Adic models of ergodic transformations, spectral theory, substitutions, and related topics, in *Representation theory and dynamical systems*, pp. 185–204, Amer. Math. Soc., Providence, RI, 1992.

440. J. VIDAL AND R. MOSSERI, Generalized Rauzy tilings: construction and electronic properties, *Materials Science and Engineering A* 294–296 (2000), 572–575.
441. L. VUILLON, Combinatoire des motifs d'une suite sturmienne bidimensionnelle, *Theoret. Comput. Sci.* 209 (1998), 261–285.
442. L. I. WADE, Certain quantities transcendental over $GF(p^n, x)$, *Duke Math. J.* 8 (1941), 701–720.
443. L. I. WADE, Transcendence properties of the Carlitz ψ-functions, *Duke Math. J.* 13 (1946), 79–85.
444. M. WALDSCHMIDT, Transcendence problems connected with Drinfeld modules, *İstanbul Üniv. Fen Fak. Mat. Derg.* 49 (1990), 57–75. International Symposium on Algebra and Number Theory (Silivri, 1990).
445. P. WALTERS, *An introduction to ergodic theory*, Springer-Verlag, New York, 1982.
446. K. WARGAN, *S-adic dynamical systems and Bratelli diagrams*, PhD thesis, George Washington University, 2001.
447. Z.-X. WEN, *Diverses études sur l'automaticité*, PhD thesis, Université Paris-Sud, 1991.
448. Z.-X. WEN, Relations polynomiales entre les traces de produits de matrices, *C. R. Acad. Sci. Paris Sér. I Math.* 318 (1994), 99–104.
449. Z.-X. WEN AND Z.-Y. WEN, Some studies of factors of infinite words generated by invertible substitutions, in *Procedings of the 5th Conference Formal Power Series and Algebraic Combinatorics, Florence*, pp. 455–466, 1993.
450. Z.-X. WEN AND Z.-Y. WEN, Local isomorphisms of invertible substitutions, *C. R. Acad. Sci. Paris Sér. I Math.* 318 (1994), 299–304.
451. Z.-X. WEN AND Z.-Y. WEN, On the leading term and the degree of the polynomial trace mapping associated with a substitution, *J. Statist. Phys.* 75 (1994), 627–641.
452. Z.-X. WEN AND Z.-Y. WEN, Some properties of the singular words of the Fibonacci word, *European J. Combin.* 15 (1994), 587–598.
453. Z.-X. WEN AND Y.-P. ZHANG, Some remarks on invertible substitutions on three letter alphabet, *Chinese Sci. Bull.* 44 (1999), 1755–1760.
454. Z.-Y. WEN, Singular words, invertible substitutions and local isomorphisms, in *Beyond quasicrystals (Les Houches, 1994)*, pp. 433–440, Springer, Berlin, 1995.
455. Z.-Y. WEN, T. JANSSEN AND F. M. DEKKING, Fibonacci chain as a periodic chain with discommensurations, *J. Phys. A* 27 (1994), 1691–1702.
456. Z.-Y. WEN, F. WIJNANDS AND J. S. W. LAMB, A natural class of generalized Fibonacci chains, *J. Phys. A* 27 (1994), 3689–3706.
457. A. WHITTEMORE, On special linear characters of free groups of rank $n \geq 4$, *Proc. Amer. Math. Soc.* 40 (1973), 383–388.
458. N. WIENER, The spectrum of an array and its application to the study of the translation properties of a simple class of arithmetical functions I: The spectrum of the array, *J. Math. Massachusetts* 6 (1927), 145–157.
459. F. WIJNANDS, Energy spectra for one-dimensional quasiperiodic potentials: bandwidth, scaling, mapping and relation with local isomorphism, *J. Phys. A* 22 (1989), 3267–3282.
460. R. F. WILLIAMS, Classification of subshifts of finite type, *Ann. of Math.* 98 (1973), 120–153; errata, ibid. 99 (1974), 380–381.
461. N. WOZNY AND L. Q. ZAMBONI, Frequencies of factors in Arnoux-Rauzy sequences, *Acta Arith.* 96 (2001), 261–278.
462. E. M. WRIGHT, Prouhet's 1851 solution of the Tarry-Escott problem of 1910, *Amer. Math. Monthly* 66 (1959), 199–201.

463. E. M. Wright, The Tarry-Escott and the "easier" Waring problems, *J. Reine Angew. Math.* 311/312 (1979), 170–173.
464. J.-Y. Yao, Critères de non-automaticité et leurs applications, *Acta Arith.* 80 (1997), 237–248.
465. S.-I. Yasutomi, On Sturmian sequences which are invariant under some substitutions, in *Number theory and its applications (Kyoto, 1997)*, pp. 347–373, Kluwer Acad. Publ., Dordrecht, 1999.
466. J. Yu, Transcendence and special zeta values in characteristic p, *Ann. of Math.* 134 (1991), 1–23.
467. L. Q. Zamboni, Une généralisation du théorème de Lagrange sur le développement en fraction continue, *C. R. Acad. Sci. Paris, Série I* 327 (1998), 527–530.
468. O. Zariski and P. Samuel, *Commutative algebra. Vol. 2*, Springer-Verlag, New York, 1975. With the cooperation of I. S. Cohen. Corrected reprinting of the 1958 edition, Graduate Texts in Mathematics, No. 28.
469. E. Zeckendorf, Représentation des nombres naturels par une somme de nombres de Fibonacci ou de nombres de Lucas, *Bull. Soc. Roy. Sci. Liège* 41 (1972), 179–182.

Index